Loss Models

WILEY SERIES IN PROBABILITY AND STATISTICS
APPLIED PROBABILITY AND STATISTICS SECTION

Established by WALTER A. SHEWHART and SAMUEL S. WILKS

A complete list of the titles in this series appears at the end of this volume.

Loss Models

From Data to Decisions

Stuart A. Klugman
Harry H. Panjer
Gordon E. Willmot

With assistance from:
Gary G. Venter

A Wiley-Interscience Publication
JOHN WILEY & SONS, INC.
New York / Chichester / Weinheim / Brisbane / Singapore / Toronto

This book is printed on acid-free paper ♾.

Published simultaneously in Canada.

Library of Congress Cataloging-in-Publication Data:

Klugman, Stuart A., 1949–
Loss models : from data to decisions / Stuart Allen Klugman, Harry H. Panjer, Gordon E. Willmot, with the assistance of Gary G. Venter.
p. cm. — (Wiley series in probability and mathematical statistics. Applied probability and statistics)
Includes bibliographical references and index.
ISBN 0-471-23884-8 (cloth : alk. paper)
1. Insurance—Statistical methods. 2. Insurance—Mathematical models. I. Panjer, Harry H. II. Willmot, Gordon E., 1957– . III. Title. IV. Series.
HG8781.K583 1998
368'.01—dc21 97-28718
CIP

Printed in the United States of America.

10 9 8 7 6

Contents

Preface

This textbook is organized around the principle that much of actuarial science consists of the construction and analysis of mathematical models which describe the process by which funds flow into and out of an insurance system. An analysis of the entire system is beyond the scope of a single text, so we have concentrated our efforts on the loss process, that is, the outflow of cash due to the payment of benefits. This makes the organization of the material somewhat natural as we first discuss the amount of a single claim, then the number of claims, then the total claims, then the use of additional information to modify our results, and finally the progress of the process over multiple time periods.

We have not assumed that the reader has any substantial knowledge of insurance systems. Insurance terms are defined when they are first used. In fact, most of the material could be disassociated from the insurance process altogether, and this book could be just another applied statistics text. What we have done is kept the examples focused on insurance, presented the material in the language and context of insurance, and tried to avoid getting into statistical methods that would have little use in actuarial practice.

What we have assumed is that the reader has a solid background in mathematical statistics (and of course the calculus necessary to do that) as provided by such texts as Hogg and Craig [65]. This would include the ability to work with probability, conditional probability, random variables, density and distribution functions, expectation, estimation and inference.

Because we view this as an applied text, the majority of exercises are numerical in nature. And because insurance data tend to be numerically unpleasant at times, many of the exercises will require more than a calculator to complete. The reader would be well-served to have a full-feature spreadsheet program available. Two of the features that will prove very useful are a solver/optimizer (a routine that solves equations and maximizes functions) and access to the gamma, incomplete gamma, and incomplete beta functions. As the need arises we will also provide information that should be sufficient should you like to write your own programs.

In addition, we have provided several DOS programs that will perform many (but not all) of the tasks discussed in the text. They are available along with instructions at the John Wiley & Sons web site (www.wiley.com).

It has been standard practice in many countries to qualify actuaries through examination. Many of the topics covered in this text have been tested on Casualty Actuarial Society exam 4B (loss distributions and credibility, the subjects of Chapters 2, 3, and 5) and Society of Actuaries exam 151 (risk theory, the subject of Chapters 4 and 6). The two societies have been kind enough to let us reproduce some of the questions and they are used with their permission. They can be identified by the notation (mxx-yy) at the end of the exercise, where xx is the exam year and yy is the question number. For years in which an exam was administered twice, m indicates the month (May or November) of the exam. Some of the questions were altered to conform with the terminology used in this text. Most were originally multiple choice questions, but the answer choices have not been given here. We make no claim that future examinations will have similar questions. They have been included so that readers will have some idea of what those who write questions for these exams have thought important in the past.[1]

With regard to notation, we have adopted $\log x$ for the natural logarithm (other bases for logarithms are clearly identified). This has been the subject of fierce debate among the authors, and we have chosen not to burden you with the logic (or lack thereof) behind this decision.

With regard to usage of this text, there are three natural divisions. Chapters 2 and 3 can be used together as a stand-alone course in model building. Chapters 4 and 6 can be used together as a stand-alone course in risk theory. Chapter 5 is a self-contained unit on credibility. Statistical concepts from Chapters 2 and 3 that are needed in Chapter 5 are repeated in that chapter.

S. A. KLUGMAN
H. H. PANJER
G. E. WILLMOT

Des Moines, Iowa
Waterloo, Ontario

[1] Readers who suspect the real reason is laziness on the part of the authors are probably correct.

Acknowledgments

This text was under construction for several years and is the product of many prior years of work. While each author will express some individual thanks later, there have been a number of individuals who have played major roles in the production of this text.

Throughout, we enjoyed the considerable support and advice of the members of the Casualty Actuarial Society's Committee on the Theory of Risk. They reviewed our first outline and then reviewed the material as it was produced. In addition, they maintained contact with the Casualty Actuarial Society's syllabus committees because it was our hope that this text will prove to be useful in their education and examination efforts. Particular thanks go to Roger Hayne, committee chair. Other members during this period were John Aquino, Bob Finger, Louise Francis, Jim Garven, Phil Heckman, Rodney Kreps, Isaac Mashitz, Glenn Meyers, Gary Patrik, John Robertson, Richard Roth, Jr., and Gary Venter.

Also prominent among the members is Gary Venter, who received special acknowledgment on the title page. Gary contributed a number of useful ideas in Chapters 2 and 5, some of which have been lifted directly from his writing. Gary was also instrumental in encouraging the Brownian motion sections at the end of Chapter 6.

The Society of Actuaries also played a role in the production of this text, in particular we thank Judy Anderson. Until this year, she was the staff actuary in charge of Intensive Seminar 152, Applied Risk Theory. Chapters 2, 3, and

4 evolved from the materials developed by us for that seminar, which also helped convince us of the need for a text on this subject.

Two individuals provided significant help with error and style corrections. Clive Keatinge read every word, worked every example and exercise, and kept reminding us that nothing less than perfection was acceptable. His frequent comments that this was unclear or that was vague helped us to sharpen our explanations. Chris Wiggenhorn provided a student viewpoint. He is not an actuary and had never studied this material before. He, too, helped us clear up that which was muddy.

Thanks go to David Dickson who supplied an improved proof which was incorporated into Chapter 6. Other actuaries who made helpful comments were Jim Broffitt and Hoque Sharif. University of Waterloo students and faculty who contributed comments and corrections include Claire, Bilodeau, Corrina Lin, Ying Liu, Ken Seng Tan, Chi-Liang Tsai, Shaun Wang, Julia Wirch, Keith Hoon Wong, Stella Wong, Esther Yang, and Hailiang Yang.

The following are individual acknowledgments by the authors.

Stuart Klugman:

It is imperative that I begin by thanking Bob Hogg. Bob hired me for my first actuarial teaching position, and as Department Chair he was unstinting in his support for the actuarial science program. He continued his support by inviting me to be a coauthor of *Loss Distributions* [66]. This book project came about as I realized that this 1984 text was in need of considerable updating. It was with Bob's blessing that I joined with some younger (in age, no one is younger in spirit than Bob) folks to work on the new version. Bob's love for his profession, his university, his students, and his colleagues set an example for me that I try (but only occasionally succeed) to emulate.

Thanks also go to my coauthors. It was their vision that turned my idea of a second edition of *Loss Distributions* into the tome you are now holding. Their belief that so much more could be done was inspiring.

Another reason this book exists is the support from John Wiley & Sons in its production. Our editor, Steve Quigley, was an enthusiastic supporter from the first time we contacted him. Lisa Van Horn provided expert production assistance and made sure that commas, semicolons, colons and dashes appeared only where they should. She also was willing to tolerate a few of our idiosyncracies. The style file which controls the look of this book was created by Amy Hendrickson. Her support, advice, and persistence in providing the LaTeX help I needed in converting our words, symbols, and pictures into the final copy were greatly appreciated.

Further thanks must go to my family. Authors often refer to the patience and support of their spouse. Because Marie wrote two books during this period, the support was more mutual and empathic than usual. More suffering was done by our sons David and Philip, who will no longer wail, "Dad, are you working on the book again? Aren't you ever going to finish?" So, as promised, here are your names in print.

Financial support for producing this book was provided to me by Drake University's Kelley Insurance Center.

A number of Drake students had to suffer through the use of early drafts of the book. I promised a mention to those who spotted errors. So thanks to Dave Frette, Sam Keller, Becky Risley, and Melody White.

Harry Panjer:

Any book is the result of the shared vision of the authors. It was clear to all three of us that we could produce a product that improved and updated the material in *Loss Distributions* and *Insurance Risk Models* [102], particularly with actuarial students as the intended audience. The book reflects a major combined effort, with much discussion, but little unresolvable disagreement amongst the authors. Each learned from the others in the process of the development of each chapter.

Special thanks are due to the many students at the University of Waterloo who have had to study from this material as it was being developed. They were our guinea pigs.

My wife Joanne contributed indirectly by wisely suggesting that this book (as well as another one I was working on) be finished before our wedding in June 1997. The authors collectively met my commitment to do so (and saved my neck) on this book by having the final draft completed in May 1997. I was not so lucky on the other book. Thanks, Jo. Our teenage sons provided no proofreading help at all but did stay out of our hair. They had their priorities. Thanks, Lucas and Lucas for your tolerance.

Gordon Willmot:

I wish to begin by thanking my co-authors for their role in developing the syllabus material for the Society of Actuaries intensive seminar on applied risk theory, because that material formed the basis for much of the present work. Their enthusiasm has been inspiring to me.

Much of the material on credibility theory has been drawn from course notes I wrote for an undergraduate actuarial course at the University of Waterloo. Special mention must be made in this regard of the contributions of Jim Hickman, who cheerfully volunteered to read the material and who provided much technical and historical input. Thanks also to Xiaodong Lin, Harry Panjer, and Jock Mackay for their input and to Nandanee Basdeo for her expert typing of the notes. Finally, I wish to mention the contributions of Stuart Klugman for the numerous discussions with him on credibility, as well as his numerous written contributions which undoubtedly have improved the presentation and synthesis of the material.

I wish to thank my family for their tolerance of my involvement in this work. To my wife Deborah I wish to express my gratitude for her support and patience. To my three daughters Rachel, Lauren, and Kristen, I wish to express my thanks for their understanding and cooperation when I was unable to spend time with them because of my involvement with the book.

S.A.K., H.H.P., G.E.W.

Loss Models

1

Introduction—A Model-Based Approach to Actuarial Science

1.1 INTRODUCTION

This book covers a number of subjects that have traditionally been accorded separate treatment in the actuarial literature.[1] These include risk theory (particularly the narrow view of risk theory as the study of the claims process), loss distributions, and credibility theory. As well as all being topics with a statistical basis, they also have been developed in the context of model-based analysis. In this chapter we argue that model-based analyses are appropriate for actuarial investigations and therefore the historical developments continue to keep the profession on the right track. We will then introduce the major areas covered in this book by placing them in this modeling context and describing briefly what details of those areas will be covered in the relevant chapters.

1.2 THE MODEL-BASED APPROACH

The model-based approach should be considered in the context of the objectives of any given problem. Many problems in actuarial science involve the building of a mathematical model that can be used to forecast or predict insurance costs in the future particularly, the short-term future.

[1] An exception is Hossack, Pollard, and Zehnwirth [68], which does provide an introduction to this very set of topics.

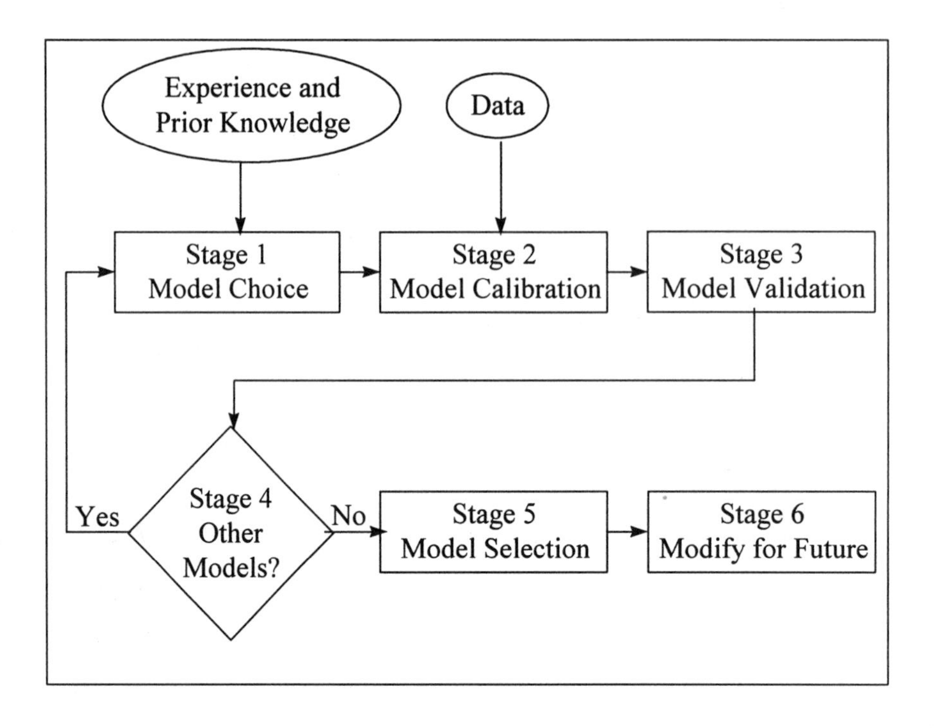

Fig. 1.1 The modeling process

A model is a simplified mathematical description which is constructed based on the knowledge and experience of the actuary combined with data from the past. The data guide the actuary in selecting the form of the model as well as in calibrating unknown quantities, usually called **parameters**. The model provides a balance between simplicity and conformity to the available data.

The simplicity is measured in terms of such things as the number of unknown parameters (the fewer the simpler); the conformity to data is measured in terms of the discrepancy between the data and the model. Model selection is based on a balance between the two criteria, namely, fit and simplicity.

1.2.1 The modeling process

The modeling process is illustrated in Figure 1.1, which describes six stages.

Stage 1 One or more models are selected based on the actuary's prior knowledge and experience and possibly on the nature and form of available data. In typical actuarial studies of mortality, the actuary may choose models containing covariate information such as age, sex, duration, policy type, medical information, and lifestyle variables. In studies of the size of insurance loss, a statistical distribution (e.g., lognormal, gamma, or Weibull) may be chosen.

Stage 2 The model is calibrated based on available data. In mortality studies, these data may be information on a set of life insurance policies. In studies of property claims, the data may be information about each of a set of actual insurance losses paid under a set of property insurance policies.

Stage 3 The fitted model is validated to determine if it adequately conforms to the data. Various diagnostic tests can be used. These may be well-known statistical tests, such as the chi-square goodness-of-fit test or the Kolomogorov–Smirnov test, or may be more qualitative in nature. The choice of test may relate directly to the ultimate purpose of the modeling exercise. In insurance-related studies, the fitted model is often required to replicate in total the losses actually experienced in the data. In insurance practice this is often referred to as **unbiasedness** of a model.

Stage 4 An opportunity is provided to consider other possible models. This is particularly useful if Stage 3 revealed that all models were inadequate. It is also possible that more than one valid model will be under consideration at this stage.

Stage 5 All valid models considered in Stages 1–4 are compared using some criteria to select between them. This may be done by using the test results previously obtained or may be done by using another criterion. In this book, we shall frequently use a likelihood ratio test as the basis of comparison. The winner will be used, although the losers may be retained for sensitivity analyses.

Stage 6 Finally, the selected model is adapted for application to the future. This could involve adjustment of parameters to reflect anticipated inflation from the time the data were collected to the period of time to which the model will be applied.

As new data are collected or the environment changes, the six stages will need to be repeated in order to improve the model.

1.2.2 The modeling advantage

The advantages of using models requires us to consider the alternative, decision making based strictly upon empirical evidence. The empirical approach assumes that the future can be expected to be exactly like a sample from the past, perhaps adjusted for trends such as inflation. Consider the following illustration.

Example 1.1 *A portfolio of group life insurance certificates consists of 1,000 employees of various ages and death benefits. Over the past five years, 14 employees died and received a total of 580,000 in benefits (adjusted for inflation because the plan relates benefits to salary). Determine the empirical estimate of next year's expected benefit payment.*

The empirical estimate for next year is then 116,000 (one-fifth of the total), which would need to be further adjusted for benefit increases. The danger, of course, is that it is unlikely that the experience of the past five years accurately reflects the future of this portfolio as there can be considerable fluctuation in such short-term results. □

It seems much more reasonable to build a model, in this case a mortality table. This table would be based on the experience of many lives, not just the 1,000 in our group. With this model we not only can estimate the expected payment for next year, but we can also measure the risk involved by calculating the standard deviation of payments or perhaps various percentiles from the distribution of payments. This is precisely the problem covered in great detail in *Actuarial Mathematics* [16]. Other advantages will be discussed in the next section.

This approach was codified by the Society of Actuaries' Committee on Actuarial Principles. In the publication"Principles of Actuarial Science" [113], Principle 3.1 states that "Actuarial risks can be stochastically modeled based on assumptions regarding the probabilities that will apply to the actuarial risk variables in the future, including assumptions regarding the future environment." The actuarial risk variables referred to are: occurrence, timing, and severity—that is, the chances of a claim event, the time at which the event occurs if it does, and the cost of settling the claim. These are discussed in Chapters 3 and 6 (occurrence and timing) and Chapter 2 (severity). Chapter 4 is devoted to putting them together and Chapter 5 discusses some special statistical techniques for rate setting.

In the following sections we give examples of how the modeling concept is used in the areas covered in this book. We also take this opportunity to begin defining some terms that will be used throughout.

1.3 LOSS DISTRIBUTIONS

In the most general sense, all of actuarial science is about loss distributions because that is precisely what an insurance agreement is all about. The policyholder is paid a random amount (the loss) at a random future time. The calculation of the amount depends on the terms of the agreement and the nature of the event when it happens. However, over time, and especially with the publication of *Loss Distributions* [66], the definition of a loss has been restricted to a single payment (although it may comprise many smaller payments). To provide an important distinction, we make the following definitions. Many of them, along with additional discussion, can be found in *Foundations of Casualty Actuarial Science* [26, Ch. 2].

Definition 1.1 *A* ***loss event*** *or* ***claim*** *is an incident in which an insured or group of insureds suffers damages which are potentially covered by their insurance contract.*

Definition 1.2 *The* ***loss*** *is the dollar amount of damage suffered by an insured or group of insureds as a result of a loss event. The loss may be zero.*

Definition 1.3 *A* ***payment event*** *is an incident in which an insured or group of insureds receives a payment as a result of a loss event covered by their insurance contract.*

Definition 1.4 *The* ***amount paid*** *is the actual dollar amount paid to the policyholder(s) as a result of a loss event or a payment event. If it is as the result of a loss event, the amount paid may be zero.*

An issue of great importance, but one that is not addressed in this text, is the timing of the payments. The process which ultimately leads to a payment involves the time until a covered event may have taken place, the additional time to determine that it is a covered event, and then more time to determine the specific amount to be paid. We assume that the loss is known immediately as is the amount paid. See *Foundations of Casualty Actuarial Science* [26, Ch. 4] for a discussion of these issues.

Not all loss events produce a payment and some loss events do not even produce a loss, as illustrated in the following example.

Example 1.2 *Consider an automobile accident in which car A suffers damages that cost 150 to repair. Another car (B) was involved, but it is unclear at the time who is at fault. There is no damage to car B. Because matters are unclear, claims are filed with the insurers of both cars. Suppose that ultimately the driver of car B is deemed to not have been at fault. Then B experienced a loss event, but the loss is for 0 as is the amount paid. If the driver of the damaged car (A) has a policy with a 100 deductible, the insurer of car A and that driver would have experienced a loss event, a 150 loss, and a 50 amount paid by the insurer. If the deductible had been 250, there would still have been a loss, but no payment event or amount paid.* □

The reason it is important to record the loss event for company B is that this company probably incurred expenses as a result of posting this potential claim. Such expenses should be reflected directly in the setting of rates for this coverage. This leads to an additional definition.

Definition 1.5 *The* ***allocated loss adjustment expense (ALAE)*** *is the amount of expense incurred directly as the result of a loss event.*

These expenses may include direct processing costs, verification costs (both of the amount of the loss and of the existence of coverage for the event), costs of defending the insured against (perhaps) invalid claims, and any other expenses that occurred specifically because of the event. The ALAE does not include indirect expenses such as the cost of creating the policy, establishing rates, or determining financial results.

The difference between the loss and the amount paid is most always attributable to specific provisions in the policy that reduce the insurer's obligation. The following definitions describe the three most common such **coverage modifications**.

Definition 1.6 *An* ***ordinary deductible*** *is a provision which states that when the loss is less than or equal to the deductible, there is no payment and when the loss exceeds the deductible, the amount paid is the loss less the deductible.*

Definition 1.7 *A* ***policy limit*** *or* ***limit*** *is a provision which states that when the loss exceeds the limit, no additional benefits will be paid.*

Definition 1.8 *A* ***coinsurance factor*** *is the proportion of any loss that is paid by the insurer after any other modifications (such as deductibles or limits) have been applied. A* ***coinsurance*** *is a provision which states that a coinsurance factor is to be applied.*

It should be noted that in the first two cases the names both refer to the existence of the provision and the numerical value of the modification. These three provisions do not exhaust the possibilities; others will be introduced later. There are other interpretations of the last two provisions. Sometimes the word **limit** refers to the maximum payment by the insurance company and sometimes the word **coinsurance** refers to the proportion paid by the policyholder. In this book we will apply the above definitions consistently.

Example 1.3 *Consider a hospitalization plan that has all three provisions, say a deductible of 100, a limit of 5,000, and a coinsurance factor of 90%. If total covered charges are 4,000, the insurance will pay* $0.9(4{,}000-100) = 3{,}510$ *while if total covered charges are 5,500, the insurance will pay* $0.9(5{,}000 - 100) = 4{,}410$. *The maximum the insurance will pay is 4,410.* □

We are now ready to define a loss distribution.

Definition 1.9 *A* ***loss distribution*** *is the probability distribution of either the loss or the amount paid from a loss event or of the amount paid from a payment event. The distribution may or may not exclude payments of zero and may or may not include ALAE.*

Because the definition is ambiguous on a number of points, the specific context must be made clear when a loss distribution is being determined or evaluated. The following definition also provides several meanings.

Definition 1.10 *The **severity** can be either the loss or amount paid random variable. Its expected value is called the **mean severity**.*

Some writers use the single term severity interchangeably for the random variable and its expected value. The context should make the meaning clear in any given situation.

While it is possible for a loss distribution to be based on the empirical data, loss distributions based on smooth parametric models have many advantages. They are illustrated in this section with an example and are discussed in detail in Chapter 2.

Example 1.4 *Your company is undertaking a study of its basic group dental policy. The policy is currently sold with a deductible of 50 per claim. One group within your company thinks the deductible should be eliminated completely in order to encourage the employees to visit the dentist more frequently and thus reduce the cost of more expensive claims. A second group wants the deductible raised to 100 in order to significantly reduce premiums. You have been asked to investigate the consequences of these two options.*

To aid in your investigation, you have directed your summer intern to pull 10 claim files at random. The amount paid on each was recorded. The results were

$$141 \quad 16 \quad 46 \quad 40 \quad 351 \quad 259 \quad 317 \quad 1511 \quad 107 \quad 567.$$

Using the empirical approach it is possible to estimate the expected amount paid per such event by just computing the sample mean of 335.5. With regard to a new deductible of 100 it is also possible to estimate the expected amount paid per such event as

$$(91 + 301 + 209 + 267 + 1461 + 57 + 517)/7 = 2{,}903/7 = 414.71.$$

We could also relate the new experience to the old as whatever sequence of events caused the 3,355 in payments under the original policy, would now cost 2,903 under the new deductible. All else being equal,[2] we can expect a reduction of $452/3{,}355 = 13.47\%$ in costs if the deductible is raised.

With regard to the new deductible of 0, it is impossible to use the empirical approach to tell anything of value. We know that the existing amounts paid would all increase by 50 each to bring the total to 3,855. But there are also a number of loss events that did not show up in our files, due to the loss not reaching the 50 deductible, typically because they would not be reported to the insurer by the policyholder. Those would now create positive amounts paid, and those amounts should be added in. So the best we can see from

[2] This is not a trivial assumption. For example, we are ignoring the fact that our policyholders may change their behavior when they discover that they are now bearing a greater portion of the loss.

the empirical viewpoint is that the cost will increase by at least $500/3{,}355 = 14.90\%$.

Another problem with the empirical approach is that we have very little idea how reliable the cost decrease and increase estimates are. We reported the answers with four significant digits, but it is hard to believe we have achieved that level of accuracy with only ten observations.

A further problem is the lack of smoothness. Suppose we consider raising the deductible to 150. The total paid would reduce to 2,553. But this is unrealistic because we certainly do not expect that with a 100 deductible every amount paid would exceed 50 (as it did in the sample) and therefore raising the deductible another 50 would lower every payment by 50. This is just an artifact of the ten items that happened to be in our sample this time.

For an even more dramatic illustration, consider imposing a limit of 1,500 on the amount paid after the 100 deductible. Empirically, there would be no change in the expected cost because our sample contained no payments in excess of this limit. Yet our customers would certainly expect a reduction in their premium if such a restriction were placed on their policy.

The model-based alternative is to hypothesize a parametric distribution as the source of the losses. For this example, suppose that losses (not amounts paid) occur according to the lognormal distribution.[3] Based on the ten observed payment amounts, we can use the method of maximum likelihood to estimate the lognormal parameters. The results are $\hat{\mu} = 5.251$ and $\hat{\sigma} = 1.119$. Based on the lognormal model, the expected amount paid per payment event with a 50 deductible is 348.97.[4] This is a different number from the empirical estimate, but is just that, an estimate. We should not have expected these two values to agree, and we also do not believe either of them precisely measures the expected amount paid. An increase in the deductible to 100 will decrease expected payments by 12.95%, and the decrease will be 23.45% if the deductible is raised to 150.

Best of all, we can now estimate the effects of removing the deductible. Because the model is for losses, the expected loss per loss event is just the mean of the lognormal distribution, $\exp(5.251+1.119^2/2) = 356.49$. We can also determine that the probability of exceeding a 50 deductible is 0.88431. Therefore, we could expect an additional $.11569/.88431 = 13.08\%$ in the number of payments. The overall effect is an increase of $1.1308(356.49)/348.97-1 = 15.52\%$.

It would also be possible to compute confidence intervals for any of these quantities. These calculations are much more involved, but will be covered in Chapter 2. If the model is correct (in the example, the lognormal distribution), the model-based approach with maximum likelihood estimators will be very close to optimal in the sense of producing the narrowest confidence intervals.

[3]This and other distributions are discussed in detail in Chapters 2 and 3. Specific information about the various distributions can be found in Appendix A.

[4]Formulas for this and all the other calculations done in this example will be explained in detail in Chapter 2.

A final advantage of the parametric model is simplicity. With one word (lognormal) and two numbers (5.251 and 1.119) we have completely described the population. □

The bulk of Chapter 2 will be devoted to methods for estimation (both theoretical and computational aspects), methods for selecting the particular parametric distribution to use, performing calculations using these models, and constructing more complex models.

The area of loss distributions has generally been restricted to a study of loss amounts. A corresponding treatment can be afforded to the random number of claims generated by a policy or collection of policies. To be meaningful the number of losses must be expressed in terms of the opportunity for loss events to occur. Examples would be number of events per car-year, per dollar of annual payroll, or per a single, specified, one-year group life insurance contract.

Definition 1.11 *The **exposure base** is the basic unit of measurement upon which premiums are determined.*

Definition 1.12 *The **frequency** is the number of losses or number of payments random variable. Its expected value is called the **mean frequency**. Unless indicated otherwise the frequency is for one exposure unit.*

For example, a common exposure base for automobile insurance is the car-year while for life insurance it is the amount of death benefit. The same concepts apply to frequency as to severity. The only difference is that the model is a discrete probability distribution, while the model for severity is more likely to be continuous. The same considerations that are covered for severity are investigated for frequency in Chapter 3. A large number of distributions that have only been recently introduced to the actuarial literature will be presented. Our list is organized in a manner similar to the presentation in *Insurance Risk Models* [102].

Example 1.5 (Example 1.4 continued) *Suppose the ten amounts paid represented all claims in one month on a particular group dental policy. A possible frequency model is that the number of such amounts paid per month has the Poisson distribution with mean number of claims,* $\lambda = 10$. *With the deductible removed the number of losses per month remains Poisson, but now* $\lambda = 11.686$. *Should the deductible be increased to 100 the frequency distribution for amounts paid is again Poisson, but with* $\lambda = 7.977$. *The formulas used to obtain these numbers are developed in Chapter 3.* □

1.4 AGGREGATE PAYMENT CALCULATIONS

While it is not necessary to do so, the separation of the insurance loss process into frequency and severity components has several advantages. Among them are the ability to modify the components separately (e.g., applying coverage modifications such as an individual claim deductible affects the severity while underwriting changes may affect the frequency), to adjust for inflation and other time-dependent factors, or to estimate the parameters of the components from separate sources of information. In addition, if the components are not kept separate, we have but one observation per year. A consequence is an increase in the complexity of the calculations.

Regardless, the goal of insurance modeling is to develop a probability distribution for the total amount paid in benefits. This allows the insurance company to manage its capital account and honor its commitments.

Definition 1.13 *The* ***aggregate payment*** *or* ***aggregate loss*** *is the total amount paid or the total losses on all claims occurring in a fixed period on a defined set of insurance contracts.*

The set or collection of contracts may include one person, one object, a line or block of business, or an entire company. Common usage does not distinguish between payments and losses. The term loss is most commonly used, but usually in reference to payments. The following definition sets up one of the two standard situations for modeling the aggregate loss.

Definition 1.14 *A* ***collective risk model*** *represents the aggregate loss as a sum, S, of a random number , N, of individual payment amounts $(X_1, \ldots, X_N)$. Therefore $S = X_1 + \cdots + X_N$. Unless stated otherwise it is assumed that all of the random variables are independent and that the X_is have identical probability distributions.*

The restrictions on the random variables are not as harsh as they might seem. In particular, the X_is can be considered to represent payment amounts recorded as the claims occur (or are paid). Thus X_1 represents the first payment in the period, X_2 the second, and on to X_N being the amount of the final payment. The actual number of policies, n, that produces these payments may or may not be known. The value of n is relevant only as it affects the distribution of N. Over a short term, there is no reason to expect the distribution of these payment amounts to change. Independence among the X_is is less likely, but is usually a reasonable approximation. A large payment early in the year may imply that some changes have occurred that will lead to larger amounts in the future as well. It is also reasonable that

N is independent of the X_is, otherwise the number of claims would tell us something about their amounts.[5]

Example 1.6 (Example 1.4 continued) *Suppose we adopt the lognormal* ($\mu = 5.167$, $\sigma = 1.182$) *model for the individual losses X and adopt the Poisson* ($\lambda = 11.686$) *model for the number of losses. Using calculation techniques developed in Chapter 4 we can compute various quantities from the distribution of S. With a deductible of 100 the expected aggregate loss is 3,118 and the standard deviation is 2,232. A few percentiles from the aggregate loss distribution are: 50th: 2,614, 75th: 4,037, 90th: 5,825, 95th: 7,231, 99th: 10,992.* □

The method used to obtain the numbers in the example is called the **recursive method**. Other methods covered in Chapter 4 are direct calculation by convolution, inversion of the characteristic function, parametric approximation, and simulation. From this list alone it is clear that a lot of attention has been given to the numerical evaluation of quantities from the collective risk model. In Chapter 4 the relative merits of each approach will be discussed and computational algorithms presented.

The other model for aggregate losses is defined as follows.

Definition 1.15 *An **individual risk model** represents the aggregate loss as the sum of the amounts paid on each component of the portfolio of risks. That is, $S = X_1 + \cdots + X_n$, where X_i is the amount paid on the* i*th contract. Unless stated otherwise, it is assumed that $X_1, \ldots, X_n$ are independent.*

Although it is usually assumed that the X_is are mutually independent random variables, it is not assumed that the X_is have identical distributions. Each contract produces losses according to its own provisions and the underwriting characteristics of the policyholder. It would not be unusual for n to be a very large number or for many of the X_is to have outcomes of amount zero. In relating this model to the collective risk model, while many of the n policies may not provide any counts toward N, some may provide several.

Example 1.7 *A group life insurance contract covers the employees of a university. During the next year, should any employee die, his or her beneficiary will receive 2.5 times the employees annual salary. For the jth employee, the loss X_j has a two-point distribution. With probability q_j (which depends on the employee's age, sex, and perhaps some other factors), $X_j = 2.5s_j$ (where s_j is the employee's salary); and with probability $p_j = 1 - q_j$, $X_j = 0$.* □

Algorithms for computing probabilities and other relevant factors under the individual risk model will be covered in Chapter 4. It turns out that

[5] This is a standard assumption required for computational convenience. Little formal analysis of possible dependence has been done.

when n is large, the amount of computational work can be considerable. One method of simplifying the work is to find a collective risk model that is similar (in its probability distribution) to the individual risk model and then do all calculations with the collective risk model. Methods for making the conversion are also discussed in Chapter 4.

1.5 CREDIBILITY

Some people or organizations purchase insurance knowing they have a greater propensity to produce claims than the insurance company believes of them. After providing insurance for some time, the company may realize that this is indeed so and would want to increase their premiums. But the evidence needs to be strong enough to prevent the insured from countering that it is just bad luck (and of course, some policyholders will indeed exhibit bad results just through bad luck rather than any high-level propensity to produce claims). Again, the key point is that good drivers are occasionally unlucky while bad drivers are often lucky. We need to be able to decide whether what we see is due to chance or to actual differences in frequency and severity.

On the other hand, members of a class of insureds may believe they deserve a lower premium. If they really do deserve a break, it should become apparent after some number of periods with low claim levels has elapsed. When (and it is probably never soon enough for the insured) has enough evidence accumulated?

In many such instances the amount of data on the individual or group is too small to allow for a firm conclusion to be drawn about the existence of differences or the extent of such differences. This leads to the definition of credibility:

Definition 1.16 ***Credibility*** *is a procedure by which data external to a particular group or individual are combined with the experience of the group or individual in order to better estimate the expected loss (or any other statistical quantity) for each group or individual.*

The external data are typically a rate that has been determined based on historical data from other, similar policies. The internal data are the experience of the individual or group itself. Often the calculation is reduced to a very simple form:

$$\hat{\mu} = Z\bar{X} + (1 - Z)\tilde{X} \tag{1.1}$$

where $\bar{X}$ is the sample mean for the group in question, $\tilde{X}$ is the externally obtained estimate, $0 \leq Z \leq 1$, and $\hat{\mu}$ is the revised estimate of the group mean. In this setting we have

Definition 1.17 *For a given problem, when (1.1) is used, the value of Z is called the* ***credibility*** *or the* ***credibility factor****.*

Example 1.8 *In a manner similar to Example 1.1, consider a group life insurance covering 1,000 employees for one year. All the employees are age 50 and a standard mortality table indicates that the probability of death for one employee in one year is 0.008. In the past year there were 5 deaths. Upon what probability should future premiums be based?*

The employees may argue for the empirical probability of $5/1{,}000 = 0.005$ citing their excellent performance with respect to not dying. The insurer may insist that this is just a random good luck and has nothing to do with superior death avoidance propensities. Using a Poisson approximation to the binomial distribution we can follow Example 5.22 and compute the credibility factor as

$$Z = \sqrt{5/1{,}082.41} = 0.068$$

and obtain the credibility estimate as

$$0.068(0.005) + 0.932(0.008) = 0.0078.$$

1.6 CONTINUOUS TIME MODELS

An insurance enterprise is not static, but operates over time in a changing environment. The timing of events is critical because the enterprise is much better off if the premiums arrive before the claims must be paid. In general, we need to be able to model and evaluate processes in which events are recorded either continuously or periodically over time. For example, we could keep track of the loss ratio (total losses paid divided by premiums earned) for a given line of business for a given accident year as estimated at the end of each subsequent calendar year. This would be an example of the latter. If the events are random, we call it a **stochastic process**.

One application of the theory of stochastic processes is modeling probabilities of a process crossing various barriers, either at any future time or during a period of interest. Issues that may be addressed include measuring the probability distribution of the time until a certain portfolio of claims reaches 10,000,000, the time until surplus becomes negative, or the time to the next major earthquake.

We will focus on processes which track the portfolio's surplus through time. This allows us to use the models developed earlier to indicate the timing of the claims payments and then the amounts of each payment. When premiums are added in, it is possible to model the progress of surplus (premiums received less claims paid) through time and to investigate the possibility of surplus becoming negative (ruin).

In Chapter 6, two widely used models will be introduced. The compound Poisson model uses the Poisson distribution to model the number of claims in any fixed time period and uses an arbitrary distribution for the amount of each claim. Analytic results can be obtained for a number of situations and approximations are available for the rest. A second model is called **Brownian motion**. In Chapter 6 it is introduced as an approximation to the compound Poisson model. However, this model has many other uses, particularly in the modeling of the value of investments. An important actuarial use of Brownian motion is in the pricing of assets, particularly options. While this application is not covered in this text, the material presented does provide an introduction to this statistical model.

1.7 THE CASE STUDY

The purpose of constructing actuarial models is not to complete a mathematical exercise, nor is it to provide an overly complex description of the insurance environment. Rather, the purpose of building and analyzing a model is to answer questions of importance regarding the operation of the portfolio or organization. This concept may be lost in the mathematics of each chapter and so we have decided to provide a reminder by solving a realistic (though not real) problem. We call it the KPWV, Inc. case study and it is described in the remainder of this section. At the conclusion of Chapters 2–5 various portions of the problem are solved. There are exercises at the end of each of these sections, but most of them do not relate to the case study, but instead provide additional insights by using problems and data sets from the literature which highlight the development of actuarial modeling.

1.7.1 The KPWV, Inc. case study

You are the pricing actuary for KPWV, Inc., a major force in worldwide reinsurance. KPWV has a reputation for underwriting most any risk and so you are not surprised when KPWV is asked to bid on excess of loss and aggregate reinsurance for a newly emerging peril, actuarial liability. This refers to amounts paid when actuaries are found guilty of negligence, failure to act professionally, or the commission of any act with financial consequences that a prudent and careful actuary would not have done. There has been little data collected on such payments, so there is no rating organization, such as the Insurance Services Office, you can turn to which will supply quantitative information.

Fortunately, your prospective client has been insuring actuaries for some time and has been able to obtain some data on these payments. This coverage has been on a "claims-made" basis, which means that when an actuary pays an annual premium for liability insurance, the insurance will cover any incident in which the initial claim against the actuary occurs in the next 12 months. It does not matter when the actuary committed the act for which the liability has been incurred, but it does matter when the claim is made against the actuary. This is convenient for you because you need not worry about the possibility that incidents which happened during the policy year will be discovered years after the fact and must still be covered by the insurance.

On each claim, your client has provided five pieces of information. The first is the policy year in which the claim was made. For example, an assignment of 1993 would mean that the claim occurred after the policy was issued in 1993, but before it expired a year later in 1994. The second is the area of practice of the actuary. The categories are life and health (LH), pension (P), and property and liability (PL). The third is the deductible associated with the policy. All of the policies written by your client have either no deductible (coded as 0) or an ordinary deductible (coded as the amount). All of the policies have a limit (coded as the maximum payment amount). The final item is the amount paid by the insurance. The data appear in Tables 1.1–1.4.

You have been informed that one problem that is typical of such data should not concern you. That is, despite the claims-made provision, after a claim is made it may be several years before the amount of liability, and hence the amount of payment by the insurance company, has been revealed. This is a major difficulty in working with liability data, but you have been assured that for each claim, an expert has projected the ultimate cost.[6]

The final piece of information you have relates to the frequency with which such claims occur. For each of the policy years, you have been provided the number of policies issued to each of the three types of actuary. As part of the insurance contract, policyholders are required to report all liability incidents, whether or not the loss exceeds the deductible. Therefore, in addition to the payment counts in the earlier table, in Table 1.5 we have the number of losses in each year.

1.7.2 Your task

As the pricing actuary, you have been asked to provide the following information. First, assume that all three types of actuary are to be treated the same. That is, both the propensity to have claims and the distribution of claim amounts are the same for all three groups. With that assumption, construct a model from which you can obtain the distribution of reinsurance payments for any one covered actuary when the reinsurance pays the excess over a specified amount (the retention) but has no limit (other than that implied by the original policy). Use that distribution to obtain the mean, standard deviation, and the 90th and 99th percentiles for various combinations of deductible and limit (in the original policy) and retention. Then repeat this calculation for a reinsurance agreement covering a group of 100 actuaries.

[6] We do not want to minimize the importance of this step in the analysis, but because it is completely outside the scope of this text we thought it best to eliminate this concern from our case study.

You are next to consider an aggregate reinsurance in which the reinsurer will pay the excess over a certain amount (the aggregate deductible) of total insurance payments in respect of all policies issued in a particular calendar year. The premium will be determined after it is known how many policies were issued and it can be assumed that all of them have the same limit and no deductible. You are again to obtain the distribution of reinsurance payments, this time for various combinations of aggregate deductible, and individual limit, assuming there are 100 actuaries covered, all with the same limit. From this distribution again obtain the mean, standard deviation, and 90th and 99th percentiles.

Next, you are to attack the issue of whether or not the three types of actuary should be treated separately. Conduct appropriate tests to determine if this is true and if the issue is differing frequency, severity, or both. Discuss the trade-offs in constructing separate models versus a single model using all the data. Credibility theory provides ideas with regard to effecting a compromise between combining the data and treating the three groups separately. Use this theory to create models for each group.

As a further step, you anticipate that your client will expect the arrangement to be experience rated. This will be done prospectively. If actual claims are low one year, the reinsurance premium for the following year should be lower, and of course the opposite will happen if actual claims are high. Use credibility theory to construct an experience rating scheme.

Finally, you want to convince your client that purchasing reinsurance is a good decision. Do this by evaluating the probability that your client will be broke (with regard to this insurance) sometime in the next five years. In particular, demonstrate that this probability will be lower if reinsurance is purchased. Or, equivalently, demonstrate that by purchasing reinsurance your client will have to devote less surplus to support this line of business.

1.7.3 Caveats

This case study was inspired by a real problem encountered by a friend of the authors. In particular, the data were collected in a form just like that in Tables 1.1–1.4, except there were no groups and the policy year was irrelevant. The problem was to fit a severity distribution to the data. We have expanded upon this problem by adding the three groups, adding frequency data, and introducing aspects involving the aggregate distribution and credibility theory. We recognize that we have made simplifications with regard to problems caused by combining different policies (classes in this case) and by combining different years. We also recognize that this problem is also more complex than many that occur in practice, but we wanted to make sure that most of the topics covered in this text appear in the case study.

It is also important to note that while the numbers in Tables 1.1–1.4 are similar to those encountered by our friend, that data were not from actuarial liability coverage. Therefore, we do not want to imply that neither the frequency of incidence nor the severity of claims bears any relationship to that which really exists for actuarial liability. Finally, should any of the three types actuary appear to be better or worse, it is merely a result of our random assignment of LH, P, and PL to the three groups of claims. We certainly would not want to imply that one group tends to be more careful and/or practices in a less risky area.

Table 1.1 Liability payments—life and health actuaries

Year	Deductible	Maximum payment	Payment
90	0	1,000,000	2,890
90	0	5,000,000	5,851
90	250,000	10,000,000	15,347
90	0	1,000,000	15,635
90	0	3,000,000	20,553
90	0	10,000,000	34,584
90	0	10,000,000	79,661
90	0	400,000	132,601
90	1,500,000	5,000,000	1,410,989
90	0	10,000,000	2,784,401
90	0	10,000,000	4,894,360
90	10,000,000	10,000,000	9,316,751
91	0	1,000,000	1,891
91	0	3,000,000	30,893
91	0	1,000,000	31,392
91	500,000	10,000,000	49,488
91	175,000	1,000,000	67,425
91	0	1,000,000	150,310
91	45,000,000	33,000,000	1,335,735
91	0	10,000,000	3,308,499
91	12,750,000	10,000,000	10,000,000
91	15,000,000	10,000,000	10,000,000
92	0	1,000,000	1,836
92	0	1,000,000	10,705
92	0	5,000,000	10,973
92	0	5,000,000	13,408
92	0	10,000,000	16,339
92	350,000	5,000,000	95,736
92	0	1,000,000	212,313
92	0	5,000,000	439,543
92	70,000,000	15,000,000	1,098,710
92	0	3,000,000	1,211,180
93	0	500,000	10,510
93	0	3,000,000	14,029
93	0	10,000,000	15,296
93	50,000	1,000,000	27,516
93	0	10,000,000	53,467
93	300,000	5,000,000	87,463
93	100,000	5,000,000	220,995
93	150,000	5,000,000	274,086
93	0	5,000,000	1,862,304
93	0	5,000,000	5,000,000

Table 1.2 Liability payments—pension actuaries, 1990–1991

Year	Deductible.	Maximum payment	Payment
90	5,000	5,000,000	10,548
90	0	1,000,000	12,959
90	0	1,000,000	13,456
90	0	1,000,000	16,148
90	0	5,000,000	20,684
90	5,000,000	2,000,000	23,691
90	75,000	3,000,000	27,196
90	0	1,000,000	28,283
90	5,000,000	2,000,000	169,616
90	12,000,000	5,000,000	268,534
90	50,000	1,000,000	1,000,000
90	500,000	10,000,000	1,033,715
90	0	3,000,000	1,363,432
90	4,500,000	5,000,000	2,205,674
90	1,500,000	30,000,000	3,148,409
90	16,000,000	10,000,000	8,652,788
90	500,000	10,000,000	8,719,031
90	55,000,000	10,000,000	9,508,586
91	0	10,000,000	1,362
91	0	10,000,000	1,883
91	50,000	3,000,000	3,394
91	100,000	1,000,000	4,246
91	0	500,000	6,992
91	0	10,000,000	10,262
91	0	1,000,000	16,452
91	0	3,000,000	20,427
91	0	3,000,000	27,494
91	0	3,000,000	30,698
91	0	1,000,000	45,743
91	100,000	5,000,000	52,023
91	100,000	1,000,000	54,481
91	10,000,000	10,000,000	164,732
91	0	1,000,000	535,593
91	1,000,000	6,000,000	1,491,732
91	30,000,000	10,000,000	2,271,437
91	200,000	5,000,000	2,732,422
91	0	10,000,000	3,130,873
91	0	10,000,000	3,622,812
91	500,000	10,000,000	4,288,766
91	0	10,000,000	4,435,099
91	0	5,000,000	5,000,000
91	10,000,000	20,000,000	5,644,894

Table 1.3 Liability payments—pension actuaries, 1992–1993

Year	Deductible.	Maximum payment	Payment
92	0	5,000,000	1,003
92	0	1,000,000	2,388
92	0	5,000,000	3,067
92	0	10,000,000	4,066
92	0	10,000,000	6,758
92	0	1,000,000	6,781
92	0	3,000,000	7,439
92	1,000	5,000,000	10,617
92	1,000	5,000,000	10,888
92	0	7,500,000	34,745
92	0	400,000	58,587
92	0	1,000,000	113,166
92	0	5,000,000	122,967
92	350,000	3,000,000	199,607
92	150,000	5,000,000	298,847
92	75,000	1,000,000	1,000,000
92	0	1,000,000	1,000,000
92	0	10,000,000	3,022,258
92	0	5,000,000	3,201,434
92	10,000,000	10,000,000	3,754,944
92	10,000,000	5,000,000	5,000,000
92	700,000	20,000,000	6,126,080
93	0	5,000,000	189
93	0	10,000,000	388
93	0	1,000,000	2,026
93	0	1,000,000	2,354
93	0	10,000,000	8,959
93	0	1,000,000	17,865
93	0	10,000,000	41,170
93	0	5,000,000	158,391
93	1,000,000	10,000,000	596,674
93	100,000	10,000,000	926,657
93	1,000,000	10,000,000	1,101,816
93	100,000	5,000,000	1,903,358
93	0	10,000,000	2,055,117
93	0	10,000,000	2,966,399

Table 1.4 Liability payments—property and liability actuaries

Year	Deductible	Maximum payment	Payment
90	0	7,500,000	60,664
90	5,000,000	5,000,000	116,134
90	0	7,500,000	576,857
91	0	1,000,000	31,698
91	50,000	10,000,000	46,427
91	0	7,500,000	119,206
91	450,000	1,000,000	405,796
91	0	5,000,000	1,519,846
91	500,000	2,000,000	2,000,000
92	0	1,000,000	505
92	250,000	5,000,000	17,833
92	0	3,000,000	20,546
92	22,000,000	10,000,000	10,000,000
93	20,000	5,000,000	4,699
93	150,000	10,000,000	6,055
93	0	1,000,000	10,950
93	25,000,000	10,000,000	244,023
93	100,000	1,000,000	255,892
93	0	5,000,000	384,222

Table 1.5 Exposures and loss counts

	Life/health		Pension		Property/liability	
Year	Exposure	Claims	Exposure	Claims	Exposure	Claims
1990	853	20	1,446	27	639	5
1991	1,105	14	1,780	35	725	8
1992	1,148	16	1,717	36	685	4
1993	1,270	21	2,065	24	864	11
Total	4,376	71	7,008	122	2,913	28

2

Loss Distributions—Models for the Amount of a Single Payment

2.1 INTRODUCTION

The purpose of insurance is to indemnify policyholders when unforeseen adverse events occur. There is tremendous variety in the events that are covered by insurance as indicated in Table 2.1.

There are a number of features common to any insurance system. They are:

1. There must be risk—a condition in which there is a possibility of an adverse deviation from an expected outcome.

2. The loss must be financial. That is, it involves a loss in value that can be measured in dollars.

3. Some or all of the risk is transferred from the insured to the insurer.

4. There is an expectation that through the pooling of risks the insurer will improve the estimation of expected total losses.

The above were taken from the introductory insurance text by Vaughan [121]. Similar statements can be found in other texts. Additional definitions of interest can be found in the Society of Actuaries' "Principles of Actuarial Science" [113]. A few of them, along with a principle (the principle number is taken from the referenced document), are repeated here.

Definition 2.1 ***Statistical regularity*** *describes phenomena such that, if a sequence of independent experiments is held under the same specified condi-*

Table 2.1 Insurance benefits

Event	Insurance	Benefit
Death	Life	Fixed payment to beneficiary
Continued life	Annuity	Fixed, periodic, payment to policyholder
Unintentional misdeed	Liability	Payment of legal fees and damages
Damage to property	Property	Payment relating to amount of damage
Inability to work	Disability	Replacement of salary
Illness	Health	Payment of medical expenses

tions, the proportion of occurrences of a given event stabilizes as the number of experiments becomes larger.

Definition 2.2 *A **mathematical model** is an abstract and simplified representation of a given phenomenon that can be expressed in mathematical terms.*

Definition 2.3 *A **stochastic model** is a mathematical model for a phenomenon displaying statistical regularity that can accurately describe the probabilities of outcomes.*

Definition 2.4 *An **actuarial risk** is a phenomenon that has economic consequences and that is subject to uncertainty with respect to one or more of the actuarial risk variables: occurrence, timing, and severity.*

Principle 3.1 *Actuarial risks can be stochastically modeled.*

One key feature of an insurance system is that all quantities can be measured in monetary units. This means that the set of real numbers will be sufficient for our purposes and that, in particular, random variables can be used to build our actuarial models.

Another feature of all insurances is the element of randomness. As noted in Definition 2.4, any stream of benefits paid to a policyholder has three components: the number of benefit payments (if any at all), the timing of the benefit payments, and the amount of each payment. Any or all of them may be random with respect to a particular policy. For example, a basic life insurance policy makes exactly one payment, for a fixed amount, but at a random time. Each of the three components is random for an automobile insurance policy.

As indicated in Definition 1.9 a loss distribution refers to the assignment of probabilities to one of these random processes. This chapter will be devoted specifically to determining loss distributions for loss or payment amounts. The determination of probability models for the component that measures the number of payments will be taken up in Chapter 3.

The discussion of loss distributions will begin with a description of two alternative methods of obtaining a loss distribution. The empirical method

Table 2.2 Grouped dental data

Amount paid	No. of payments
0–25	30
25–50	31
50–100	57
100–150	42
150–250	65
250–500	84
500–1,000	45
1,000–1,500	10
1,500–2,500	11
2,500–4,000	3

is easy to use, though not always available. The parametric approach has many appealing features, although it is more difficult. The remainder of this chapter is devoted to a thorough discussion of the parametric approach. In particular, the various problems that arise due to the nature of insurance data are addressed. These include the variety of parametric estimation methods, data from policies with coverage modifications, data from multiple sources, tests of hypotheses, and more complicated (for example, bivariate) situations.

Throughout the chapter, two examples will be used to illustrate the methods being used. The first is Example 1.4. As a reminder, the data are basic dental claims on a policy with a deductible of 50. The 10 payments observed were

$$141 \quad 16 \quad 46 \quad 40 \quad 351 \quad 259 \quad 317 \quad 1{,}511 \quad 107 \quad 567.$$

This example will be referred to as the *individual data dental example.* The second is the following.

Example 2.1 *Consider the same setting—that is, dental payments—only now assume we were able to collect a larger amount of data. In particular the data were summarized as in Table 2.2.*

This example will be referred to as the *grouped data dental example.* □

2.2 EMPIRICAL ESTIMATION

Of the two most common estimation procedures, empirical estimation is by far the simplest. As such, it should not be ignored and, in particular, when there are a very large number of observations, the method may well be the most accurate. In this section, a formal definition is given and then a few specific empirical estimators are introduced. Given that we will be presented with more than one means of solving a particular problem, it is imperative that we have some method of evaluating a particular estimator. This will be done in Section 2.3, and then the empirical estimators will be evaluated according to these criteria.

An important point is that quality is a property of the estimator and not of the estimate. We are interested in the quality of the method, not the quality of any particular outcome from using the method. Use of a high-quality estimator does not ensure that realized outcomes will be consistent with estimated outcomes. This point is stated nicely in a Society of Actuaries' principles draft ([114], Discussion to Principle 4) regarding the level of adequacy of a provision for a block of life risk obligations (that is, the probability that the company will have enough money to meet its contractual obligations).

> Adequacy estimates are prospective, but the actuarial model is generally validated against past experience. It is incorrect to conclude on the basis of subsequent experience that the actuarial assumptions were inappropriate at the time the estimate was made or that the level of adequacy was overstated.

2.2.1 Definition

The purpose of any estimation process is to use the results of a sample to learn about the population from which it was taken. We will assume throughout that all of our samples are random samples.

Definition 2.5 *Let $X_1, \ldots, X_n$ be random variables such that they are mutually independent and have the same probability distribution. Then the collection of random variables is said to be a* ***random sample****.*

One consequence of this definition is that the joint distribution of the n random variables is easy to obtain:

$$F_{X_1,\ldots,X_n}(x_1, \ldots, x_n) = F_X(x_1) \cdots F_X(x_n)$$

where $F_X(x)$ is the common distribution function for the random sample. It should also be noted that the definition says nothing about an actual sample or an actual population. However, if there is a population and a sample is drawn such that each member of the population has an equal chance of being drawn and the identity of any one sampled member has no impact on the identity of any other sampled member, then the above definition holds.

Our goal is to learn as much as possible about $F(x)$ based on the random sample.[1] While it may be that we are interested in just a few features of

[1] When necessary, the subscript on a distribution or density function will indicate the particular random variable to which it belongs. When it is obvious from the context, the subscript will be left off.

the population (for example, its mean or 90th percentile), all of the methods discussed here begin by estimating the population distribution and then obtaining the quantities of interest. One advantage of this approach is that if we change our mind about the quantities of interest, no additional estimation work may be needed.

The empirical approach estimates $F(x)$ by the empirical distribution.

Definition 2.6 *The* ***empirical distribution*** *is obtained from a sample by placing probability* $1/n$ *at each observation. More formally, the* ***empirical cumulative distribution function*** *(cdf) is*

$$F_n(x) = \frac{\text{number of } x_j \leq x}{n}.$$

The empirical cdf is a step function that increases by $1/n$ at each data point. It is a discrete distribution placing probability at at most n values. The probability function[2] is

$$f_n(x) = \frac{\text{number of } x_j = x}{n}.$$

Any time we have a cdf or pf, there is an associated random variable. For notation purposes, it is convenient to define the random variable which has the empirical cdf as its cdf.

Definition 2.7 *The* ***empirical random variable*** *is a random variable which has* $F_n(x)$ *as its cdf. It will be denoted* $\hat{X}$.

Therefore, we could just as easily write $F_{\hat{X}}(x)$ for the empirical cdf, however, in this case the notation used above is much simpler.

Example 2.2 (individual data dental example) *The graph of the empirical distribution for this data is presented in Figure 2.1.* □

The graph in Figure 2.1 shows vertical bars where there should just be open space. The true empirical cdf is a sequence of horizontal bars with the left-hand point of each bar included in the graph and the right-hand point excluded. We think it is easier to visualize the graph with the vertical lines inserted.

For grouped data it is impossible to determine the empirical cdf because the individual data points are not available. However, it is possible to approximate

[2] In this text the terms **probability function** and **probability density function** and the abbreviations pf and pdf will be used interchangeably. The context should make it clear if the random variable is discrete, continuous, or partly discrete and partly continuous. The pdf will usually be denoted as a function, such as $f(x)$. In the special case where the random variable is discrete and takes on probability only on integers, the pf can be written p_n.

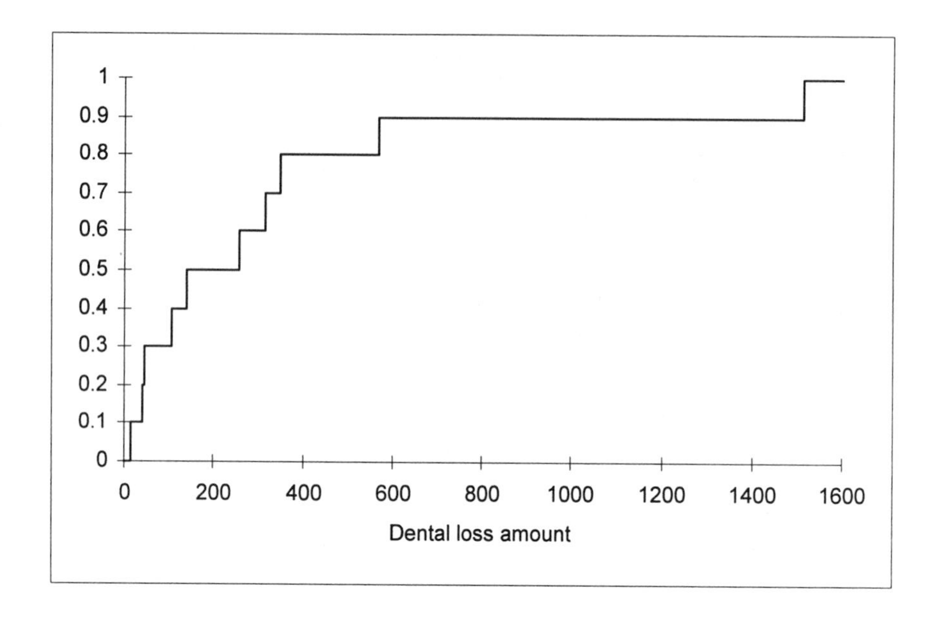

Fig. 2.1 Empirical distribution function of individual dental loss amounts

it. The following definition also provides some standard notation for grouped data sets.

Definition 2.8 *Let $c_0 < c_1 < \cdots < c_r$ be the boundaries for a grouped data set. Let n_j be the number of observations in the interval $(c_{j-1}, c_j]$, $j = 1, \ldots, r$. It is possible that $c_r = \infty$. The empirical cdf can be obtained at the boundaries as $F_n(c_j) = \dfrac{1}{n}\sum_{i=1}^{j} n_i$. The graph formed by connecting the empirical cdf at these points by straight lines is called the* **ogive** *and is an approximation to the empirical cdf. The formal definition is*

$$\tilde{F}_n(x) = \begin{cases} 0, & x \le c_0 \\ \dfrac{(c_j - x)F_n(c_{j-1}) + (x - c_{j-1})F_n(c_j)}{c_j - c_{j-1}}, & c_{j-1} < x \le c_j \\ 1, & x > c_r. \end{cases}$$

Note that the ogive is not defined for $x > c_{r-1}$ if $c_r = \infty$ (unless $n_r = 0$). Also note that the ogive must be a piecewise linear function. That means the derivative exists everywhere except at the boundaries and where it exists, it is just the slope of the straight line segment.

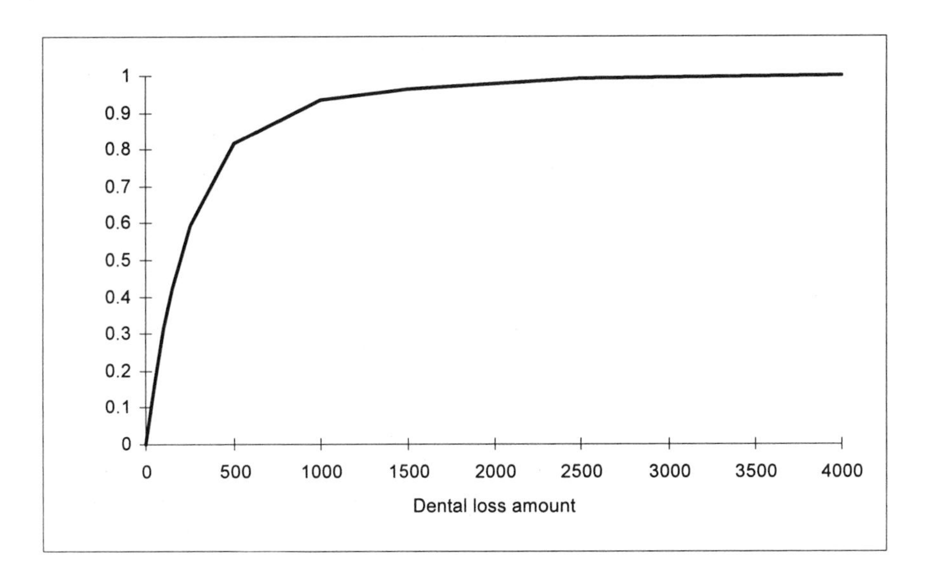

Fig. 2.2 Ogive of grouped dental loss amounts

Definition 2.9 *The derivative (where it exists) of the ogive is an empirical approximation to the density function and is called a* ***histogram****. The formal definition is*

$$\tilde{f}_n(x) = \begin{cases} 0, & x \le c_0 \\ \dfrac{F_n(c_j) - F_n(c_{j-1})}{c_j - c_{j-1}} = \dfrac{n_j}{n(c_j - c_{j-1})}, & c_{j-1} < x \le c_j \\ 0, & x > c_r. \end{cases} \tag{2.1}$$

Example 2.3 (grouped data dental example) *The ogive and histogram for the grouped dental example are (to highlight the beginning part of the data, values past 1,500 were left off) presented in Figures 2.2 and 2.3.* □

Note that as an approximation to the pdf the histogram will be positive with an area of one (unless $c_k = \infty$ in which case there is no way to represent the probability for the last group). It is important to note (as is clear from (2.1)) that it is the area, not the height, of the histogram bars that is proportional to the number in the group.

2.2.2 Specific empirical estimators

From the empirical distribution or the ogive, we can obtain estimates of any desired feature of the population distribution. Throughout we assume a random sample from the population has yielded the observations $x_1, \ldots, x_n$. Any inference we make is for the population that produced these values. The key

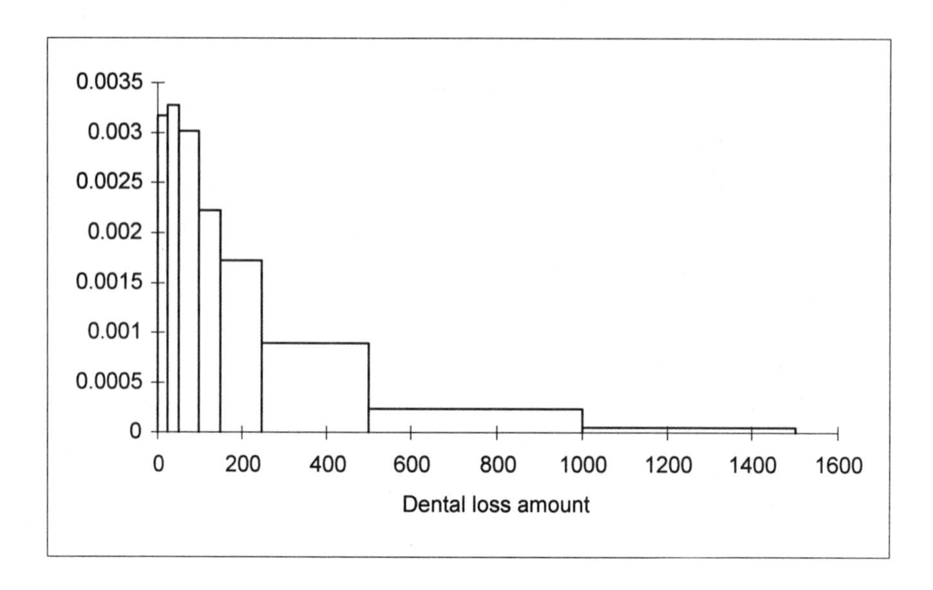

Fig. 2.3 Histogram of grouped dental loss amounts

to empirical estimation is to compute for the empirical distribution the quantity we would like to know for the population that was sampled. Specific cases are presented in this subsection.

2.2.2.1 Empirical estimate of the mean For individual data the mean of the empirical distribution is easy to obtain. It is

$$\hat{\mu} = \bar{x} = \frac{1}{n}\sum_{j=1}^{n} x_j.$$

This is the familiar sample mean which exemplifies the basis for empirical estimation: To determine a particular feature of the population, determine that same feature from the sample. Throughout, an estimate or estimator will be indicated by a circumflex (or "hat").

2.2.2.2 Empirical estimate of other moments Population moments (if they exist) are defined as follows.

Definition 2.10 *The* k**th raw moment** *(also called the* k**th moment about the origin***) is*

$$\mu_k' = E(X^k).$$

The k**th central moment** *is*

$$\mu_k = E[(X-\mu)^k].$$

It is customary to denote μ'_1 by the simpler μ. Although most applications use whole number values for the moments, there is nothing in these definitions to prevent us from using fractional or negative moments. There is no assurance that the sums or integrals that define these moments converge. When the sum or integral does not converge we say that the moment does not exist. The moment is said to be "infinite" if the sum or integral approaches infinity.

In addition to the mean, certain moments have acquired specific names and additional symbols.

Definition 2.11 *The* ***variance*** *is* $\sigma^2 = \mu_2 = \mu'_2 - \mu^2$.
The ***standard deviation*** *is* $\sigma = \sqrt{\sigma^2}$.
The ***coefficient of variation*** *is* σ/μ.
The ***skewness*** *is* $\gamma_1 = \mu_3/\sigma^3$.
The ***kurtosis*** *is* $\gamma_2 = \mu_4/\sigma^4$.
The k***th factorial moment*** *is* $\mu_{(k)} = E[X(X-1)\cdots(X-k+1)]$. *Here* k *must be an integer.*

For any symmetric distribution the skewness is zero. Positively skewed distributions tend to have most of the probability on small values, but the remaining probability is stretched over a long range of larger values. The histogram in Figure 2.3 is an example of a positively skewed distribution. The kurtosis measures the nature of the spread of the values around the mean. A small kurtosis (it is always non-negative) indicates a sharp peak at the middle. A large kurtosis indicates a much slower drop-off. While the kurtosis is similar to the variance in that it measures spread, it is more effective at distinguishing those distributions that place additional probability on larger values. A random variable with a normal distribution has a kurtosis of three, regardless of its parameters.

The relationships between the central and raw moments are

$$\begin{aligned}
\mu_3 &= \mu'_3 - 3\mu'_2\mu + 2\mu^3 \\
\mu_4 &= \mu'_4 - 4\mu'_3\mu + 6\mu'_2\mu^2 - 3\mu^4.
\end{aligned}$$

The empirical estimates are

$$\hat{\mu}'_k = E(\hat{X}^k) = \frac{1}{n}\sum_{j=1}^{n} x_j^k$$

and

$$\hat{\mu}_k = E[(\hat{X} - \hat{\mu})^k] = \frac{1}{n}\sum_{j=1}^{n}(x_j - \bar{x})^k.$$

Notice that the sample variance is defined by dividing the sum of squares by the sample size. This is slightly different than the usual division by $n-1$. An explanation for the difference is provided in the next subsection.

For grouped data, provided that $c_r < \infty$, the histogram can be integrated to yield the appropriate moments. The raw moments are

$$\hat{\mu}'_k = \sum_{j=1}^{r} \int_{c_{j-1}}^{c_j} x^k \frac{n_j}{n(c_j - c_{j-1})} dx = \sum_{j=1}^{r} \frac{n_j(c_j^{k+1} - c_{j-1}^{k+1})}{n(k+1)(c_j - c_{j-1})}. \tag{2.2}$$

This is similar to applying a uniform distribution of deaths assumption when interpolating in a life table.

Another set of moments is extremely useful for insurance calculations. As indicated in Definition 1.7, it is common to limit the benefit that will be paid by the insurance. If Y is the amount paid random variable it is related to the loss random variable X by $Y = \min(X, u) = (X \wedge u)$, where u is the policy limit and the symbol "$\wedge$" is interpreted to mean "minimum." This leads to the following definition.

Definition 2.12 *The* k**th limited moment** *of* X *is*

$$E[(X \wedge u)^k] = E(Y^k) = E\{[\min(X, u)]^k\}.$$

The first limited moment is called the ***limited expected value (LEV)*** *and is denoted by* $E(X \wedge u)$.

If X takes on only non-negative values all limited moments with $k \geq 0$ must exist. Furthermore, $\lim_{u \to \infty} E(X \wedge u) = E(X)$ if it exists. Formulas for calculating the limited moments in the discrete and continuous cases are

$$E[(X \wedge u)^k] = \sum_{x_j < u} x_j^k f(x_j) + u^k \sum_{x_j \geq u} f(x_j)$$

and

$$E[(X \wedge u)^k] = \int_0^u x^k f(x) dx + u^k \int_u^\infty f(x) dx. \tag{2.3}$$

In both cases the second term can be replaced by $u^k[1 - F(u)]$.

Empirical estimates are obtained as before. For individual data, the formula is

$$E[(\hat{X} \wedge u)^k] = \frac{1}{n}\left(\sum_{x_j < u} x_j^k + \sum_{x_j \geq u} u^k\right). \tag{2.4}$$

For grouped data, the histogram can be used. The formula is (where u is such that $c_{j-1} \leq u \leq c_j$)

$$\begin{aligned}\tilde{E}[(\hat{X} \wedge u)^k] &= \sum_{i=1}^{j-1} \int_{c_{i-1}}^{c_i} x^k \frac{n_i}{n(c_i - c_{i-1})} dx + \int_{c_{j-1}}^{u} x^k \frac{n_j}{n(c_j - c_{j-1})} dx \\ &\quad + \int_u^{c_j} u^k \frac{n_j}{n(c_j - c_{j-1})} dx + \sum_{i=j+1}^{r} \int_{c_{i-1}}^{c_i} u^k \frac{n_i}{n(c_i - c_{i-1})} dx\end{aligned}$$

$$= \sum_{i=1}^{j-1} \frac{n_i(c_i^{k+1} - c_{i-1}^{k+1})}{n(k+1)(c_i - c_{i-1})} + \frac{n_j(u^{k+1} - c_{j-1}^{k+1})}{n(k+1)(c_j - c_{j-1})}$$
$$+ \frac{n_j u^k (c_j - u)}{n(c_j - c_{j-1})} + \sum_{i=j+1}^{r} \frac{n_i u^k}{n}.$$

For the special case where $k = 1$ the above simplifies to

$$\tilde{E}(\hat{X} \wedge u) = \sum_{i=1}^{j-1} \frac{n_i(c_i + c_{i-1})}{2n} + \frac{n_j(2uc_j - c_{j-1}^2 - u^2)}{2n(c_j - c_{j-1})} + \sum_{i=j+1}^{r} \frac{n_i u}{n}. \qquad (2.5)$$

In reports that use grouped data, it is common to include the sample mean for the observations in each group. Let a_i be the sample mean for the ith group (so $c_{i-1} < a_i \le c_i$). Then the empirical limited expected value can be calculated exactly at the boundaries by

$$E(\hat{X} \wedge c_j) = \sum_{i=1}^{j} \frac{n_i a_i}{n} + \sum_{i=j+1}^{r} \frac{n_i c_j}{n}.$$

It is not at all clear what the best way is to interpolate between boundaries. One idea is suggested in Exercise 2.3.

Example 2.4 (dental example) *Determine the empirical mean, standard deviation, skewness, and kurtosis from the two samples. Also, estimate the limited expected value at* 400 *for each sample.*

For the individual data example we have

$$\begin{aligned}
\hat{\mu}_1' &= (141 + \cdots + 567)/10 = 335.5 \\
\hat{\mu}_2' &= (141^2 + \cdots + 567^2)/10 = 2.9307 \times 10^5 \\
\hat{\mu}_3' &= (141^3 + \cdots + 567^3)/10 = 3.7287 \times 10^8 \\
\hat{\mu}_4' &= (141^4 + \cdots + 567^4)/10 = 5.3463 \times 10^{11} \\
\hat{\mu} &= 335.5 \\
\hat{\sigma} &= \sqrt{293{,}068.3 - 335.5^2} = 424.86 \\
\hat{\gamma}_1 &= (1.5343 \times 10^8)/(7.6691 \times 10^7) = 2.0006 \\
\hat{\gamma}_2 &= (1.9415 \times 10^{11})/(3.2583 \times 10^{10}) = 5.9586.
\end{aligned}$$

From (2.4) the LEV at 400 is

$$\begin{aligned}
E(\hat{X} \wedge 400) &= (141 + 16 + 46 + 40 + 351 + 259 + 317 + 400 + 107 + 400)/10 \\
&= 207.7.
\end{aligned}$$

For the grouped data example

$$\hat{\mu}_1' = \frac{1}{378(2)} \left[\frac{30(25^2 - 0^2)}{25 - 0} + \cdots + \frac{3(4{,}000^2 - 2{,}500^2)}{4{,}000 - 2{,}500} \right] = 353.34$$

$$
\begin{aligned}
\hat{\mu}_2' &= \frac{1}{378(3)}\left[\frac{30(25^3-0^3)}{25-0}+\cdots+\frac{3(4{,}000^3-2{,}500^3)}{4{,}000-2{,}500}\right]=3.5768\times10^5\\
\hat{\mu}_3' &= \frac{1}{378(4)}\left[\frac{30(25^4-0^4)}{25-0}+\cdots+\frac{3(4{,}000^4-2{,}500^4)}{4{,}000-2{,}500}\right]=6.5863\times10^8\\
\hat{\mu}_4' &= \frac{1}{378(5)}\left[\frac{30(25^5-0^5)}{25-0}+\cdots+\frac{3(4{,}000^5-2{,}500^5)}{4{,}000-2{,}500}\right]=1.6261\times10^{12}\\
\hat{\mu} &= 353.34\\
\hat{\sigma} &= \sqrt{357{,}680.2-353.34^2}=482.53\\
\hat{\gamma}_1 &= (3.6771\times10^8)/(1.1235\times10^8)=3.2730\\
\hat{\gamma}_2 &= (9.1635\times10^{11})/(5.4210\times10^{10})=16.904.
\end{aligned}
$$

From (2.5) the LEV at 400 is

$$
\begin{aligned}
\tilde{E}(\hat{X}\wedge 400) &= \{30(12.5)+31(37.5)+57(75)+42(125)+65(200)\\
&\qquad +84[2(400)(500)-250^2-400^2]/2(250)\\
&\qquad +(45+10+11+3)(400)\}/378\\
&= 215.56.
\end{aligned}
$$

□

2.2.2.3 *Percentiles* Knowing all of the percentiles is equivalent to knowing the cdf. The formal definition of a percentile follows.

Definition 2.13 *The* 100***p*th percentile** *of a distribution $F(x)$ is any number, π_p such that*

$$F(\pi_p-)\le p\le F(\pi_p)$$

where $F(\pi_p-)=\lim_{h\downarrow 0}F(\pi_p-h)$.

If the cdf is strictly increasing (such as the ogive illustrated in Example 2.3) the solution will be unique. If the cdf is constant for any interval (as for any discrete distribution and any empirical distribution), the values for which it is constant are all acceptable as a particular percentile.

The empirical estimator of a percentile is simply the corresponding percentile from the empirical distribution.

Example 2.5 (dental example) *Determine the empirical estimates of the 45th and 80th percentiles from the two samples.*

For the individual data example we have $F_{10}(141)=0.5$ and $F_{10}(141-)=0.4$ and so $\hat{\pi}_{0.45}=141$. Also, $F_{10}(351)=0.8$ and $F_{10}(567-)=0.8$ and so $351<\hat{\pi}_{0.8}<567$.

For the grouped data example we have

$$
\begin{aligned}
0.45 &= \tilde{F}_{378}(\hat{\pi}_{0.45})=\frac{(250-\hat{\pi}_{0.45})(160/378)+(\hat{\pi}_{0.45}-150)(225/378)}{250-150}\\
\hat{\pi}_{0.45} &= 165.54
\end{aligned}
$$

and

$$\begin{aligned} 0.8 &= \tilde{F}_{378}(\hat{\pi}_{0.8}) = \frac{(500 - \hat{\pi}_{0.8})(225/378) + (\hat{\pi}_{0.8} - 250)(309/378)}{500 - 250} \\ \hat{\pi}_{0.8} &= 480.36. \end{aligned}$$

In both these cases, the percentile can be found graphically by locating the percentage (0.45 and 0.80 in the above example) on the vertical axis of the appropriate figure (empirical cdf or ogive) and then finding the percentile on the horizontal axis. □

For individual data the method described above is unsatisfactory because it does not provide a unique answer in some cases and for the others there are several percentiles with the same value. A number of ways have been suggested for smoothing out the process.[3] We prefer the following one.

Definition 2.14 *The **smoothed empirical estimate** of a percentile is found by*

$$\begin{aligned} \hat{\pi}_p &= (1-h)x_{(g)} + hx_{(g+1)}, \text{ where} \\ g &= [(n+1)p] \text{ and } h = (n+1)p - g. \end{aligned}$$

Here $[\cdot]$ *indicates the greatest integer function and* $x_{(1)} \le x_{(2)} \le \cdots \le x_{(n)}$ *are the order statistics from the sample.*

Unless there are two or more data points with the same value, the percentiles will be unique. One feature of this process is that $\hat{\pi}_p$ cannot be obtained for $p < 1/(n+1)$ or $p > n/(n+1)$. This seems reasonable as we should not expect to be able to infer the value of large or small percentiles from small samples. We will use the smoothed version whenever an empirical percentile estimate is called for.

Example 2.6 (individual data dental example) *Determine the smoothed empirical estimates of the 45th and 80th percentiles.*

For the 45th percentile,

$$\begin{aligned} g &= [11(0.45)] = 4, \ h = 4.95 - 4 = 0.95, \\ \hat{\pi}_{0.45} &= 0.05(107) + 0.95(141) = 139.3. \end{aligned}$$

For the 80th percentile,

$$\begin{aligned} g &= [11(0.8)] = 8, \ h = 8.8 - 8 = 0.8, \\ \hat{\pi}_{0.8} &= 0.2(351) + 0.8(567) = 523.8. \end{aligned}$$

□

[3] Hyndman and Fan [70] present nine different methods. They recommend a slight modification of the one presented here, using $g = [p(n + \frac{1}{3}) + \frac{1}{3}]$ and $h = p(n + \frac{1}{3}) + \frac{1}{3} - g$.

2.2.3 Interval estimation

All of the estimators discussed to this point have been **point estimators**. That is, the estimation process produces a single value that represents our best attempt to determine the value of the unknown population quantity. While that value may be a good one, we do not expect it to exactly match the true value. A more useful statement is often provided by an **interval estimator**. Instead of a single value, the result of the estimation process is a range of possible numbers, any of which is likely to be the true value. A specific type of interval estimator is the confidence interval.

Definition 2.15 *A* $100(1-\alpha)\%$ ***confidence interval*** *for a parameter* θ *is a pair of values* L *and* U *computed from a random sample such that* $\Pr(L \le \theta \le U) \ge 1-\alpha$ *for all* θ.

Note that this definition does not uniquely specify the interval. Because the definition is a probability statement and must hold for all θ, it says nothing about whether or not a particular interval encloses the true value of θ from a particular population. Instead, the **level of confidence**, $1-\alpha$, is a property of the method used to obtain L and U and not of the particular values obtained. The proper interpretation is that if we use a particular interval estimator over and over on a variety of samples, about $100(1-\alpha)\%$ of the time our interval will enclose the true value.

Constructing confidence intervals is usually very difficult. For example, we know that if a population has a normal distribution with unknown mean and variance, a $100(1-\alpha)\%$ confidence interval for the mean uses

$$\begin{aligned} L &= \bar{X} - t_{\alpha/2,n-1}s/\sqrt{n} \\ U &= \bar{X} + t_{\alpha/2,n-1}s/\sqrt{n} \end{aligned} \tag{2.6}$$

where $s = \sqrt{\sum_{i=1}^{n}(X_i - \bar{X})^2/(n-1)}$ and $t_{\alpha,b}$ is the $100(1-\alpha)$th percentile of the t distribution with b degrees of freedom. But it takes a great deal of effort to verify that this is correct (see, for example, [65, p. 214]).

However, there is a method for constructing approximate confidence intervals that is often accessible. Suppose we have a point estimator $\hat{\theta}$ of parameter θ such that $E(\hat{\theta}) \doteq \theta$, $Var(\hat{\theta}) \doteq v(\theta)$, and $\hat{\theta}$ has approximately a normal distribution. With all these approximations, we have that approximately

$$1-\alpha \doteq \Pr\left(-z_{\alpha/2} \le \frac{\hat{\theta}-\theta}{\sqrt{v(\theta)}} \le z_{\alpha/2}\right) \tag{2.7}$$

and solving for θ produces the desired interval. Sometimes this is difficult to do (due to the appearance of θ in the denominator) and so replace $v(\theta)$ in (2.7) with $v(\hat{\theta})$ to obtain a further approximation

$$1-\alpha \doteq \Pr\left(\hat{\theta} - z_{\alpha/2}\sqrt{v(\hat{\theta})} \le \theta \le \hat{\theta} + z_{\alpha/2}\sqrt{v(\hat{\theta})}\right) \tag{2.8}$$

where z_α is the $100(1-\alpha)th$ percentile of the standard normal distribution.

Example 2.7 *Use (2.8) to construct an approximate 95% confidence interval for the mean of a normal population with unknown variance.*

Use $\hat{\theta} = \bar{X}$ and then note that $E(\hat{\theta}) = \theta$, $Var(\hat{\theta}) = \sigma/\sqrt{n}$, and $\hat{\theta}$ does have a normal distribution. The confidence interval is then $\bar{X} \pm 1.96s/\sqrt{n}$. Because $t_{.025,n-1} > 1.96$, this approximate interval must be narrower than the exact interval given by (2.6). That means that our level of confidence is something less than 95%. □

Example 2.8 *Use (2.7) and (2.8) to construct approximate 95% confidence intervals for the mean of a Poisson distribution. Obtain intervals for the particular case where $n = 25$ and $\bar{x} = 0.12$.*

For the first interval

$$0.95 \doteq \Pr\left(-1.96 \le \frac{\bar{X} - \theta}{\sqrt{\theta/n}} \le 1.96\right)$$

is true if and only if

$$|\bar{X} - \theta| \le 1.96\sqrt{\theta/n}$$

which is equivalent to

$$(\bar{X} - \theta)^2 \le 3.8416\theta/n$$

or

$$\theta^2 - \theta(2\bar{X} + 3.8416/n) + \bar{X}^2 \le 0.$$

Solving the quadratic produces the interval

$$\bar{X} + \frac{1.9208}{n} \pm \frac{1}{2}\sqrt{\frac{15.3664\bar{X} + 3.8416^2/n}{n}}$$

and for this problem the interval is 0.197 ± 0.156.

For the second approximation the interval is $\bar{X} \pm 1.96\sqrt{\bar{X}/n}$ and for the example it is 0.12 ± 0.136. This interval extends below zero (which is not possible for the true value of θ). This is because (2.8) is too crude an approximation in this case. □

The intervals created to this point have been parametric, at least in the sense that the normal distribution was employed as an approximation. There is one situation in which we can form a confidence interval that is truly nonparametric. That is, the stated level of confidence is correct regardless of the population distribution. The situation is the construction of a confidence interval for a population percentile. The following theorem provides the required result.

Theorem 2.1 *Let $X_1, \ldots, X_n$ be a random sample of size n where each X_j has a continuous distribution. Let $X_{(1)} \le X_{(2)} \le \cdots \le X_{(n)}$ be the order statistics from the sample. Let $1 \le a < b \le n$ be two integers. Then the interval $(X_{(a)}, X_{(b)})$ is a $100(1-\alpha)\%$ confidence interval for π_p, the population pth percentile where*

$$1 - \alpha = \Pr(a \le B < b)$$

where B is a random variable having the binomial distribution with parameters n and p.

Proof: The event $\{X_{(a)} < \pi_p < X_{(b)}\}$ is equivalent to the event which requires at least a of the sample items to be less than π_p and at least $n-b+1$ of the sample items to be greater than π_p. Define "success" to be a sample item being less than π_p. The event is thus equivalent to there being at least a successes, but at most $b-1$ successes. By the definition of percentile (and the fact that each X_j is a continuous random variable) the probability of success is p. Therefore we have a binomial sampling situation, and the probability of the event is as stated in the theorem. □

It should be noted that because a and b must be integers, there is a limit to the possible values for $1-\alpha$. This is illustrated in the following example.

Example 2.9 (individual data dental example) *Construct an interval with at least 90% confidence for the 70th percentile.*

The answer is not unique, but it makes sense to construct the interval by using as few of the order statistics as possible. The probabilities for the binomial distribution with $n = 10$ and $p = 0.7$ are given in Table 2.3. The fastest way to get the probability to 0.9 is by adding $0.26683 + 0.23347 + 0.20012 + 0.12106 + 0.10292 = 0.92440$. This is $\Pr(5 \le B < 10)$ and so the general interval for a sample of size 10 is $(X_{(5)}, X_{(10)})$. For this data set the interval is $(141, 1{,}511)$ and the confidence level is actually 92.44%. Note that had $\Pr(B = 10)$ been used to obtained the desired probability, the value of b would be infinity and the upper limit for the confidence interval would be the largest possible value of the random variable. Similarly, if it turns out that $a = 0$, the lower limit is the smallest possible value for the random variable. □

For large sample sizes this process becomes cumbersome as the binomial probabilities become numerous and painful to compute. In that setting we can appeal to the normal distribution as an effective approximation to the binomial. The following example illustrates the process.

Example 2.10 *Determine the order statistics that form the boundaries for a 90% confidence interval for the 70th percentile when the sample size is 750.*

The variable B has the binomial distribution with parameters 750 and .7. This can be approximated by a normal distribution with mean $750(0.7) = 525$

Table 2.3 Probabilities from the binomial(10, 0.7) distribution

i	$\Pr(B = i)$	i	$\Pr(B = i)$
0	0.00001	6	0.20012
1	0.00014	7	0.26683
2	0.00145	8	0.23347
3	0.00900	9	0.12106
4	0.03676	10	0.02825
5	0.10292		

and variance $750(0.7)(0.3) = 157.5$ for a standard deviation of 12.55. We then have

$$\begin{aligned} 0.9 &= \Pr(a \le B < b) \\ &= \Pr\left(\frac{a - 0.5 - 525}{12.55} < Z < \frac{b - 0.5 - 525}{12.55}\right) \end{aligned}$$

where Z has the standard normal distribution and the continuity correction has been employed. A symmetric 90% interval is obtained by setting

$$\frac{a - 0.5 - 525}{12.55} = -1.645$$

and

$$\frac{b - 0.5 - 525}{12.55} = 1.645.$$

The solutions are $a = 504.86$ and $b = 546.14$. In order to ensure at least 90% confidence, the interval is $(X_{(504)}, X_{(547)})$. □

You are asked to derive the general formula for a and b when n, p, and $1 - \alpha$ are arbitrary, in Exercise 2.16. One particular use for this result is in simulation studies where the goal is to estimate a particular population percentile. This result allows us to determine the number of simulations needed to achieve a desired level of accuracy. This is explored in detail in Chapter 4.

2.3 EVALUATING AN ESTIMATOR

Now that we have the ability to estimate quantities based on samples, it is essential that we be able to assess the quality of our work. In general, there are four types of error that we can make:

1. We have sampled from a population other than the one intended.

2. We have selected a model for or made assumptions about the population that are not true.

3. Our sample is not representative of the population, due to chance selection.

4. The estimation method itself is imperfect.

The first type of error can occur when the sampling is conducted over a lengthy time period. For example, data on automobile claims would be inaccurate if it included experience from a period before a recent change in the driving environment (such as a change in the speed limit). Another possibility is a faulty sampling scheme. For example, claims data may not include a provision for those that have occurred but have yet to be reported (e.g., calendar year mortality experience collected on January 1 may miss some December deaths).

The second type of error is not possible with empirical estimates, because no assumptions are made beyond the randomness of the sample. We will see how this error can arise in the next section.

The third type of error is one that can only be minimized, not eliminated. Any time we settle for examining a sample in place of the population the possibility of error arises. There is no assurance that the members of the sample will accurately reflect the population and there would be no way we could even know (at the time) that we have drawn such a sample.

The fourth type of error is usually evaluated along with the third type. The idea here is that for some estimators, even if we were to sample the entire population, we may still be in error. An example of such an estimator would be one that requires the data be grouped and then the population mean be estimated by (2.2).

In this section we will deal only with the combined effect of the third and fourth types of error, called **sampling error** or **estimation error**. We will use the term "sampling error" to avoid confusion with other kinds of error. This leads to the following definitions.

Definition 2.16 *A* ***point estimator*** *is a function of the values obtained from a random sample. As such, an estimator itself is a random variable, with its own distribution (called the* ***sampling distribution****).*

Definition 2.17 *A* ***point estimate*** *is the numerical realization of an estimator based on a particular random sample.*

We will drop the modifier "point" when it is clear that we are not referring to an interval estimator. Although not discussed here, we should again note that the limits of a confidence interval as described in Subsection 2.2.3 are also random variables. We presume that the purpose of constructing the estimator and then using the estimate is to produce a good guess at the value of some feature of the population. For notation, we typically use Greek letters for population features and place circumflexes above them to denote estimators. Being a random variable, it is also common to denote an estimator by an uppercase Roman letter and denote the corresponding estimate with a lowercase Roman letter. The following example illustrates these concepts.

Example 2.11 *Consider the mean and variance. They are clearly population features. Possible estimators are*

$$\hat{\mu} = \bar{X} = \sum_{j=1}^{n} X_j/n, \qquad \hat{\sigma}^2 = S_n^2 = \frac{1}{n}\sum_{j=1}^{n}(X_j - \bar{X})^2$$

and the corresponding estimates are

$$\hat{\mu} = \bar{x} = \sum_{j=1}^{n} x_j/n, \qquad \hat{\sigma}^2 = s_n^2 = \frac{1}{n}\sum_{j=1}^{n}(x_j - \bar{x})^2.$$

Alternative estimators are

$$\hat{\mu} = \hat{\pi}_{0.5}, \qquad \hat{\sigma}^2 = S_{n-1}^2 = \frac{1}{n-1}\sum_{j=1}^{n}(X_j - \bar{X})^2.$$

With the notation defined as above, there is no way to distinguish the estimator from the estimate when using Greek letters. The first set of estimators are the empirical estimators. The second set uses the empirical estimator of the median to estimate the mean and uses a more common choice for the denominator when estimating the variance. □

We must once again make it clear that the only quantity that can be evaluated with regard to quality is the estimator, not the estimate. All of the measures used here apply to the random variable, trying to indicate how well we might do if we adopt that particular procedure. In any particular application, the estimate itself may turn out to be good or bad, something which may be revealed in the future but cannot be apparent in advance or at the time the estimation process is completed.

The measures of estimator quality introduced here are the standard ones covered in most mathematical statistics texts. See Hogg and Craig [65] for example. For all of the definitions to follow, θ (possibly a vector) will indicate the parameter and $\hat{\theta}$ the estimator.

The first measure indicates how well the estimator does on average. If a good estimator is used repeatedly on various, similar, situations, the errors should cancel and there will be no tendency to over- or underestimate the parameter.

Definition 2.18 *The* ***bias*** *of an estimator,* $\hat{\theta}$, *is*

$$b_\theta(\hat{\theta}) = E(\hat{\theta}) - \theta.$$

An estimator for which the bias is identically zero ($b_\theta(\hat{\theta}) = 0$ for all θ) is said to be ***unbiased.***

Note that the bias is a function of the parameter value. The degree to which the estimator is off, on average, may depend on the particular value for the parameter. All else being equal, we would prefer that an estimator be unbiased.

Example 2.12 (Example 2.11 continued) *Determine the bias of each of the four estimators. Do so first with no assumptions on the population, then under the assumption that the population has a distribution function of $F_X(x) = x, \quad 0 < x < 1$. Note that in this very artificial example, we have assumed that the population distribution (and therefore its mean of 1/2 and variance of 1/12) is known.*

With no specific assumptions the following calculations can be done

$$E(\bar{X}) = E\left(\frac{1}{n}\sum_{j=1}^{n} X_j\right) = \frac{1}{n}\sum_{j=1}^{n} E(X_j) = \mu, \quad b_\mu(\bar{X}) = 0$$

$$\begin{aligned} E(S_n^2) &= \frac{1}{n}E\left[\sum_{j=1}^{n}(X_j - \bar{X})^2\right] = \frac{1}{n}\left[E\left(\sum_{j=1}^{n} X_j^2\right) - E(n\bar{X}^2)\right] \\ &= (\sigma^2 + \mu^2) - \frac{1}{n^2}\sum_{i=1}^{n}\sum_{j=1}^{n} E(X_i X_j) \\ &= (\sigma^2 + \mu^2) - \frac{1}{n^2}\left[n(n-1)\mu^2 + n(\sigma^2 + \mu^2)\right] = \frac{n-1}{n}\sigma^2 \\ b_{\sigma^2}(S_n^2) &= -\frac{\sigma^2}{n} \end{aligned}$$

$$E(S_{n-1}^2) = E\left(\frac{n}{n-1}S_n^2\right) = \sigma^2, \quad b_{\sigma^2}(S_{n-1}^2) = 0$$

and so regardless of the distribution of the population, $\bar{X}$ and S_{n-1}^2 are unbiased and S_n^2 has a negative bias. The expected value of the sample median depends on the population distribution. If n is odd we can write $n = 2m + 1$ and the pdf of the sample median (see, for example, Hogg and Craig [65, p. 159]), letting p represent $\hat{\pi}_{0.5}$, is

$$f_{\hat{\pi}_{.5}}(p) = \frac{n!}{m!^2} f_X(p)[F_X(p)]^m[1 - F_X(p)]^m.$$

For the distribution in question, $f_X(p) = 1$ and so

$$E(\hat{\pi}_{0.5}) = \int_0^1 p\frac{n!}{m!^2}(p^m)(1-p)^m dp$$

$$= \frac{n!}{m!^2}\frac{\Gamma(m+2)\Gamma(m+1)}{\Gamma(2m+3)} \quad \text{(noting the beta function)}$$

$$= \frac{n!(m+1)!m!}{m!^2(n+1)!} = \frac{1}{2} = \pi_{0.5}$$

and we see that $\hat{\pi}_{0.5}$ is unbiased. This result is not typical for a sample median, it is due to the symmetry of the density. □

Although S_n^2 is biased, we see that as the sample size increases, the bias decreases, and in the limit is zero. This prompts the following definition.

Definition 2.19 *Let $\hat{\theta}_n$ be an estimator based on a sample of size n. An estimator is* ***asymptotically unbiased*** *if $\lim_{n\to\infty} b_\theta(\hat{\theta}_n) = 0$ for all θ.*

For an estimator to be useful, it not only should be accurate on average, but also should come close to the true parameter value, at least most of the time. The accuracy should improve with the sample size. In particular, if the sample is infinite (so in effect, we are sampling the entire population), we should expect our estimator to be perfect.

The weakest such statement (that is, it is the one that is easiest to satisfy) is given by the following definition.

Definition 2.20 *An estimator is* ***consistent*** *(often called, in this context,* ***weakly consistent****) if for all $\delta > 0$ and any θ,*

$$\lim_{n\to\infty} \Pr(|\hat{\theta}_n - \theta| < \delta) = 1.$$

A sufficient (although not necessary) condition for weak consistency is that the estimator be asymptotically unbiased and $Var(\hat{\theta}_n) \to 0$.

Example 2.13 (Example 2.11 continued) *Determine which of the estimators of the mean are consistent. Do this under the assumption that the population has the uniform distribution ($F(x) = x, 0 < x < 1$).*

It is clear from the previous development that both the sample mean and median are unbiased.

With regard to the variance we have $Var(\bar{X}) = \sigma^2/n \to 0$ as $n \to \infty$, and so as long as the population variance exists the sample mean will be consistent for the population mean. Also note that with regard to the median

$$\begin{aligned} Var(\hat{\pi}_{.5}) &= \int_0^1 p^2 \frac{n!}{m!^2}(p^m)(1-p)^m dp - \frac{1}{4} \\ &= \frac{n!(m+2)!m!}{m!^2(2m+3)!} - \frac{1}{4} \\ &= \frac{m+2}{2(2m+3)} - \frac{1}{4} \to 0 \text{ as } m \to \infty. \end{aligned}$$

□

While consistency is nice, most estimators have this property. What would be truly impressive is an estimator that is not only correct on average, but comes very close most of the time and, in particular, comes closer than rival estimators. One measure, for a finite sample, is motivated by the definition of consistency. The quality of an estimator could be measured by the probability that it gets within δ of the true valuet—that is, by measuring $\Pr(|\hat{\theta}_n - \theta| < \delta)$. But the choice of δ is arbitrary and we prefer measures that cannot be altered to suit the investigator's whim. Then we might consider $E(|\hat{\theta}_n - \theta|)$, the average absolute error. But we know that working with absolute values often presents unpleasant mathematical challenges, and so the following has become widely accepted as a measure of accuracy.

Definition 2.21 *The* ***mean squared error (MSE)*** *of an estimator is*

$$MSE_\theta(\hat{\theta}) = E[(\hat{\theta} - \theta)^2].$$

Note that the MSE is a function of the true value of the parameter. An estimator may perform extremely well for some values of the parameter, but poorly for others.

Example 2.14 *Consider the estimator $\hat{\theta} = 5$ of an unknown parameter θ. The MSE is $(5 - \theta)^2$, which is very small when θ is near 5 but becomes poor for other values. Of course this estimate is both biased and inconsistent.* □

A result that follows directly from the various definitions is

$$MSE_\theta(\hat{\theta}) = E\{[\hat{\theta} - E(\hat{\theta}) + E(\hat{\theta}) - \theta]^2\} = Var(\hat{\theta}) + [b_\theta(\hat{\theta})]^2. \tag{2.9}$$

If we restrict attention to only unbiased estimators, the best such could be defined as follows.

Definition 2.22 *An estimator, $\hat{\theta}$, is called a* ***uniformly minimum variance unbiased estimator (UMVUE)*** *if it is unbiased and for any true value of θ there is no other unbiased estimator that has a smaller variance.*

Because we are looking only at unbiased estimators it would have been equally effective to make the definition in terms of MSE. We could also generalize the definition by looking for estimators that are uniformly best with regard to MSE, but the previous example indicates why that is not feasible. There are a few theorems that can assist with the determination of UMVUEs. However, such estimators are difficult to determine. On the other hand, it is still a useful criterion for comparing two alternative estimators.

Example 2.15 *In Example 2.12 it was demonstrated that both the sample mean and median are unbiased for the mean of a uniform distribution. Which has the smallest MSE for a sample of size 11?*

For the sample mean, the MSE is the variance which is $(1/12)/n = 1/132$. For the sample median it is $7/26 - 1/4 = 1/52$ (using the calculation from Example 2.13 with $m = 5$) and so the sample mean is to be preferred in this instance. □

All of the material discussed to this point comes under the heading of what is called "classical" or "frequentist" statistics. Some basic assumptions of this approach are that the population exists, the parameter is a fixed value that happens to be unknown, and the analyst knows no more than what is revealed by a sample of numbers taken from the population. Inferences are made by making further assumptions concerning the nature of the sampling process and/or the population and then the laws of probability are applied.

There is an alternative view of the estimation process that is called "Bayesian" after the theorem attributed to The Reverend Thomas Bayes. This approach assumes that we already have an idea of what the parameter might be, that this idea can be expressed as a probability distribution, and that the data are fixed with no further numbers worthy of consideration. Once again, the laws of probability are applied. The details of this process will be taken up in Section 2.8.

2.4 PARAMETRIC ESTIMATION

So far, empirical estimation looks pretty good. However, it has a number of drawbacks, which will be illustrated later in this chapter. That means we need something better. One such approach is called **parametric**. There are some others, but with the exception of a brief introduction in Section 2.11, they will not be presented.

2.4.1 Definition

Earlier we obtained quantities from the population (or its distribution) that described various features. There are certain cases when the process can be reversed.

Definition 2.23 *A* ***parametric family of distributions*** *is a collection of distribution functions where the identity of a particular member is indexed by a fixed number of variables called* **parameters**. *More formally, the family is*

$$\{F(x;\theta) : \theta \in \Theta\}$$

where θ is a scalar or vector and Θ is the set of all possible parameter values. Also, the random variable may be multivariate and thus x may also be a vector.

If it is true that the population distribution is one of the members of a parametric family, it is sufficient to be able to obtain the value of θ. That will determine the distribution. Then any quantity of interest can be determined. Parametric inference is thus reduced to four steps.

1. Determine which parametric family describes the population.
2. Determine the value of the parameter.[4]
3. Determine the value of the quantity of interest.
4. Assess the accuracy of the value determined in Step 3.

2.4.2 Methods based on matching sample and population quantities

Methods of estimating parameters fall into two categories. One is to set up a system of equations where the number of equations equals the number of parameters. We then hope that there is exactly one admissible solution to the equations, which then becomes the estimator. The equations are selected so that certain facts that we would like be true are indeed true. The other set of methods optimizes some criterion that is relevant for our purposes.[5]

There are two popular methods in the first category: the method of moments and percentile matching. The method of moments is based on ensuring that the parametric model has the same moments as the empirical model.

Definition 2.24 *If a parametric family has* r *parameters, the* ***moment equations*** *are*

$$\mu_j' = \frac{1}{n}\sum_{i=1}^{n} x_i^j, \qquad j = 1, \ldots, r$$

where $\mu_j' = E(X^j|\theta)$ is a function of the unknown parameter vector θ. The ***method of moments estimator*** *is the solution to these equations.*

Example 2.16 (individual data dental example) *Determine the method of moments estimates for the exponential, gamma, and Pareto distributions.*[6] *For each case, estimate the mean, standard deviation, and the probability that a single claim exceeds 500.*

For the exponential distribution, the equation is $\theta = 335.5$, so $\hat{\theta} = 335.5$. The mean is θ which is estimated by 335.5, the standard deviation is also θ and is estimated by 335.5. $\Pr(X > 500) = \exp(-500/\theta) = 0.22530$.

[4] The singular "parameter" will be used even if there is more than one (that is, a vector of parameters).

[5] The Bayesian method is sufficiently different that it will be discussed in a separate section (Section 2.8).

[6] Moments, along with density and distribution functions, can be found for all named distributions in Appendix A. The collection of distributions will be more formally introduced in Section 2.7.

For the gamma distribution, $\alpha\theta = 335.5$ and $\alpha(\alpha+1)\theta^2 = 293{,}068.3$. The solutions are $\hat{\alpha} = 0.62357$ and $\hat{\theta} = 538.03$. The mean is $\alpha\theta$ which is estimated by $(0.62357)(538.03) = 335.5$. The standard deviation is $\alpha^{1/2}\theta$ which is estimated by 424.86. $\Pr(X > 500) = 1 - \Gamma(\alpha, 500/\theta)$, which is estimated by 0.22593.[7]

For the Pareto distribution, $\theta/(\alpha-1) = 335.5$ and $2\theta^2/[(\alpha-1)(\alpha-2)] = 293{,}068.3$. The solutions are $\hat{\alpha} = 5.3131$ and $\hat{\theta} = 1,447.1$. The mean is $\theta/(\alpha-1)$, which is estimated by 335.5 and the standard deviation is $\theta\alpha^{1/2}/[(\alpha-1)(\alpha-2)^{1/2}]$, which is estimated by 424.88. $\Pr(X > 500) = [\theta/(500+\theta)]^\alpha$, which is estimated by 0.20663. □

It should be noted that for the gamma and Pareto examples the method of moments estimates of the first two moments match the empirical estimates. This should be clear from the definition of the method of moments. For the exponential case, with only one parameter, only the mean matches. As a reminder, the empirical estimate of $\Pr(X > 500)$ is 0.2.

The second popular method is called **percentile matching**. Here the sample and model percentiles are forced to be equal at r arbitrarily selected points. The equations are

$$p_j = F(\hat{\pi}_{p_j}; \theta), \qquad j = 1, \ldots, r.$$

Example 2.17 (Example 2.16 continued) *Estimate the parameters by using percentile matching for the exponential and Pareto distributions. Use the 70th percentile for the exponential distribution and the 40th and 70th percentiles for the Pareto distribution.*

The percentiles are $\hat{\pi}_{0.7} = 0.3(317) + 0.7(351) = 340.8$ and $\hat{\pi}_{0.4} = 0.6(107) + 0.4(141) = 120.6$.

For the exponential distribution the equation is $0.7 = 1 - \exp(-340.8/\theta)$ and the solution is $\hat{\theta} = 283.06$.

For the Pareto distribution the two equations are $0.7 = 1 - [\theta/(\theta + 340.8)]^\alpha$ and $0.4 = 1 - [\theta/(\theta + 120.6)]^\alpha$. The second equation can be solved for $\alpha = \log(0.6)/\log[\theta/(\theta + 120.6)]$. This can be substituted into the first equation to produce

$$\log(0.3) = \frac{\log(0.6)}{\log[\theta/(\theta + 120.6)]} \log[\theta/(\theta + 340.8)].$$

An iterative method such as bisection or Newton–Raphson can be used to obtain the solution $\hat{\theta} = 424.5$ and then the previous equation evaluated to give $\hat{\alpha} = 2.0428$. □

[7] The incomplete gamma function is discussed in more detail later in this chapter and in Appendix A.

2.4.3 Evaluating parametric estimators

When evaluating parametric estimators it is important to recognize that most evaluations are done under an assumption that the population conforms to the parametric family selected and that only the value of the parameter is uncertain. In reality, the model is likely to be at least a bit wrong and thus there is an additional source of error. Models that perform well even when the population does not conform precisely to a member of the parametric family are said to be **robust**.

Example 2.18 *Assume that the population is exponential and the task is to estimate the probability that a random observation exceeds 500. Evaluate the empirical and method of moments estimators for a sample of size n.*

We begin by assuming that the true parameter is θ. The empirical estimator is the proportion of observations in the sample that exceed 500. The number of observations exceeding 500 has a binomial distribution with mean $n[1-F(500)]$. Hence the proportion of observations over 500 is an unbiased estimator for $1-F(500)$. The variance of the proportion is $[1-F(500)]F(500)/n \to 0$ as $n \to \infty$ and hence this estimator is also consistent.

For the method of moments the estimator is $\exp(-500/\bar{X})$. It is difficult to find moments for this particular random variable. An approximation can be obtained from the Taylor series expansion of the estimator about the true value θ.

$$e^{-500/\bar{X}} = e^{-500/\theta}[1+(\bar{X}-\theta)(500/\theta^2)+(\bar{X}-\theta)^2(250{,}000/\theta^4-1000/\theta^3)/2+\cdots] \tag{2.10}$$

The expected value of the right-hand side is

$$e^{-500/\theta}[1+0+(\theta^2/n)(250{,}000/\theta^4-1000/\theta^3)/2+\cdots]$$

where the missing terms also have a factor of n in the denominator. Therefore the last term and all missing terms go to zero as n goes to infinity and the method of moments estimator is asymptotically unbiased.

To evaluate the variance we need the expected value

$$Var(e^{-500/\bar{X}}) = E\{[e^{-500/\bar{X}} - E(e^{-500/\bar{X}})]^2\}.$$

Approximating $E(e^{-500/\bar{X}})$ by $e^{-500/\theta}$ and using (2.10) we have

$$\begin{aligned} Var(e^{-500/\bar{X}}) &\doteq E\left\{\left[e^{-500/\theta}(\bar{X}-\theta)(500/\theta^2)+\cdots\right]^2\right\} \\ &= e^{-1000/\theta}(250{,}000/\theta^4)Var(\bar{X})+\ldots \\ &= e^{-1000/\theta}(250{,}000/\theta^4)(\theta^2/n)+\cdots. \end{aligned}$$

Then, we have, approximately

$$nVar(e^{-500/\bar{X}}) \doteq e^{-1000/\theta}(250{,}000/\theta^2).$$

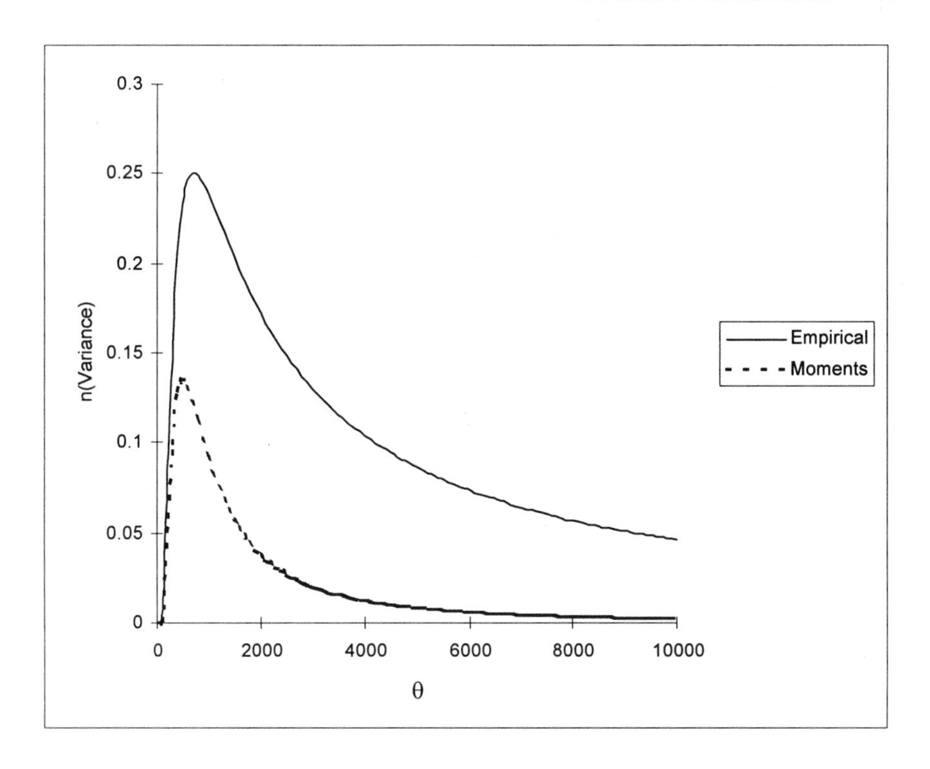

Fig. 2.4 Variance of two estimators of an exponential probability

Figure 2.4 shows a plot of n times the variance[8] for each estimator for various values of θ. It is clear that the moment-based estimator is superior (when the population is truly exponential). □

The only problem with the above analysis is that the moment estimator had the advantage of being evaluated under the assumption that the exponential distribution was correct. Suppose instead that the population was Weibull. It is no more difficult to evaluate the variance of the empirical estimator as we continue to need only the population probability of exceeding 500. Evaluating the moment-based estimator is much more difficult. One possible approach is a simulation analysis. The basic procedure is as follows:

1. Select the Weibull parameters for the population distribution.

[8] Most any reasonable estimator has $Var(\hat{\theta}) \to 0$ as $n \to \infty$. However, it is often the case that $nVar(\hat{\theta})$ tends to a constant. This constant becomes a measure of the quality of the estimator.

Table 2.4 Simulation study of two estimators

Method	Mean	Variance	MSE
Empirical	0.22801	0.007152	0.007153
Moment	0.22522	0.003602	0.003607

2. Generate a simulated sample of size n from this Weibull distribution.[9]
3. Evaluate $e^{-500/\bar{x}}$ for this sample.
4. Repeat the process a large number of times.
5. An estimate of the bias is the difference between the sample mean of the simulated set of probability estimates and the true probability under the Weibull distribution.
6. An estimate of the mean squared error is obtained by subtracting the true Weibull probability from each simulated estimate, squaring the difference, adding the squares, and then dividing by n.

If our goal is to compare the moment based estimator to the empirical estimator, it may be more fair to also simulate these quantities for the empirical estimator, even though we can obtain them exactly. This is so that in case our simulations produce some unusual samples, these neither favor nor harm our evaluation.

Example 2.19 *Evaluate the exponential based moment estimator of the probability of exceeding* 500 *when the population has a Weibull distribution with parameters* $\tau = 1.1$ *and* $\theta = 350$ *for a sample of size 25.*

Ten thousand samples of size 25 were generated. Key quantities are presented in Table 2.4. The true Weibull probability is $\exp[-(500/350)^{1.1}] = 0.22754$. The variance (and mean squared error) for the empirical estimator is $(0.22754)(0.77246)/25 = 0.007031$. These compare favorably with the simulated quantities, providing some evidence that the simulation is valid. □

The above analysis still requires some assumption about the true population distribution. What would be best would be to evaluate our estimators against the population from which the sample was taken. At first thought this would appear to be impossible. If we knew what the population looked like, sampling would be unnecessary! A somewhat recent statistical development has found a way around this problem. It is called the **bootstrap** (an excellent

[9] As simulation is not an integral part of this text we will devote limited space to the generation of random variables. An elementary method of doing this is given in Appendix H.

introductory reference is the book by Efron and Tibshirani [36]) and is based heavily on the empirical distribution. The steps of a bootstrap analysis are quite simple.

1. Generate a random number from the integers $1, 2, \ldots n$, where n is the original sample size. Let j be this number.

2. Obtain the jth member of the original sample.

3. Repeat the first two steps n times. Note that the same value from the original sample may be selected more than once.

4. Compute the estimator being evaluated from these n new values (called a bootstrap sample and a bootstrap estimate).

5. Repeat steps 1 through 4 a large number (at least 250) times. This is referred to as the number of bootstrap samples.

Note that the first three steps are simply the drawing of a random sample with replacement.

The large number of bootstrap estimates obtained can be considered as a random sample from the sampling distribution of the estimator being evaluated. It can be treated in the same manner as the simulated quantities discussed earlier. Thus the sample mean from the sample of bootstrap estimates is a good indicator of the expected value of the estimator, and the variance from the sample is a good indicator of the variance of the estimator. The MSE is not available because that does require the true value from the population. Bootstrap theory is devoted to demonstrating that this procedure provides a reliable indicator of the properties of the estimator in question with regard to the population from which the sample was drawn.

Example 2.20 *Obtain bootstrap estimates of the sampling mean and variance of the empirical and moment estimators of the probability of observing a value in excess of 500 from the population that generated the individual data dental example.*

The number of bootstrap samples was varied from 1,000 to 5,000. The results are summarized in Table 2.5. As expected, the bootstrap estimate of the mean of the empirical estimator is very close to .2, the value that would be obtained with an infinite bootstrap sample. Similarly, an infinite sample would yield a variance of 0.016. For the moment estimate we do not know what the result should be. It is clear that the mean is slightly larger and the variance slightly smaller. For the particular population that generated these ten numbers it appears that there is little difference in the two estimators. □

While numerical procedures exist for solving systems of non-linear equations, we do not recommend using the method of moments or percentile matching unless the equations can be reduced to no more than one equation in one

Table 2.5 Bootstrap study of two estimators

Method	Sample size	Mean	Variance
Empirical	1,000	0.1972	0.01735
Empirical	2,000	0.2004	0 .01563
Empirical	5,000	0.1999	0.01566
Moment	1,000	0.2147	0.01550
Moment	2,000	0.2131	0.01481
Moment	5,000	0.2168	0.01441

unknown (at which point a simple search method such as bisection could be used). The lack of accuracy of these methods does not justify a significant numerical effort.

2.4.4 Optimization-based estimators

The major problem with some equation-based estimators is their inability to fit well over the full range of observed values (concentrating instead on matching a few select characteristics). In order to fit well throughout, we must not insist that the empirical and model distributions agree exactly in some way. Instead we ask that they be close in some measurable way. We will look at three ways of doing this. One comes under the general name of minimum distance and relates data and model at specific locations. The second has no recognized name. It requires separation of the data into intervals and then compares a feature of the model to the data in each interval. The third is maximum likelihood.

All of these methods require the ability to maximize or minimize functions of one or more variables. In most applications the technique of setting all partial derivatives equal to zero and solving the resulting equations will prove to be difficult. Many numerical methods have been devised for optimizing functions. Most numerical analysis texts provide an introduction. Numerous algorithms for implementing them can be found in the *Numerical Recipes* series (e.g., [103]). As well, many spreadsheet programs include an optimization tool. Finally, as an all-purpose tool we recommend the simplex method. A thorough discussion with detailed algorithms can be found in *Sequential Simplex Optimization* [123], and programs are provided in [103]. The algorithm is described in Appendix C. Its major advantage is that derivatives are not required. However, convergence may be slow. Regardless of the method used, we strongly recommend that the proposed solution be checked by evaluating the function at nearby points to verify that a maximum or minimum has been achieved. Many "canned" programs will often announce success when in fact the optimum point has not been obtained. Finally, be aware that any numerical routine is only able to determine local optima. It is impossible to know for sure that a superior value does not exist some distance away from the proposed solution.

The following definition of the general minimum distance estimator is taken from Klugman and Parsa [77].

Definition 2.25 *For a parametric family, let $F(x;\theta)$ be the cdf and let $G(x;\theta)$ be any function of x that is uniquely related to F. That is, if we know F and θ we can obtain G and if we know G and θ we can obtain F. Let $G_n(x)$ be obtained from the empirical cdf in the same manner. Then, if it exists, the value of θ that minimizes*

$$Q(\theta) = \sum_{j=1}^{k} w_j [G(c_j;\theta) - G_n(c_j)]^2 \tag{2.11}$$

is called a ***minimum distance estimator*** *of θ. The values of $c_1 < c_2 < \cdots < c_k$ and $w_1, \ldots, w_k \geq 0$ are arbitrarily selected.*[10]

If grouped data were used, the c_j will usually be the class boundaries. If individual data is used, the boundary values are temporarily set for estimation purposes. Two specific examples of this type of estimator are

1. *Minimum cdf*: $G(x;\theta) = F(x;\theta)$. This is also known as the Cramér–von Mises estimator.

2. *Minimum LEV*: $G(x;\theta) = E(X \wedge x;\theta)$

The second estimator confines each term in the sum (2.11) to observations in a particular interval.

Definition 2.26 *Let the range of possible values be partitioned as $c_0 < c_1 < \cdots < c_k$ where c_0 is the smallest possible value in the model and c_k is the largest possible value. Let $G_j(\theta)$ be any function that depends only upon θ, $F(c_{j-1};\theta)$, $F(c_j;\theta)$, and $f(x;\theta)$ for $c_{j-1} < x < c_j$. Let G_j be the same quantity for the empirical distribution. Then, if it exists, the value of θ that minimizes*

$$Q(\theta) = \sum_{j=1}^{k} w_j [G_j(\theta) - G_j]^2$$

is called a ***minimum interval-based distance estimator*** *of θ. The weights $w_1, \ldots, w_k \geq 0$ are arbitrary.*

Two specific examples of this type of estimator are as follows:

[10] It is not mandatory that the criterion involve squares. We could have used absolute value or some other loss function. Squaring has the usual advantage of mathematical tractability. This also applies to Definition 2.26.

1. *Minimum modified chi-square*: $G_j(\theta) = n[F(c_j;\theta) - F(c_{j-1};\theta)]$, $w_j = 1/G_j$.

2. *Minimum layer average severity (LAS)*: $G_j(\theta) = E(X \wedge c_j;\theta) - E(X \wedge c_{j-1};\theta)$.

You are asked in Exercise 2.24 to verify that the minimum LAS specification meets the conditions of the definition. The main advantage of these two estimators over their minimum distance counterparts is that errors are somewhat more independent. That is, if it is difficult to match the cdf at one point, that difficulty will be perpetuated to higher arguments when using minimum distance. The modification in the chi-square based procedure refers to the use of the empirical value for the weight rather than the model value.[11]

These estimators are particularly useful for grouped data because they only require empirical information at selected values. The following example illustrates the ease with which these problems can be set up on a spreadsheet.

Example 2.21 (grouped data dental example) *Determine the minimum cdf and minimum modified chi-square estimates of θ for the exponential distribution. For the minimum cdf case use weights of* 1.

For minimum cdf estimation the quantity to minimize is

$$Q(\theta) = (1 - e^{-25/\theta} - 30/378)^2 + \cdots + (1 - e^{-4000/\theta} - 378/378)^2.$$

Using a spreadsheet minimization routine produces the estimate $\hat{\theta} = 281.5852$. The final result of the optimization appears in Table 2.6.

For minimum modified chi-square estimation the quantity to minimize is

$$\begin{aligned} Q(\theta) &= [378(1 - e^{-25/\theta}) - 30]^2/30 + \cdots \\ &\quad + [378(e^{-1500/\theta} - e^{-2500/\theta}) - 11]^2/11 + [378e^{-2500/\theta} - 3]^2/3. \end{aligned}$$

Note that we have combined the last two groups (2,500 to 4,000 and 4,000 to infinity) because with no observations above 4,000 the contribution of the final interval to Q would be undefined. Again, using a spreadsheet we obtain the estimator, $\hat{\theta} = 274.7305$. The result is in Table 2.7. □

The methods presented so far have a feature that is both an advantage and a disadvantage. By allowing great flexibility in the choice of weights and by allowing some flexibility with regard to the function to match, these methods allow for a great deal of latitude on the part of the analyst. If the analyst has a thorough knowledge of the purpose of the estimation exercise, a criterion

[11]The chi-square test statistic and its use in hypothesis tests will be discussed in Section 2.9.

Table 2.6 Minimum cdf estimation for dental data

x	$F_n(x)$	$F(x;\hat{\theta})$	$[F_n(x) - F(x;\hat{\theta})]^2$
25	0.07937	0.08496	0.000031
50	0.16138	0.16269	0.000002
100	0.31217	0.29892	0.000176
150	0.42328	0.41298	0.000106
250	0.59524	0.58845	0.000046
500	0.81746	0.83063	0.000173
1,000	0.93651	0.97131	0.001211
1,500	0.96296	0.99514	0.001035
2,500	0.99206	0.99986	0.000061
4,000	1.00000	1.00000	0.000000
$\hat{\theta} = 281.5852$			$Q = 0.002842$

Table 2.7 Minimum modified chi-square estimation for dental data

Interval	G_j	$G_j(\hat{\theta})$	$[G_j - G_j(\hat{\theta})]^2/G_j$
0–25	30	32.8787	0.27623
25–50	31	30.0189	0.03105
50–100	57	52.4317	0.36613
100–150	42	43.7073	0.06940
150–250	65	66.8066	0.05021
250–500	84	90.9089	0.56826
500–1,000	45	51.3238	0.88867
1,000–1,500	10	8.3161	0.28356
1,500–2,500	11	1.5658	8.09129
2,500–∞	3	0.0422	2.91616
$\hat{\theta} = 274.7305$			$Q = 13.54097$

can be selected that will allow the data and model to match in the way that is most important. For example, if the objective is to price a reinsurance agreement in which we are to be responsible for paying losses in the one to five million dollar range, then minimum LAS estimation with most of the weight on the intervals that span the key range would seem the best choice. On the other hand, such flexibility allows an unscrupulous (or perhaps just ill-informed) analyst to obtain results that are not appropriate.

The final method to be presented in this section does not leave room for arbitrary decisions. Furthermore, it is the best method, at least with regard to statistical properties. It is called **maximum likelihood estimation**; the process, the estimator, and the estimate are all identified by the abbreviation **mle**. The philosophy is quite simple. Let the likelihood function be the probability of observing what was observed given a hypothetical value of the parameter. The most reasonable estimate of the true parameter value is the one that yields the highest probability of obtaining what was actually obtained. Assuming independent observations we have the following definition.

Definition 2.27 *The* ***likelihood function*** *for a set of n independent observations is*

$$L(\theta) = \prod_{j=1}^{n} L_j(\theta)$$

where $L_j(\theta)$ is the contribution of the jth observation to the likelihood. If the jth observation is an event with positive probability (such as from a discrete distribution or from an interval), then the contribution is that probability. If the jth observation is a value from a continuous distribution, the contribution is the pdf at that value.

Two cases where the likelihood function is easy to write are

1. *Individual data*: $L(\theta) = \prod_{j=1}^{n} f(x_j;\theta)$

2. *Grouped data*: $L(\theta) = \prod_{j=1}^{k} [F(c_j;\theta) - F(c_{j-1};\theta)]^{n_j}$.

Note that if there is a policy limit of u, the contribution of a loss that exceeds the limit (so the amount paid was u) is $1 - F(u;\theta)$ and not $f(u;\theta)$. That is because the underlying distribution is for losses, not amounts paid. When u is paid, all that is known about the loss is that it was at or above the limit.

There is no guarantee that the likelihood function will have a maximum. That is, $L(\theta)$ may increase as θ goes to a boundary such as zero or infinity. In addition, there is the possibility of local maxima. For most applications it is not possible to perform the maximization by analytic means, so again, numerical methods must be employed.

Maximum likelihood estimators enjoy a number of properties that, collectively, are not shared by any other parametric estimators. Among them are (specific statements of these results are presented in the next section) the following.

1. They are asymptotically unbiased.

2. Among estimators that have asymptotic normal distributions, they have the smallest asymptotic variance.

3. The maximum likelihood estimate of a function of the parameter is that same function of the maximum likelihood estimate (mle) of the parameter. That is, the mle is invariant under parameter transformation.

4. An explicit formula for the asymptotic variance of the estimator can be derived.

Item 3 means that no matter how we parameterize a distribution we will get the same answer when it comes to estimating the quantity of interest. Item 4 means that we will always be able to approximate the variance of the estimator (under the condition that the true distribution is a member of the chosen family).

Example 2.22 (individual data dental example) *Determine the method of moments and maximum likelihood estimates of the parameters for the lognormal distribution. Use them to estimate the mean of the lognormal population. Conduct a simulation study to see which estimator is superior when the population is lognormal with* $\mu = 5$ *and* $\sigma = 1$.

If the goal were just to estimate the mean, we would know without formulas that the moment estimator is the sample mean. With regard to the parameter estimates, the moment equations are

$$\begin{aligned} 335.5 &= \exp(\mu + \sigma^2/2) \\ 293{,}068.3 &= \exp(2\mu + 2\sigma^2). \end{aligned}$$

The solutions are $\hat{\mu} = 5.33716$ and $\hat{\sigma} = 0.97822$. The estimate of the mean is $\exp(5.33716 + 0.97822^2/2) = 335.5$. The likelihood function is

$$L(\mu, \sigma) = \prod_{j=1}^{10} \frac{1}{x_j \sigma \sqrt{2\pi}} \exp\left[-\frac{(\log x_j - \mu)^2}{2\sigma^2}\right].$$

The logarithm of the likelihood function and its partial derivatives are

$$\begin{aligned} l(\mu, \sigma) &= -\sum_{j=1}^{10} \log x_j - 10 \log \sigma - (10/2) \log 2\pi - \sum_{j=1}^{10} (\log x_j - \mu)^2/2\sigma^2 \\ \partial l/\partial \mu &= \sum_{j=1}^{10} (\log x_j - \mu)/\sigma^2 \\ \partial l/\partial \sigma &= -10/\sigma + \sum_{j=1}^{10} (\log x_j - \mu)^2/\sigma^3. \end{aligned}$$

Setting the partial derivatives equal to zero yields the maximum likelihood estimates

$$\hat{\mu} = \frac{1}{10} \sum_{j=1}^{10} \log x_j, \quad \hat{\sigma} = \sqrt{\frac{1}{10} \sum_{j=1}^{10} (\log x_j - \hat{\mu})^2}.$$

For this example the estimates are $\hat{\mu} = 5.07491$ and $\hat{\sigma} = 1.30055$. You may have noted that these numbers differ from those presented in Example 1.4. The ten numbers were the amounts paid with a deductible of 50. The

Table 2.8 Simulation study of two estimators of the lognormal mean

Method	Mean	Variance	MSE
Moment	245.14	10,430.45	10,430.65
Maximum likelihood	248.85	8,849.95	8,867.21

model found in that earlier example was for the loss itself. In this example, we are finding a model for the amount paid. The estimate of the mean is $\exp(5.07491 + 1.30055^2/2) = 372.65$.

The simulation results presented in Table 2.8 are based on 1,000 samples of size 10. While there is a slight bias for this sample size (the true mean is $\exp(5.5) = 244.69$), the maximum likelihood estimator is clearly superior with regard to variance and mean squared error. □

When data are grouped there is a particularly effective method of maximizing the likelihood function. It is an analog of the well-known Newton–Raphson method that is called the **method of scoring**. Begin by letting $P_j(\theta)$ be the probability that an observation falls in the jth group. In the simple case presented earlier it is equal to $F(c_j;\theta) - F(c_{j-1};\theta)$. However, we will see in Section 2.10 that it can take on more complicated forms. The logarithm of the likelihood function (the **loglikelihood**) is

$$l(\theta) = \sum_{j=1}^{k} n_j \log[P_j(\theta)].$$

Assume for now that the parameter is scalar. Then the maximum likelihood estimate will satisfy

$$dl/d\theta = l'(\theta) = \sum_{j=1}^{k} n_j P_j'(\theta)/P_j(\theta) = 0.$$

Newton–Raphson iteration for the root would use

$$\theta^* = \theta - l'(\theta)/l''(\theta)$$

with

$$l''(\theta) = \sum_{j=1}^{k} n_j \frac{P_j(\theta)P_j''(\theta) - P_j'(\theta)^2}{P_j(\theta)^2}$$

where θ^* is the new value of the parameter.

The main problem with this approach is that the second derivative is often difficult to obtain. The modification provided by the scoring method is to temporarily view n_j as a random variable and replace it by its expectation in the formula for the second derivative. It has a binomial distribution with expected value $nP_j(\theta)$. With this replacement we have

$$E[l''(\theta)] = n\sum_{j=1}^{k} \frac{P_j(\theta)P_j''(\theta) - P_j'(\theta)^2}{P_j(\theta)} = n\sum_{j=1}^{k} P_j''(\theta) - n\sum_{j=1}^{k} \frac{P_j'(\theta)^2}{P_j(\theta)}.$$

It does not appear that any savings have accrued. However, taking the second derivative outside the sum yields

$$E[l''(\theta)] = n\frac{d^2}{d\theta^2}\sum_{j=1}^{k} P_j(\theta) - n\sum_{j=1}^{k} \frac{P_j'(\theta)^2}{P_j(\theta)} = -n\sum_{j=1}^{k} \frac{P_j'(\theta)^2}{P_j(\theta)}$$

because the sum of the $P_j(\theta)$s is one, regardless of θ and so its second derivative is zero. The iteration formula for the method of scoring is then

$$\theta^* = \theta + \frac{\sum_{j=1}^{k} n_j P_j'(\theta)/P_j(\theta)}{n\sum_{j=1}^{k} P_j'(\theta)^2/P_j(\theta)}. \tag{2.12}$$

This formula requires only first derivatives (note that these must be taken with respect to the parameter as it appears in the distribution function). If analytic derivatives are not available, approximate derivatives could be used. However, while (2.12) will converge when $P_j'(\theta)$ is approximated, convergence will not be to the true maximum, but rather to the point where the approximate derivative of the loglikelihood is zero.

This approach is easily extended to the multiparameter case. The general formulas are

$$\begin{aligned}
\boldsymbol{\theta}^* &= \boldsymbol{\theta} + [\mathbf{I}(\boldsymbol{\theta})]^{-1}\mathbf{S}(\boldsymbol{\theta}) \\
\mathbf{S}(\boldsymbol{\theta})_r &= \sum_{j=1}^{k} n_j \frac{\partial P_j(\boldsymbol{\theta})/\partial\theta_r}{P_j(\boldsymbol{\theta})} \\
\mathbf{I}(\boldsymbol{\theta})_{rs} &= n\sum_{j=1}^{k} \frac{[\partial P_j(\boldsymbol{\theta})/\partial\theta_r][\partial P_j(\boldsymbol{\theta})/\partial\theta_s]}{P_j(\boldsymbol{\theta})}
\end{aligned} \tag{2.13}$$

where $\mathbf{S}(\boldsymbol{\theta})_r$ indicates the rth element of the vector $\mathbf{S}(\boldsymbol{\theta})$ and $\mathbf{I}(\boldsymbol{\theta})_{rs}$ indicates the (r, s)th element of the matrix $\mathbf{I}(\boldsymbol{\theta})$.

A case that occurs fairly often with insurance data is that of several samples with different groupings. Let n_j be the sample size for the jth sample and let n_{ij} be the number of observations that fall in the ith group for the jth sample. Let $P_{ij}(\boldsymbol{\theta})$ be the probability of an observation from the jth sample falling in the ith group. You are asked in Exercise 2.36 to prove that the formulas for the method of scoring become

$$
\begin{aligned}
\boldsymbol{\theta}^* &= \boldsymbol{\theta} + [\mathbf{I}(\boldsymbol{\theta})]^{-1}\mathbf{S}(\boldsymbol{\theta}) \\
\mathbf{S}(\boldsymbol{\theta})_r &= \sum_{j=1}^{J}\sum_{i=1}^{k_j} n_{ij}\frac{\partial P_{ij}(\boldsymbol{\theta})/\partial\theta_r}{P_{ij}(\boldsymbol{\theta})} \\
\mathbf{I}(\boldsymbol{\theta})_{rs} &= \sum_{j=1}^{J} n_j \sum_{i=1}^{k_j} \frac{[\partial P_{ij}(\boldsymbol{\theta})/\partial\theta_r][\partial P_{ij}(\boldsymbol{\theta})/\partial\theta_s]}{P_{ij}(\boldsymbol{\theta})}
\end{aligned}
\tag{2.14}
$$

where J is the number of samples and k_j is the number of groups in the jth sample.

Example 2.23 (grouped data dental example) *Perform one iteration of the method of scoring to estimate the lognormal parameters. Use $\mu = 5$ and $\sigma = 1$ as starting values.*

For the jth group we have

$$
\begin{aligned}
P_j(\mu,\sigma) &= \Phi[(\log c_j - \mu)/\sigma] - \Phi[(\log c_{j-1} - \mu)/\sigma] \\
\partial P_j(\mu,\sigma)/\partial\mu &= -\frac{1}{\sigma}\phi[(\log c_j - \mu)/\sigma] + \frac{1}{\sigma}\phi[(\log c_{j-1} - \mu)/\sigma] \\
\partial P_j(\mu,\sigma)/\partial\sigma &= -\frac{\log c_j - \mu}{\sigma^2}\phi[(\log c_j - \mu)/\sigma] \\
&\quad + \frac{\log c_{j-1} - \mu}{\sigma^2}\phi[(\log c_{j-1} - \mu)/\sigma]
\end{aligned}
$$

where $\Phi(x)$ is the standard normal cdf and $\phi(x)$ is the standard normal pdf. Derivatives of the distribution function with respect to the parameters for numerous distributions are given in Appendix A. The calculations are displayed in spreadsheet form in Table 2.9. The remaining calculations are

$$
\begin{aligned}
\mathbf{S} &= \begin{bmatrix} 55.2818 \\ 172.7856 \end{bmatrix}, \quad \mathbf{I} = \begin{bmatrix} 365.1746 & 7.6842 \\ 7.6842 & 669.6065 \end{bmatrix} \\
\mathbf{I}^{-1} &= \begin{bmatrix} 2.7391 \times 10^{-3} & -3.1433 \times 10^{-5} \\ -3.1433 \times 10^{-5} & 1.4938 \times 10^{-3} \end{bmatrix}, \quad \boldsymbol{\theta}^* = \begin{bmatrix} 5.1460 \\ 1.2564 \end{bmatrix}
\end{aligned}
$$

The new values are $\mu = 5.1460$ and $\sigma = 1.2564$. Repeated iterations produce the maximum likelihood estimates $\hat{\mu} = 5.1418$ and $\hat{\sigma} = 1.2308$. □

2.5 EVALUATING OPTIMIZATION ESTIMATORS

As noted in the previous section, maximum likelihood estimators have the smallest variance among estimators that have asymptotic normal distributions. In this section we make that statement precise and show how it can be used to evaluate the quality of maximum likelihood estimators without resorting to simulation. We also investigate similar properties for minimum distance estimators.

Table 2.9 One iteration of the scoring method

Interval	n_j	P_j	$P_j^{(\mu)}$	$P_j^{(\sigma)}$	$\frac{n_j P_j^{(\mu)}}{P_j}$	$\frac{n_j P_j^{(\sigma)}}{P_j}$	$\frac{n P_j^{(\mu)} P_j^{(\mu)}}{P_j}$	$\frac{n P_j^{(\mu)} P_j^{(\sigma)}}{P_j}$	$\frac{n P_j^{(\sigma)} P_j^{(\sigma)}}{P_j}$
0–25	30	0.0374	-0.0817	0.1455	-65.43	116.53	67.32	-119.91	213.57
25–50	31	0.1009	-0.1391	0.0947	-42.75	29.11	72.49	-49.36	33.61
50–100	57	0.2082	-0.1483	-0.0945	-40.60	-25.86	39.93	25.43	16.20
100–150	42	0.1578	-0.0299	-0.1499	-7.96	-39.92	2.14	10.74	53.87
150–250	65	0.1947	0.0507	-0.1773	16.92	-59.20	4.99	-17.45	61.05
250–500	84	0.1888	0.1574	-0.0501	70.06	-22.32	49.63	-15.81	5.04
500–1,000	45	0.0840	0.1261	0.1084	67.54	58.04	71.56	61.49	52.84
1,000–1,500	10	0.0179	0.0372	0.0598	20.82	33.48	29.26	47.05	75.67
1,500–2,500	11	0.0080	0.0201	0.0427	27.66	58.78	19.09	40.56	86.19
2,500–4,000	3	0.0019	0.0056	0.0151	9.01	24.14	6.41	17.16	45.94
4,000–	0	0.0005	0.0018	0.0058	0.00	0.00	2.36	7.78	25.63
Totals	378	1	0	0	55.28	172.79	365.17	7.68	669.61

2.5.1 The variance of maximum likelihood estimators

The key is a theorem that can be found in most mathematical statistics books. The particular version stated here and its multi-parameter generalization is taken from Rohatgi [109, p. 384] and stated without proof. Recall that $L(\theta)$ is the likelihood function and $l(\theta)$ its logarithm. All of the results assume that the population has a distribution that is a member of the chosen parametric family.

Theorem 2.2 *Assume that the pf $f(x;\theta)$ satisfies the following (for θ in an interval containing the true value, and replace integrals by sums for discrete variables):*

(i) $\log f(x;\theta)$ *is three times differentiable with respect to θ.*

(ii) $\int \frac{\partial}{\partial\theta} f(x;\theta)dx = 0$. *This allows the derivative to be taken outside the integral and so we are just differentiating the constant 1.*

(iii) $\int \frac{\partial^2}{\partial\theta^2} f(x;\theta)dx = 0$. *This is the same concept for the second derivative.*

(iv) $-\infty < \int f(x;\theta)\frac{\partial^2}{\partial\theta^2}\log f(x;\theta)dx < 0$. *This establishes that the indicated integral exists and that the location where the derivative is zero is a maximum.*

(v) There exists a function $H(x)$ such that $\int H(x)f(x;\theta)dx < \infty$ with $\left|\frac{\partial^3}{\partial\theta^3}\log f(x;\theta)\right| < H(x)$. This makes sure that the population is not too strange with regard to extreme values.

Then the following results hold

(a) As $n \to \infty$, the probability that the likelihood equation ($L'(\theta) = 0$) has a solution goes to one.

(b) As $n \to \infty$ the distribution of the maximum likelihood estimator $\hat{\theta}_n$ converges to a normal distribution with mean θ and variance such that $I(\theta)Var(\hat{\theta}_n) \to 1$ where

$$\begin{aligned} I(\theta) &= -nE\left[\frac{\partial^2}{\partial\theta^2}\log f(X;\theta)\right] = -n\int f(x;\theta)\frac{\partial^2}{\partial\theta^2}\log f(x;\theta)dx \\ &= nE\left[\left(\frac{\partial}{\partial\theta}\log f(X;\theta)\right)^2\right] = n\int f(x;\theta)\left(\frac{\partial}{\partial\theta}\log f(x;\theta)\right)^2 dx. \end{aligned}$$

□

For any z, the last statement is to be interpreted as

$$\lim_{n\to\infty} \Pr\left(\frac{\hat{\theta}_n - \theta}{[I(\theta)]^{-1/2}} < z\right) = \Phi(z)$$

and therefore $[I(\theta)]^{-1}$ is a useful approximation for $Var(\hat{\theta}_n)$. The quantity $I(\theta)$ is called the **information** (sometimes more specifically, **Fisher's information**). It follows from this result that the mle is asymptotically unbiased and consistent. The conditions in statements *(i)–(v)* are often referred to as "mild regularity conditions." A skeptic would translate this statement as "conditions that are almost always true but are often difficult to establish, so we'll just assume they hold in our case." Their purpose is to ensure that the density function is fairly smooth with regard to changes in the parameter and that there is nothing unusual about the density itself.

The results stated above assume that the sample consists of independent and identically distributed random observations. A more general version of the result uses the logarithm of the likelihood function:

$$I(\theta) = -E\left[\frac{\partial^2}{\partial\theta^2} l(\theta)\right] = E\left[\left(\frac{\partial}{\partial\theta} l(\theta)\right)^2\right].$$

An intermediate case occurs when the observations are independent, but not identical. In that case, let $L_j(\theta)$ be the contribution of the jth observation to the likelihood function and let $l_j(\theta) = \log L_j(\theta)$. Then $l(\theta) = \sum_{j=1}^{n} l_j(\theta)$. The logarithm of the likelihood function is then

$$I(\theta) = -\sum_{j=1}^{n} E\left[\frac{\partial^2}{\partial\theta^2} l_j(\theta)\right] = E\left[\left(\frac{\partial}{\partial\theta}\sum_{j=1}^{n} l_j(\theta)\right)^2\right].$$

If there is more than one parameter, the only change is that the vector of maximum likelihood estimates now has an asymptotic multivariate normal distribution. The covariance matrix[12] of this distribution is obtained from the inverse of the matrix with rsth element

$$\begin{aligned}
\mathbf{I}(\boldsymbol{\theta})_{rs} &= -E\left[\frac{\partial^2}{\partial\theta_s\partial\theta_r} l(\boldsymbol{\theta})\right] = -nE\left[\frac{\partial^2}{\partial\theta_s\partial\theta_r}\log f(X;\boldsymbol{\theta})\right] \\
&= E\left[\frac{\partial}{\partial\theta_r} l(\boldsymbol{\theta}) \frac{\partial}{\partial\theta_s} l(\boldsymbol{\theta})\right] = nE\left[\frac{\partial}{\partial\theta_r}\log f(X;\boldsymbol{\theta})\frac{\partial}{\partial\theta_s}\log f(X;\boldsymbol{\theta})\right].
\end{aligned}$$

The first expression on each line is always correct. The second expression assumes that the likelihood is the product of n identical pfs. When the ob-

[12] For any multivariate random variable the covariance matrix has the variances of the individual random variables on the main diagonal and covariances on the off-diagonal positions.

servations are independent but do not have identical distributions,

$$\mathbf{I}(\boldsymbol{\theta})_{rs} = -\sum_{j=1}^{n} E\left[\frac{\partial^2}{\partial\theta_s\partial\theta_r} l_j(\boldsymbol{\theta})\right].$$

This matrix is often called the **information matrix**. It should look familiar as it is the same matrix that was used in the method of scoring (recall that (2.13) is based on $E\left[\frac{\partial^2}{\partial\theta_s\partial\theta_r} l(\boldsymbol{\theta})\right]$). The only difference is that the true asymptotic variance requires the true parameter values while the method of scoring inserts the maximum likelihood estimates. Because we will never know the true values, this is as close as we are likely to be able to get to evaluating the variance of our estimator.

The information matrix also forms the Rao–Cramér lower bound (see, for example, Hogg and Craig [65, pp. 370–373]). That is, under the usual conditions, no unbiased estimator has a smaller variance than that given by the inverse of the information. Therefore, at least asymptotically, no unbiased estimator is more accurate than the mle.

Example 2.24 (individual data dental example) *Estimate the covariance matrix of the maximum likelihood estimator for the lognormal distribution.*

The first partial derivatives were obtained in Example 2.22,

$$\partial l/\partial\mu = \sum_{j=1}^{10}(\log x_j - \mu)/\sigma^2 \text{ and } \partial l/\partial\sigma = -10/\sigma + \sum_{j=1}^{10}(\log x_j - \mu)^2/\sigma^3.$$

The second partial derivatives are

$$\begin{aligned}
\partial^2 l/\partial\mu^2 &= -10/\sigma^2 \\
\partial^2 l/\partial\sigma\partial\mu &= -2\sum_{j=1}^{10}(\log x_j - \mu)/\sigma^3 \\
\partial^2 l/\partial\sigma^2 &= 10/\sigma^2 - 3\sum_{j=1}^{10}(\log x_j - \mu)^2/\sigma^4.
\end{aligned}$$

The expected values are ($\log X$ has a normal distribution with mean μ and standard deviation σ)

$$\begin{aligned}
E[\partial^2 l/\partial\mu^2] &= -10/\sigma^2 \\
E[\partial^2 l/\partial\sigma\partial\mu] &= 0 \\
E[\partial^2 l/\partial\sigma^2] &= -20/\sigma^2.
\end{aligned}$$

Changing the signs and inverting produces an estimate of the covariance matrix (it is an estimate because Theorem 2.2 only provides the covariance matrix in the limit). Here it is

$$\begin{bmatrix} \sigma^2/10 & 0 \\ 0 & \sigma^2/20 \end{bmatrix}.$$

We now further approximate the covariance matrix by inserting the maximum likelihood estimate of σ as found in Example 2.22:

$$\begin{bmatrix} 0.16914 & 0 \\ 0 & 0.084572 \end{bmatrix}.$$

□

The zeroes off the diagonal indicate that the two parameter estimates are asymptotically uncorrelated. For the particular case of the lognormal distribution, that is also true for any sample size. One thing we could do with this information is construct approximate 95% confidence intervals for the true parameter values. These would be 1.96 standard deviations either side of the estimate:

$$\begin{aligned} \mu &: \quad 5.07491 \pm 1.96(0.16914)^{1/2} = 5.07491 \pm 0.80608 \\ \sigma &: \quad 1.30055 \pm 1.96(0.084572)^{1/2} = 1.30055 \pm 0.56999. \end{aligned}$$

If we had been unable to obtain the expected values needed to get the information matrix (either due to our lack of skill at integrating or the lack of an analytic anti-derivative), an approximation can be constructed by inserting the second derivatives of the logarithm of the likelihood function directly into the matrix. For the lognormal distribution, using the parameter estimates, it turns out that the numbers are unchanged. (You are asked to show that this is always true in Exercise 2.51.) This good fortune is unlikely to hold in future examples or in the exercises. (An example appears in Exercise 2.52.) As the ultimate in approximation, in case we cannot even take the derivatives, we can use an approximate differentiation formula.

Example 2.25 (Example 2.24 continued) *Approximate the covariance matrix by each of the two less refined methods suggested above.*

Prior to taking the expected values, the elements of the information matrix are computed as

$$\begin{aligned} \partial^2 l/\partial\mu^2 &= -10/\hat{\sigma}^2 = -5.91216 \\ \partial^2 l/\partial\sigma\partial\mu &= -2\sum_{j=1}^{10}(\log x_j - \hat{\mu})/\hat{\sigma}^3 = 0 \\ \partial^2 l/\partial\sigma^2 &= 10/\hat{\sigma}^2 - 3\sum_{j=1}^{10}(\log x_j - \hat{\mu})^2/\hat{\sigma}^4 = -11.82431. \end{aligned}$$

Changing the sign and inverting the matrix gives the exact same values as were obtained using the expected value.

For the crudest level of approximation the second derivatives are approximated. A reasonable approximation is

$$\frac{\partial^2 f(\boldsymbol{\theta})}{\partial\theta_i\partial\theta_j} \doteq \frac{1}{h_i h_j}[f(\boldsymbol{\theta}+h_i\mathbf{e}_i/2+h_j\mathbf{e}_j/2)-f(\boldsymbol{\theta}+h_i\mathbf{e}_i/2-h_j\mathbf{e}_j/2)$$
$$-f(\boldsymbol{\theta}-h_i\mathbf{e}_i/2+h_j\mathbf{e}_j/2)+f(\boldsymbol{\theta}-h_i\mathbf{e}_i/2-h_j\mathbf{e}_j/2)]$$

where $\mathbf{e}_i$ is a vector with all zeros except for a one in the ith position and $h_i = \theta_i/10^v$ where v is one-third the number of significant digits used in calculations.

For this example, with 12 significant digits we have $h_1 = 0.000507491$ and $h_2 = 0.000130055$. The first approximation is

$$\begin{aligned}
\partial^2 l/\partial\mu^2 &\doteq \frac{l(5.075418, 1.300553) - 2l(5.074910, 1.300553) + l(5.0744403, 1.300553)}{(0.000507491)^2} \\
&= \frac{-67.5663823 - 2(-67.5663816) + (-67.5663823)}{(0.000507491)^2} \\
&= -5.912129.
\end{aligned}$$

The other two approximations are

$$\begin{aligned}
\partial^2 l/\partial\sigma\partial\mu &\doteq 0.000000 \\
\partial^2 l/\partial\sigma^2 &\doteq -11.824258.
\end{aligned}$$

We see that here the approximation works pretty well. □

For grouped data there is no need to approximate the expected value because it is taken as part of the scoring formula. It may be necessary to approximate the derivatives. For the grouped data example given earlier (Example 2.23), at convergence, the approximate covariance matrix turns out to be

$$\begin{bmatrix} 0.0041259 & -0.0000793 \\ -0.0000793 & 0.0022986 \end{bmatrix}.$$

In the grouped data case the two estimators are almost, but not exactly, uncorrelated.

2.5.2 Functions of maximum likelihood estimators

Without the following theorem, all this work would be of little value. That is because we are typically not interested in the parameters themselves, but rather in a function of the parameters. The following theorem is taken from Rao [104, p. 321].

Theorem 2.3 *Let $\mathbf{X}_n = (X_{1n}, \ldots, X_{kn})'$ be a multivariate random variable of dimension k based on a sample of size n. Assume that $\mathbf{X}_n$ is asymptotically normal with mean $\boldsymbol{\theta}$ and covariance matrix $\boldsymbol{\Sigma}/n$, where neither $\boldsymbol{\theta}$ nor $\boldsymbol{\Sigma}$ depend on n. Let g be a function of k variables that is totally differentiable. Let $G_n = g(X_{1n}, \ldots, X_{kn})$. Then G_n is asymptotically normal with mean $g(\boldsymbol{\theta})$ and covariance matrix $(\partial\mathbf{g})'\boldsymbol{\Sigma}(\partial\mathbf{g})$, where $\partial\mathbf{g}$ is the vector of first derivatives, that is, $\partial\mathbf{g} = (\partial g/\partial\theta_1, \ldots, \partial g/\partial\theta_k)'$ and it is to be evaluated at $\boldsymbol{\theta}$, the true parameter of the original random variable.* □

This is a specific case of a method for approximating moments of functions of random variables. The general method is called the delta method. For our purposes $\mathbf{X}_n$ is the vector of maximum likelihood estimators of the parameters and $\boldsymbol{\theta}$ is the true value. As usual, we approximate the result by inserting parameter estimates where necessary.

Example 2.26 (individual data dental example) *Approximate the variance of the maximum likelihood estimator of the mean using the lognormal distribution. Compare this to the variance of the method of moments estimator.*

The function in question is $g(\mu, \sigma) = \exp(\mu + \sigma^2/2)$. The two derivatives are $\exp(\mu + \sigma^2/2)$ and $\sigma\exp(\mu + \sigma^2/2)$. Inserting the maximum likelihood estimates yields 372.64 and 484.64. The approximate variance of the estimator is

$$\begin{bmatrix} 372.64 & 484.64 \end{bmatrix} \begin{bmatrix} 0.16914 & 0 \\ 0 & 0.084572 \end{bmatrix} \begin{bmatrix} 372.64 \\ 484.64 \end{bmatrix} = 43{,}351.$$

An approximate 95% confidence interval is

$$372.64 \pm 1.96(43{,}351)^{1/2} = 372.64 \pm 408.09.$$

The method of moments estimator is the sample mean which has variance $Var(X)/10$. This can be estimated as $200{,}564.5/10 = 20{,}056.45$ using the unbiased estimator for the variance. □

Why does the method of moments estimator appear to be so superior? The answer lies in the estimates themselves. The moment estimate of σ was 0.97822 while the mle was 1.30055. The mle indicates a population with a much larger variance and therefore any estimates from that population would be expected to have a larger sampling variance. A fairer test would be to use the mle parameters to evaluate the moment estimator. With those parameters the population variance is 614,784 and the variance of the sample mean is 61,478.4. Similarly, if the approximate variance of the mle were evaluated using the moment estimates, the variance estimate would be much smaller.

The interval above is a confidence interval for the mean of the population. Of more interest may be the actual value resulting from next year's claims.

Example 2.27 (Example 2.26 continued) *Suppose we know that there will be 100 claims next year. Determine a 95% prediction interval for the total payment.*

Let $S = X_1 + \cdots + X_{100}$, where X_i is the amount of the ith claim. Then $E(S) = 100E(X)$ is estimated by $100(372.64) = 37{,}264$. The squared error of this estimate is given by

$$\begin{aligned}
E[(S - 100e^{\hat{\mu}+\hat{\sigma}^2/2})^2] &= E\{[(S - 100e^{\mu+\sigma^2/2}) + 100(e^{\mu+\sigma^2/2} - e^{\hat{\mu}+\hat{\sigma}^2/2})]^2\} \\
&= E[(S - 100e^{\mu+\sigma^2/2})^2] \\
&\quad +10{,}000E[(e^{\mu+\sigma^2/2} - e^{\hat{\mu}+\hat{\sigma}^2/2})^2] \\
&\quad +200E[(S - 100e^{\mu+\sigma^2/2})(e^{\mu+\sigma^2/2} - e^{\hat{\mu}+\hat{\sigma}^2/2})] \\
&= Var(S) + 10{,}000Var(e^{\hat{\mu}+\hat{\sigma}^2/2}) \\
&= 100Var(X) + 10{,}000(43,351) \\
&= 100(e^{2\mu+2\sigma^2} - e^{2\mu+\sigma^2}) + 10{,}000(43,351) \\
&= 61{,}478{,}393 + 433{,}510,000 \\
&= 494{,}988{,}393.
\end{aligned}$$

The third line follows because the third term is the product of independent variables, each with an expected value of zero. They are independent because S depends only upon future observations while $\hat{\mu}$ and $\hat{\sigma}$ depend only upon past observations. In order to evaluate $Var(X)$, parameter estimates were inserted. A 95% prediction interval is $37{,}264 \pm 1.96\sqrt{494{,}988{,}393}$ or $37{,}264 \pm 43{,}607$.□

We must emphasize again that if the population is lognormal, the maximum likelihood estimator of the mean will have a smaller variance than the method of moments estimator. The true variances of these estimators depend on the true values of the parameters and would be evaluated at that common set of values.

Example 2.28 (grouped data dental example) *Estimate the parameters for the Pareto distribution, approximate the covariance matrix, and construct a 95% confidence interval for the population mean.*

Using the method of scoring, the parameter estimates are $\hat{\alpha} = 3.8275$ and $\hat{\theta} = 948.52$. The approximate covariance matrix is

$$\begin{bmatrix} 0.97058 & 290.01 \\ 290.01 & 90{,}384 \end{bmatrix}.$$

The estimate of the mean is $\hat{\theta}/(\hat{\alpha} - 1) = 948.52/2.8275 = 335.46$. The derivatives with respect to α and θ respectively are $-\theta/(\alpha - 1)^2$ and $1/(\alpha - 1)$.

When evaluated at the parameter estimates they produce the values -118.64 and 0.35367. The approximate variance of the mle of the mean is

$$\begin{bmatrix} -118.64 & 0.35367 \end{bmatrix} \begin{bmatrix} 0.97058 & 290.01 \\ 290.01 & 90{,}384 \end{bmatrix} \begin{bmatrix} -118.64 \\ 0.35367 \end{bmatrix} = 629.51$$

and an approximate 95% confidence interval is

$$335.46 \pm 1.96(629.51)^{1/2} = 335.46 \pm 49.18. \qquad \square$$

2.5.3 Minimum distance estimation

A similar set of results is available for minimum distance estimators. The following result is from [77].

Theorem 2.4 *If the minimum distance estimator is an implicit function of random variables to which the Central Limit Theorem can be applied then* $\sqrt{n}(\widehat{\boldsymbol{\theta}} - \boldsymbol{\theta})$ *has an asymptotic multivariate normal distribution with mean vector* $\mathbf{0}$ *and covariance matrix* $\mathbf{A}^{-1}\mathbf{B}\boldsymbol{\Sigma}\mathbf{B}'\mathbf{A}^{-1}$ *where the matrices are obtained as follows.*

$$\begin{aligned} \mathbf{A}_{rs} = \frac{\partial^2 Q}{\partial\theta_r \partial\theta_s} &= 2\sum_{j=1}^{k} w_j \frac{\partial G(c_j;\boldsymbol{\theta})}{\partial\theta_r}\frac{\partial G(c_j;\boldsymbol{\theta})}{\partial\theta_s} \\ &\quad + 2\sum_{j=1}^{k} w_j [G(c_j;\boldsymbol{\theta}) - G_n(c_j)] \frac{\partial^2 G(c_j;\boldsymbol{\theta})}{\partial\theta_r \partial\theta_s} \\ \mathbf{B}_{rs} = \frac{\partial^2 Q}{\partial\theta_r \partial G_n(c_s)} &= -2w_s \frac{\partial G(c_s;\boldsymbol{\theta})}{\partial\theta_r} \end{aligned}$$

and $\boldsymbol{\Sigma}$ *is the asymptotic covariance matrix of* $\sqrt{n}(\mathbf{G}_n - \boldsymbol{\mu})$ *where* $\mathbf{G}_n$ *is the vector of the* $G_n(c_j)$ *and* $\boldsymbol{\mu}$ *is its asymptotic mean.*

A fair amount of effort is required to obtain these quantities, examples for both minimum cdf and minimum LEV are given in [77]. With regard to functions of the parameters, Theorem 2.3 still applies.

The main reason for going through the exercise would be to justify the use of a minimum distance estimator in place of maximum likelihood. Although we know that the mle will have smaller variances, for a large sample size the variance of the minimum distance estimator may be sufficiently small that we are willing to trade the loss of accuracy for the increased flexibility.

2.6 ADVANTAGES OF PARAMETRIC ESTIMATION

We have spent a great deal of effort covering the mechanics of parametric estimation as well as determining methods of evaluating estimators. While many of the concepts involved in making comparisons have already been discussed, we use this section to organize and expand those ideas.

2.6.1 Accuracy

It has already been established that when the population follows the selected parametric family, maximum likelihood estimators are superior to other competitors. It is likely that even when the population is slightly unlike the selected family, maximum likelihood estimators will still do well. However, this is a big risk. If this were the only advantage of parametric estimation, it would be worthwhile only when we had a fair degree of confidence in the nature of the population. But there are many other reasons to choose parametric estimators. These are outlined in the remaining subsections.

2.6.2 Inferences can be made beyond the population that generated the data

The purpose of an actuarial model is not only to represent the past, but also to represent the future. The future will differ from the past in ways that are not predictable (e.g., random changes in the loss producing environment) and those that are (e.g., planned changes in the benefit structure or in the characteristics of those insured). Even random changes can be estimated, such as using a forecasted inflation rate. Once the rate has been set, the change is now planned, not random. It is essential that we be able to use our model to investigate the impact of planned changes as well as to perform "what-if" analysis on potential random changes.

It is not always possible to perform a successful adjustment to a new population when empirical estimators are being used. The following two examples illustrate this point.

Example 2.29 (individual data dental example) *The data set consists of amounts paid with a deductible of 50. Estimate the expected amount paid per payment after 10% inflation is imposed on all losses. Attempt to do this both empirically and parametrically.*

Empirically we can obtain the new amount paid from the ten previous amounts paid:

$$\begin{aligned} 1.1(141+50)-50 &= 160.1 \\ 1.1(16+50)-50 &= 22.6 \end{aligned}$$

$$55.6,\ 49.0,\ 391.1,\ 289.9,\ 353.7,\ 1{,}667.1,\ 122.7,\ 628.7.$$

These numbers represent what would be paid after inflation has affected the losses. We could use the sample mean of 374.05 as our estimate, but that quantity must overstate the true value. The smallest possible amount paid that could have been recorded by this empirical analysis is $1.1(0+50)-50=5$. Amounts paid under 5 arise from losses in the 45.45–50 range. (A loss of 45.45 when inflated by 10% is right at the deductible of 50.). These losses do not and cannot appear in our data set. Also, inflation affects the frequency of payment in this case. There will be more payments due to it being easier to exceed the deductible. An empirical estimator is impossible to obtain.

For a parametric solution, consider the exponential distribution with the parameter estimated by the method of moments. To solve this problem we must let the exponential distribution model the amount of the loss, not the amount of the payment. If X is the loss random variable and Y is the payment random variable, then for the model we obtain

$$\begin{aligned}
E(Y) &= E(X-50|X>50) \\
&= \int_{50}^{\infty}(x-50)f(x|X>50)dx \\
&= \int_{50}^{\infty}(x-50)\frac{f(x)}{1-F(50)}dx \\
&= \int_{50}^{\infty}(x-50)\frac{\theta^{-1}\exp(-x/\theta)}{\exp(-50/\theta)}dx \\
&= -(x-50)\exp[-(x-50)/\theta]-\theta\exp[-(x-50)/\theta]|_{50}^{\infty} \\
&= \theta=\bar{Y}=335.5.
\end{aligned}$$

With inflation the amount paid is $1.1X-50|1.1X>50$ and the expected payment is

$$\begin{aligned}
E(1.1X-50|1.1X>50) &= \int_{500/11}^{\infty}(1.1x-50)\frac{(335.5)^{-1}\exp(-x/335.5)}{\exp[-(500/11)/335.5]}dx \\
&= 369.05.
\end{aligned}$$

Also note that prior to inflation the probability that a loss produced a payment was

$$\Pr(X>50)=\exp(-50/335.5)=0.86154.$$

After inflation it is

$$\Pr(1.1X>50)=\exp[-(500/11)/335.5]=0.87329.$$ □

In the parametric solution we followed a specific set of steps. We first postulated a model for *losses*, including those below 50. However, the only data available were on the *amount paid.* We then used that model to study the amount paid under the new scenario. This will be a standard approach and will be expanded upon in Section 2.9.

Example 2.30 *The data in Table 2.10*[13] *represent 217 general liability payments from policies with a limit of 300,000. Estimate the percentage change in*

[13] Copyright, Insurance Services Office, Inc., 1994. The data in Tables 2.12 and 2.27 are also Copyright, Insurance Services Office, Inc., 1994.

Table 2.10 General liability payments

Payment	Number	Average
0–2,500	41	1,389
2,500–7,500	48	4,661
7,500–12,500	24	9,991
12,500–17,500	18	15,482
17,500–22,500	15	20,232
22,500–32,500	14	26,616
32,500–47,500	16	40,278
47,500–67,500	12	56,414
67,500–87,500	6	74,985
87,500–125,000	11	106,851
125,000–225,000	5	184,735
225,000–300,000	4	264,025
at 300,000	3	300,000

the average payment that will result from (a) 10% inflation and (b) imposition of a deductible of 1,000. In both cases the limit remains at 300,000.

Empirically we have enough information to estimate the mean prior to the modifications. It is

$$[41(1,389) + 48(4,661) + \cdots + 3(300{,}000)]/217 = 33,648.$$

However, with 10% inflation it is impossible to determine the effect on the 4 payments that were in the 225,000–300,000 group. Some of them may hit the limit after being inflated. With regard to the deductible, we do not know how many of the 41 payments in the first group will be affected or the magnitude of the effect. Again, empirical methods fail to perform. We should note, however, that if we use the histogram as the density function, the calculations become possible (see Exercise 2.53).

For a parametric solution, consider the lognormal distribution with parameters estimated by matching the 30th and 70th percentiles. The percentiles are

$$\begin{aligned} 2{,}500 + (65.1 - 41)5{,}000/48 &= 5{,}010 \\ 22{,}500 + (151.9 - 146)10{,}000/14 &= 26{,}714. \end{aligned}$$

The resulting equations are

$$\begin{aligned} 0.3 &= \Phi[(\log 5{,}010 - \mu)/\sigma] \\ 0.7 &= \Phi[(\log 26{,}714 - \mu)/\sigma] \end{aligned}$$

or

$$\begin{aligned} -0.52440 &= (8.51919 - \mu)/\sigma \\ 0.52440 &= (10.19294 - \mu)/\sigma \end{aligned}$$

and the solution is $\hat{\sigma} = 1.595871$ and $\hat{\mu} = 9.356065$. We note that the expected payment is $E(X \wedge 300,000) = 33,960.11$. With 10% inflation the desired quantity is

$$\begin{aligned} E[\min(1.1X, 300{,}000)] &= E(1.1X \wedge 300{,}000) \\ &= E[1.1(X \wedge 272{,}727)] \\ &= 1.1E(X \wedge 272{,}727) \\ &= 1.1(33{,}354.59) \\ &= 36{,}690.05. \end{aligned}$$

The increase due to inflation is 36,690.05/33,960.11 − 1 = 0.0804, an 8.04% increase.

Imposing a deductible of 1,000 changes the payment variable. If we want to obtain the expected payment per loss, the random variable of interest is

$$Y = \begin{cases} 0, & X \leq 1{,}000, \\ X - 1{,}000, & 1{,}000 < X < 300{,}000, \\ 299{,}000, & X \geq 300{,}000 \end{cases}$$

and the expected value is

$$\begin{aligned} E(Y) &= \int_{1{,}000}^{300{,}000} (x - 1{,}000) f(x) dx + 299{,}000[1 - F(300{,}000)] \\ &= \int_0^{300{,}000} x f(x) dx - \int_0^{1000} x f(x) dx \\ &\quad -1000[F(300{,}000) - F(1{,}000)] + 299{,}000[1 - F(300{,}000)] \\ &= \int_0^{300{,}000} x f(x) dx + 300{,}000[1 - F(300{,}000)] \\ &\quad - \int_0^{1{,}000} x f(x) dx - 10{,}000[1 - F(1{,}000)] \\ &= E(X \wedge 300{,}000) - E(X \wedge 1{,}000) \\ &= 33{,}960.11 - 973.63 = 32{,}986.48. \end{aligned}$$

The reduction due to the deductible is 1 − 32,986.48/33,960.11 = 0.0287, or 2.87%. □

The examples make clear that parametric models provide flexibility that is not present in their empirical counterparts. We see that grouping does not present an obstacle nor does the imposition of inflation. The second of the examples points out the usefulness of the limited expected value. The following theorem confirms that observation.

Theorem 2.5 *Let X be the random variable for the loss. With inflation at rate r, a deductible of d, a limit of u, and coinsurance of α, the amount paid (per loss) random variable, Y, is*

$$Y = \begin{cases} 0, & X \leq d/(1+r), \\ \alpha[(1+r)X - d], & d/(1+r) < X < u/(1+r), \\ \alpha(u-d), & X \geq u/(1+r). \end{cases}$$

Then the expected amount paid per loss is

$$E(Y) = \alpha(1+r)\{E[X \wedge u/(1+r)] - E[X \wedge d/(1+r)]\}.$$

Proof: The proof is provided for a continuous random variable. A similar argument would apply for any other random variable.

$$\begin{aligned} E(Y) &= \int_{d/(1+r)}^{u/(1+r)} [\alpha(1+r)x - \alpha d]f(x)dx + \alpha(u-d)\{1 - F[u/(1+r)]\} \\ &= \alpha(1+r)\int_0^{u/(1+r)} xf(x)dx - \alpha(1+r)\int_0^{d/(1+r)} xf(x)dx \\ &\quad -\alpha d\{F[u/(1+r)] - F[d/(1+r)]\} \\ &\quad +\alpha(u-d)\{1 - F[u/(1+r)]\} \\ &= \alpha(1+r)\left\{\int_0^{u/(1+r)} xf(x)dx + [u/(1+r)]\{1 - F[u/(1+r)]\}\right. \\ &\quad \left. - \int_0^{d/(1+r)} xf(x)dx - [d/(1+r)]\{1 - F[d/(1+r)]\}\right\} \\ &= \alpha(1+r)\{E[X \wedge u/(1+r)] - E[X \wedge d/(1+r)]\}. \end{aligned}$$

□

If we wanted the expected payment per payment, we need only to understand that the random variable in question is now $Y|X > d/(1+r)$.

Corollary 2.6 *The expected payment per payment is*

$$E[Y|X > d/(1+r)] = \alpha(1+r)\frac{E[X \wedge u/(1+r)] - E[X \wedge d/(1+r)]}{1 - F[d/(1+r)]}.$$

□

From Theorem 2.5 we see that the difference of LEVs is often a useful quantity. The following result provides an alternative method of computing model LEVs that may be especially useful for obtaining differences.

Theorem 2.7 *If* $\Pr(X < 0) = 0$, *then* $E(X \wedge x) = \int_0^x [1 - F(y)]dy$.

Proof: For the case where X has a continuous distribution,

$$\begin{aligned} E(X \wedge x) &= \int_0^x yf(y)dy + x[1 - F(x)] \\ &= -y[1 - F(y)]|_0^x + \int_0^x [1 - F(y)]dy + x[1 - F(x)] \\ &= \int_0^x [1 - F(y)]dy \end{aligned}$$

where the second line follows from integration by parts with $u = y$, $dv = f(y)dy$, $du = dy$, and $v = -[1 - F(y)]$. The same result follows for discrete or mixed distributions. □

Corollary 2.8 $E(X \wedge u) - E(X \wedge d) = \int_d^u [1 - F(y)]dy$. □

Theorem 2.5 indicates that for most common modifications the limited expected value is sufficient. However, in the literature you will find that two other quantities are sometimes used.

Definition 2.28 *The* ***loss elimination ratio*** *for a deductible of d is the relative reduction in the expected payment given the imposition of a deductible. More formally,*

$$LER_X(d) = E[\min(X, d)]/E(X) = E(X \wedge d)/E(X)$$

provided that $E(X)$ *and* $E(X \wedge d)$ *exist.*

In general, the term loss elimination ratio can refer to the reduction in the expected payment under any set of modifications. An interesting, but not necessarily useful, observation is that $LER_X(d)$ satisfies all of the properties of a distribution function.

In the definition, it was noted that the moments must exist before the LER is defined. At this point some words are in order about the existence of moments. When $E(X)$ fails to exist, it could be either because $\lim_{u \to \infty} \int_d^u xf(x)dx$ does not converge or because $\lim_{d \to 0} \int_d^u xf(x)dx$ does not converge. If the second limit exists, $E(X \wedge x)$ will still exist. This makes the limited expected value a bit more useful than the loss elimination ratio, because it will exist, even for heavy-tailed distributions. If the first limit exists but the second does not, both $E(X)$ and $E(X \wedge x)$ will fail to exist, but their difference will. The concept of tail weight will be discussed further in the next section.

The second quantity is useful in describing the behavior of the loss random variable with regard to large losses. This will be expanded upon in the next section.

Definition 2.29 *The* ***mean excess loss*** *for a deductible of* d *is the expected loss in excess of the deductible, conditioned on the loss exceeding the deductible.*[14] *From Corollary 2.6 it is*

$$e_X(d) = e(d) = E(X - d|X > d) = \frac{E(X) - E(X \wedge d)}{1 - F_X(d)}.$$

Of course, if $E(X)$ is infinite due to X having a heavy tail, the mean excess loss will also be infinite. If there is an existence problem near zero, the numerator can be evaluated as $\int_d^\infty (x-d)f(x)dx$.

Example 2.31 *Determine the mean excess loss function for the Pareto distribution.*

From Appendix A we have

$$\begin{aligned} e_X(d) &= \frac{\dfrac{\theta}{\alpha-1} - \dfrac{\theta}{\alpha-1} + \dfrac{\theta}{\alpha-1}\left(\dfrac{\theta}{d+\theta}\right)^{\alpha-1}}{1 - 1 + \left(\dfrac{\theta}{d+\theta}\right)^{\alpha}} \\ &= \frac{\theta}{\alpha-1}\frac{d+\theta}{\theta} = \frac{d+\theta}{\alpha-1} \end{aligned}$$

provided that $\alpha > 1$. Note that the function is a straight line. □

The final result for this discussion indicates how the limited second moment can be used to calculate the variance of a modified loss.

Theorem 2.9 *Let* X *be the random variable for the loss. With a deductible of* d*, a limit of* u*, and coinsurance of* α*, the amount paid (per loss) random variable,* Y*, is*

$$Y = \begin{cases} 0, & X \le d, \\ \alpha(X-d), & d < X < u, \\ \alpha(u-d), & X \ge u. \end{cases}$$

Then the variance of the amount paid per loss is

$$\begin{aligned} E(Y^2) - [E(Y)]^2 &= \alpha^2 \left\{E[(X \wedge u)^2] - E[(X \wedge d)^2]\right. \\ &\quad -2dE(X \wedge u) + 2dE(X \wedge d) \\ &\quad \left.-[E(X \wedge u) - E(X \wedge d)]^2\right\}. \end{aligned} \tag{2.15}$$

[14] In the biostatistics literature this quantity is called the **mean residual life**. When applied to the age at death random variable it is the complete expectation of life, denoted $\mathring{e}_d$. This is the motivation for the use of the symbol e.

The variance of the amount paid per payment is found by dividing the first term of (2.15) by $1 - F_X(d)$ *and dividing the second term by* $[1 - F_X(d)]^2$.

Proof: The proof is left as Exercise 2.57. □

Example 2.32 (Example 2.30 continued) *Estimate the variance of the loss with a deductible of 1,000 and a limit of 300,000.*

The two required additional numbers are $E[(X \wedge 1{,}000)^2]$ and $E[(X \wedge 300{,}000)^2]$. For the lognormal distribution, in general, we have

$$\begin{aligned}
E[(X \wedge x)^2] &= \int_0^x \frac{t^2}{t\sigma\sqrt{2\pi}} e^{-(\log t - \mu)^2/2\sigma^2} dt + x^2\{1 - \Phi[(\log x - \mu)/\sigma]\} \\
&= \int_{-\infty}^{(\log x - \mu)/\sigma} \frac{e^{y\sigma+\mu}}{\sigma\sqrt{2\pi}} e^{-y^2/2} \sigma e^{y\sigma+\mu} dy \\
&\quad + x^2\{1 - \Phi[(\log x - \mu)/\sigma]\} \\
&= \int_{-\infty}^{(\log x - \mu)/\sigma} \frac{1}{\sqrt{2\pi}} e^{-(y^2 - 4\sigma y - 4\mu)/2} dy \\
&\quad + x^2\{1 - \Phi[(\log x - \mu)/\sigma]\} \\
&= \int_{-\infty}^{(\log x - \mu)/\sigma} \frac{1}{\sqrt{2\pi}} e^{-(y-2\sigma)^2/2} e^{2\mu+2\sigma^2} dy \\
&\quad + x^2\{1 - \Phi[(\log x - \mu)/\sigma]\} \\
&= e^{2\mu+2\sigma^2}\Phi[(\log x - \mu - 2\sigma^2)/\sigma] + x^2\{1 - \Phi[(\log x - \mu)/\sigma]\}.
\end{aligned}$$

Then, using the parameter estimates $\hat{\mu} = 9.356065$ and $\hat{\sigma} = 1.595871$, we have

$$E[(X \wedge 1{,}000)^2] = 962{,}490 \text{ and } E[(X \wedge 300{,}000)^2] = 4{,}581{,}050{,}352.$$

Finally,

$$\begin{aligned}
Var(Y) &= 4{,}581{,}050{,}352 - 962{,}490 - 2{,}000(33{,}960.11) \\
&\quad + 2{,}000(973.63) - (33{,}960.11 - 973.63)^2 \\
&= 3{,}426{,}007{,}039.
\end{aligned}$$

□

The calculation of the cdf or the LEV for many distributions often involves integrals that must be evaluated numerically. Two specific cases that show up regularly are the incomplete beta (denoted in this text by $\beta(a, b; x)$) and gamma (denoted $\Gamma(\alpha; x)$) integrals. They are defined formally in Subsection 2.7.3 and are discussed in detail at the beginning of Appendix A. These functions are also available in most spreadsheet and statistics programs.

2.6.3 Interpolation and parsimony

For the most part we believe that the progression of loss probabilities is smooth. That is, as the possible amount increases there should be a slight change in probability, and this trend should be steadily increasing or decreasing over reasonably large ranges of values. The ultimate in smoothness would be provided by a continuous random variable with a continuous pdf. This is true for all of the models in Appendix A.

For individual data the empirical cdf is not even differentiable; the model represents a discrete random variable. For grouped data the ogive is continuous but the histogram is not. Also it must be kept in mind that the histogram is a complex model, involving $k-1$ parameters, one for each group less one because the probabilities must add to one.

This property of the histogram is an example of a lack of parsimony. A typical dictionary definition of parsimony is[15]

1. Unusual or excessive frugality; extreme economy or stinginess.
2. Adoption of the simplest assumption in the formulation of a theory or in the interpretation of data, especially in accordance with the rule of Ockham's razor.

For modeling purposes we take a narrower view.

Definition 2.30 *One model is* ***more parsimonious*** *than another if it can be completely specified using a smaller number of objects. These objects are usually the parameters of the model.*

The principle of parsimony states that unless there is a very good reason to do otherwise, the more parsimonious model should be used. One reason is that the fewer parameters there are to estimate, the more accurately each one can be estimated. We have a fixed amount of data and want to use it as efficiently as possible. Also, models with a large number of parameters tend to perform poorly when used for prediction.

Example 2.33 *Consider the regression problem of fitting a curve to a collection of ten (x, y) pairs. We know that as long as the ten x values are unique there will be a unique ninth degree polynomial that passes through them without error. With this polynomial and the ten x values we can reproduce the ten y values. But when confronted with a new, eleventh, x value, how well do we think the ninth degree polynomial will do? It has to take many twists and turns to successfully hit all ten points and is likely to provide useless predictions for any other value. The two curves in Figure 2.5 provide an illustration.* □

[15] American Heritage Electronic Dictionary [3].

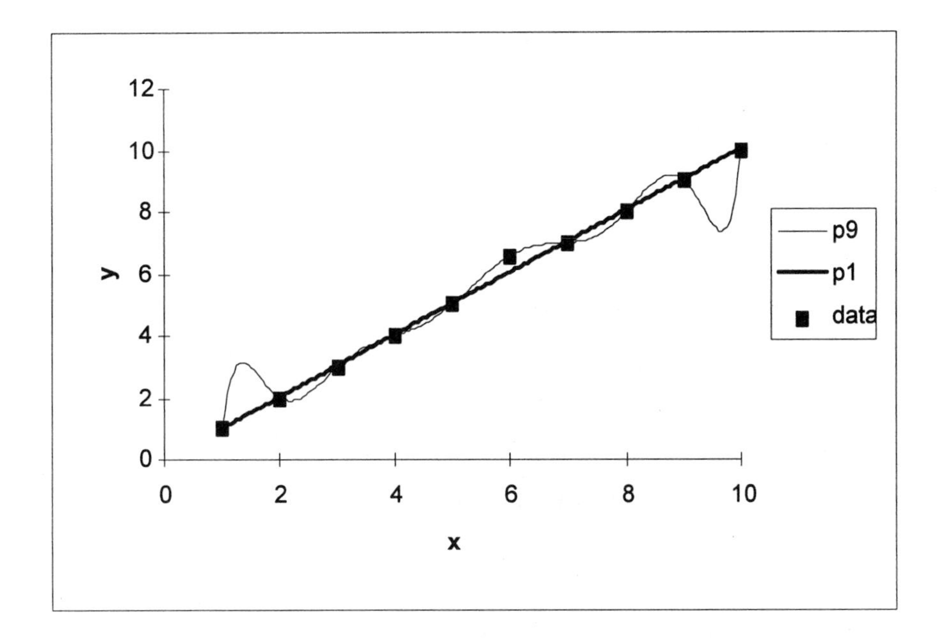

Fig. 2.5 Two polynomial regressions: p9 = ninth degree; p1 = first degree

Our goal is to create simple models that adequately (but not necessarily perfectly) capture the essence of our data.

We illustrate these ideas with two examples.

Example 2.34 (individual data dental example) *Determine the effect on expected payment amounts of raising the deductible from 50 to 60.*

This problem can be solved empirically. With a deductible of 50 the expected payment is 335.5. Raising the deductible to 60 reduces each amount paid by 10. We then discard any negative payments. But there were no such payments here, so the sample mean drops by 10 to 325.5. The empirical analysis implies that raising the deductible by 10 will reduce average payments by 10 and will have no impact on the number of payments.

Now consider the exponential model estimated in Example 2.29. From Theorem 2.5 and Appendix A we have that the expected payment per payment with a deductible of 50 is 335.5 while for a deductible of 60 it is

$$\frac{E(X) - E(X \wedge 60)}{1 - F(60)} = \frac{335.5 - [335.5(1 - e^{-60/335.5})]}{e^{-60/335.5}} = 335.5$$

This result is due to the memoryless property of the exponential distribution. With the deductible at 50 the proportion of losses that produce payments is $\exp(-50/335.5) = 0.86154$ while with a deductible of 60 it is $\exp(-60/335.5) = 0.83624$. Thus there will be a $(0.86154 - 0.83624)/0.86154 =$

Table 2.11 General liability payments

Payment	Number
0–2,500	25
2,500–7,500	22
7,500–12,500	10
12,500–17,500	14
17,500–22,500	7
22,500–32,500	6
32,500–47,500	9
47,500–67,500	3
67,500–87,500	2
87,500–125,000	6
125,000–225,000	2
225,000–300,000	2
at 300,000	1

0.029366 reduction in payments. This is less than the empirical reduction of 10,which is $10/335.5 = 0.029806$. The reduction should be less because those future losses that will be between 50 and 60 will result in a saving of less than 10. □

The following example illustrates how parsimonious models assist in projections.

Example 2.35 *The data behind Example 2.30 was split into two pieces by randomly selecting 109 of the 217 observations. The grouped data appears in Table 2.11. Estimate the 70th percentile from the ogive and from the lognormal distribution as estimated by maximum likelihood. Compare the estimates to the 70th percentile of the remaining 108 observations in order to see how well the estimators predict a future result.*

The empirical estimate of the 70th percentile is

$$17,500 + (5.3)(5{,}000)/7 = 21{,}286$$

while the maximum likelihood estimates of the lognormal parameters (writing the likelihood function as in Definition 2.27 and noting that the one observation at 300,000 belongs to the group $[300{,}000, \infty)$) are $\hat{\mu} = 9.13169$ and $\hat{\sigma} = 1.65556$. The 70th percentile of this distribution is found from

$$\begin{aligned}
\Phi\left(\frac{\log \pi_{0.7} - 9.13169}{1.65556}\right) &= 0.7 \\
\frac{\log \pi_{0.7} - 9.13169}{1.65556} &= 0.52440
\end{aligned}$$

$$\hat{\pi}_{0.7} = 22{,}024.$$

The 70th percentile from the remaining 108 observations is 32,000. While neither model performed particularly well, the two-parameter model did come closer than the 12-parameter model did. □

2.6.4 Hypothesis tests

Sometimes our questions are of the "yes/no" type rather than a search for a numerical quantity. If the question concerns parameters, we can formulate a statistical hypothesis test. The test is conducted by obtaining a test statistic and a critical region. One of the big advantages of likelihood based procedures is that we have access to the likelihood ratio test. The following Theorem is taken from Rao [104, p. 350] and describes the test.

Theorem 2.10 *Let the population random variable have r-dimensional parameter $\boldsymbol{\theta} = (\theta_1, \ldots, \theta_r)'$. Let the null hypothesis be that k restrictions of the form $R_j(\boldsymbol{\theta}) = 0$, $j = 1, \ldots, k$ hold where each $R_j(\boldsymbol{\theta})$ has continuous first partial derivatives. Let the alternative be that there are no restrictions on the parameter values. Let $L(\boldsymbol{\theta}; x)$ be the likelihood function, where x represents all the data in the random sample. The restricted mle of $\boldsymbol{\theta}$ is the value that maximizes the likelihood function subject to the k restrictions. Let $\boldsymbol{\theta}^*$ denote this value. Let $\widehat{\boldsymbol{\theta}}$ be the unrestricted mle. Finally, let the likelihood ratio test statistic be*

$$2[\log L(\widehat{\boldsymbol{\theta}}; x) - \log L(\boldsymbol{\theta}^*; x)].$$

Then, provided that (i) through (v) from Theorem 2.2 hold, as the sample goes to infinity, the test statistic (under the null hypothesis) has a chi-square distribution with k degrees of freedom. □

The following example illustrates this process.

Example 2.36 *The study that produced Table 2.10 also included amounts paid from policies with a 500,000 limit. Those results appear in Table 2.12. Do the two populations have the same, or different, lognormal distributions?*

Under the null hypothesis of identical distributions we fit a single lognormal distribution to the combined data set. The likelihood function is simply the product of the likelihood functions from the two separate data sets. The parameter estimates are $\hat{\mu} = 9.36545$ and $\hat{\sigma} = 1.64653$ with a loglikelihood of $-1{,}874.72$. When fit separately the results for the 300,000 limit are $\hat{\mu} = 9.29376$ and $\hat{\sigma} = 1.62713$ with a loglikelihood of -498.29 and for the 500,000 limit are $\hat{\mu} = 9.39218$ and $\hat{\sigma} = 1.65247$ with a loglikelihood of $-1{,}376.12$. The likelihood ratio statistic is

$$2[-498.29 - 1,376.12 - (-1{,}874.72)] = 0.62.$$

Table 2.12 General liability payments, 500,000 limit

Payment	Number
0–2,500	101
2,500–7,500	132
7,500–12,500	61
12,500–17,500	50
17,500–22,500	29
22,500–32,500	50
32,500–47,500	28
47,500–67,500	40
67,500–87,500	22
87,500–125,000	27
125,000–175,000	19
175,000–225,000	6
225,000–325,000	10
325,000–500,000	4
at 500,000	5

The null hypothesis has two restrictions (equal μ and equal σ) and so the critical value is based on the chi-square distribution with two degrees of freedom. At a 5% significance level the critical value is 5.99 and the null hypothesis is accepted. The p-value is 0.733.

The conclusion is that the two populations could have the same lognormal distribution. □

This example illustrates another advantage of maximum likelihood estimation. When data are collected from independent samples from different populations (with regard to data modification), it is possible to estimate the parameters under the assumption that the two populations (with the modifications removed) have the same distribution. This is because the loglikelihood is just the sum of the individual loglikelihoods. Thus if we can assume that purchasers of the two limits have the same loss distribution, we can use the 217 observations at the 300,000 limit to aid in establishing expected costs for those who purchase a 500,000 limit.

2.6.5 Scale parameters/families

While not precisely an advantage of parametric models, scale parameters provide an easy method of handling inflationary changes. The definitions are as follows:

Definition 2.31 *A parametric family of distributions is called a* **scale family** *if for any random variable X in the family, the random variable $Y = cX$ is also a member of that family.*

Definition 2.32 *A parametric family of distributions possesses a* **scale parameter** *if when X is a member of the family, the random variable $Y = cX$ is also a member of the family and, furthermore, the parameters of Y are identical to those of X except for the scale parameter which is multiplied by c.*

Example 2.37 *Show that the lognormal distribution represents a scale family but does not have a scale parameter.*

The cdf of Y is

$$\begin{aligned} F_Y(y) &= \Pr(Y \leq y) = \Pr(cX \leq y) = \Pr(X \leq y/c) = \Phi\left(\frac{\log(y/c) - \mu}{\sigma}\right) \\ &= \Phi\left(\frac{\log y - \log c - \mu}{\sigma}\right) \end{aligned}$$

and therefore Y has the lognormal distribution with parameters $\mu + \log c$ and σ. □

Example 2.38 *Show that the Pareto distribution has a scale parameter.*

The cdf of Y is

$$\begin{aligned} F_Y(y) &= \Pr(X \leq y/c) = 1 - \left(\frac{\theta}{\theta + y/c}\right)^\alpha \\ &= 1 - \left(\frac{c\theta}{c\theta + y}\right)^\alpha \end{aligned}$$

and therefore Y has the Pareto distribution with parameters α and $c\theta$. □

The main advantage of being in a scale family is that when inflation is applied (provided it applies uniformly) the resulting distribution is known and available for further analysis. In the next section we will find that a scale parameter can also be used as a device for modeling uncertainty about inflation.

All of the distributions in Appendix A are scale families. With the exception of the lognormal and inverse Gaussian, all have a scale parameter. The first two could be parameterized with a scale parameter, but we have adopted the more conventional representations.

2.7 AN INVENTORY OF PARAMETRIC DISTRIBUTIONS

By now we hope you are convinced that there are many advantages to using parametric estimation. To be successful we must be able to choose a family of models that is close enough to the population to be useful. Our chances of being successful will be enhanced if we possess a large collection of distributions that is rich in terms of possible shapes. In this section we present over twenty distributions and then indicate a way to increase that number to over two hundred. Another way to add flexibility is introduced in Section 2.11.

2.7.1 Desirable characteristics of parametric families

There are a number of characteristics that our collection of distributions should have. They are

1. Place probability on all non-negative real numbers.

2. Be smooth.

3. Some should have a mode at zero, some should not.

4. Some should have moderate tails, some should have heavy tails.

The first characteristic represents the fact that in most loss processes there is no maximum possible loss. Of course in reality there is always a maximum such as all the money in the world, the net worth of a corporation, or the value of the object (e.g., house or car) insured against physical damage. In the last case the maximum is more critical because the maximum loss is known and is not a matter for speculation. This is different from a contractual imposition of a maximum payment. In that case the distribution should be confined to the interval $(0, c)$. The beta and generalized beta distributions provide examples of this type of distribution. They are described in Appendix A, but are not discussed further in this section.

There is one model that does not start at zero, namely, the single parameter Pareto. It makes some sense to start probability above zero when losses below a certain amount are not important. This would be the case when there is a deductible and there are no plans to lower it. Sometimes the deductible may be a large number, but we begin our model at a smaller number. This allows inflation to be incorporated, but still ignores data that has no bearing on the questions being asked.

Smoothness implies that the transition with regard to probability is steady and continuous as the loss amount changes. This can be translated as a pdf that is continuous. There is one situation in which this may not be the most appropriate model for losses, or at least amounts paid. When there is some discretion over the benefit to be provided by the insurance, it is not unusual for the amount to be a round figure. If a car is completely destroyed in an accident, the adjuster is more likely to offer a round figure such as 1,400 rather than 1,412. When the settlement includes compensation for suffering and inconvenience, round numbers become especially likely. This is a difficult problem because it is not always clear from the data which round numbers are popular, nor is it clear how to model them. It also means that inflation is unlikely to apply uniformly. An incident that yields a 100,000 liability settlement may still yield a 100,000 settlement after 5% inflation, but a few years later after there has been 20% inflation the settlement for the same incident may be for 125,000.

There have been three suggestions for finding smooth models for data that have these concentrations. The first is to use the individual losses as recorded and let the model provide the smoothing. The second and third require that the data be smoothed by grouping prior to estimation. The second method is to create the groups so that the boundaries are unpopular loss values and the popular values occur near the middle of such intervals. The third method is to create narrow groups (width of one dollar) around popular values. The second method might produce boundaries such as (0, 2,500], (2,500, 7,500], and (7,500, 12,500] where the popular values are 5,000 and 10,000. The third method would produce boundaries of (0, 4,999], (4,999, 5,000], (5,000, 9,999], and (9,999, 10,000]. The advantage of the third method is that no arbitrary decisions need to be made with regard to the boundaries. On the other hand, more groups are needed. Regardless of the method used, it should be understood that the real loss process is not as smooth as the model we are fitting. These techniques are attempts to help us get the best possible smooth model.

At first glance it might appear that distributions with a mode at zero make little sense. While that may be true they tend to be effective. The main reason is that the exact distribution of losses from zero to some small number is of very little interest. The impact on the mean or on the imposition of coverage modifications is small.

The issue of the shape of the right-hand tail is a critical one for the actuary. If the possibility of large losses were remote, there might be little interest in the purchase of insurance. On the other hand, many loss processes produce large claims with enough regularity to inspire people and companies to purchase insurance and to create problems for the provider of insurance. For any distribution that extends probability to infinity, the issue is one of how quickly the density function approaches zero as the loss approaches infinity. The slower this happens, the more probability is pushed onto higher values and we then say the distribution has a heavy, thick, or fat tail. We avoid the adjective "long" because in insurance a long tail usually means that it takes a long time from the issuance of the policy until all claims are settled. Before developing our collection of distributions we devote the next subsection to a further exploration of these issues.

2.7.2 Tail behavior

Various functions related to the density function may be examined to assess the likelihood of a large loss. For example, the **survival function**

$$S(x) = 1 - F(x) = \int_x^\infty f(t)dt$$

gives the probability of the loss exceeding the value x, and (relatively) large values of $S(x)$ for given x indicate heavy tailed behavior. Also, the density function itself can be used to measure tail weight. For some distributions, $S(x)$ and/or $f(x)$ are simple functions of x and we can immediately determine what the tail looks like. If they are complex functions of x, we may be able to obtain a simple function that has similar behavior as x gets large. We use the notation

$$a(x) \sim b(x), \qquad x \to \infty$$

to mean $\lim_{x\to\infty} a(x)/b(x) = 1$.

This concept can be used to compare the tail behavior of two random variables, X and Y. If $S_X(x) \sim cS_Y(x)$ (or, equivalently from L'Hôpital's rule, $f_X(x) \sim cf_Y(x)$), then the two variables have proportional tails and therefore are similar. If the ratio goes to zero, then X has the lighter tail, while if it goes to infinity, then Y has the lighter tail.

Example 2.39 *Compare the tail weights of the lognormal, gamma, and Pareto distributions using the concept introduced above.*

For gamma versus Pareto, the limit of the ratio of their density functions is

$$\lim_{x\to\infty} \frac{x^{\alpha-1}e^{-x/\theta}}{(x+\lambda)^{-\tau-1}} = \lim_{x\to\infty} \exp[(\tau+1)\log(x+\lambda) + (\alpha-1)\log x - x/\theta].$$

Because x goes to infinity much faster than $\log x$, the above exponent has a limit of negative infinity. Thus the original limit is zero and the Pareto distribution has the heavier tail. For lognormal versus gamma, we need

$$\lim_{x\to\infty} \frac{x^{-1}\exp\left[-\frac{1}{2\sigma^2}(\log x - \mu)^2\right]}{x^{\alpha-1}e^{-x/\theta}}$$
$$= \lim_{x\to\infty} \exp\left[-\frac{1}{2\sigma^2}(\log x - \mu)^2 - \alpha\log x + x/\theta\right].$$

The exponent goes to infinity and so does the original limit. Thus the lognormal distribution has the heavier tail. A similar calculation indicates that the Pareto has a heavier tail than the lognormal. □

A quantity that can assist us in evaluating tail weight is defined as follows.

Definition 2.33 *The ratio*

$$\lambda(x) = \frac{f(x)}{S(x)}, \quad x \geq 0$$

is called the ***force of mortality, failure rate,*** *or* ***hazard rate.***

If X is the associated random variable, then

$$\begin{aligned}
\lambda(x) &= \frac{\frac{d}{dx}\Pr(X \leq x)}{\Pr(X > x)} \\
&= \lim_{h \to 0} \frac{\Pr(X \leq x+h) - \Pr(X \leq x)}{h \Pr(X > x)} \\
&= \lim_{h \to 0} \frac{\Pr(x < X \leq x+h)}{h \Pr(X > x)} \\
&= \lim_{h \to 0} \frac{1}{h} \Pr(x < X \leq x+h | X > x). \qquad (2.16)
\end{aligned}$$

Thus, $\lambda(x)h$ may be interpreted as the probability of "failure" at x given "survival" to x. Intuitively, if $\lambda(x)$ becomes small, immediate "failure" is less likely and the distribution is heavy-tailed. Conversely, if $\lambda(x)$ becomes large, the distribution is light-tailed.

Note that

$$\lambda(t) = -\frac{d}{dt} \log S(t)$$

and integrating both sides over t from 0 to x gives

$$\int_0^x \lambda(t)dt = -\log S(x) + \log S(0).$$

Since $S(0) = 1$, this means that

$$S(x) = e^{-\int_0^x \lambda(t)dt}, \quad x \geq 0$$

and $\lambda(x)$ uniquely characterizes the distribution.

Example 2.40 *Determine the asymptotic behavior of the failure rate of the gamma distribution.*

For this distribution, the density is

$$f(x) = \frac{x^{\alpha-1}e^{-x/\theta}}{\theta^\alpha \Gamma(\alpha)}, \quad x > 0$$

where the gamma function, $\Gamma(\alpha)$, is defined later in this section. The survival function $S(x)$ is complicated if α is not a positive integer. But $\lim_{x\to\infty} S(x) =$

$\lim_{x\to\infty} f(x) = 0$, so by L'Hôpital's rule,

$$\begin{aligned}
\lim_{x\to\infty}\frac{f(x)}{S(x)} &= \lim_{x\to\infty}\frac{f'(x)}{-f(x)} \\
&= \lim_{x\to\infty}\frac{[(\alpha-1)x^{\alpha-2}-\theta^{-1}x^{\alpha-1}]e^{-x/\theta}}{-x^{\alpha-1}e^{-x/\theta}} \\
&= \lim_{x\to\infty}\left(\frac{1}{\theta}-\frac{\alpha-1}{x}\right) \\
&= \theta^{-1}.
\end{aligned}$$

Thus $S(x) \sim \theta f(x)$ and so the tail behavior of the gamma survival function is the same as that for the pdf (with α increased by one). □

Example 2.41 *Determine the failure rate for the Pareto distribution.*

For this distribution, the survival function is

$$S(x) = \left(\frac{\theta}{\theta+x}\right)^{\alpha}, \quad x \ge 0$$

and so the failure rate is

$$\lambda(x) = -\frac{d}{dx}\ln S(x) = \frac{\alpha}{\theta+x}, \quad x \ge 0.$$

In this case, $\lambda(x)$ is strictly decreasing from $\lambda(0) = \alpha/\theta$ *to* $\lambda(\infty) = 0$. □

If, as in the above example, $\lambda(x)$ is decreasing, then we say that $F(x)$ has a **decreasing failure rate** (DFR). If $\lambda(x)$ is increasing, then $F(x)$ has an **increasing failure rate** (IFR). A DFR distribution has a heavier tail than an IFR distribution.

As before, it is difficult to examine $\lambda(x)$ if $S(x)$ is complicated. But,

$$[\lambda(x)]^{-1} = \frac{\int_x^\infty f(t)dt}{f(x)} = \int_0^\infty \frac{f(x+y)dy}{f(x)}.$$

Clearly, if $f(x+y)/f(x)$ is an increasing function of x for any fixed $y \ge 0$ (i.e. $f(x)$ is said to be **log-convex**), then $[\lambda(x)]^{-1}$ is increasing in x and $F(x)$ has a DFR. Conversely, if $f(x)$ is **log-concave** then $F(x)$ has an IFR.

Example 2.42 *Determine if the gamma distribution is IFR or DFR.*

The density is

$$f(x) = \frac{x^{\alpha-1}e^{-x/\theta}}{\theta^{\alpha}\Gamma(\alpha)}, \quad x \ge 0.$$

A general expression for $S(x)$ is complicated but

$$\frac{f(x+y)}{f(x)} = \left(1+\frac{y}{x}\right)^{\alpha-1} e^{-y/\theta}.$$

Viewed as a function of x for fixed $y \geq 0$, it is clear that $f(x+y)/f(x)$ is increasing in x if $\alpha \leq 1$ and decreasing in x if $\alpha \geq 1$. Thus, $F(x)$ has a DFR if $\alpha \leq 1$ and an IFR if $\alpha \geq 1$ (when $\alpha = 1$ the distribution is exponential, has a constant failure rate, and thus is both IFR and DFR). We also know that $\lim_{x\to\infty}\lambda(x) = \theta^{-1}$ from Example 2.40 so the failure rate increases to θ^{-1} if $\alpha \geq 1$ and decreases to θ^{-1} if $\alpha \leq 1$. □

Note also from (2.16) that

$$\lambda(x) = \lim_{h\to 0}\frac{1}{h}\left[1-\frac{S(x+h)}{S(x)}\right]$$

so that "$F(x)$ has a DFR" is equivalent to "$S(x)$ is log-convex," i.e., $S(x+y)/S(x)$ is increasing in x for fixed $y \geq 0$. Similarly, "$F(x)$ has an IFR" is equivalent to "$S(x)$ is log-concave."

One other function which is useful in analyzing thickness of tails is the mean excess loss (see Definition 2.29), also called the mean residual lifetime. It is

$$e_X(x) = e(x) = E(X-x|X>x) = \frac{\int_x^\infty (t-x)f(t)dt}{S(x)}. \tag{2.17}$$

Intuitively, if $e(x)$ is large for large x, then the distribution has a heavy-tail since the expected excess loss $X - x$ is large. Conversely, if $e(x)$ is small for large x, then the distribution has a light-tail.

There is a simpler form than (2.17) for $e(x)$, which follows by integration by parts. Let $u = (t-x)$ and $dv = f(t)dt = -S'(t)dt$. Then $v = -S(t)$ and $du = dt$, and so

$$\begin{aligned} e(x) &= \left.\frac{-(t-x)S(t)}{S(x)}\right|_{t=x}^{\infty} + \int_x^\infty \frac{S(t)}{S(x)}dt \\ &= -\lim_{t\to\infty}(t-x)\frac{S(t)}{S(x)} + \int_x^\infty \frac{S(t)}{S(x)}dt. \end{aligned}$$

The mean of the distribution is $e(0) = E(X) = \int_0^\infty yf(y)dy$. If $e(x) < \infty$, then $\lim_{t\to\infty}\int_t^\infty yf(y)dy = 0$. Now, for $t \geq x$

$$\begin{aligned} 0 \leq (t-x)S(t) &= (t-x)\int_t^\infty f(y)dy \\ &\leq \int_t^\infty (y-x)f(y)dy = \int_t^\infty yf(y)dy - xS(t). \end{aligned}$$

Thus

$$0 \leq \lim_{t\to\infty}(t-x)S(t) \leq \lim_{t\to\infty}\left\{\int_t^\infty yf(y)dy - xS(t)\right\} = 0-0=0$$

and so $\lim_{t\to\infty}(t-x)S(t)=0$. But this means that

$$e(x)=\int_x^\infty \frac{S(t)}{S(x)}dt, \quad x\geq 0 \tag{2.18}$$

as long as $e(0)=E(X)<\infty$.

Note that

$$1=\int_0^\infty \frac{S(x)}{e(0)}dx$$

so that

$$f_e(x)=\frac{S(x)}{e(0)}, \quad x\geq 0$$

is a continuous pdf. The survival function corresponding to $f_e(x)$ is

$$S_e(x)=\int_x^\infty f_e(t)dt=\frac{\int_x^\infty S(t)dt}{e(0)}, \quad x\geq 0. \tag{2.19}$$

The failure rate or force of mortality corresponding to $f_e(x)$ is

$$\begin{aligned}\lambda_e(x) &= \frac{f_e(x)}{S_e(x)}\\ &= \frac{S(x)}{\int_x^\infty S(t)dt}\\ &= [e(x)]^{-1}\end{aligned} \tag{2.20}$$

from (2.18). Thus,

$$\begin{aligned}S(x) &= e(0)f_e(x)\\ &= e(0)\lambda_e(x)S_e(x)\\ &= \frac{e(0)}{e(x)}e^{-\int_0^x [e(t)]^{-1}dt}, \quad x\geq 0\end{aligned}$$

which demonstrates that $e(x)$ uniquely characterizes the distribution $F(x)$.

There is a close relationship between $e(x)$ and $\lambda(x)$. First note that if the indicated limits exist, then by L'Hôpital's rule,

$$\begin{aligned}\lim_{x\to\infty} e(x) &= \lim_{x\to\infty}\frac{\int_x^\infty S(t)dt}{S(x)}\\ &= \lim_{x\to\infty}\frac{-S(x)}{-f(x)}\\ &= \lim_{x\to\infty}\frac{1}{\lambda(x)}\end{aligned}$$

and so the behavior of $e(x)$ is easily established from that of $\lambda(x)$. This is useful if $S(x)$, and hence $e(x)$, is complicated. Also, if $e(x)$ is monotone increasing (nondecreasing), then we say that $F(x)$ has an **increasing mean residual lifetime** (IMRL). Conversely, if $e(x)$ is monotone decreasing (nonincreasing), then we say that $F(x)$ has a **decreasing mean residual lifetime** (DMRL). From (2.18),

$$e(x) = \int_0^\infty \frac{S(x+y)}{S(x)} dy$$

and since $\lambda(x)$ decreasing or DFR is equivalent to "$S(x)$ is log-convex," it is clear that $e(x)$ is increasing. That is, DFR $\Rightarrow$ IMRL and similarly, IFR $\Rightarrow$ DMRL.

Example 2.43 *Determine the asymptotic behavior of $e(x)$ for the gamma distribution.*

An expression for $e(x)$ is complicated, but

$$\lim_{x\to\infty} e(x) = \lim_{x\to\infty} \frac{1}{\lambda(x)} = \theta.$$

Also, for $\alpha \le 1$, $\lambda(x)$ is decreasing, implying that $e(x)$ is increasing to its limit θ. For $\alpha \ge 1$, $\lambda(x)$ is increasing, implying that $e(x)$ is decreasing to its limit θ. □

Note that from (2.18) and (2.19),

$$\frac{e(x)}{e(0)} = \frac{S_e(x)}{S(x)}. \tag{2.21}$$

If $e(x) \ge e(0)$ (i.e. if $F(x)$ has a DFR, implying that $F(x)$ has an IMRL), then $S_e(x) \ge S(x)$. Thus,

$$\int_0^\infty S_e(x)dx \ge \int_0^\infty S(x)dx. \tag{2.22}$$

But, $E(X) = \int_0^\infty S(x)dx$, and

$$\begin{aligned}\int_0^\infty S_e(x)dx &= \int_0^\infty x f_e(x)dx \\ &= \int_0^\infty x\frac{S(x)}{e(0)}dx.\end{aligned}$$

Let $u = S(x)$, $dv = x$, then integration by parts gives

$$\begin{aligned}\int_0^\infty xS(x)dx &= \frac{1}{2}x^2S(x)\Big|_0^\infty + \frac{1}{2}\int_0^\infty x^2 f(x)dx \\ &= \frac{1}{2}\lim_{x\to\infty} x^2S(x) + \frac{1}{2}E(X^2).\end{aligned}$$

Now, if $E(X^2) < \infty$, then

$$\lim_{x\to\infty} \int_x^\infty t^2 f(t)dt = 0$$

and

$$0 \le \lim_{x\to\infty} x^2 S(x) = \lim_{x\to\infty} x^2 \int_x^\infty f(t)dt \le \lim_{x\to\infty} \int_x^\infty t^2 f(t)dt = 0$$

implying that

$$\lim_{x\to\infty} x^2 S(x) = 0.$$

Thus, if $E(X^2) < \infty$,

$$\int_0^\infty S_e(x)dx = \frac{E(X^2)}{2E(X)}.$$

But this implies that (2.22) is equivalent to $E(X^2) \ge 2[E(X)]^2$, or $Var(X) \ge [E(X)]^2$. Taking square roots, this is equivalent to $SD(X) = \sqrt{Var(X)} \ge E(X)$. Reversing the inequalities, IFR $\Rightarrow$ DMRL $\Rightarrow e(x) \le e(0) \Rightarrow SD(X) \le E(X)$.

Evaluation of the moments of X provides another method to assess heavy-tailed behavior. If the distribution has only a finite number of moments (for example, Pareto or Burr), then the distribution is heavy tailed. If only a finite number of moments exist, then the moment generating function $E(e^{tX})$ does not exist (i.e., $E(e^{tX}) = \infty$ for any $t > 0$). The converse is not true, however. For example, the moment generating function of the lognormal distribution does not exist, but all moments exist.

2.7.3 Methods of creating new families from existing ones

To increase the number of distributional families at our disposal we can employ one or more of the techniques discussed in this subsection. Through examples and exercises, most of the distributions in Appendix A will be developed.

2.7.3.1 Multiplication by a constant This transformation is equivalent to applying inflation uniformly across all loss levels and is known as a change of scale.

Theorem 2.11 *Let X be a continuous random variable with pdf $f_X(x)$ and cdf $F_X(x)$ with $F_X(0) = 0$. Let $Y = \theta X$ with $\theta > 0$. Then*

$$F_Y(y) = F_X(y/\theta), \quad f_Y(y) = \frac{1}{\theta} f_X(y/\theta), \qquad y > 0.$$

Proof:

$$F_Y(y) = \Pr(Y \le y) = \Pr(\theta X \le y) = \Pr(X \le y/\theta) = F_X(y/\theta)$$

$$f_Y(y) = \frac{d}{dy}F_Y(y) = \frac{1}{\theta}f_X(y/\theta).$$

□

Corollary 2.12 *θ is a scale parameter.* □

The following example illustrates this process.

Example 2.44 *Let X have pdf $f(x) = e^{-x}$, $x > 0$. Determine the cdf and pdf of $Y = \theta X$.*

$$\begin{aligned} F_X(x) &= 1 - e^{-x}, \quad F_Y(y) = 1 - e^{-y/\theta} \\ f_Y(y) &= \frac{1}{\theta}e^{-y/\theta}. \end{aligned}$$

We recognize this as the exponential distribution. □

2.7.3.2 Raising to a power

Theorem 2.13 *Let X be a continuous random variable with pdf $f_X(x)$ and cdf $F_X(x)$ with $F_X(0) = 0$. Let $Y = X^{1/\tau}$. Then if $\tau > 0$,*

$$F_Y(y) = F_X(y^\tau), \quad f_Y(y) = \tau y^{\tau-1} f_X(y^\tau), \qquad y > 0$$

while if $\tau < 0$,

$$F_Y(y) = 1 - F_X(y^\tau), \quad f_Y(y) = -\tau y^{\tau-1} f_X(y^\tau). \tag{2.23}$$

Proof: If $\tau > 0$

$$F_Y(y) = \Pr(X \le y^\tau) = F_X(y^\tau)$$

while if $\tau < 0$

$$F_Y(y) = \Pr(X \ge y^\tau) = 1 - F_X(y^\tau).$$

The pdf follows by differentiation. □

It is more common to keep parameters positive and so when τ is negative, create a new parameter $\tau^* = -\tau$. Then (2.23) becomes

$$F_Y(y) = 1 - F_X(y^{-\tau^*}), \quad f_Y(y) = \tau^* y^{-\tau^*-1} f_X(y^{-\tau^*}).$$

Then drop the * for future use of this positive parameter.

When $\tau > 0$ the resulting distribution is called **transformed**, when $\tau = -1$ it is called **inverse**, and when $\tau < 0$ it is called **inverse transformed**. To create the distributions in Appendix A and to retain θ as a scale parameter the base distribution should be raised to a power before being multiplied by θ.

Example 2.45 *Suppose X has the exponential distribution. Determine the cdf of the inverse, transformed, and inverse transformed exponential distributions.*

The inverse exponential with no scale parameter has cdf

$$F(y) = 1 - [1 - e^{-1/y}] = e^{-1/y}.$$

With the scale parameter added it is $F(y) = e^{-\theta/y}$.

The transformed exponential with no scale parameter has cdf

$$F(y) = 1 - \exp(-y^{\tau}).$$

With the scale parameter added it is $F(y) = 1 - \exp[-(y/\theta)^{\tau}]$. This distribution is more commonly known as the **Weibull distribution.**

The inverse transformed exponential with no scale parameter has cdf

$$F(y) = 1 - [1 - \exp(-y^{-\tau})] = \exp(-y^{-\tau}).$$

With the scale parameter added it is $F(y) = \exp[-(\theta/y)^{\tau}]$. This distribution is the **inverse Weibull.** □

One other basic distribution that was briefly introduced in Example 2.40 has pdf $f(x) = x^{\alpha-1}e^{-x}/\Gamma(\alpha)$. When a scale parameter is added, this becomes the **gamma distribution.** It has inverse and transformed versions that can be created using the results in this section. Unlike the distributions introduced to this point, this one does not have a closed form cdf. The best we can do is define notation for the function.

Definition 2.34 *The* ***incomplete gamma function*** *with parameter α is denoted and defined by*

$$\Gamma(\alpha; x) = \frac{1}{\Gamma(\alpha)} \int_0^x t^{\alpha-1}e^{-t}dt$$

while the ***gamma function*** *is denoted and defined by*

$$\Gamma(\alpha) = \int_0^\infty t^{\alpha-1}e^{-t}dt.$$

Appendix A provides details on numerical methods of evaluating these quantities. Furthermore, most spreadsheet programs and many statistical and numerical analysis programs have these functions built in.

2.7.3.3 Exponentiation

Theorem 2.14 *Let X be a continuous random variable with pdf $f_X(x)$ and cdf $F_X(x)$ with $f_X(x) > 0$ for all real x. Let $Y = \exp(X)$. Then*

$$F_Y(y) = F_X(\log y), \quad f_Y(y) = \frac{1}{y} f_X(\log y).$$

Proof: $F_Y(y) = \Pr(e^X \leq y) = \Pr(X \leq \log y) = F_X(\log y)$. □

Example 2.46 *Let X have the normal distribution with mean μ and variance σ^2. Determine the cdf and pdf of $Y = e^X$.*

$$F_Y(y) = \Phi\left(\frac{\log y - \mu}{\sigma}\right)$$

$$f_Y(y) = \frac{1}{y}\phi\left(\frac{\log y - \mu}{\sigma}\right) = \frac{1}{y\sigma\sqrt{2\pi}} \exp\left[-\frac{1}{2}\left(\frac{\log y - \mu}{\sigma}\right)^2\right].$$ □

We could try to add a scale parameter by creating $W = \theta Y$, but this adds no value, as is demonstrated in Exercise 2.73. This example created the lognormal distribution (the name has stuck even though expnormal would seem more descriptive).

The next example uses the incomplete beta function. Because it is an increasing function, it also defines a distribution, the beta distribution. That distribution and a numerical approximation are given in Appendix A.

Definition 2.35 *The* ***incomplete beta function*** *is given by*

$$\beta(a, b; x) = \frac{\Gamma(a+b)}{\Gamma(a)\Gamma(b)} \int_0^x t^{a-1}(1-t)^{b-1}dt, \qquad 0 \leq x \leq 1,\ a, b > 0.$$

Example 2.47 *A random variable X has the t distribution with parameter r if it has pdf*

$$f(x) = \frac{\Gamma\left(\frac{r+1}{2}\right)}{\sqrt{\pi r}\,\Gamma\left(\frac{r}{2}\right)} \frac{1}{(1 + x^2/r)^{(r+1)/2}}, \qquad -\infty < x < \infty,\ r > 0.$$

A general version of the log-t distribution can be obtained as $Y = \exp(aX + b)$. Determine its pdf and cdf.

Using the same arguments as in the previous theorems in this subsection we have $f_Y(y) = (ya)^{-1} f_X[(\log y - b)/a]$. Then

$$f_Y(y) = \frac{\Gamma\left(\frac{r+1}{2}\right)}{ya\sqrt{\pi r}\,\Gamma\left(\frac{r}{2}\right)\left[1 + \frac{1}{r}\left(\frac{\log y - b}{a}\right)^2\right]^{(r+1)/2}}, \qquad y > 0.$$

To obtain the cdf of Y we first obtain it for X. If $x \leq 0$,

$$\begin{aligned}
F_X(x) &= \int_{-\infty}^{x} \frac{\Gamma\left(\frac{r+1}{2}\right)}{\sqrt{\pi r}\,\Gamma\left(\frac{r}{2}\right)} \frac{1}{(1+t^2/r)^{(r+1)/2}} dt \\
&= \frac{\Gamma\left(\frac{r+1}{2}\right)}{\sqrt{\pi r}\,\Gamma\left(\frac{r}{2}\right)} \int_0^{r/(r+x^2)} s^{(r+1)/2} \sqrt{r} \frac{1}{2} \left(\frac{1-s}{s}\right)^{-1/2} \frac{1}{s^2} ds \\
&= \frac{\Gamma\left(\frac{r+1}{2}\right)}{2\Gamma\left(\frac{1}{2}\right)\Gamma\left(\frac{r}{2}\right)} \int_0^{r/(r+x^2)} s^{r/2-1}(1-s)^{-1/2} ds \\
&= \frac{1}{2}\beta[r/2, 1/2; r/(r+x^2)].
\end{aligned}$$

By symmetry of the integrand, if $x \geq 0$, $F_X(x) = 1 - \frac{1}{2}\beta[r/2, 1/2; r/(r+x^2)]$. We then have

$$F_Y(y) = F_X[(\log y - b)/a] = \begin{cases} \frac{1}{2}\beta\left[\frac{r}{2}, \frac{1}{2}; \frac{r}{r + \left(\frac{\log y - b}{a}\right)^2}\right], & 0 < y \leq e^b, \\ 1 - \frac{1}{2}\beta\left[\frac{r}{2}, \frac{1}{2}; \frac{r}{r + \left(\frac{\log y - b}{a}\right)^2}\right], & y \geq e^b. \end{cases}$$

The second line is accomplished by the transformation $s = r/(r+t^2)$. In the third line the relationship $\Gamma(1/2) = \sqrt{\pi}$ was used. □

2.7.3.4 Mixing

Theorem 2.15 *Let X have pdf $f_{X|\Theta}(x|\theta)$ and cdf $F_{X|\Theta}(x|\theta)$ where θ is a parameter of X. Let θ be a realization of the random variable Θ with pdf $f_\Theta(\theta)$. Then the unconditional pdf of X is*

$$f_X(x) = \int f_{X|\Theta}(x|\theta) f_\Theta(\theta) d\theta$$

where the integral is taken over all values of θ with positive probability.

Proof: The integrand is, by definition, the joint density of X and Θ. The integral is then the marginal density. □

There are two physical interpretations of mixing, both of which are versions of parameter uncertainty. Suppose θ is a scale parameter and we have determined that the appropriate loss distribution has pdf $f_X(x;\theta^*)$ and cdf $F_X(x;\theta^*)$ where θ^* is a particular parameter value. For next year we believe losses will be inflated to cX, but we are not sure about the value of c. Suppose we can quantify our uncertainty by specifying a probability distribution for C. The cdf of CX is then

$$\begin{aligned} F_{CX}(y) &= \int_0^\infty \Pr(CX \le y|C=c) f_C(c)dc \\ &= \int_0^\infty F_X(y;c\theta^*) f_C(c)dc \end{aligned}$$

and the pdf is

$$\begin{aligned} f_{CX}(y) &= \int_0^\infty f_X(y;c\theta^*) f_C(c)dc \\ &= \int_0^\infty f_X(y;\theta) f_C(\theta/\theta^*)\frac{1}{\theta^*}d\theta \\ &= \int_0^\infty f_X(y;\theta) g(\theta)d\theta. \end{aligned}$$

The second line used the transformation $\theta = c\theta^*$. The third line is just the definition of $g(\theta) = f_C(\theta/\theta^*)(1/\theta^*)$.

Care must be taken with this interpretation of mixing. Consider the following example.

Example 2.48 *Individual automobile physical damage losses have the exponential distribution with $\theta = 500$. Suppose the inflation factor for the year ahead has the inverse exponential distribution with $\theta = 1$. Determine the unconditional distribution of next year's losses.*

The pdf is

$$\begin{aligned} f(x) &= \int_0^\infty \frac{1}{500c} e^{-x/500c} \frac{1}{c^2} e^{-1/c} dc \\ &= \frac{1}{500} \int_0^\infty c^{-3} \exp\left[-\frac{1}{c}(1 + x/500)\right] dc \\ &= \frac{1}{500} \int_0^\infty (1 + x/500)^{-3} y^3 \exp(-y)(1 + x/500) y^{-2} dy \\ &= \frac{1}{500}(1 + x/500)^{-2} \int_0^\infty y \exp(-y) dy \\ &= 500/(500 + x)^2. \end{aligned}$$

This is a Pareto distribution with parameters $\alpha = 1$ and $\theta = 500$. □

However, the Pareto distribution in this example is not the loss distribution for each individual accident. The derivation assumes that each accident has its loss amount inflated by a factor that is unique to that loss, but likely to be different from the factor that is applied to other losses. But the truth is that inflation will impact every loss in the same way and so observations will still have the exponential distribution. The conclusion would be valid if the exponential distribution were the model for the sum of all payments for a year. Then with uncertainty, the total obligation would have the Pareto distribution.

An alternative approach is to consider that all loss events are the result of a realization of a particular random variable, but that the parameter θ may vary from event to event. In the above example this variation may be related to the value of the car. That is, all insured cars have the exponential distribution, but the potential cost of a loss varies in an inverse exponential manner. Then the population of losses does have the Pareto distribution.

The following example is taken from Hayne [57].

Example 2.49 *In the valuation of warranties on automobiles it is important to recognize that the number of miles driven varies from driver to driver. It is also the case that for a particular driver, the number of miles varies from year to year. Suppose the number of miles for a randomly selected driver has the inverse Weibull distribution but that the year to year variation in the scale parameter has the transformed gamma distribution with the same value for τ. What is the distribution for the number of miles driven in a randomly selected year by a randomly selected driver?*

Using the parameterizations from Appendix A, the inverse Weibull has parameters θ and τ while the transformed gamma distribution has parameters τ, γ and α (γ has been substituted for θ) and the marginal density is

$$f(x) = \int_0^\infty \frac{\tau\theta^\tau}{x^{\tau+1}} e^{-(\theta/x)^\tau} \frac{\tau\theta^{\tau\alpha-1}}{\gamma^{\tau\alpha}\Gamma(\alpha)} e^{-(\theta/\gamma)^\tau} d\theta$$

$$
\begin{aligned}
&= \frac{\tau^2}{\gamma^{\tau\alpha}\Gamma(\alpha)x^{\tau+1}}\int_0^\infty \theta^{\tau+\tau\alpha-1}\exp[-\theta^\tau(x^{-\tau}+\gamma^{-\tau})]d\theta \\
&= \frac{\tau^2}{\gamma^{\tau\alpha}\Gamma(\alpha)x^{\tau+1}}\int_0^\infty \{y^{1/\tau}(x^{-\tau}+\gamma^{-\tau})^{-1/\tau}\}^{\tau+\tau\alpha-1}e^{-y} \\
&\quad \times y^{\tau^{-1}-1}\tau^{-1}(x^{-\tau}+\gamma^{-\tau})^{-1/\tau}dy \\
&= \frac{\tau}{\gamma^{\tau\alpha}\Gamma(\alpha)x^{\tau+1}(x^{-\tau}+\gamma^{-\tau})^{\alpha+1}}\int_0^\infty y^\alpha e^{-y}dy \\
&= \frac{\tau\Gamma(\alpha+1)}{\gamma^{\tau\alpha}\Gamma(\alpha)x^{\tau+1}(x^{-\tau}+\gamma^{-\tau})^{\alpha+1}} \\
&= \frac{\tau\alpha\gamma^\tau x^{\tau\alpha-1}}{(x^\tau+\gamma^\tau)^{\alpha+1}}.
\end{aligned}
$$

The third line is obtained by the transformation $y=\theta^\tau(x^{-\tau}+\gamma^{-\tau})$. The final lines uses the fact that $\Gamma(\alpha+1)=\alpha\Gamma(\alpha)$. This is an **inverse Burr distribution**. In an exercise you are asked to derive the inverse Burr distribution by transformation. □

As a final insight into mixing, consider the case of mixing on the scale parameter. It was shown earlier that this is equivalent to multiplying X by an independent random variable C. If X has already been parameterized to the best of our ability then it is convenient to assume that $E(C)=1$. We then have

$$E(CX)=E(C)E(X)=E(X)$$

and

$$
\begin{aligned}
Var(CX) &= E(C^2X^2)-[E(CX)]^2 \\
&= E(C^2)E(X^2)-[E(C)]^2[E(X^2)-Var(X)] \\
&= E(X^2)Var(C)+Var(X) \\
&> Var(X).
\end{aligned}
$$

Thus the addition of uncertainty is reflected in an increased variance.

2.7.4 Special and limiting cases

Many of the distributions presented in this section and in Appendix A are special cases of others. For example, a Weibull distribution with $\tau=1$ and θ arbitrary is an exponential distribution. Through this process, many of our distributions can be organized into three groupings as illustrated in Figures 2.6 and 2.7. The transformed beta family indicates two special cases of a different nature. The paralogistic and inverse paralogistic distributions are created by setting the two nonscale parameters of the Burr and inverse Burr distributions equal to each other rather than to a specified value. Some ad-

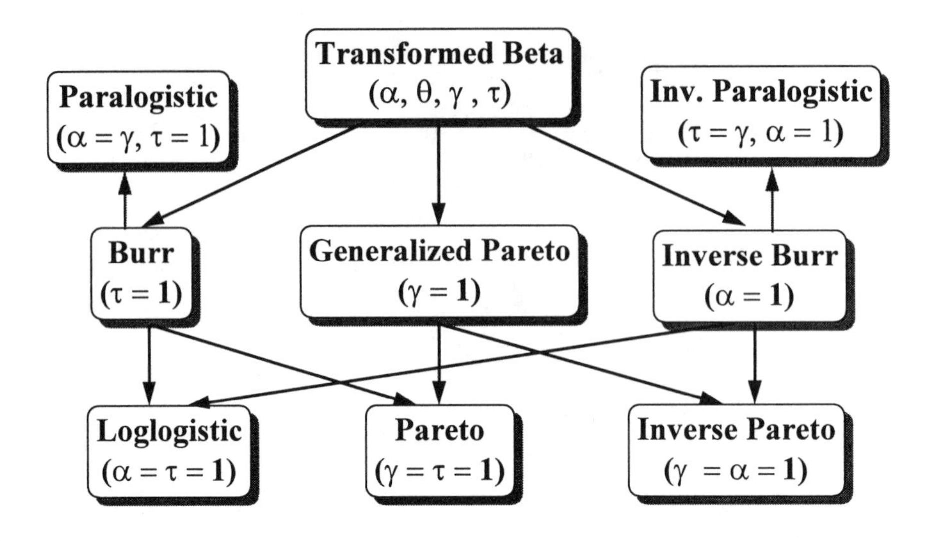

Fig. 2.6 Transformed beta family

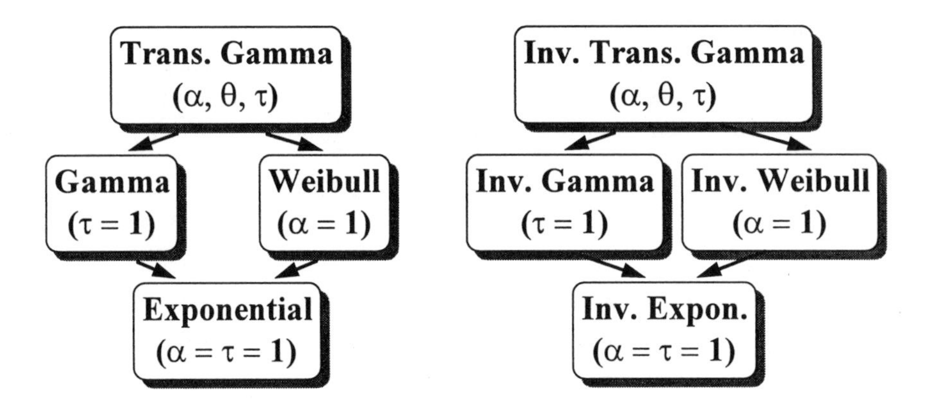

Fig. 2.7 Transformed and inverse transformed gamma families

ditional distributions useful for modeling large claims only can be found in *Practical Analysis of Extreme Values* [11].

It is also the case that some distributions are limiting cases of others. This means that as two or more parameters go to zero or infinity in a special way the result is a member of another family.

Example 2.50 *Show that the transformed gamma distribution is a limiting case of the transformed beta distribution as $\theta \to \infty$, $\alpha \to \infty$, and $\theta/\alpha^{1/\gamma} \to \xi$, a constant.*

The demonstration relies on two facts concerning limits:

$$\lim_{\alpha\to\infty} \frac{e^{-\alpha}\alpha^{\alpha-1/2}(2\pi)^{1/2}}{\Gamma(\alpha)} = 1 \tag{2.24}$$

and

$$\lim_{a\to\infty} \left(1+\frac{x}{a}\right)^{a+b} = e^x. \tag{2.25}$$

The limit in (2.24) is known as Stirling's formula and provides an approximation for the gamma function. The limit in (2.25) is a standard result found in most calculus texts.

To ensure that the ratio $\theta/\alpha^{1/\gamma}$ goes to a constant, it is sufficient to force it to be constant as α and θ become larger and larger. This can be accomplished by substituting $\xi\alpha^{1/\gamma}$ for θ in the transformed beta pdf and then letting $\alpha \to \infty$. The first steps, which also include using Stirling's formula to replace two of the gamma function terms, are

$$\begin{aligned} f(x) &= \frac{\Gamma(\alpha+\tau)\gamma x^{\gamma\tau-1}}{\Gamma(\alpha)\Gamma(\tau)\theta^{\gamma\tau}(1+x^\gamma\theta^{-\gamma})^{\alpha+\tau}} \\ &= \frac{e^{-\alpha-\tau}(\alpha+\tau)^{\alpha+\tau-1/2}(2\pi)^{1/2}\gamma x^{\gamma\tau-1}}{e^{-\alpha}\alpha^{\alpha-1/2}(2\pi)^{1/2}\Gamma(\tau)(\xi\alpha^{1/\gamma})^{\gamma\tau}(1+x^\gamma\xi^{-\gamma}\alpha^{-1})^{\alpha+\tau}} \\ &= \frac{e^{-\tau}\left(\dfrac{\alpha+\tau}{\alpha}\right)^{\alpha+\tau-1/2}\gamma x^{\gamma\tau-1}}{\Gamma(\tau)\xi^{\gamma\tau}\left[1+\dfrac{(x/\xi)^\gamma}{\alpha}\right]^{\alpha+\tau}}. \end{aligned}$$

The two limits

$$\lim_{\alpha\to\infty}(1+\tau/\alpha)^{\alpha+\tau-1/2} = e^\tau$$

and

$$\lim_{\alpha\to\infty}\left[1+\frac{(x/\xi)^\gamma}{\alpha}\right]^{\alpha+\tau} = e^{(x/\xi)^\gamma}$$

can be substituted to yield

$$\lim_{\alpha\to\infty} f(x) = \frac{\gamma x^{\gamma\tau-1}e^{-(x/\xi)^\gamma}}{\Gamma(\tau)\xi^{\gamma\tau}}$$

which is the pdf of the transformed gamma distribution. □

With a similar argument, the inverse transformed gamma distribution is obtained by letting τ go to infinity instead of α (see Exercise 2.83).

Because the Burr distribution is a transformed beta distribution with $\tau = 1$, its limiting case is the transformed gamma with $\tau = 1$ (using the parameterization in the previous example), which is the Weibull distribution. Similarly the inverse Burr has the inverse Weibull as a limiting case.

As a final illustration of a limiting case, consider the transformed gamma distribution as parameterized above. Let $\gamma^{-1}\sqrt{\xi^\gamma} \to \sigma$ and $\gamma^{-1}(\xi^\gamma\tau - 1) \to \mu$. If this is done by letting $\tau \to \infty$ (so both γ and ξ must go to zero), the limiting distribution will be lognormal.

The distinction between special and limiting cases is important with regard to likelihood ratio tests. When one distribution is a special case of another we can perform a likelihood ratio test to determine if the data provides evidence for the more complex distribution. The theory of likelihood ratio tests does not apply to limiting cases. Nevertheless, our search for parsimonious models should lead us to choose the limiting case over the more general model unless the data provides compelling evidence (such as a significantly larger value for the likelihood function). The notion of limiting cases also tells us why it is possible for maximum likelihood estimates not to exist. If the limiting case is a viable model, we should expect to see the likelihood function for the more general model increase as the parameters go to infinity in the right way.

Example 2.51 *Consider the data in Table 2.10. Test the hypothesis that a Burr distribution provides a better model when compared to the Pareto distribution.*

Maximum likelihood estimation produced the following values. For the Pareto distribution $\alpha = 1.22448$, $\theta = 14{,}810.5$, and the loglikelihood is -499.306. For the Burr distribution $\alpha = 2.08426$, $\theta = 35{,}723.4$, $\gamma = 0.829453$, and the loglikelihood is -498.410. Twice the difference is 1.792. This is not significant at any reasonable level with respect to a chi-square distribution with 1 degree of freedom. Thus the Pareto model (null hypothesis) is accepted as reasonable for this data (at least as compared to the Burr). □

2.7.5 Tail behavior in the transformed beta family

It is to be expected that with four parameters, the transformed beta distribution can assume a variety of shapes. Here we investigate those shapes by determining the relationships between the parameters and the tail behavior. For the right-hand tail we look at the existence of positive moments as well as the behavior of $1 - F(x)$ as $x \to \infty$. We then provide the same analysis for the left-hand tail (that is, behavior for small claims).

2.7.5.1 Right tail behavior For the transformed beta distribution the kth moment exists, for positive values of k, provided $k < \alpha\gamma$. Thus for all members of this family only a finite number of moments exist. It is clear, however, that only the parameters α and γ are involved in determining the behavior at the right end. To further investigate this behavior, consider the ratio

$[1-F(x)]/x^{-\alpha\gamma}$. As $x \to \infty$ both the numerator and denominator go to zero and so the limit can be found by L'Hôpital's rule.

$$\begin{aligned}
\lim_{x\to\infty} \frac{1-F(x)}{x^{-\alpha\gamma}} &= \lim_{x\to\infty} \frac{-f(x)}{-\alpha\gamma x^{-\alpha\gamma-1}} \\
&= \lim_{x\to\infty} \frac{\Gamma(\alpha+\tau)\gamma x^{\gamma\tau-1}\theta^{-\gamma\tau}}{\Gamma(\alpha)\Gamma(\tau)[1+(x/\theta)^{\gamma}]^{\alpha+\tau}\alpha\gamma x^{-\alpha\gamma-1}} \\
&= \lim_{x\to\infty} \frac{\Gamma(\alpha+\tau)}{\Gamma(\alpha)\Gamma(\tau)\theta^{\gamma\tau}\alpha} \left[\frac{\theta^{\gamma}x^{\gamma}}{\theta^{\gamma}+x^{\gamma}}\right]^{\alpha+\tau}.
\end{aligned}$$

As $x \to \infty$ the term in brackets goes to θ^{γ} and therefore

$$\lim_{x\to\infty} \frac{1-F(x)}{x^{-\alpha\gamma}} = \frac{\Gamma(\alpha+\tau)\theta^{\alpha\gamma}}{\Gamma(\alpha)\Gamma(\tau)\alpha}$$

which is a constant and so the probability in the right-hand tail is approximately proportional to $x^{-\alpha\gamma}$. This reinforces the conclusion that behavior in the right tail is tied to the product $\alpha\gamma$. As well, as the product gets larger, the tail gets shorter (becomes zero faster).

With regard to the limiting distributions, as α goes to infinity we obtain the transformed gamma distribution. This distribution has all positive moments and so all of its members have lighter tails than members of the transformed beta family. As τ goes to infinity we obtain the inverse transformed gamma distribution. The same limit ($\alpha\gamma$) applies to the existence of moments and so the right tail is similar to that for the transformed beta distribution. The inverse transformed gamma distribution is obtained as $\tau \to \infty$ and so all members of this family have heavy right tails whose behavior is governed by the product of the two nonscale parameters.

Example 2.52 *Illustrate the right-hand tail behavior of the transformed beta family by plotting the pdf of the generalized Pareto distribution. Keep both τ and the mode of the distribution constant and then let α vary, including the case where α goes to infinity (a gamma distribution, see Exercise 2.76).*

By using the generalized Pareto distribution we have set $\gamma = 1$. We have arbitrarily set $\tau = 3$ and the mode at 100 for this example. The plot (on log–log axes) appears in Figure 2.8. It is clear that changing α affects only the right-hand tail and that the gamma distribution is a limiting case. □

2.7.5.2 Left tail behavior Left tail behavior can be investigated by looking at the existence of negative moments. For the transformed beta distribution more negative moments exist as $\gamma\tau$ gets larger, so it would appear these parameters control the shape of the distribution for small loss values. With regard to probability in the left tail, the requisite limit is

$$\lim_{x\to 0} \frac{F(x)}{x^{\gamma\tau}} = \lim_{x\to 0} \frac{f(x)}{\gamma\tau x^{\gamma\tau-1}}$$

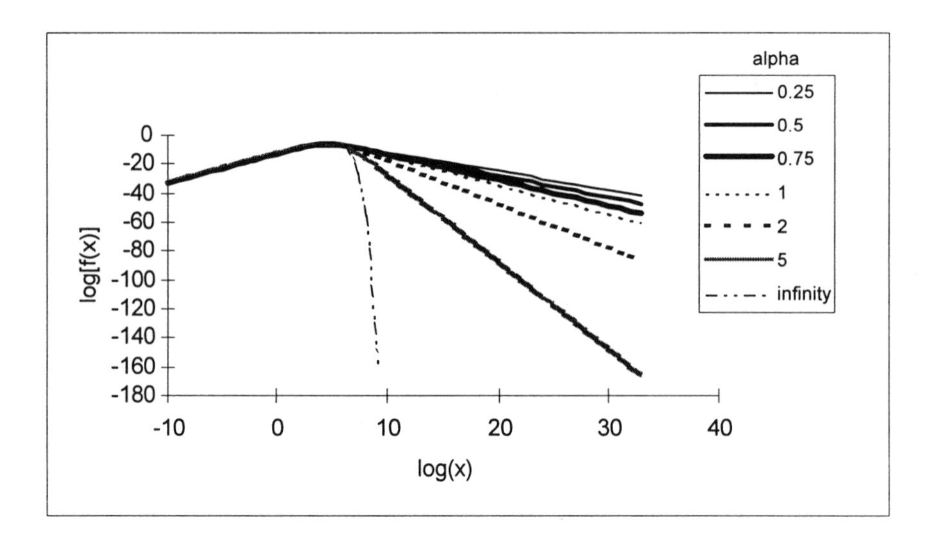

Fig. 2.8 Generalized Pareto pdfs with a mode of 100 and varying α

$$
\begin{aligned}
&= \lim_{x\to 0} \frac{\Gamma(\alpha+\tau)\gamma x^{\gamma\tau-1}\theta^{-\gamma\tau}}{\Gamma(\alpha)\Gamma(\tau)[1+(x/\theta)^{\gamma}]^{\alpha+\tau}\gamma\tau x^{\gamma\tau-1}} \\
&= \lim_{x\to 0} \frac{\Gamma(\alpha+\tau)}{\Gamma(\alpha)\Gamma(\tau)\tau\theta^{\gamma\tau}}\left[\frac{\theta^{\gamma}}{x^{\gamma}+\theta^{\gamma}}\right]^{\alpha+\tau} \\
&= \frac{\Gamma(\alpha+\tau)}{\Gamma(\alpha)\Gamma(\tau)\tau\theta^{\gamma\tau}}.
\end{aligned}
$$

The probability in the left-hand tail is approximately proportional to $x^{\gamma\tau}$. Again, the thickness of the tail is governed by the product, and larger values indicate a thinner tail.

It is interesting to note that the parameter γ affects the behavior of both tails. As it increases, both tails get lighter and lighter. So in some sense this parameter controls the spread around the middle while the parameters α and τ control the right and left ends, respectively. As a scale parameter, θ plays no role in setting the shape of the distribution.

Example 2.53 *Illustrate the left-hand tail behavior of the transformed beta family by plotting the pdf of the generalized Pareto distribution. Keep α constant as well as the mean of the distribution and then let τ vary, including the case where τ goes to infinity (an inverse gamma distribution).*

By using the generalized Pareto distribution we have set $\gamma = 1$. We have arbitrarily set $\alpha = 3$ and the mean at 100 for this example. The plot (on log–log axes) appears in Figure 2.9. It is clear that changing τ affects only the left-hand tail and that the inverse gamma distribution is a limiting case.□

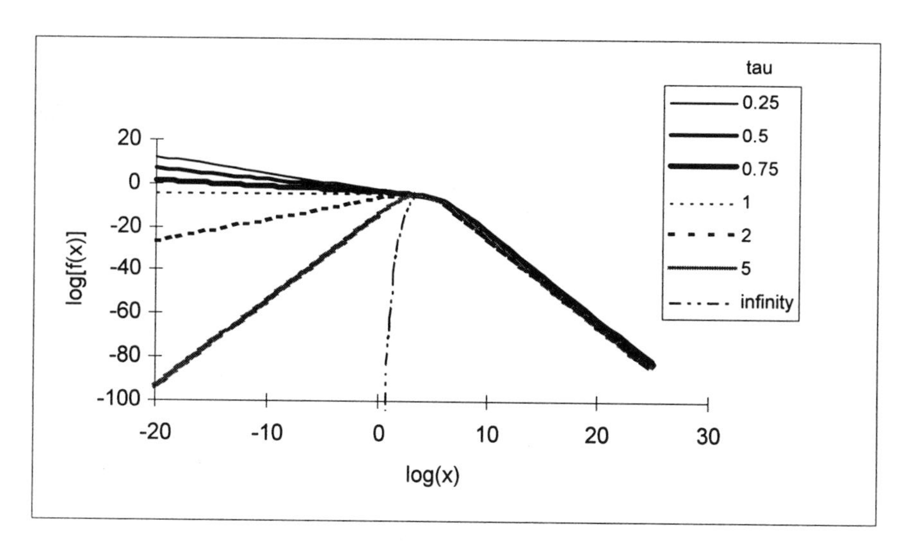

Fig. 2.9 Generalized Pareto pdfs with a mean of 100 and varying τ

2.7.6 Two-point mixtures

There is a special type of mixture that deserves attention. At first glance it appears to be simpler than the mixture from the previous subsection. A generalization of the two-point mixture is provided in the following definition.

Definition 2.36 *A random variable Y is an* ***n-point mixture*** *of the random variables* $X_1, X_2, \ldots, X_n$ *if its cdf is given by*

$$F_Y(y) = a_1 F_{X_1}(y) + a_2 F_{X_2}(y) + \cdots + a_n F_{X_n}(y) \tag{2.26}$$

where all $a_i \geq 0$ *and* $a_1 + a_2 + \cdots + a_n = 1$.

This is equivalent to using a mixing distribution that is discrete, placing probability a_i on the random variable X_i. Although mathematically the n-point mixture is easier to work with (using a sum instead of an integral) there is a major problem – there will be an enormous number of parameters to estimate. Consider a situation where each of the X_i has a different Burr distribution. There will be $3n$ Burr parameters and $n-1$ mixing parameters to estimate, for a total of $4n-1$.

However, a two-point mixture is not unreasonable. If the two original distributions each have two parameters, there will be a total of five parameters overall. With over twenty choices for each of X_1 and X_2, we can create over 200 new distributions using just two-point mixtures. That should be sufficient for most shapes that occur in practice.

To motivate the use of two-point mixtures consider a loss process in which there are two distinct events yielding insurance payments. An example might be a hospital's general liability coverage. One process is carelessness in patient handling, which would include lost items such as dentures and eyeglasses as well as small incidents such as wheelchair collisions. The other process takes place in the operating room, where surgeons may create events with costly consequences. The first process would have a large value of a_1 due to the higher frequency of such incidents, but the mean would be much smaller. Of course, if we could separate the data into these two components we should simply model them separately. As with all our models, it is not necessary to have a physical motivation in order to use the distribution.

A final computational point is worth mentioning. Maximizing over quantities that are constrained to the interval from zero to one is difficult because most iterative methods have a possibility of leading us to a new point that is outside this range. One way around this problem is to replace a_j in (2.26) with $e^{b_j}/(1+e^{b_j})$. Now the parameter b_j can take on any real number as its value. For a two-point mixture we have $a_1 = e^b/(1+e^b)$ and $a_2 = 1/(1+e^b)$.

Example 2.54 *For models involving general liability insurance, the Insurance Services Office has had some success with a mixture of two Pareto distributions. They also found that five parameters were not necessary. The distribution they selected has cdf*

$$F(x) = 1 - a\left(\frac{\theta_1}{\theta_1 + x}\right)^{\alpha} - (1-a)\left(\frac{\theta_2}{\theta_2 + x}\right)^{\alpha+2}.$$

Note that the shape parameters in the two Pareto distributions must differ by 2. The second distribution has a lighter tail. This might be a model for the frequent, small claims while the first distribution covers the large, but infrequent claims. This distribution has only four parameters, making maximization easier and bringing some parsimony to the modeling process. □

Finally, it is possible to do hypothesis tests to determine if a two-point mixture is better than a single distribution. While it might appear that a formal likelihood ratio test can be used, we again face the problem that the null hypothesis is on the boundary of the parameter space ($a_1 = 1$). However, we still think the likelihood ratio method provides a good decision rule. It is still necessary that the null hypothesis be that the model is in the family of X_1. A likelihood ratio test could be used in Example 2.54 to determine if all four parameters should have been used.

2.8 BAYESIAN ESTIMATION

All of the previous discussion on estimation has assumed a frequentist approach. That is, the population distribution is fixed, but unknown, and our decisions are concerned not only with the sample we obtained from the population, but also with the possibilities attached to other samples that might have been obtained. The Bayesian approach assumes that only the data count and it is the population that is variable. For parameter estimation the following definitions describe the process and then Bayes' theorem provides the solution.

2.8.1 Definitions and Bayes' Theorem

Definition 2.37 *The* ***prior distribution*** *is a probability distribution over the space of possible parameter values. It is denoted* $\pi(\theta)$ *and represents our opinion concerning the relative chances that various values of* θ *are the true value.*

As earlier, the parameter θ may be scalar- or vector-valued. Determination of the prior distribution has always been one of the barriers to the widespread acceptance of Bayesian methods. It is almost certainly the case that your experience has provided some insights about possible parameter values before the first data point has been observed. (If you have no such opinions, perhaps the wisdom of the person who assigned this task to you should be questioned.) The difficulty is translating this knowledge into a probability distribution. An excellent discussion about prior distributions and the foundations of Bayesian analysis can be found in Lindley [81], and for a discussion about issues surrounding the choice of Bayesian versus frequentist methods, see Efron [35]. The book by Klugman [76] contains more detail on the Bayesian approach along with several actuarial applications. A good source for a thorough mathematical treatment of Bayesian methods is the text by Berger [12].

Due to the difficulty of finding a prior distribution that is convincing (you will have to convince others that your prior opinions are valid) and the possibility that you may really have no prior opinion, the definition of prior distribution has been loosened.

Definition 2.38 *An* ***improper prior distribution*** *is one for which the probabilities (or pdf) are non-negative, but their sum (or integral) is infinite.*

A great deal of research has gone into the determination of a so-called **noninformative** or **vague** prior. Its purpose is to reflect minimal knowledge. Universal agreement on the best way to construct a vague prior does not exist. However, there is agreement that the appropriate noninformative prior for a scale parameter is $\pi(\theta) = 1/\theta, \quad \theta > 0$. Note that this is an improper prior.

For a Bayesian analysis, the model is no different than before.

Definition 2.39 *The* ***model distribution*** *is the probability distribution for the data as collected, given a particular value for the parameter. Its pdf is denoted* $f_{\mathbf{X}|\Theta}(\mathbf{x}|\theta)$*, where vector notation for* $\mathbf{x}$ *is used to remind us that all the data appears here. Also note that this is identical to the likelihood function and so that name may also be used at times.*

We use concepts from multivariate statistics to obtain two more definitions. In both cases, as well as in the rest of this section, integrals should be replaced by sums if the distributions are discrete.

Definition 2.40 *The* ***joint distribution*** *has pdf*

$$f_{\mathbf{X},\Theta}(\mathbf{x},\theta) = f_{\mathbf{X}|\Theta}(\mathbf{x}|\theta)\pi(\theta).$$

Definition 2.41 *The* ***marginal distribution*** *of* $\mathbf{x}$ *has pdf*

$$f_{\mathbf{X}}(\mathbf{x}) = \int f_{\mathbf{X}|\Theta}(\mathbf{x}|\theta)\pi(\theta)d\theta.$$

If the prior distribution is discrete, the integral should be replaced by a sum.

The final two quantities of interest are the following.

Definition 2.42 *The* ***posterior distribution*** *is the conditional probability distribution of the parameters given the observed data. It is denoted* $\pi_{\Theta|\mathbf{X}}(\theta|\mathbf{x})$.

Definition 2.43 *The* ***predictive distribution*** *is the conditional probability distribution of a new observation* y *given the data* $\mathbf{x}$. *It is denoted* $f_{Y|\mathbf{X}}(y|\mathbf{x})$.[16]

These last two items are the key output of a Bayesian analysis. The posterior distribution tells us how our opinion about the parameter has changed once we have observed the data. The predictive distribution tells us what the next observation might look like given the information contained in the data (as well as, implicitly, our prior opinion). Bayes' theorem tells us how to compute the posterior distribution.

Theorem 2.16 *The posterior distribution can be computed as*

$$\pi_{\Theta|\mathbf{X}}(\theta|\mathbf{x}) = \frac{f_{\mathbf{X}|\Theta}(\mathbf{x}|\theta)\pi(\theta)}{\int f_{\mathbf{X}|\Theta}(\mathbf{x}|\theta)\pi(\theta)d\theta} \tag{2.27}$$

while the predictive distribution can be computed as

$$f_{Y|\mathbf{X}}(y|\mathbf{x}) = \int f_{Y|\Theta}(y|\theta)\pi_{\Theta|\mathbf{X}}(\theta|\mathbf{x})d\theta \tag{2.28}$$

where $f_{Y|\Theta}(y|\theta)$ *is the pdf of the new observation, given the parameter value. In both formulas the integrals are replaced by sums for a discrete distribution.* □

[16] In this section and in any subsequent Bayesian discussions, we reserve $f(\cdot)$ for distributions concerning observations (such as the model and predictive distributions) and $\pi(\cdot)$ for distributions concerning parameters (such as the prior and posterior distributions). The arguments will usually make it clear which particular distribution is being used. To make matters explicit, we also employ subscripts to enable us to keep track of the random variables.

The following example illustrates the above definitions and results. The setting, though not the data, is taken from Meyers [89].

Example 2.55 *The following amounts were paid on a hospital liability policy.*

$$125 \quad 132 \quad 141 \quad 107 \quad 133 \quad 319 \quad 126 \quad 104 \quad 145 \quad 223.$$

The amount of a single payment has the single-parameter Pareto distribution with $\theta = 100$ and α unknown. The prior distribution has the gamma distribution with $\alpha = 2$ and $\theta = 1$. Determine all of the relevant Bayesian quantities.

The prior density has a gamma distribution and is

$$\pi(\alpha) = \alpha e^{-\alpha}, \qquad \alpha > 0$$

while the model is (evaluated at the data points)

$$f_{\mathbf{X}|A}(\mathbf{x}|\alpha) = \frac{\alpha^{10}(100)^{10\alpha}}{\left(\prod_{j=1}^{10} x_j^{\alpha+1}\right)} = \alpha^{10}e^{-3.801121\alpha-49.852823}.$$

The joint density of $\mathbf{x}$ and A is (again evaluated at the data points)

$$f_{\mathbf{X},A}(\mathbf{x},\alpha) = \alpha^{11}e^{-4.801121\alpha-49.852823}.$$

The posterior distribution of α is

$$\pi_{A|\mathbf{X}}(\alpha|\mathbf{x}) = \frac{\alpha^{11}e^{-4.801121\alpha-49.852823}}{\int_0^\infty \alpha^{11}e^{-4.801121\alpha-49.852823}d\alpha} = \frac{\alpha^{11}e^{-4.801121\alpha}}{(11!)(1/4.801121)^{12}}. \tag{2.29}$$

There is no need to evaluate the integral in the denominator. Because we know that the result must be a probability distribution the denominator is just the appropriate normalizing constant. A look at the numerator reveals that we have a gamma distribution with $\alpha = 12$ and $\theta = 1/4.801121$.

The predictive distribution is

$$\begin{aligned} f_{Y|\mathbf{X}}(y|\mathbf{x}) &= \int_0^\infty \frac{\alpha 100^\alpha}{y^{\alpha+1}} \frac{\alpha^{11}e^{-4.801121\alpha}}{(11!)(1/4.801121)^{12}} d\alpha \\ &= \frac{1}{y(11!)(1/4.801121)^{12}} \int_0^\infty \alpha^{12}e^{-(0.195951+\log y)\alpha}d\alpha \\ &= \frac{1}{y(11!)(1/4.801121)^{12}} \frac{(12!)}{(0.195951+\log y)^{13}} \\ &= \frac{12(4.801121)^{12}}{y(0.195951+\log y)^{13}}, \qquad y > 100. \end{aligned} \tag{2.30}$$

While this density function may not look familiar, you are asked to show in Exercise 2.93 that $\log Y - \log 100$ has the Pareto distribution. □

2.8.2 Inference and prediction

In one sense the analysis is complete. We begin with a distribution that quantifies our knowledge about the parameter and/or the next observation and we end with a revised distribution. But we have a suspicion that your boss may not be satisfied if you produce a distribution in response to his/her request. No doubt a specific number, perhaps with a margin for error, is what is desired. The usual Bayesian solution is to pose a loss function.

Definition 2.44 *A* ***loss function*** $l_j(\hat{\theta}_j, \theta_j)$ *describes the penalty paid to the investigator when* $\hat{\theta}_j$ *is the estimate when* θ_j *is the true value of the* j*th parameter.*

It would also be possible to have a multidimensional loss function $l(\widehat{\boldsymbol{\theta}}, \boldsymbol{\theta})$ which allowed the loss to depend simultaneously on the errors in the various parameter estimates.

Definition 2.45 *The* ***Bayes estimate*** *for a given loss function is the one that minimizes the expected loss given the posterior distribution of the parameter in question.*

The three most commonly used loss functions are defined as follows.

Definition 2.46 *For* ***squared-error loss*** *the loss function is (all subscripts are dropped for convenience)* $l(\hat{\theta}, \theta) = (\hat{\theta} - \theta)^2$. *For* ***absolute loss*** *it is* $l(\hat{\theta}, \theta) = |\hat{\theta} - \theta|$. *For* ***zero–one loss*** *it is* $l(\hat{\theta}, \theta) = 0$ *if* $\hat{\theta} = \theta$ *and is* 1 *otherwise.*

The following theorem indicates the Bayes estimates for these three common loss functions.

Theorem 2.17 *For squared-error loss the Bayes estimate is the mean of the posterior distribution, for absolute loss it is the median, and for zero–one loss it is the mode.* □

Note that there is no guarantee that the posterior mean exists or that the posterior median or mode will be unique.

Example 2.56 (Example 2.55 continued) *Determine the three Bayes estimates of* α.

The mean of the posterior gamma distribution is $\alpha\theta = 12/4.801121 = 2.499416$. The median must be determined numerically, it is 2.430342 while the mode is $(\alpha - 1)\theta = 11/4.801121 = 2.291132$. Note that the α used here is the parameter of the posterior gamma distribution, not the α for the single-parameter Pareto distribution, that we are trying to estimate. □

The Bayesian equivalent of a confidence interval is easy to construct. The following definition will suffice.

Definition 2.47 *The points* $a < b$ *define a* $100(1-\alpha)\%$ ***credibility interval*** *for* θ_j, *provided that* $\Pr(a \le \Theta_j \le b|\mathbf{x}) \ge 1-\alpha$.

The use of the term credibility has no relationship to its use in actuarial analyses as developed in Chapter 5. The inequality is present for the case where the posterior distribution of θ_j is discrete. Then it may not be possible for the probability to be exactly $1-\alpha$. This definition does not produce a unique solution. The following theorem indicates one way to produce a unique interval.

Theorem 2.18 *If the posterior random variable* $\theta_j|\mathbf{x}$ *is continuous and unimodal, then the* $100(1-\alpha)\%$ *credibility interval with smallest width* $(b-a)$ *is the unique solution to*

$$\int_a^b \pi_{\Theta_j|\mathbf{X}}(\theta_j|\mathbf{x})d\theta_j = 1-\alpha$$

$$\pi_{\Theta|\mathbf{X}}(a|\mathbf{x}) = \pi_{\Theta|\mathbf{X}}(b|\mathbf{x}).$$

This interval is a special case of a highest posterior density (HPD) credibility set. □

The following example may clarify the theorem.

Example 2.57 (Example 2.55 continued) *Determine the shortest 95% credibility interval for the parameter* α. *Also determine the interval that places 2.5% probability at each end.*

The two equations from Theorem 2.18 are

$$\Pr(a \le A \le b|\mathbf{x}) = \Gamma(12; 4.801121b) - \Gamma(12; 4.801121a) = .95$$

$$a^{11}e^{-4.801121a} = b^{11}e^{-4.801121b}.$$

Numerical analysis methods can be used to find the solution $a = 1.1832$ and $b = 3.9384$. The width of this interval is 2.7552.

Placing 2.5% probability at each end yields the two equations

$$\Gamma(12; 4.801121b) = 0.975, \quad \Gamma(12; 4.801121a) = 0.025.$$

This solution requires either access to the inverse of the incomplete gamma function or the use of root-finding techniques with the incomplete gamma function itself. The solution is $a = 1.2915$ and $b = 4.0995$. The width is 2.8080, wider than the first interval. Figure 2.10 shows the difference in the two intervals. The solid vertical bars represent the HPD interval. The total area to the left and right of these bars is 0.05. Any other interval must also

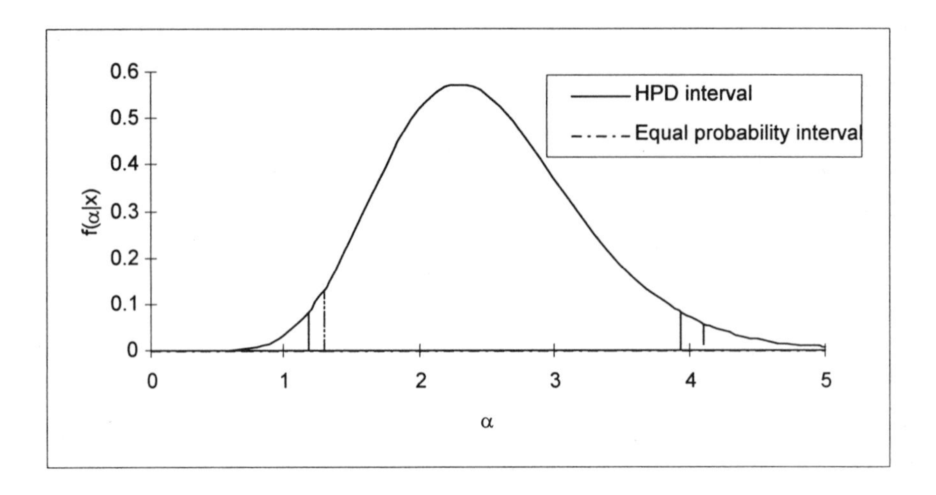

Fig. 2.10 Two Bayesian credibility intervals

have this probability. To create the interval with 0.025 probability on each side, both bars must be moved to the right. To subtract the same probability on the right end that is added on the left end, the right limit must be moved a greater distance, because the posterior density is lower over that interval than it is on the left end. This must lead to a wider interval. □

The following definition provides the equivalent result for any posterior distribution.

Definition 2.48 *For any posterior distribution the* $100(1-\alpha)\%$ ***HPD credibility set*** *is the set of parameter values* C *such that*

$$\Pr(\theta_j \in C) \geq 1 - \alpha \tag{2.31}$$

$$C = \{\theta_j : \pi_{\Theta_j|\mathbf{X}}(\theta_j|\mathbf{x}) \geq c\} \text{ for some } c$$

where c *is the largest value for which the inequality (2.31) holds.*

This set may be the union of several intervals (which can happen with a multimodal posterior distribution). This definition produces the set of minimum total width that has the required posterior probability. Construction of the set is done by starting with a high value of c and then lowering it. As it decreases, the set C gets larger as does the probability. The process continues until the probability reaches $1 - \alpha$. It should be obvious to see how the definition can be extended to the construction of a simultaneous credibility set for a vector of parameters, $\boldsymbol{\theta}$.

Sometimes it is the case that while computing posterior probabilities is difficult, computing posterior moments may be easy. We can then use the

Bayesian Central Limit Theorem. The following is a paraphrase from Berger [12, p. 224].

Theorem 2.19 *If $\pi(\theta)$ and $f_{\mathbf{X}|\Theta}(\mathbf{x}|\theta)$ are both twice differentiable in the elements of θ and other commonly satisfied assumptions hold, then the posterior distribution of Θ given $\mathbf{X} = \mathbf{x}$ is asymptotically normal.* □

The "commonly satisfied assumptions" are like those in Theorem 2.2. As in that theorem, it is possible to do further approximations. In particular, the asymptotic normal distribution also results if the posterior mode is substituted for the posterior mean and/or if the posterior covariance matrix is estimated by inverting the matrix of second partial derivatives of the negative logarithm of the posterior density.

Example 2.58 (Example 2.55 continued) *Construct a 95% credibility interval for α using the Bayesian Central Limit Theorem.*

The posterior distribution has a mean of 2.499416 and a variance of $\alpha\theta^2 = 0.520590$. Using the normal approximation the credibility interval is $2.499416 \pm 1.96(0.520590)^{1/2}$, which produces $a = 1.0852$ and $b = 3.9136$. This interval (with regard to the normal approximation) is HPD due to the symmetry of the normal distribution.

The approximation is centered at the posterior mode of 2.291132 (see Example 2.56). The second derivative of the negative logarithm of the posterior density (from (2.29)) is

$$-\frac{d^2}{d\alpha^2}\log\left[\frac{\alpha^{11}e^{-4.801121\alpha}}{(11!)(1/4.801121)^{12}}\right] = \frac{11}{\alpha^2}.$$

The variance estimate is the reciprocal. Evaluated at the modal estimate of α we get $(2.291132)^2/11 = 0.477208$ for a credibility interval of $2.29113 \pm 1.96(0.477208)^{1/2}$, which produces $a = 0.9372$ and $b = 3.6451$. □

The same concepts can apply to the predictive distribution. However, the Central Limit Theorem does not help here because the predictive sample has only one member. The only potential use for it is that for a large original sample size, we can replace the true posterior distribution in (2.28) with a multivariate normal distribution.

Example 2.59 (Example 2.55 continued) *Construct a 95% highest density prediction interval for the next observation.*

It is easy to see that the predictive density function (2.30) is strictly decreasing. Therefore the region with highest density runs from $a = 100$ to b. The value of b is determined from

$$0.95 = \int_{100}^{b} \frac{12(4.801121)^{12}}{y(0.195951 + \log y)^{13}} dy$$

$$= \int_0^{\log(b/100)} \frac{12(4.801121)^{12}}{(4.801121+x)^{13}} dx$$

$$= 1 - \left[\frac{4.801121}{4.801121+\log(b/100)}\right]^{12}$$

and the solution is $b = 390.1840$. It is interesting to note that the mode of the predictive distribution is 100 (because the pdf is strictly decreasing) while the mean is infinite (with $b = \infty$ and an additional y in the integrand, after the transformation, the integrand is like $e^x x^{-13}$, which goes to infinity as x goes to infinity). □

2.8.3 Computational issues

It should be obvious by now that all Bayesian analyses proceed by taking integrals or sums. So at least conceptually it is always possible to do a Bayesian analysis. However, only in rare cases are the integrals or sums easy to do, and that means most Bayesian analyses will require numerical integration. While one-dimensional integrations are easy to do to a high degree of accuracy, multidimensional integrals are much more difficult to approximate. A great deal of effort has been expended in the past ten years with regard to solving this problem. A number of ingenious methods have been developed. Some of them are summarized in Klugman [76], while a more recent technique is used in Carlin and Klugman [23].

There is another way which completely avoids computational problems. This is illustrated using the example (in an abbreviated form) from Meyers [89], which also employed this technique. The example also shows how a Bayesian analysis is used to estimate a function of parameters.

Example 2.60 *Data were collected on 100 losses in excess of 100,000. The single-parameter Pareto distribution is to be used with $\theta = 100{,}000$ and α unknown. The objective is to estimate the layer average severity for the layer from 1,000,000 to 5,000,000. For the observations, $\sum_{j=1}^{100} \log x_j = 1{,}208.4354$.*

The model density is

$$\begin{aligned} f_{\mathbf{X}|A}(\mathbf{x}|\alpha) &= \prod_{j=1}^{100} \frac{\alpha(100{,}000)^\alpha}{x_j^{\alpha+1}} \\ &= \exp\left[100\log\alpha + 100\alpha\log 100{,}000 - (\alpha+1)\sum_{j=1}^{100}\log x_j\right] \\ &= \exp(100\log\alpha - 100\alpha/1.75 - 1{,}208.4354). \end{aligned}$$

The density appears in column three of Table 2.13. To prevent computer overflow, the value 1,208.4354 was not subtracted prior to exponentiation.

This makes the entries proportional to the true density function. The prior density is given in the second column. It was chosen based on a belief that the true value is in the range 1 to 2.5 and is more likely to be near 1.5 than at the ends. The posterior density is then obtained using (2.27). The elements of the numerator are found in column four. The denominator is no longer an integral but a sum. The sum is at the bottom of column four and then the scaled values are in column five.

We can see from column five that the posterior mode is at $\alpha = 1.7$, as compared to the maximum likelihood estimator of 1.75. The posterior mean of α could be found by adding the product of columns one and five. Here we are interested in a layer average severity. For this problem it is

$$\begin{aligned} \mathrm{LAS}(\alpha) &= E(X \wedge 5{,}000{,}000) - E(X \wedge 1{,}000{,}000) \\ &= \frac{100{,}000^{\alpha}}{\alpha - 1}\left(\frac{1}{1{,}000{,}000^{\alpha-1}} - \frac{1}{5{,}000{,}000^{\alpha-1}}\right), \quad \alpha \neq 1 \\ &= 100{,}000\,(\log 5{,}000{,}000 - \log 1{,}000{,}000)\,, \quad \alpha = 1. \end{aligned}$$

Values of $\mathrm{LAS}(\alpha)$ for the sixteen possible α values appear in column six. The last two columns are then used to obtain the posterior expected values of the layer average severity. The point estimate is the posterior mean, 18,827. The posterior standard deviation is

$$\sqrt{445{,}198{,}597 - 18{,}827^2} = 9{,}526.$$

We can also use columns five and six to construct a credibility interval. Discarding the first five rows and the last four rows eliminates 0.0406 of posterior probability. That leaves [5,992, 34,961] as a 96% credibility interval for the layer average severity. Part of Meyers' paper was the observation that even with a fairly large sample, the accuracy of the estimate is poor.

The discrete approximation to the prior distribution could be refined by using many more than sixteen values. This adds little to the spreadsheet effort. The number was kept small here only for display purposes. □

2.9 SELECTING AND VALIDATING A MODEL

To this point our search for a parametric model has allowed us to postulate too many models. We have a large collection of distributions from which to choose and a variety of ways to estimate the parameters. What we need now is a method to narrow our selection to a single model and a unique parameter estimate. With regard to selecting the model, there are two approaches. One is a yes/no mechanism. That is, for each proposed model, a decision is made that it is acceptable or unacceptable. Any acceptable model could then be used, with some nonstatistical criteria guiding the decision. The major drawbacks are that there may be no acceptable models (an unacceptable outcome)

Table 2.13 Bayesian estimation of a layer average severity

α	$\pi(\alpha)$	$f_{\mathbf{X}\mid A}(\mathbf{x}\mid\alpha)$	$\pi(\alpha) f_{\mathbf{X}\mid A}(\mathbf{x}\mid\alpha)$	$\pi_{A\mid\mathbf{X}}(\alpha\mid\mathbf{x})$	$\mathrm{LAS}(\alpha)$	$\pi_{A\mid\mathbf{X}}(\alpha\mid\mathbf{x})\mathrm{LAS}(\alpha)$	$\pi_{A\mid\mathbf{X}}(\alpha\mid\mathbf{x})\mathrm{LAS}(\alpha)^2$
1.0	0.0400	1.52E-25	6.10E-27	0.0000	160,944	0	6,433
1.1	0.0496	6.93E-24	3.44E-25	0.0000	118,085	2	195,201
1.2	0.0592	1.37E-22	8.13E-24	0.0003	86,826	29	2,496,935
1.3	0.0688	1.36E-21	9.33E-23	0.0038	63,979	243	15,558,906
1.4	0.0784	7.40E-21	5.80E-22	0.0236	47,245	1,116	52,737,840
1.5	0.0880	2.42E-20	2.13E-21	0.0867	34,961	3,033	106,021,739
1.6	0.0832	5.07E-20	4.22E-21	0.1718	25,926	4,454	115,480,050
1.7	0.0784	7.18E-20	5.63E-21	0.2293	19,265	4,418	85,110,453
1.8	0.0736	7.19E-20	5.29E-21	0.2156	14,344	3,093	44,366,353
1.9	0.0688	5.29E-20	3.64E-21	0.1482	10,702	1,586	16,972,802
2.0	0.0640	2.95E-20	1.89E-21	0.0768	8,000	614	4,915,383
2.1	0.0592	1.28E-20	7.57E-22	0.0308	5,992	185	1,106,259
2.2	0.0544	4.42E-21	2.40E-22	0.0098	4,496	44	197,840
2.3	0.0496	1.24E-21	6.16E-23	0.0025	3,380	8	28,650
2.4	0.0448	2.89E-22	1.29E-23	0.0005	2,545	1	3,413
2.5	0.0400	5.65E-23	2.26E-24	0.0001	1,920	0	339
	1.0000		2.46E-20	1.0000		18,827	445,198,597

or too many acceptable models, some of which are likely to violate our goal of parsimony.

The second approach is to rank the models and then select the one that appears at the top of the list. This makes much more sense, but we will still examine the yes/no approach, partly because it eventually leads to a ranking criterion.

Regardless of the approach taken, keep in mind that your own expertise should not be ignored. Even if you are not taking a formal Bayesian approach, your opinions about which models are likely to be successful are valuable. For example, if in the past a Weibull distribution has consistently provided a useful model, but this time the data indicate that a transformed gamma distribution is slightly better, it may be advisable to use the Weibull anyway. Or, suppose a client needs some help with the estimation process but believes that the lognormal distribution is the one to use. In a close call between the lognormal and another model, there is little reason to promote the alternative. Of course, if the data indicate that the generalized Pareto distribution is far better, than you must explain your choice and describe this new (to your client) model.

With regard to parameter estimation, we support maximum likelihood estimation unless there is a good reason to do otherwise. Some of those arguments were provided earlier in this chapter. We should also point out that we are recommending a generic approach to estimation. That is, whether you favor maximum likelihood, minimum modified chi-square, or some other method, that method will be used regardless of the model being considered. Statisticians have devoted a great deal of energy to studying specific distributions in an attempt to learn how they should best be handled (for example, there is an entire book [4] devoted to the Pareto distribution), and it is certainly true that one could match the estimation and testing strategy to the model. With our large number of models, we would rather adopt a consistent strategy.

2.9.1 Testing the acceptability of a model

The issue can be phrased as a hypothesis test. The null hypothesis is that the model is acceptable, whereas the alternative is that it is not. In the parametric case this can be formally stated as

$$\begin{aligned} H_0 &: F_X(x) = F(x;\theta) \quad \text{for some } \theta \in \Theta \\ H_1 &: F_X(x) \neq F(x;\theta) \quad \text{for any } \theta \in \Theta. \end{aligned}$$

Here $F_X(x)$ is the population cdf and $F(x;\theta)$ is the cdf for a member of the selected parametric family. This is a situation where classical hypothesis testing is less than ideal. In most testing situations the purpose of sampling and testing is to convince others that the alternative hypothesis is true. To be convincing we control the probability of a Type I error—that is, of selecting H_1 when H_0 is true. Here we are hoping to select the null hypothesis (and thereby convince others that our model is a good one), but we cannot control the probability that our decision was in error.

Furthermore, we face a serious problem when the sample size is large. We know for a fact that the null hypothesis is false! It is extremely unlikely that any process as complex as the one that produces insurance claims yields a population that can be described with a few parameters. What we do hope is that we can find a simple model that is an excellent approximation to the population; one that is useful for the calculations we subsequently plan to perform. With a large sample size our hypothesis test will be so powerful that the false nature of the null hypothesis will be detectable. The result is that there is a good chance that all models will be rejected.

2.9.1.1 Informal tests The above indicates that we may actually prefer an informal test, one that uses judgment rather than statistics. Two informal procedures can be very helpful. The first is to create at least one useful picture. Our favorite is an overlay plot of the histogram of the data and the pdf of the model. A good model should have a density that looks similar to the histogram. If the data were not grouped, it must be grouped in order to construct the histogram. Another popular graph is a q–q plot. This one requires individual data. The observations need to be ranked so that $x_1 \leq x_2 \leq \cdots \leq x_n$. The n points to be plotted are $(j/(n+1),\ F(x_j;\theta))$. If the model is a good one, the points will lie very close to the line from $(0,0)$ to $(1,1)$.

The second informal approach is to compute some relevant numbers from the sample and compare them to the same numbers from the model. These could be cumulative probabilities, limited expected values, layer average severities, or anything else that might be considered important.

Example 2.61 *In Example 2.36 a lognormal distribution was fit to the data in Table 2.10. The maximum likelihood estimates were* $\hat{\mu} = 9.29376$ *and* $\hat{\sigma} = 1.62713$*. Evaluate the quality of this model by comparing the histogram to the pdf and by computing layer average severities for each class.*

Due to the skewness of this data, it is best to present the graph in two pieces. The first (Figure 2.11) covers the range 0–32,500 while the second (Figure 2.12) covers the range 32,500–300,000. We see that the fit is pretty good throughout. In particular, the pdf tends to go through both the horizontal and vertical portions of the histogram about in the middle of each segment.

The layer average severities appear in Table 2.14. The empirical layer average severities are found by (i) taking the entries in the third column of Table 2.10, subtracting the lower limit of the interval, and then multiplying the difference by the entry in the second column, then (ii) taking the width of the interval times the sum of the entries in the second column for all subsequent rows, and then (iii) dividing the sum of the first two items by 217, the sample size. The model layer average severities are found using the limited expected

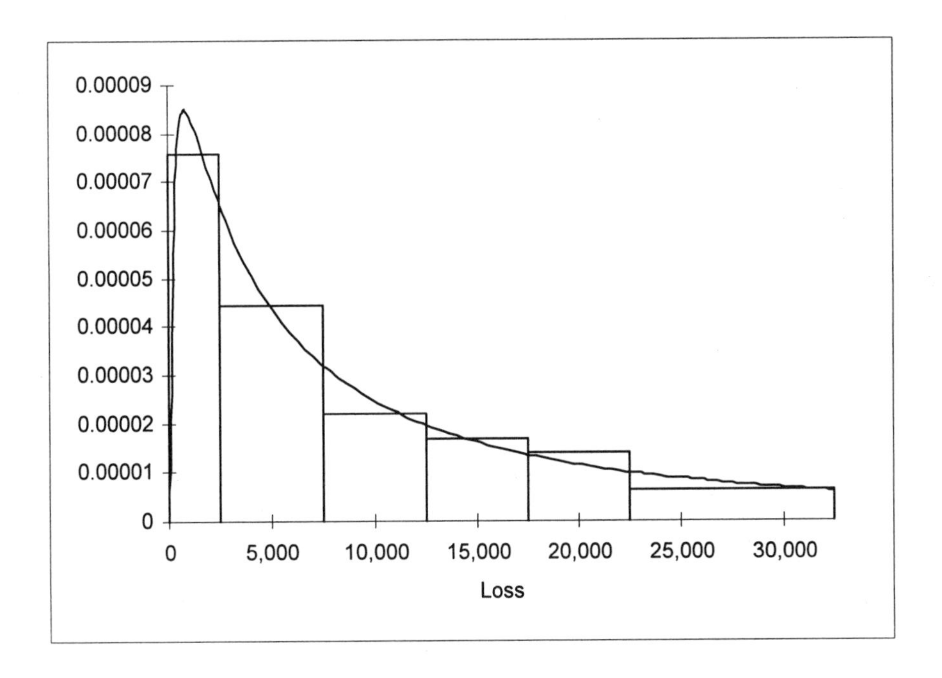

Fig. 2.11 Lognormal pdf and histogram for small losses

value formula in Appendix A. They appear to be fairly close to each other, again indicating a good fit for the lognormal model. □

Example 2.62 *In Example 2.56 the Bayesian estimate of the single-parameter Pareto parameter α for the data in Example 2.55 was* 2.499416. *Evaluate the quality of this model by constructing a q-q plot.*

The plot is in Figure 2.13. It appears that there is a systematic departure from the desired straight line and thus the single-parameter Pareto model is questionable for these data. However, with so few data points, it is hard to be certain about this conclusion. □

It should be noted that comparisons of the data and model only work as described here when the data are complete in the sense that no deductibles, limits, or similar modifications prevented us from recovering all losses. Methods to deal with this problem are given in Section 2.10.

2.9.1.2 Formal tests While these judgments are certainly helpful, a formal test may be more persuasive. While specific tests have been developed for particular models, the most commonly used generic test is the chi-square goodness-of-fit test. This test requires that the data be grouped, but as in

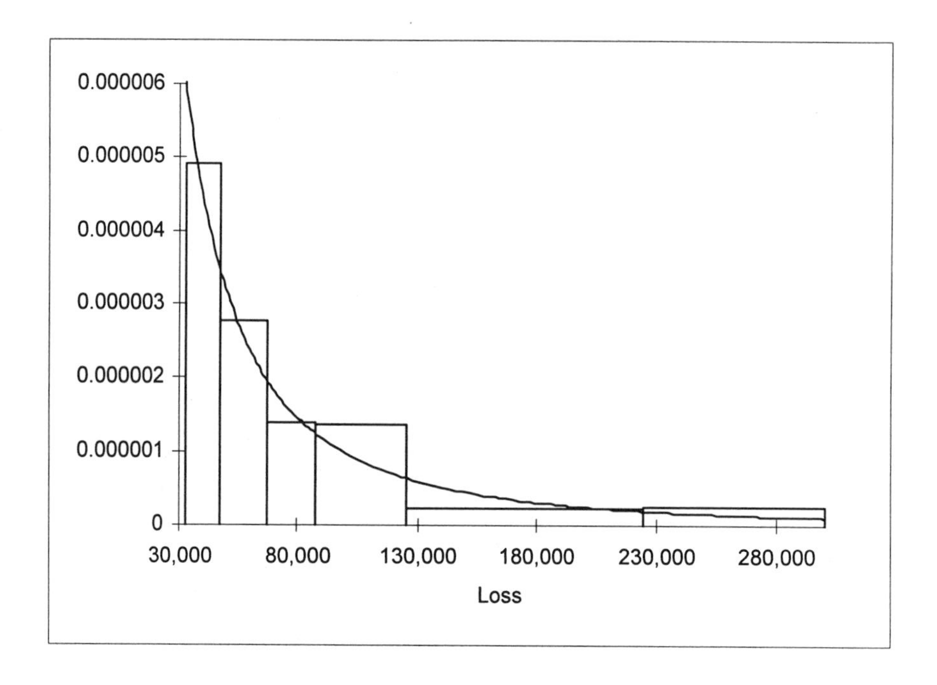

Fig. 2.12 Lognormal pdf and histogram for large losses

the histogram comparison, that does not mean the parameters need to have been estimated from the grouped version of the data.

For the jth group the test is based on n_j, the number of observations in that group and

$$E_j = n\,\Pr(X \in j\text{th group}; \hat{\theta})$$

where n is the sample size and the probability is that for a single observation falling in the jth group, given that θ has its estimated value. In words, E_j is the expected number of observations in the jth group given that the model is correct and the parameter has its estimated value. The test statistic is

$$Q = \sum_{j=1}^{k} \frac{(n_j - E_j)^2}{E_j}$$

and the null hypothesis is rejected if Q exceeds $\chi^2_{d,\alpha}$, where $d = k - r - 1$ is the number of degrees of freedom (recall that r is the number of parameters) and α is the significance level. Then the critical value is the number that makes $\Pr(\chi^2 > \chi^2_{d,\alpha}) = \alpha$, where χ^2 has the chi-square distribution with d degrees of freedom. The p-value for the test is $\Pr(\chi^2 > Q)$.

Table 2.14 Lognormal layer average severities

Payment	Empirical LAS	Lognormal LAS
0–2,500	2,290	2,275
2,500–7,500	3,427	3,450
7,500–12,500	2,672	2,615
12,500–17,500	2,229	2,114
17,500–22,500	1,825	1,773
22,500–32,500	2,892	2,858
32,500–47,500	3,408	3,198
47,500–67,500	3,166	3,084
67,500–87,500	2,327	2,285
87,500–125,000	3,055	3,056
125,000–225,000	4,602	4,550
225,000–300,000	1,756	1,909
300,000–	—	7,679

One common convention for the test to be valid (that is, the probability of making a Type I error really is α) is that $E_j \geq 5$ for all cells.[17] If that is not so, the boundaries must be expanded and the test redone, or adjacent groups combined. The theory that supports this test insists that the parameter be estimated by a particular method. According to Moore [91], estimation by either maximum likelihood or minimum chi-square is satisfactory.

Example 2.63 (Example 2.61 continued) *Perform the chi-square goodness-of-fit test to see if the lognormal model is appropriate.*

The results from the test appear in Table 2.15. Note that the last two groups were combined in order to make the expected count exceed five. A typical calculation is

$$\begin{aligned} E_4 &= 217\left[\Phi\left(\frac{\log(17{,}500) - 9.29376}{1.62713}\right) - \Phi\left(\frac{\log(12{,}500) - 9.29376}{1.62713}\right)\right] \\ &= 217(0.615109 - 0.534216) = 17.55. \end{aligned}$$

The test statistic is 4.51. There are 9 degrees of freedom (12 cells, less 2 parameters estimated, less 1) and the critical value for a 5% significance level

[17] Moore [92] cites a number of other rules. Among them are: (1) All cells with an expected frequency of at least 1 and 80% of the cells with an expected frequency of at least 5. (2) An average count per cell of at least four when testing at the 1% significance level and an average count of two when testing at the 5% level. (3) At least three cells, a sample size of at least ten, and the ratio of the square of the sample size to the number of cells must be at least ten.

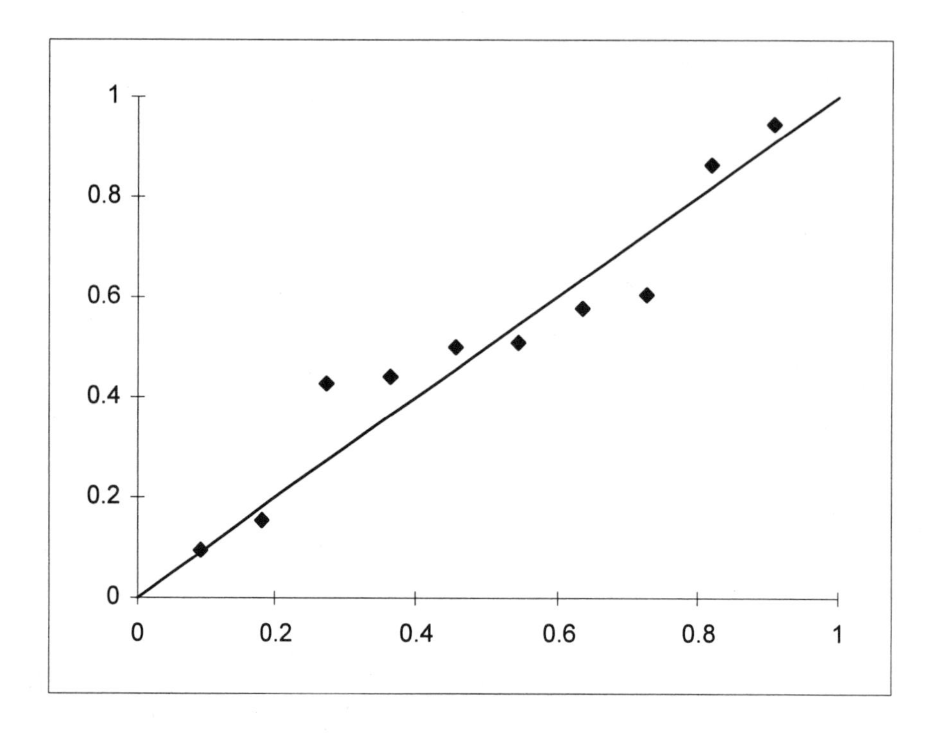

Fig. 2.13 q–q plot for the single parameter Pareto distribution

is 16.92 and so the null hypothesis is accepted. That is, there is no evidence in this sample of size 217 to reject the lognormal distribution as a model. The p-value is 0.8748 and so this model would be accepted for any commonly used significance level. (Recall that the null hypothesis is accepted by anyone whose significance level is less than the p-value.) □

The results of the goodness-of-fit test agree with the visual evidence presented earlier. The lognormal model does check out as being useful for these data.

The only time this test is impractical is when there are very few observations. In that setting it is not possible to have enough groups to capture the shape of the data and the model and at the extreme it may not be possible to have a positive number of degrees of freedom. For example, with ten observations, careful grouping could create two groups with our minimum expected count of five. But even with just one parameter to estimate there will be zero degrees of freedom, and therefore no chi-square test can be done. An alternative test works specifically with individual data. It is the Kolmogorov–Smirnov test. The test statistic is easy to compute, it is the maximum absolute difference between the model cdf and the empirical cdf.

Table 2.15 Chi-square goodness-of-fit test

Interval	Observed	Expected	Chi-square
0–2,500	41	39.75	0.04
2,500–7,500	48	49.17	0.03
7,500–12,500	24	27.00	0.33
12,500–17,500	18	17.55	0.01
17,500–22,500	15	12.48	0.51
22,500–32,500	14	16.70	0.44
32,500–47,500	16	14.77	0.10
47,500–67,500	12	11.18	0.06
67,500–87,500	6	6.71	0.07
87,500–125,000	11	7.22	1.98
125,000–225,000	5	7.68	0.94
225,000–	7	6.79	0.01
		Total	4.51

That is,

$$D = \sup_x \left| F_n(x) - F(x; \hat{\theta}) \right| .$$

Because the empirical cdf is discontinuous, we require a supremum (least upper bound). This just means that at one of the data points, the model cdf must be compared with the empirical cdf both just before and just after the jump. A pleasant feature of this test is that the maximum must occur at one of the data points. While detailed tables are available for this test approximate critical values (good for $n \geq 15$) are given in Table 2.16. It is important to recognize that the critical values in this table only apply when the parameters of the hypothesized distribution are specified by the null hypthesis. When the parameters are estimated from the data, the critical values must be reduced. There is no easy way to make the adjustment (such as reducing the degrees of freedom as is done in the chi-square test). Stephens [116] provides tables for use in testing a number of particular distributions. He also suggests a general approach: Use only half the data to estimate the parameters, but then use the entire data set to conduct the test. In this case, the critical values in Table 2.16 can be applied (at least asymptotically). For the examples and exercises in this text we will use the tabled values as we have nothing else at hand.

Example 2.64 *Test the suitability of the single-parameter Pareto model in Example 2.62 using the Kolmogorov–Smirnov test.*

A graph comparing the empirical and model cdfs appears in Figure 2.14. It should be easy to see that the largest difference occurs just prior to the third jump. The easiest way to compute the value of the test statistic is with a table

Table 2.16 Kolmogorov–Smirnov critical values

Significance level (α)	Critical value
0.20	$1.07/\sqrt{n}$
0.10	$1.22/\sqrt{n}$
0.05	$1.36/\sqrt{n}$
0.01	$1.63/\sqrt{n}$

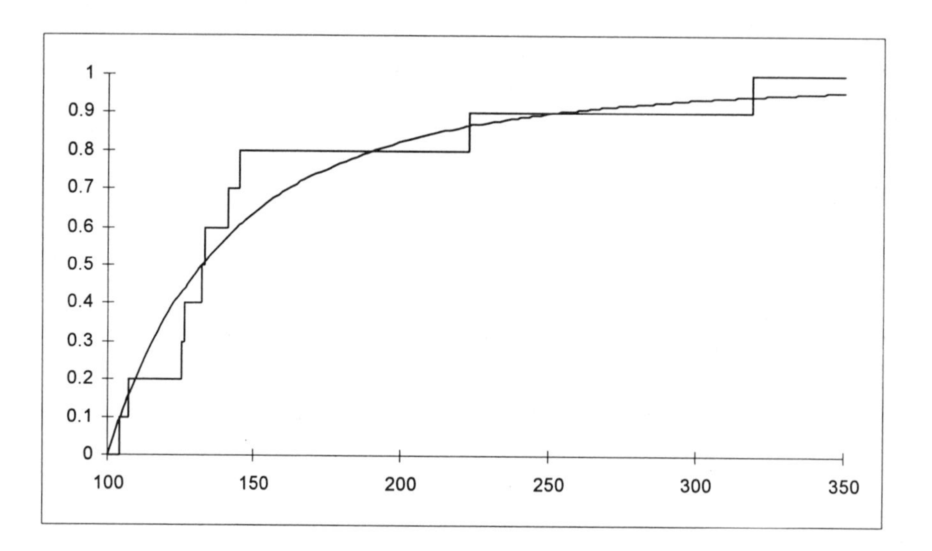

Fig. 2.14 Empirical versus model cdf

such as Table 2.17. Here the empirical cdf simply increases by $1/n$ at each data point. Then the model cdf is computed at each data point. For example, in the third row the model value of 0.4275 is compared with the empirical value before (0.2) and after (0.3) the jump. The largest difference occurs before the jump and is 0.2275. This is the maximum overall and becomes the test statistic. At a significance level of 10% the critical value is $1.22/\sqrt{10} = 0.3858$. The null hypothesis is accepted and thus there is no evidence in the data to reject the single-parameter Pareto model. Nevertheless, the q–q plot we constructed earlier makes this model suspect. A larger sample may be needed to resolve this issue. □

The major problem with the Kolmogorov–Smirnov test is its lack of power. As a result, for small samples it is unlikely that a model will be rejected. On the other hand, for large samples the more powerful chi-square goodness-of-fit test is available.

Table 2.17 Kolmogorov–Smirnov test statistic

Observation	Empirical cdf−	Empirical cdf+	Model cdf	Maximum difference
104	0.0	0.1	0.0934	0.0934
107	0.1	0.2	0.1556	0.0556
125	0.2	0.3	0.4275	0.2275
126	0.3	0.4	0.4388	0.1388
132	0.4	0.5	0.5004	0.1004
133	0.5	0.6	0.5097	0.0903
141	0.6	0.7	0.5763	0.1237
145	0.7	0.8	0.6049	0.1951
223	0.8	0.9	0.8653	0.0653
319	0.9	1.0	0.9449	0.0551
				$D = 0.2275$

2.9.2 Ranking and selecting models

Recall that our objective is to select a single model that we can claim adequately represents the population. The easiest way to do this would be to assign a numerical value to each proposed model and then select the one with the best score. We have already encountered a number of ways to assign a score. A partial list is as follows:

1. Value of the likelihood function at its maximum (large is good).

2. Value of the chi-square goodness-of-fit test statistic (small is good).

3. Value of the Kolmogorov–Smirnov test statistic (small is good).

4. The p-value from the chi-square goodness-of-fit test (large is good).

5. The value from any optimization based estimation, such as minimum cdf, minimum MSE, or minimum LAS (small is good).

In general it is reasonable to perform the ranking using the same method that was used to estimate the parameters. However, this is not necessary.

All but one of the suggestions has a problem that must be resolved. When one model is a special case of another (for example, Pareto and Burr) and the criterion matches the estimation method, the more complex model must come out at least as good as the simpler one. This is because the special case is always available as a potential candidate for the minimum (or maximum) for the more complex case. This is the same phenomenon that applies when additional variables are added to a multiple regression. The coefficient of determination (R^2) cannot decrease, even if the new variable is not related to the dependent variable. In regression a significant improvement is required before we can accept a new variable. The same applies here. There are two ways to resolve this issue.

Before discussing this resolution, we note that the exception is the p-value for the chi-square goodness-of-fit test. It automatically corrects for the increase in complexity by reducing the degrees of freedom and therefore a more complex model needs a significant reduction in the test statistic before the p-value becomes larger. It should also be noted that using the test statistic itself causes additional problems when groups are combined to achieve the required minimum of five expected counts. Then, even with the same number of parameters, two models may involve a different number of degrees of freedom. Based on this, we prefer using p-values and not test statistics to rank models.

Example 2.65 *In Table 2.18 the general liability payments from Table 2.10 have been reproduced. A large number of models have been fit by maximum likelihood. The models, the negative logarithm of the likelihood function, the chi-square test statistic, degrees of freedom, and p-value all appear in Table 2.19. Using this information, select the best model according to the p-value criterion.*

Because the phrase "negative logarithm of the likelihood function" is awkward, we will write either "negative loglikelihood," "−loglikelihood," or "NLL." In constructing Table 2.19, there was no convergence for a few of the attempted distributions. For example, when iterating to maximize the likelihood function for the transformed gamma distribution it became apparent after a few hundred iterations of the simplex method that α was heading toward infinity while θ and τ were heading toward zero. This indicates that some two-parameter distribution which is a limiting case will do just as well. While this limiting distribution may not be in our inventory, it does indicate that three parameters may be more than are needed.

From Table 2.19, it is clear that on the basis of p-value the lognormal model is the winner. The next best choice is the Burr distribution, but it would be hard to justify using three parameters when the result is both a lower p-value and a higher value of the test statistic itself. □

2.9.2.1 Likelihood ratio test When one model is a special case of another we can use the likelihood ratio test (introduced in Section 2.6.4). The null hypothesis is that the simpler model is appropriate, the alternative is the more complex model. The test statistic is twice the difference of the logarithm of the likelihoods. It is to be compared to a chi-square critical value based on a number of degrees of freedom equal to the difference in the number of parameters. In Example 2.51 it was noted that for a particular data set a Burr distribution was not a significant improvement over the Pareto.

A second possibility is that one model may be a limiting case of another one. For example, in Section 2.7.4 it was shown that the three-parameter

Table 2.18 General liability payments

Payment	Number	Average
0–2,500	41	1,389
2,500–7,500	48	4,661
7,500–12,500	24	9,991
12,500–17,500	18	15,482
17,500–22,500	15	20,232
22,500–32,500	14	26,616
32,500–47,500	16	40,278
47,500–67,500	12	56,414
67,500–87,500	6	74,985
87,500–125,000	11	106,851
125,000–225,000	5	184,735
225,000–300,000	4	264,025
at 300,000	3	300,000

transformed gamma distribution is a limiting case of the four-parameter transformed beta distribution. Another example appears in Chapter 3, where it is observed that the Poisson distribution is a limiting case of the negative binomial distribution. In these cases the likelihood ratio test may still be used, but the test statistic need not have the approximate chi-square distribution. Self and Liang [110] show that the correct approximate distribution is a mixture of chi-square distributions. In this text we will use the chi-square distribution with degrees of freedom equal to the difference in the number of parameters, even for limiting cases. We recognize that this is a further approximation, but does simplify the decision process.

When the two models are unrelated, but have a different number of parameters (for example, exponential versus Pareto), the likelihood ratio test can still be applied, but we must understand that it is no longer a hypothesis test in the formal sense, but just a reasonable decision rule. One way to see that the test is no longer precise is that the test statistic could be negative (that is, the unrelated model with more parameters produces a smaller value of the likelihood function at its maximum). Therefore the test statistic could not have a chi-square distribution.

Finally, we note that similar methods do not exist for measuring the effect of an extra parameter when using other decision variables such as minimum cdf.

Example 2.66 *Use the information in Table 2.19 to determine the best model using the likelihood value as the criteria and a 5% significance level for all tests.*

Table 2.19 Maximum likelihood fits to liability data

Model	No. of parameters	NLL	χ^2	*df*	*p*-value
Exponential	1	548.72	81.02	8	<0.0001
Inverse exponential	1	520.27	49.06	8	<0.0001
Lognormal	2	498.29	4.51	9	0.8744
Inverse Gaussian	2	502.26	12.95	9	0.1648
Pareto	2	499.31	6.37	9	0.7028
Inverse Pareto	2	500.09	7.52	9	0.5831
Loglogistic	2	499.93	7.51	9	0.5847
Gamma	2	507.84	16.38	8	0.0372
Inverse gamma	2	509.80	26.77	8	0.0008
Weibull	2	501.63	8.16	8	0.4183
Inverse Weibull	2	506.72	20.27	9	0.0163
Paralogistic	2	499.79	7.30	9	0.6055
Inverse Paralogistic	2	500.01	7.58	9	0.5767
Burr	3	498.41	4.79	8	0.7793
Inverse Burr	3	499.01	5.33	8	0.7220
Generalized Pareto	3	498.62	5.00	8	0.7580

Within a fixed number of parameters, the choice is the model with the smallest negative logarithm of the likelihood function. For one parameter the winner is the inverse exponential at 520.27, for two parameters it is the lognormal at 498.29, and for three parameters it is the Burr at 498.41. Moving in a forward direction we first test inverse exponential versus lognormal. This is not a formal test as the inverse exponential is not a special case. The test statistic is $2(520.27 - 498.29) = 43.96$. There is one degree of freedom and so the critical value is 3.84. The null hypothesis is rejected and so the lognormal distribution is selected. Once again, note that this is not a formal test and so the 5% significance is not the Type I error probability for the procedure just used. We then test lognormal versus Burr. The test statistic is $2(498.29 - 498.41) = -0.24$, which clearly favors the null hypothesis (being negative) and so the lognormal is again our choice. □

2.9.2.2 *Penalized likelihood values* There is an alternative to a formal hypothesis test that eliminates the distinction between special cases and non-special cases. There are a number of such methods that have been suggested (indicating that this problem has yet to be resolved to everyone's satisfaction). We present one here, called the Schwartz Bayesian Criterion (SBC), introduced by Schwartz [111].[18] Such methods take the likelihood value and then adjust it in some way to reflect the sample size and the number of parameters. The SBC takes the logarithm of the likelihood function[19] and subtracts $r\log(n/2\pi)$, where r is the number of estimated parameters and n is the sample size. Note that as the sample size increases the penalty for an extra parameter increases. This is a difference from the likelihood ratio test. It is not clear which is better. On the one hand, with a large sample we should be able to successfully estimate a few more parameters and thus can justify a more complex model. On the other hand, hypothesis tests become more powerful as the sample size increases and alternative hypotheses tend to be selected even if they are only slightly more accurate descriptions of the population.

Example 2.67 *Using the information in Table 2.19, select the best model according to the SBC.*

For models with the same number of parameters the winner is still the one with the smallest negative loglikelihood. Applying the penalty produces the following three scores (the penalty is $r\log(217/2\pi) = 3.542r$):

Model	−Loglikelihood	Penalty	Score
Inverse exponential	520.27	3.54	523.81
Lognormal	498.29	7.08	505.37
Burr	498.41	10.63	509.04

The penalty is added here because we are dealing with the negative loglikelihood. Once again the lognormal distribution is our choice. We recall from earlier that the visual and tabular checks on this model were also acceptable, so it would be our choice for this population. □

2.9.3 Sensitivity

We recognize that model-building is partly science, partly art, and partly luck. The luck comes in being fortunate enough to have obtained a sample that really does look like the population. To guard against the possibility that we were unlucky, it may be wise to consider alternative models. It may well be that the model that was only second or third best at fitting our data is

[18] There are other information based decision rules. Section 3 of the paper by Brockett [17] promotes the Akaike Information Criterion. In a discussion to that paper, Carlin provides support for the SBC.

[19] In this context the SBC is being applied in a classical estimation setting. When doing a Bayesian analysis the logarithm of the product of the prior density and the model density (evaluated at the point estimates of the parameters) is used instead of the likelihood.

in reality the one that best describes the population. This may be especially true when we have limited data from a heavy-tailed population. Our fit will depend strongly on the lower values, but we are concerned with our ability to forecast the occasional large loss. Models with different tails may appear to perform equally well at matching our data. The only reasonable solution is to continue our investigation using both models. This assumes, as must be the case, that the purpose of the model fitting is not to report the name of the model, but to use the model to answer the real question (for example, the expected cost of a particular reinsurance arrangement). If we carry out the analyses with each model, we will learn either that it doesn't make much difference or that it does. In the former case we have little to worry about. In the latter case we have to admit that our data set was not up to the task of identifying the model with enough accuracy to know which answer is right. We must then report some uncertainty about the answer and use the two results to bound the possibilities.

Similarly, we know that our parameters have been estimated. Additional sensitivity measures can be introduced by constructing confidence intervals for true parameter values and conducting our analysis using the endpoints of the intervals in order to see how much estimation error there is in the analysis.

Example 2.68 *Continuing with the data set in Table 2.18, use the model to estimate the expected cost for the layer from 100,000 to 250,000. Conduct a sensitivity analysis by both varying the lognormal parameters and using the second best two-parameter distribution, the Pareto.*

The results appear in Table 2.20. The cost of the layer is computed from $E(X\wedge 250{,}000)-E(X\wedge 100{,}000)$. The extreme cases for the lognormal distribution were obtained by using the variances from the inverse of the information matrix as obtained from the scoring method. The two 95% confidence intervals were obtained as

$$\begin{aligned}\mu &: \quad 9.293762 \pm 2\sqrt{0.01303204}\\ \sigma &: \quad 1.627131 \pm 2\sqrt{0.008659418}.\end{aligned}$$

It should be clear from the table that for this calculation the Pareto alternative makes very little difference. However, the results are very sensitive to the lognormal parameters. Our sample size of 217 was not sufficient to narrow the possibilities for the parameter values. □

The confidence intervals for the lognormal parameters make a great deal of sense. Because the estimators are nearly independent, the probability that both enclose the true value is simply the product of the individual probabilities, in this case $(0.95)(0.95) = 0.9025$. The Pareto distribution is an entirely different matter. From the estimated covariance matrix we observe a correlation coefficient of 0.905 between the estimators of α and θ. One method

Table 2.20 Sensitivity analysis

Model	Parameters	Expected cost
Lognormal	$\mu = 9.293762,\ \sigma = 1.627131$	7,172
Lognormal	$\mu = 9.065446,\ \sigma = 1.441019$	3,170
Lognormal	$\mu = 9.522078,\ \sigma = 1.813243$	12,695
Pareto	$\alpha = 1.224477,\ \theta = 14,810.47$	7,127

of constructing a confidence region for correlated estimators is to include in the region all parameter values which would not be rejected by the likelihood ratio test. That is, consider a null hypothesis that specifies the two (in the Pareto case) parameter values versus an alternative of arbitrary values. The null hypothesis will be accepted at the 5% level if twice the difference of the loglikelihood evaluated at the null hypothesis value and evaluated at the mle is less than 5.99. In other words, any pair of parameters for which the loglikelihood is within 3 of the maximum would be in the region.

Example 2.69 *Construct a 95% confidence region for the Pareto parameters using the likelihood ratio test.*

The contours of the loglikelihood function appear in Figure 2.15. The second innermost contour is at -502.3, which is three less than the value at the maximum. The correlation in the estimator is evident in the shape of the contours. Assuming independence, reasonable extreme values for 97.5% (for 95% simultaneous confidence under independence) confidence are, for α, $1.224477 - 2.241(0.198198) = 0.78032$ and for θ, $14{,}810.47 + 2.241(3{,}863.75) = 23{,}469$, which, taken simultaneously, are well outside the confidence region. This is not surprising. The mean of a Pareto distribution is $\theta/(\alpha - 1)$ and so a larger estimated value of α would be expected to be accompanied by a larger, not smaller, value of θ in order to keep the model mean near the mean of the data.

2.10 MODELS FOR INCOMPLETE DATA

Although the title of this section might appear innocuous, it represents the culmination of our loss distributions discussion. The next section of this chapter includes a few additional points and topics. The issues that remain are (1) How do incomplete data alter the estimation process? and (2) How do we answer questions with regard to situations that are like those that generate incomplete data?

To make this more precise, incomplete data means that specific observations either are lost or are not recorded exactly. There are two main ways in which data can be incomplete.

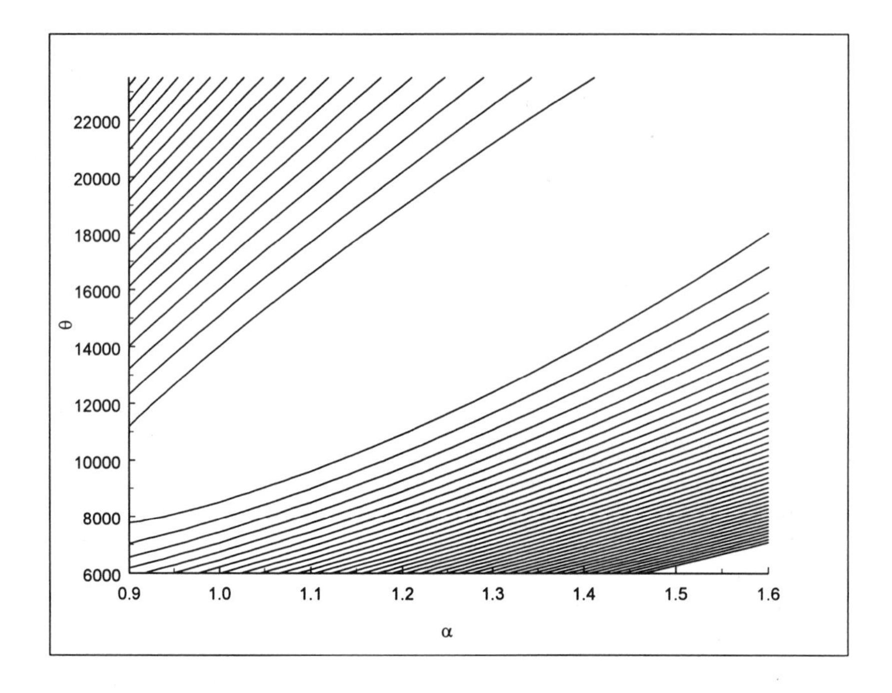

Fig. 2.15 Contours of the loglikelihood function

Definition 2.49 *Data are said to be* ***truncated*** *when observations that fall in a given set are excluded. Data are said to be* ***truncated from below*** *when the set is all numbers less than a specific value. Data are said to be* ***truncated from above*** *when the set is all numbers greater than a specific value.*

Definition 2.50 *Data are said to be* ***censored*** *when the number of observations that fall in a given set are known, but the specific values of the observations are unknown. Data are said to be* ***censored from below*** *when the set is all numbers less than a specific value. Data are said to be* ***censored from above*** *when the set is all numbers greater than a specific value.*

A deductible is an example of truncation from below, while a limit is an example of censoring from above. From the above definitions it would appear that grouping of data to form a frequency table is censoring. While that may be strictly true, we reserve this term for cases where the nature of the process (as opposed to how we chose to record the outcomes) makes determination of the specific value impossible.

Any analysis must start with the model. That means one or more random variables and the process that it represents. The next step is to use the data, however it was collected, to estimate the parameters of the model. The final step is to answer the questions that prompted the model building exercise.

Throughout this section we use two specific random variables and reserve certain capital letters for them. The first (X) is used for building a model. The second random variable (Y) is used for answering the question; it relates directly to the quantity of interest, the amount to be paid. Note that we do not require a random variable to represent the data as collected.

Definition 2.51 *The* ***ground-up*** *loss random variable is what would be paid if there were no modifications. It will be denoted by X. In the language of Chapter 1, it represents the loss.*

Definition 2.52 *The* ***amount paid per payment*** *random variable is what is actually paid on the policy. This random variable is undefined when due to coverage modifications there is a loss, but no benefits are paid. It will be denoted by Y. In Chapter 1 this was simply called the amount paid.*

Definition 2.53 *The* ***amount paid per loss*** *random variable is the actual amount paid, but on a per loss basis. When coverage modifications lead to no payment, this random variable is defined and has a value of zero. It will be denoted by Y^*.*

Because the process is always to go from data to model to answer, we need to have the ability to express the random variables Y and Y^* in terms of X. With the great variety of policies available, no general formula can be written. The following example provides an illustration which will be carried through this section.

Example 2.70 *A policy has an ordinary deductible of 100 and a policy limit of 1,000. Observations were made for 10 randomly selected payments. They are*

$$307 \quad 376 \quad 900 \quad 900 \quad 346 \quad 900 \quad 900 \quad 900 \quad 567 \quad 17.$$

Determine the expected cost per payment, given that a payment is made, if the coverage is changed to a deductible of 200 and a limit of 2,000. Calculations should be done using a Weibull distribution as the model.

The observations are of Y, the amount paid per payment (we have no information about those losses that were below 100). Begin by identifying the relationship between Y and the ground-up loss, X:

$$Y = \begin{cases} \text{undefined}, & X \leq 100 \\ X - 100, & 100 < X \leq 1{,}000 \\ 900, & X > 1{,}000. \end{cases}$$

The distribution function for Y is

$$F_Y(y) = \begin{cases} 0, & y = 0 \\ \dfrac{F_X(y+100) - F_X(100)}{1 - F_X(100)}, & 0 < y < 900 \\ 1, & y \geq 900. \end{cases}$$

and the density function is

$$f_Y(y) = \begin{cases} \dfrac{f_X(y+100)}{1 - F_X(100)}, & 0 \leq y < 900 \\ \dfrac{1 - F_X(1,000)}{1 - F_X(100)}, & y = 900 \\ 0, & y > 900. \end{cases}$$

The third line in the density expression is actually $\Pr(Y = 900)$. The random variable Y here is partially continuous and is discrete at 900. This example will be continued later in this section. □

2.10.1 Estimation with incomplete data

All estimation methods become somewhat ad hoc in the presence of incomplete data. That is because the estimation process must account for the specific nature of the modifications. In the case of maximum likelihood estimation, the likelihood function is unique, but may be difficult to write.

2.10.1.1 Maximum likelihood estimation As always, the key is to write the likelihood function. What must be done is to go through the observations item by item and record the probability (or density function if continuous) of making that observation. These probabilities must be calculated using the model (X) selected.

Example 2.71 (Example 2.70 continued) *Estimate the Weibull parameters by maximum likelihood.*

For the ground-up model, the contribution to the likelihood function for the five observations that were at the limit is

$$f_Y(900) = \frac{1 - F_X(1{,}000)}{1 - F_X(100)} = \frac{\exp\left[-(1{,}000/\theta)^\tau\right]}{\exp\left[-(100/\theta)^\tau\right]}$$

while for the five observations that were above the deductible, but below the limit, the contribution for an observation of x is

$$f_Y(x) = \frac{f_X(x+100)}{1-F_X(100)} = \frac{\dfrac{\tau(x+100)^{\tau-1}}{\theta^\tau}\exp\left[-\left(\dfrac{x+100}{\theta}\right)^\tau\right]}{\exp\left[-(100/\theta)^\tau\right]}.$$

In the numerator, 100 is added to the amount paid to restore the ground-up loss. The denominators appear in these expressions to reflect the fact that the observations were conditional on the ground-up loss exceeding the deductible of 100. The simplex method yields the parameter estimates $\hat{\tau} = 0.700744$ and $\hat{\theta} = 1{,}199.09$. □

Now suppose the observations in the example were from a policy with a deductible of 100, a coinsurance of 75%, and a limit of 1,300. (This keeps the maximum payment at $0.75(1{,}300-100) = 900$). For estimating the parameters of a ground-up model, the amounts paid would have to be restored by dividing by 0.75 and then adding 100.

2.10.1.2 Other estimation methods The other methods must be adapted as needed to the data format. Consider the following example.

Example 2.72 (Example 2.70 continued) *Estimate the Weibull parameters by percentile matching, using the 25th percentile and the proportion of observations at the limit.*

The 25th percentile is the 2.75th order statistic, that is,

$$0.25(307) + 0.75(346) = 336.25.$$

The two equations for the ground-up model are

$$\begin{aligned} 0.25 = F_Y(336.25) &= \frac{1-\exp[-(436.25/\theta)^\tau] - 1 + \exp[-(100/\theta)^\tau]}{1 - 1 + \exp[-(100/\theta)^\tau]} \\ &= 1 - \frac{\exp[-(436.25/\theta)^\tau]}{\exp[-(100/\theta)^\tau]} \end{aligned}$$

and

$$0.5 = \Pr(Y = 900) = \frac{1 - 1 + \exp[-(1{,}000/\theta)^\tau]}{1 - 1 + \exp[-(100/\theta)^\tau]}.$$

The solution is $\hat{\tau} = 0.83349$ and $\hat{\theta} = 1{,}283.2$. □

2.10.2 Shifted models

When data is collected from a policy with a deductible, losses below d are not observed. When the goal is to predict payments from a policy with a deductible, probabilities for ground-up losses below d are not interesting. Up to now our strategy has been to use whatever data is available to build a ground-up model. That means we are extrapolating from d to zero, a dangerous endeavor. An alternative is to shift the origin. While formulas similar to those in Example 2.70 may be created, it is simpler to invoke the shift of the origin and then act as if we had a ground-up model.

Example 2.73 (Example 2.70 continued) *Estimate the Weibull parameters by maximum likelihood and percentile matching for a model shifted at 100.*

The data as presented have already been shifted at 100. That is, 100 was subtracted from each *loss*. Therefore the contribution to the likelihood function for each value below 900 is simply $f_X(x)$, while for each value at 900 it is $1 - F_X(900)$. The likelihood function is maximized at $\hat{\tau} = 0.822013$ and $\hat{\theta} = 1{,}373.88$.

For percentile matching the equations are

$$0.25 = F_X(336.25) = 1 - \exp[-(336.25/\theta)^\tau]$$

and

$$0.5 = 1 - F_X(900) = \exp[-(900/\theta)^\tau].$$

The solution is $\hat{\tau} = 0.893195$ and $\hat{\theta} = 1{,}356.60$. □

If the goal had been to create a model that is shifted at 75, the first step would have been to add 25 to each observation (this represents restoring the 100 deductible and then subtracting the 75). With regard to the shift point, the effective deductible is 25 and the limit is 925. These numbers would replace the 100 and 1,000 in the earlier examples. The maximum payment is still 900.

It may appear at first that the shifted approach is superior because no extrapolation is done. However, in the original case, provided that we never perform calculations concerning a deductible less than 100, the extrapolated probabilities are never used. On the other hand, if we want to induce inflation or estimate the effect of reducing the deductible, we are forced to extrapolate. In that case the shifted model cannot be used.

2.10.3 Truncation from above

While censoring from below is rarely encountered, there is one case where truncation from above is common. Consider the time from when a policy is issued to when a claim is reported. In some types of insurance (liability in particular) the lag can be lengthy. For example, it may be several years after being exposed to asbestos that an employee develops symptoms of asbestos-caused disease. Now consider policies issued in 1983. As of December 31,

Table 2.21 Medical malpractice report lags

Lag in months	No. of claims	Lag in months	No. of claims
0–6	4	84–90	11
6–12	6	90–96	9
12–18	8	96–102	7
18–24	38	102–108	13
24–30	45	108–114	5
30–36	36	114–120	2
36–42	62	120–126	7
42–48	33	126–132	17
48–54	29	132–138	5
54–60	24	138–144	8
60–66	22	144–150	2
66–72	24	150–156	6
72–78	21	156–162	2
78–84	17	162–168	0

1994 any claim that takes more than twelve years from issue to report will not appear in the records.

Example 2.74 *This example is taken from Accomando and Weissner [2]. This data are also analyzed in [77]. Reporting of medical malpractice claims for policies issued in 1975 were recorded with times to report grouped in intervals of six months. At the time the data were collected, fourteen years had elapsed. The data appear in Table 2.21. Estimate the parameters of a Weibull distribution by maximum likelihood. Then determine a 95% confidence interval for the number of unreported claims.*

For the group covering times from c to d the contribution to the likelihood is

$$\frac{F_X(d) - F_X(c)}{F_X(168)}.$$

The denominator indicates that observations were made conditioned on them being less than 168. Maximization by the scoring method is available for this problem because the above expression is easily differentiated with regard to the parameters of the Weibull distribution. The quantity $P_i(\theta)$ as used in (2.12) is the ratio above, so scoring does become a bit more involved when data are incomplete.

The Weibull parameter estimates are $\hat{\tau} = 1.71268$ and $\hat{\theta} = 67.3002$. Scoring also produces the covariance matrix of the estimators:

$$\begin{bmatrix} 0.00503498 & 0.0143151 \\ 0.0143151 & 4.18651 \end{bmatrix}.$$

According to the Weibull distribution, the probability that a claim is reported as of time 168 is

$$F(168) = 1 - e^{-(168/\theta)^\tau}.$$

If N is the unknown total number of claims, the number observed by time 168 (463) is the result of binomial sampling, and thus on an expected value basis we obtain

$$N[1 - e^{-(168/\theta)^\tau}] = 463.$$

Solving for the unreported number of claims yields the prediction

$$N - 463 = \frac{463}{1 - e^{-(168/\theta)^\tau}} - 463$$

and inserting the parameter estimates yields the value 3.88. The confidence interval is obtained by using Theorem 2.3. The required first derivatives (evaluated at the estimates) are

$$\frac{\partial \dfrac{463}{1 - e^{-(168/\theta)^\tau}} - 463}{\partial\tau} = -\frac{463e^{-(168/\theta)^\tau}(168/\theta)^\tau \log(168/\theta)}{\left[1 - e^{-(168/\theta)^\tau}\right]^2} = -17.1331$$

$$\frac{\partial \dfrac{463}{1 - e^{-(168/\theta)^\tau}} - 463}{\partial\theta} = \frac{463\tau e^{-(168/\theta)^\tau}(168/\theta)^\tau}{\theta\left[1 - e^{-(168/\theta)^\tau}\right]^2} = 0.476616$$

and the variance of the estimated number of unreported claims is

$$\begin{bmatrix} -17.1331 & 0.476616 \end{bmatrix} \begin{bmatrix} 0.00503498 & 0.0143151 \\ 0.0143151 & 4.18651 \end{bmatrix} \begin{bmatrix} -17.1331 \\ 0.476616 \end{bmatrix} = 2.19521.$$

Then an approximate 95% confidence interval is

$$3.88 \pm 1.96\sqrt{2.19521} = 3.88 \pm 2.90.$$

It should be noted that this interval is for the expected number of unreported claims, not the actual number. The actual number will vary from 3.88 due to both estimation error (reflected in this interval) and process error (reflected in the uncertainty of the report process). □

2.10.4 Other deductibles

The inclusion of a deductible in a policy is usually done to serve one of two purposes. The first is to lower administrative costs by removing small claims from the insurance process. The second is to provide some inducement to the insured to avoid incidents that lead to claims. If only the first objective is desired, it may be more reasonable to provide full compensation for large claims.

Definition 2.54 *A* ***franchise deductible*** *is a provision under which the insurance pays nothing when the loss is below the deductible amount, but pays everything prior to imposing limits and deductibles when the loss exceeds the deductible amount.*

This presents no additional problems with regard to estimation. For a ground-up model, the deductible does not have to be restored. In Example 2.70 the 100 would already be added to the data points. However, if a shifted model were to be fitted, the 100 would have to be subtracted from the amount paid. With regard to imposing this requirement on the random variable X, the relationships are (assuming a franchise deductible of 100 and a policy limit of 1,000)

$$Y = \begin{cases} \text{undefined}, & X \leq 100 \\ X, & 100 < X \leq 1{,}000 \\ 1{,}000, & X > 1{,}000 \end{cases}$$

$$F_Y(y) = \begin{cases} 0, & y < 100 \\ \dfrac{F_X(y) - F_X(100)}{1 - F_X(100)}, & 100 \leq y < 1{,}000 \\ 1, & y \geq 1{,}000 \end{cases}$$

$$f_Y(y) = \begin{cases} 0, & y < 100 \\ \dfrac{f_X(y)}{1 - F_X(100)}, & 100 \leq y < 1{,}000 \\ \dfrac{1 - F_X(1{,}000)}{1 - F_X(100)}, & y = 1{,}000 \\ 0, & y > 1{,}000. \end{cases}$$

There is another type of deductible that is a compromise between the ordinary and the franchise deductible. The larger the claim, the less of the deductible becomes the responsibility of the policyholder.

Definition 2.55 *A* ***disappearing deductible*** *provides no benefits when the loss is below d and pays the entire loss when the loss exceeds d′. In between, the payment is a linear function of the loss.*

The following equations make this explicit with regard to the ground-up loss. A policy limit of u has also been included.

$$Y = \begin{cases} \text{undefined}, & X \le d \\ d' \dfrac{X-d}{d'-d}, & d < X \le d' \\ X, & d' < X \le u \\ u, & X > u \end{cases}$$

$$F_Y(y) = \begin{cases} \dfrac{F_X[y(d'-d)/d' + d] - F_X(d)}{1 - F_X(d)}, & 0 \le y < d' \\ \dfrac{F_X(y) - F_X(d)}{1 - F_X(d)}, & d' \le y < u \\ 1, & y \ge u \end{cases}$$

$$f_Y(y) = \begin{cases} \dfrac{(d'-d) f_X[y(d'-d)/d' + d]/d'}{1 - F_X(d)}, & 0 \le y < d' \\ \dfrac{f_X(y)}{1 - F_X(d)}, & d' \le y < u \\ \dfrac{1 - F_X(u)}{1 - F_X(d)}, & y = u \\ 0, & y > u. \end{cases}$$

2.10.5 Calculations with modifications

Once a model has been fitted to the data, the data are no longer needed. What we are saying, in effect, is that we believe the model is a better representation of the truth than the data. In order to do calculations, the essential element is to be aware of the process that has been modeled (usually the ground-up loss or a shifted loss). Then relationships between Y and Y^* and X such as those developed earlier can be exploited.

Example 2.75 (Example 2.71 continued) *Using the Weibull parameters as estimated by maximum likelihood, determine the expected payment per payment and per loss for an ordinary deductible of 200 and a limit of 2,000. Do this for both the ground-up model ($\tau = 0.700744$ and $\theta = 1{,}199.09$) and the model that is shifted at 100 ($\tau = 0.822013$ and $\theta = 1{,}373.88$).*

When a ground-up model is used we can use Theorem 2.5 and its corollary to determine the required expected values. For the expected payment per loss we have

$$\begin{aligned} E(Y^*) &= E[X \wedge 2{,}000] - E[X \wedge 200] \\ &= 928.53 - 169.63 = 758.90. \end{aligned}$$

For the expected payment per payment we must divide by the probability of making a payment, $1 - F_X(200) = 0.75196$, to obtain $E(Y) = 1{,}009.23$.

From the shifted model it is not possible to obtain the expected payment per loss as the model provides no information about the probability of losses being below 100. For the expected payment per payment, with regard to the shifted model the deductible is 100 and the limit 1900. Then

$$\begin{aligned} E(Y) &= \frac{E(X \wedge 1{,}900) - E(X \wedge 100)}{1 - F_X(100)} \\ &= \frac{991.52 - 93.88}{1 - 0.109557} = 1{,}008.08. \end{aligned}$$

It is not surprising that the two numbers are different. We have used two different models to describe the population, at least one of which may provide a poor description. Recall that at no time in this example did we check either of these models with regard to the quality of fit. □

If the difference in the two modeling processes is still unclear, the following example may provide some useful pictorial insights.

Example 2.76 *Dental payments with a deductible of 100 were collected. Fit lognormal models for both ground-up losses and losses shifted at 100.*

The data appear as histograms in Figures 2.16 and 2.17[20]. In fitting the ground-up model the data appear as loss amounts and thus the histogram begins at 100. To match the model to the histogram, the lognormal density function is chopped off below 100 and the remainder proportionally increased so the area below it is one. Because with regard to shifting the data is essentially ground-up, no adjustments need to be made in the second figure.

It should be apparent from the two pictures that the ground-up lognormal model provides an excellent fit while the shifted lognormal does not. The reason is that in the real population of dental losses there is a mode between 0 and 100. The ground-up model can recognize this, while the shifted model cannot. It would appear that if we want a shifted model, we should choose from those distributions that allow a mode at zero. □

[20]These figures were created using the software available with the text. See the Preface for more information.

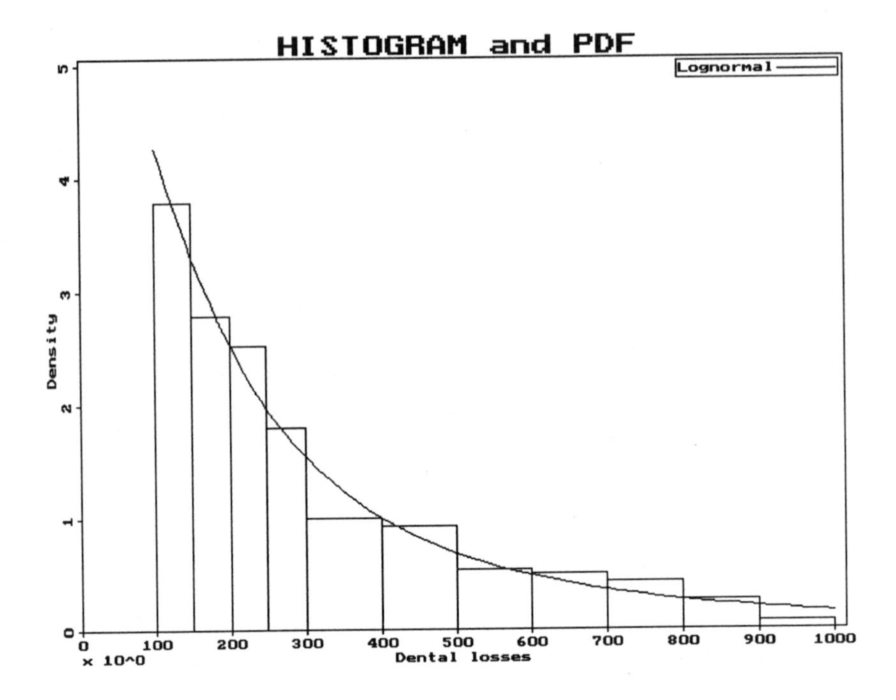

Fig. 2.16 Ground-up model for dental losses

There is essentially no limit to what we can compute with regard to modifications. Consider the following example.

Example 2.77 *Using the ground-up Weibull model determined in Example 2.71 ($\tau = 0.700744$ and $\theta = 1{,}199.09$), determine the expected payment per loss and per payment if uniform inflation of 10%, coinsurance of 80%, a disappearing deductible with $d = 110$ and $d' = 275$, and a limit of 1,980 are imposed.*

With regard to the expected payment per loss, the random variable is

$$Y^* = \begin{cases} 0, & X < 100 \\ 0.8\frac{275}{165}(1.1X - 110), & 100 \leq X < 250 \\ 0.8(1.1)X, & 250 \leq X < 1{,}800 \\ 0.8(1{,}980), & X \geq 1{,}800. \end{cases}$$

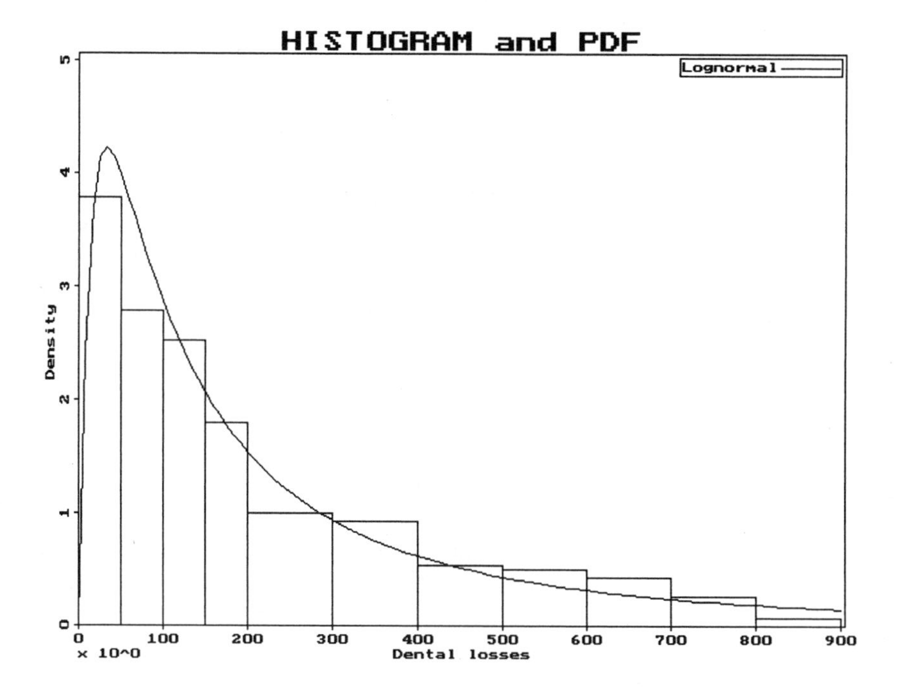

Fig. 2.17 Shifted model for dental losses

The expected value is

$$\begin{aligned}E(Y^*) &= \int_{100}^{250} \left(\tfrac{22}{15}x - \tfrac{2{,}200}{15}\right) f(x)dx + \tfrac{22}{25}\int_{250}^{1{,}800} xf(x)dx \\ &\quad +1{,}584[1 - F(1{,}800)].\end{aligned}$$

The two integrals could be evaluated directly using the inverse Pareto density function. An alternative is to add and subtract the appropriate quantities in order to identify various limited expected values. It turns out that

$$\begin{aligned}E(Y^*) &= \tfrac{22}{25}E(X \wedge 1800) + \tfrac{44}{75}E(X \wedge 250) - \tfrac{22}{15}E(X \wedge 100) \\ &= \tfrac{22}{25}878.21 + \tfrac{44}{75}206.32 - \tfrac{22}{15}90.30 = 761.43. \qquad (2.32)\end{aligned}$$

For the expected payment per payment we need to divide by the probability that a payment is made. This is when inflated losses exceed 110, or when uninflated losses exceed 100. That is, we need to divide by $1 - F(100) = 0.839131$, for an expected value of 907.40. □

2.11 OTHER MODELS

This section introduces three additional models for loss data. The first model is useful when the number of observations is large. In that setting the empirical distribution is likely to be pretty good and it is possible to smooth it somewhat without going as far as using a parametric family. The second is similar to the two-point mixtures introduced earlier. This time the two models are split, each being applied to a specific range of losses. The final set of models is useful when there are two correlated loss variables.

2.11.1 Nonparametric smoothing

The many advantages of parametric modeling have been discussed earlier. One situation in which these advantages diminish is when there are a large number of observations. In that situation the empirical distribution function is likely to be fairly smooth and just a touch of smoothing will complete the task. The smoothing is needed in order to make the density function look nice (say as compared to a histogram derived from an ogive). One way of doing this is through kernel density estimation. There are two common types of kernel density estimators.

Definition 2.56 *A **location-based kernel density estimator** is*

$$f_K(x;\theta) = \frac{1}{n}\sum_{j=1}^{n} K[(x - x_j)/\theta]$$

where $K(\cdot)$ can be any legitimate probability density function.

The quantity $\theta > 0$ is not a parameter to be estimated, but rather controls the amount of smoothing. Large values will lead to more smoothing. The function itself is often symmetric. An example would be $K(x) = 1,\ -1/2 < x < 1/2$, and zero otherwise. The effect of this process is to replace each observation with a density function and then create a mixture of all n of them. For an example of an actuarial application to graduation see the papers by Gavin, Haberman, and Verrall ([45] and [46]). They considered the kernel functions $K(x) = \exp(-x^2/2)/\sqrt{2\pi}$ and $K(x) = \exp(-|x|)/2$. A major drawback for loss distributions is that probability will be extended into the negative numbers. The second method avoids this problem. A version of this estimator was suggested by Carriere [24].

Definition 2.57 *A **scale-based kernel density estimator** is*

$$f_K(x;\theta) = \frac{1}{n}\sum_{j=1}^{n} K(x; x_j/\mu)$$

where $K(x;\theta)$ is a probability density function from a family with a scale parameter θ and $\mu = E(X)/\theta$ where X is a random variable with density function $K(x;\theta)$.

Example 2.78 *Determine the formula for a scale based kernel density estimator that uses a Pareto distribution as the kernel.*

We have $K(x;\theta) = \alpha\theta^\alpha(x+\theta)^{-\alpha-1}$ and $\mu = E(X)/\theta = [\theta/(\alpha-1)]/\theta = 1/(\alpha-1)$. Then

$$K(x;x_j/\mu) = K[x;(\alpha-1)x_j] = \alpha[(\alpha-1)x_j]^\alpha[x+(\alpha-1)x_j]^{-\alpha-1}.$$

The parameter α controls the amount of smoothing. A high value of α indicates a Pareto distribution with a light tail, and therefore less smoothing will be done. □

The example in [24] used the gamma distribution with $\mu = \alpha$. There is more smoothing as α decreases to zero. The following theorem shows that this model has the same mean as the empirical distribution.

Theorem 2.20 *Let X_K denote the random variable with density function $f_K(x;\theta)$. Then $E(X_K) = \bar{x}$.*

Proof:

$$E(X_K) = \frac{1}{n}\sum_{j=1}^{n}\mu(x_j/\mu) = \bar{x}.$$ □

The next theorem shows that the variance will not be preserved.

Theorem 2.21 *Let $\delta = E(X^2)/\theta^2$ where X has density function $K(x;\theta)$. Then*

$$Var(X_K) = \hat{\mu}_2 + \left(\frac{\delta}{\mu^2} - 1\right)\hat{\mu}_2'.$$

Proof:

$$E(X_K^2) = \frac{1}{n}\sum_{j=1}^{n}\delta(x_j/\mu)^2 = \delta\hat{\mu}_2'/\mu^2$$

$$Var(X_K) = \delta\hat{\mu}_2'/\mu^2 - \bar{x}^2 = \hat{\mu}_2 + \left(\frac{\delta}{\mu^2} - 1\right)\hat{\mu}_2'.$$ □

Because $\delta > \mu^2$, the model variance will always be larger than the empirical variance. For the gamma distribution we have $\mu = \alpha$ and $\delta = \alpha(\alpha+1)$, and so

$$Var(X_K) = \hat{\mu}_2 + \hat{\mu}_2'/\alpha.$$

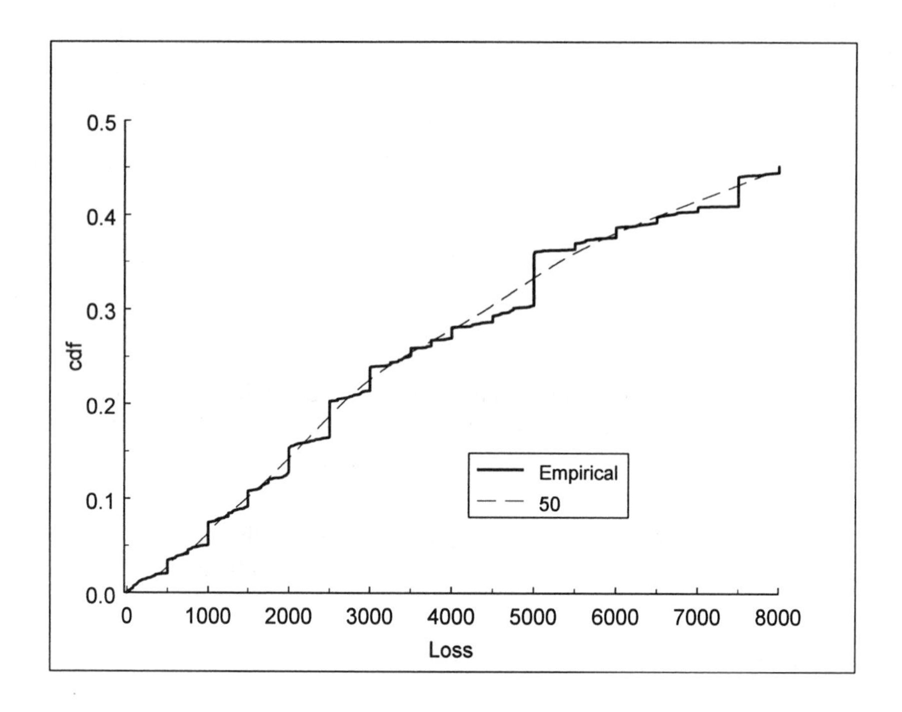

Fig. 2.18 Empirical and kernel ($\alpha = 50$) cdfs for small losses

The extra variance decreases as α goes to infinity. This is consistent with the earlier observations that larger values of α provide less smoothing.

Example 2.79 *The liability losses discussed earlier in this chapter (see Tables 2.10, 2.12, and 2.27) were combined. Then all values from policies with a limit below 100,000 were excluded along with all losses at or above 100,000. Determine a model for these data using the gamma distribution as a kernel.*

While the 1,307 data points are not listed here, the empirical cdf (Figure 2.18) reveals clustering at round numbers. In fact, there were 19 losses at 500, 33 at 1,000, 22 at 1,500, 33 at 2,000, 50 at 2,500, 33 at 3,000, 72 at 5,000, and 41 at 7,500. The figure also shows the model cdf using $\alpha = 50$. In Figure 2.19 the density function using various values of α (17, 50 and 100) is shown. The first value is based on a recommendation from [24] to try $\alpha = \sqrt{n}/(\hat{\mu}'_4/\hat{\mu}'^2_2 - 1)^{1/2}$. It is clear that the larger values of α provide less smoothing and larger concentrations at the round loss values. Similar results obtain for losses between 8,000 and 100,000. □

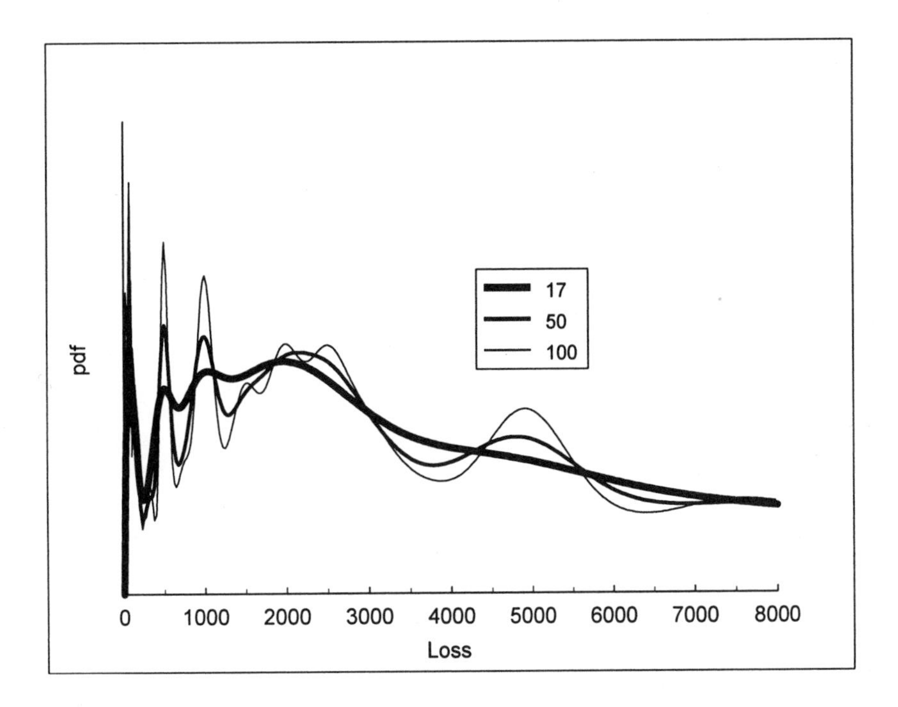

Fig. 2.19 Kernel ($\alpha = 17, 50, 100$) pdfs for small losses

2.11.2 Splicing

This approach is similar to mixing in that it might be believed that two or more separate processes are responsible for generating the losses. With mixing, the various processes operate on subsets of the population. Once the subset is identified, a simple loss model suffices. For splicing, the processes differ with regard to the loss amount. That is, one model governs the behavior of losses in some interval of possible losses while other models cover the other intervals. The following definition will make this precise.

Definition 2.58 *An **n-component spliced model** has a density function that can be expressed as follows.*

$$f_X(x) = \begin{cases} a_1 f_1(x), & b_1 < x < c_1 \\ \vdots & \vdots \\ a_n f_n(x), & b_n < x < c_n. \end{cases}$$

For $j = 1, \ldots, n$ each $f_j(x)$ must be a legitimate density function with all probability on the interval (b_j, c_j). Also, $a_1 + \cdots + a_n = 1$.

It was not necessary to use density functions and coefficients, but this is one way to ensure that the result is a legitimate density function. When using parametric models the motivation for splicing is that the tail behavior may be inconsistent with the behavior for small losses. For example, experience (based on knowledge beyond that available in the current, perhaps small, data set) may indicate that the tail follows the Pareto distribution, but there is a positive mode more in keeping with the lognormal or inverse Gaussian distributions. A second instance is when there is a large amount of data below some value, but a limited amount of information elsewhere. We may want to use the empirical distribution (or a smoothed version of it) up to a certain point and a parametric model beyond that value.

Example 2.80 *Indicate how to estimate a 2-component spliced model where losses up to 10,000 follow a smoothed empirical distribution and losses beyond 10,000 follow the Pareto distribution.*

If a scale-based kernel density estimator is used, all data points theoretically contribute to losses below 10,000. However, it is likely that the really large losses will not make much of a contribution and so can be ignored. One problem is that the area under the smoothed density and between 0 and 10,000 is not necessarily a good estimator of the true probability. The best estimate is probably the proportion of the sample that is below 10,000. The smoothed empirical density would then be proportionately scaled to make the area match this proportion.

For the Pareto portion, all of the data above 10,000 plus perhaps some of the data below could be used. Maximum likelihood estimation of the ground-up Pareto distribution from this truncated sample would produce the parameters. The portion of the Pareto curve from 10,000 to infinity would then have to be multiplied by a constant so that the area matches the proportion of the sample in excess of 10,000. □

The only problem with an approach like that used above is that the transition from one curve to another may not be smooth. One could employ techniques similar to those used in creating splines, or just accept the lack of smoothness.

2.11.3 Bivariate models

At times a bivariate distribution with dependent variables is the appropriate model. One such situation is a joint life annuity or insurance. Here the timing of the payments depends on the first or second death of two individuals. Because these individuals are often related (typically spouses), the times of death will be dependent. As another example, in casualty insurances it is common to record the expenses that are directly related to the payment of the loss (the allocated loss adjustment expenses, ALAE). The loss and the

Table 2.22 Twenty-four losses with ALAE

Loss	ALAE	Loss	ALAE
1,500	301	11,750	2,530
2,000	3,043	12,500	165
2,500	415	14,000	175
2,500	4,940	14,750	28,217
4,500	395	15,000	2,072
5,000	25	17,500	6,328
5,750	34,474	19,833	212
7,000	50	30,000	2,172
7,000	10,593	33,033	7,845
7,500	50	44,887	2,178
9,000	406	62,500	12,251
10,000	1,174	210,000	7,357

ALAE are usually strongly positively correlated. In this subsection, three methods of constructing a bivariate model will be introduced. All of them use the following small data set.

Example 2.81 *The loss and ALAE were recorded for twenty-four claims (Table 2.22). Determine a model for the joint distribution. Then determine the expected total cost under each of the following situations: (1) A policy limit of 50,000 and it is assumed that the ALAE on losses in excess of 50,000 is the same as that for losses of exactly 50,000. (2) A policy limit of 50,000 and it is assumed that the ALAE is related to the loss, not the amount paid. (3) A per loss reinsurance where the reinsurer pays the excess of each loss over 50,000 with a maximum payment of 50,000 per loss. However, any ALAE is paid at the same proportion as is paid on the loss.*

Let X be the loss random variable and let Y be the ALAE random variable. Then the three quantities requested in the example can be obtained as follows.

$$\begin{aligned}(1)\quad & \int_0^\infty \int_0^{50{,}000} (x+y) f(x,y)\,dx\,dy + 50{,}000 \Pr(X > 50{,}000) \\ & + \Pr(X > 50{,}000) \int_0^\infty y f(50{,}000, y)\,dy / f_X(50{,}000)\end{aligned}$$

$$(2)\quad \int_0^\infty \int_0^{50{,}000} (x+y) f(x,y)\,dx\,dy + \int_0^\infty \int_{50{,}000}^\infty (50{,}000 + y) f(x,y)\,dx\,dy$$

$$\begin{aligned}(3)\quad & \int_0^\infty \int_{50{,}000}^{100{,}000} \left(x - 50{,}000 + \frac{x - 50{,}000}{x} y \right) f(x,y)\,dx\,dy \\ & + \int_0^\infty \int_{100{,}000}^\infty \left(50{,}000 + \frac{50{,}000}{x} y \right) f(x,y)\,dx\,dy\end{aligned}$$

□

In most settings these integrals are likely to be very difficult to evaluate. Numerical integration may work in some cases (particularly if the integrals can be reduced to one-dimensional evaluations). Simulation is another possibility, one which is used in the specific examples that follow.

2.11.3.1 *Bivariate densities* One option is to work directly with a bivariate density function. Unfortunately, it is very difficult to find a parametric distribution that is flexible and yet has convenient marginal distributions. For example, while bivariate versions of the gamma and Pareto distributions exist, they tend to have restrictions (such as the gamma marginals having identical shape parameters, α). Three sources of such collections are the books by Hutchinson and Lai ([69]), Johnson and Kotz ([72]), and Mardia ([87]). Here the only distribution we will work with is the bivariate lognormal. It is easy to use, provided the data is the actual amounts and there are no modifications.

The definition of a bivariate lognormal distribution is that the logarithms of the two variables have the bivariate normal distribution (see, for example, Hogg and Craig [65, pp. 117–120]). If there are no modifications, maximum likelihood estimation of the five parameters is easy. The means and variances of $\log X$ and $\log Y$ are estimated by the respective sample means and variances (divide by the sample size to obtain the mle) and the correlation is estimated by the sample correlation. In addition, if the model fits, the logarithms should follow a simple linear regression pattern.

Example 2.82 (Example 2.81 continued) *Estimate the parameters for the bivariate lognormal model and the requested expected values. Also, estimate the expected ALAE for a given loss.*

After taking logarithms, the five estimates are

$$\begin{aligned}\hat{\mu}_X &= 9.293716\\ \hat{\mu}_Y &= 7.117804\\ \hat{\sigma}^2_X &= 1.268426\\ \hat{\sigma}^2_Y &= 4.027530\\ \hat{\rho}_{XY} &= 0.322037.\end{aligned}$$

The scatterplot of the logarithms (Figure 2.20) reveals the linear trend, although the relationship is not strong. The regression line is $\log Y = 1.784672 + 0.573843 \log X$ with a standard error of 1.899959 (also estimated by maximum likelihood). Then, given $X = x$ we obtain

$$\begin{aligned}Y &= \exp(1.784672 + 0.573843 \log x + Z)\\ &= 5.957626x^{0.573843}e^Z\end{aligned}$$

where Z has the normal distribution with mean zero and standard deviation 1.899959. Then

$$E(Y|X = x) = 5.957626x^{.573843}e^{1.899959^2/2} = 36.219372x^{.573843}.$$

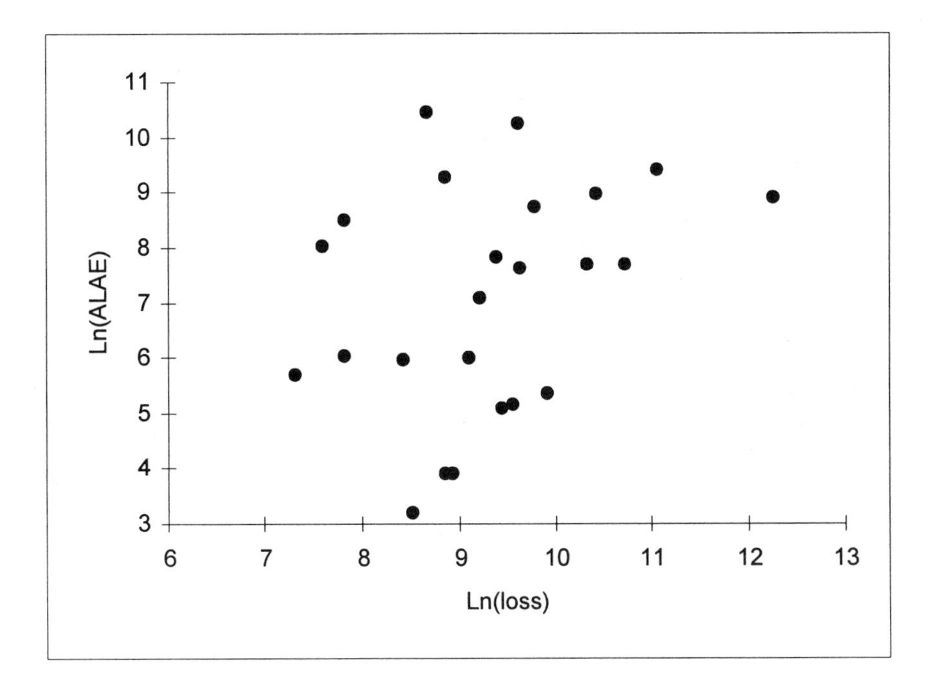

Fig. 2.20 Scatterplot of logarithms of losses and ALAE

With regard to the three expected values, while some of the integrals are relatively easy to evaluate for this distribution, some are not. Instead of trying to solve the problem in pieces, one simulation was done for them all. The process is simple. For each simulation, generate a pair of observations from the bivariate normal distribution.[21] Then exponentiate each number to change them to the lognormal distribution. Next determine the loss and ALAE according to the terms of the coverage and add them. Repeat this process many times and average the results. A 95% confidence interval for accuracy of the simulation can be constructed by using twice the standard deviation of the numbers divided by the square root of the number of simulations. After 5,000 simulations the three confidence intervals were $25{,}913 \pm 5{,}434$; $29{,}541 \pm 5{,}801$; and $3{,}301 \pm 1{,}671$. The large variances in the population should have indicated to us that a large number of simulations would be required to achieve a reasonable level of accuracy. □

2.11.3.2 *Regression models* The notion of regression can be extended to models other than normal/lognormal. The two models postulated are any

[21] A brief introduction to simulation as well as an alogorithm for the bivariate normal distribution is given in Appendix H.

model for X and then a conditional model for Y. Suppose Y has a scale parameter θ and that α represents all other parameters. Then let the conditional distribution for $Y|X=x$ have nonscale parameter(s) α and scale parameter $\theta = g(x)$. We would then have $E(Y|X=x) \propto g(x)$.

Example 2.83 *Let X have the Pareto distribution with parameters α and θ. Let the conditional distribution of Y have the gamma distribution with nonscale parameter δ (this one is usually denoted α) and scale parameter $\gamma + \beta \log x$. Determine the joint density function of X and Y and indicate how values could be simulated. Also indicate how estimation might proceed.*

The marginal density function of X is

$$f_X(x) = \frac{\alpha\theta^\alpha}{(x+\theta)^{\alpha+1}}.$$

The conditional density function of $Y|X=x$ is

$$f_{Y|X}(y|x) = \frac{y^{\delta-1}e^{-y/(\gamma+\beta\log x)}}{\Gamma(\delta)(\gamma+\beta\log x)^\delta}$$

with conditional expected value (regression) $E(Y|X=x) = \delta(\gamma + \beta \log x)$. The joint density is the product

$$f(x,y) = \frac{\alpha\theta^\alpha y^{\delta-1}e^{-y/(\gamma+\beta\log x)}}{\Gamma(\delta)(\gamma+\beta\log x)^\delta(x+\theta)^{\alpha+1}}.$$

Values could be simulated by first simulating a value for X from the Pareto distribution and then using that value to establish the parameters for the gamma distribution. The value of Y would then be simulated from that gamma distribution. With regard to estimation, if there are no modifications, the likelihood function is just the product of the joint density function and then the simplex method could be used to maximize over the five parameters. If that proves difficult (or as a way to get starting values), the values of α and θ can be estimated from the marginal data on X.[22] Then least squares regression could be used to estimate $\delta\gamma$ and $\delta\beta$. Then the likelihood function could be maximized over the one remaining variable, δ. □

Note that in this scheme the two distributions are arbitrary as is the regression function that links the two variables. This provides a great deal of flexibility, yet analysis by simulation will be straightforward. The biggest difficulty occurs when data are grouped or modified. Suppose, for example, that there is a policy limit on the loss with the ALAE related to the loss (item (2)

[22] In Exercise 2.127 you will be asked to prove that the marginal mle and the mle from maximizing over all the parameters at once must be the same.

in Example 2.81). For losses below the limit, the contribution to the likelihood is just the density function $f(x,y)$. For losses above the limit (say u) the contribution for an ALAE of y is $\int_u^\infty f(x,y)dx$. In this regression setting the integral is likely to be difficult to do. It will simplify if the roles of X and Y are reversed. That is, we are given the marginal distribution for Y and the conditional distribution for X. Then

$$\int_u^\infty f(x,y)dx = \int_u^\infty f_{X|Y}(x|y)f_Y(y)dx = f_Y(y)\Pr(X > u|Y = y).$$

Example 2.84 (Example 2.83 continued) *Reverse the roles of X and Y and determine the contribution to the likelihood function when there is a limit of 10,000, the loss exceeds the limit, and the ALAE is 2,000. To keep a similar relationship between the two variables, change the regression function to $g(y) = \exp(\gamma + \beta y)$.*

Y has the Pareto distribution and so

$$f_Y(y) = \frac{\alpha\theta^\alpha}{(2{,}000 + \theta)^{\alpha+1}}.$$

The conditional distribution of X is gamma with parameters δ and $\exp(\gamma + 2{,}000\beta)$. The probability is

$$\Pr(X > 10{,}000|Y = 2{,}000) = 1 - \Gamma[\delta; 10{,}000/\exp(\gamma + 2{,}000\beta)].$$

The contribution to the likelihood function is

$$\frac{\alpha\theta^\alpha\{1 - \Gamma[\delta; 10{,}000/\exp(\gamma + 2{,}000\beta)]\}}{(2{,}000 + \theta)^{\alpha+1}}.$$

□

We close this subsection with an analysis of the recurring example.

Example 2.85 (Example 2.81 continued) *For the data in this example fit a model that provides a Pareto marginal distribution for X and a conditional Pareto distribution for Y where the scale parameter is $g(x) = \exp(\gamma + \beta \log x)$. Determine the expected value of Y given $X = x$.*

The density function, using two Pareto distributions is

$$f(x,y) = \frac{\alpha\theta^\alpha}{(x + \theta)^{\alpha+1}} \frac{\delta e^{\delta(\gamma+\beta\log x)}}{(y + e^{\gamma+\beta\log x})^{\delta+1}}.$$

From the marginal data on losses, the maximum likelihood estimates are $\hat{\alpha} = 2.44612$ and $\hat{\theta} = 32{,}248.8$. Maximization over the remaining three parameters produces $\hat{\delta} = 0.848435$, $\hat{\gamma} = 1.10673$, and $\hat{\beta} = 0.620045$. The expected value does not exist because $\hat{\delta} \leq 1$. Note that if it did, the regression is of exactly the same form as for the bivariate lognormal distribution. □

2.11.3.3 Copulas Another way to create a bivariate distribution from marginal information is via a **copula**. Instead of specifying one marginal and one conditional density function, this approach requires the two marginal distribution functions. They are then combined through a third function, called a copula. This function must itself be a legitimate bivariate distribution function over the unit square with uniform marginals. Denote the two marginal distribution functions $F_X(x)$ and $F_Y(y)$ and the copula by $C(u,v)$. The bivariate distribution function created by the three is then $F_{X,Y}(x,y) = C[F_X(x), F_Y(y)]$. A simple but fairly useless example is the copula $C(u,v) = uv$. This creates the bivariate distribution function $F_{X,Y}(x,y) = F_X(x)F_Y(y)$, which is just the case for independent variables.

A good general introduction is [48]. The paper by Frees, Carriere, and Valdez, [44], works with Frank's copula. In that paper they show how to write the likelihood function under various truncations and censoring. We will be content to continue our example.

Example 2.86 (Example 2.81 continued) *Consider a model with Pareto marginals combined via Frank's copula. Estimate the parameters.*

Frank's copula is

$$C(u,v) = \log_\alpha \left[1 + \frac{(\alpha^u - 1)(\alpha^v - 1)}{\alpha - 1}\right] \tag{2.33}$$

where the parameter α controls the degree of association between the two variables. Values less than one indicate a positive association, values greater than one indicate an inverse association, and one indicates independence. If we let β and θ be the parameters of the marginal Pareto distribution for X (where θ is the scale parameter) and let γ and τ be the parameters for Y (with τ the scale parameter), the bivariate distribution function is

$$F(x,y) = \log_\alpha \left\{1 + \frac{[\alpha^{1-(1+x/\theta)^{-\beta}} - 1][\alpha^{1-(1+y/\tau)^{-\gamma}} - 1]}{\alpha - 1}\right\}.$$

Taking partial derivatives with respect to x and y provides the joint density function

$$f(x,y) = \frac{\begin{array}{c}(\alpha-1)\dfrac{\beta\gamma}{\theta\tau}\alpha^{2-(1+x/\theta)^{-\beta}-(1+y/\tau)^{-\gamma}} \\ \times(1+x/\theta)^{-\beta-1}(1+y/\tau)^{-\gamma-1}\log\alpha\end{array}}{\{\alpha - 1 + [\alpha^{1-(1+x/\theta)^{-\beta}} - 1][\alpha^{1-(1+y/\tau)^{-\gamma}} - 1]\}^2}.$$

Starting values for the four Pareto parameters were obtained by finding the maximum likelihood estimates for the two marginal distributions. Simplex maximization yields the estimates $\hat{\alpha} = 0.133024$, $\hat{\beta} = 2.59889$, $\hat{\theta} = 36{,}141.4$, $\hat{\gamma} = 0.759943$, and $\hat{\tau} = 803.839$. The positive association is apparent and could be tested. One way is to use the likelihood ratio test. With all five parameters the logarithm of the likelihood function is −483.126, while with independence (note that the required answer is the sum of the logarithms of the marginal likelihoods) it is −484.496. Twice the difference is 2.740, which does not exceed the chi-square 5% critical value of 3.841. For the Schwartz Bayesian Criterion the difference (which is 1.370) must exceed $\log(24/2\pi) = 1.340$. It does, and so the two methods lead to opposite conclusions. At least based on this small sample, there is some, but not convincing, evidence of dependence.□

A number of results concerning Frank's copula can be found in the paper by Genest [47]. Two are presented here.

To simulate an (X, Y) pair, begin by simulating the X value from the marginal distribution. This can be done using the standard inversion technique. Follow this by simulating a value of Y from the conditional distribution of Y given $X = x$. To do this first note that the distribution function is

$$F_{Y|x}(y|x) = \frac{\frac{d}{dx}F(x,y)}{f_X(x)}.$$

For Frank's copula we have

$$\begin{aligned}\frac{d}{dx}F(x,y) &= f_X(x)\frac{d}{dx}C(u,v)|_{u=F_X(x),v=F_Y(y)} \\ &= \frac{f_X(x)\alpha^{F_X(x)}[\alpha^{F_Y(y)}-1]}{\alpha-1+[\alpha^{F_X(x)}-1][\alpha^{F_Y(y)}-1]}.\end{aligned}$$

To simulate a conditional value of Y, obtain a uniform(0,1) random number, r, and solve the equation

$$\frac{\alpha^{F_X(x)}[\alpha^{F_Y(y)}-1]}{\alpha-1+[\alpha^{F_X(x)}-1][\alpha^{F_Y(y)}-1]} = r$$

for $F_Y(y)$ to obtain

$$\alpha^{F_Y(y)} = 1 + \frac{r(\alpha-1)}{\alpha^{F_X(x)}(1-r)+r}$$

or

$$F_Y(y) = \frac{1}{\log\alpha}\log\left[1+\frac{r(\alpha-1)}{\alpha^{F_X(x)}(1-r)+r}\right].$$

The right-hand side is a number and then the distribution function of Y can be inverted to solve for the simulated value.

The regression function can be found from

$$\begin{aligned}E(Y|X=x) &= \int 1 - F_{Y|x}(y|x)dy \\ &= \int 1 - \frac{\alpha^{F_X(x)}[\alpha^{F_Y(y)}-1]}{\alpha-1+[\alpha^{F_X(x)}-1][\alpha^{F_Y(y)}-1]}dy\end{aligned}$$

but it is likely that the integral will have to be done numerically.

Table 2.23 Trended hurricane losses

Year	Loss (10^3)	Year	Loss (10^3)	Year	Loss (10^3)
1964	6,766	1964	40,596	1975	192,013
1968	7,123	1949	41,409	1972	198,446
1971	10,562	1959	47,905	1964	227,338
1956	14,474	1950	49,397	1960	329,511
1961	15,351	1954	52,600	1961	361,200
1966	16,983	1973	59,917	1969	421,680
1955	18,383	1980	63,123	1954	513,586
1958	19,030	1964	77,809	1954	545,778
1974	25,304	1955	102,942	1970	750,389
1959	29,112	1967	103,217	1979	863,881
1971	30,146	1957	123,680	1965	1,638,000
1976	33,727	1979	140,136		

2.12 EXERCISES

Section 2.2

2.1 The data in Table 2.23 are from *Loss Distributions* [66, page 128]. It represents the total damage done by 35 hurricanes between the years 1949 and 1980. The losses have been adjusted for inflation (using the Residential Construction Index) to be in 1981 dollars. The entries represent all hurricanes for which the trended loss was in excess of 5,000,000.

The federal government is considering funding a program that would provide 100% payment for all damages for any hurricane causing damage in excess of 5,000,000. You have been asked to make some preliminary estimates.

(a) Estimate the mean, standard deviation, coefficient of variation, and skewness for the population of hurricane losses.

(b) Estimate the 50th, 75th, 90th, and 95th percentiles for the population of hurricane losses.

(c) Determine a nonparametric 90% confidence interval for the 75th percentile.

(d) Estimate the first and second limited moments at 500,000,000.

2.2 The data in Table 2.24 are from *Practical Risk Theory for Actuaries* [27, page 76]. It represents fire losses in the United Kingdom (measured in pounds). There were 16,536 losses in the data set with an average loss of 7,009. As is common in grouped data sets, the sample mean was computed for each class.

Table 2.24 United Kingdom fire losses (in £)

Upper limit	Number	Average	Upper limit	Number	Average
100	4,319	41	12,800	329	10,760
140	795	118	18,100	273	15,060
200	910	167	25,600	214	21,510
280	962	237	36,200	172	30,510
400	1,097	336	51,200	136	42,740
570	1,121	475	72,410	108	60,500
800	1,046	673	102,400	88	85,720
1,130	969	957	250,000	117	155,700
1,600	843	1,350	500,000	47	336,900
2,260	805	1,900	750,000	12	635,380
3,200	694	2,700	1,000,000	4	845,190
4,530	602	3,810	2,000,000	8	1,277,010
6,400	480	5,390	3,000,000	3	2,579,420
9,050	382	7,560			

(a) Estimate the mean, standard deviation, coefficient of variation, and skewness for the population of fire losses.

(b) Estimate the 50th, 75th, 90th, and 95th percentiles for the population of fire losses.

(c) Estimate the first and second limited moments at 350,000.

2.3 Consider the following method of developing a quadratic interpolation formula for the limited expected value from grouped data: Rather than assume the density function is a horizontal line throughout the interval in question, assume it is a straight line. That is, $f(x) = r + sx$, $c_{j-1} \leq x \leq c_j$. Determine the coefficients r and s so that the probability in that range matches the empirical probability (n_j/n) and that the average observation in that range matches the empirical average (a_j). Use that function along with the integral for the continuous case (2.3) to complete the approximation. Use this technique to estimate the LEV at 350,000 for the fire loss example.

2.4 The cdf of a random variable is $F(x) = 1 - x^{-2}$, $x \geq 1$. Determine the mean, median, and mode of this random variable. (89-24)

2.5 There have been 30 claims recorded in a random sampling of claims. There were 2 claims for 2,000, 6 for 4,000, 12 for 6,000, and 10 for 8,000. Determine the empirical skewness coefficient. (89-27)

2.6 The severities of individual claims have the Pareto[23] distribution with parameters $\alpha = 8/3$, and $\theta = 8{,}000$. Use the Central Limit Theorem to approximate the probability that the sum of 100 independent claims will exceed 600,000. (89-44)

2.7 The severity distribution of individual claims has pdf $f(x) = 2.5x^{-3.5}$, $x \geq 1$. Determine the coefficient of variation. (M92-15)

2.8 A random sample of 20 observations has been ordered as follows:

12	16	20	23	26	28	30	32	33	35
36	38	39	40	41	43	45	47	50	57

Determine the 60th sample percentile using the smoothed empirical estimate. (N92-11)

2.9 The severities of individual claims have the gamma distribution (see Appendix A) with parameters $\alpha = 5$ and $\theta = 1{,}000$. Use the Central Limit Theorem to approximate the probability that the sum of 100 independent claims exceeds 525,000. (N92-31)

2.10 The following twenty wind losses (in millions of dollars) were recorded in one year.

1	1	1	1	1	2	2	3	3	4
6	6	8	10	13	14	15	18	22	25

(a) Determine the sample 75th percentile using the smoothed empirical estimate. (M93-30)

(b) Construct on ogive based on using class boundaries at 0.5, 2.5, 8.5, 15.5, and 29.5. (M93-31)

(c) Construct a histogram using the same boundaries as in part (b). (M93-31)

2.11 Claim sizes are for 100, 200, 300, 400, or 500. The true probabilities for these values are 0.05, 0.20, 0.50, 0.20, and 0.05 respectively. Determine the skewness and kurtosis for this distribution. (M93-34)

2.12 A sample of size 50 has a sample mean of 100 and a sample standard deviation of 75.

(a) Construct an approximate 95% confidence interval for the population mean.

[23] In Appendix A three Pareto distributions are listed. They are the generalized Pareto (3 parameters), the Pareto (2 parameters) and the single-parameter Pareto (1 parameter). In this text, when no modifier is present, the two-parameter version is to be used.

(b) Suppose it is known that the population has the exponential distribution with mean μ. In this case the population variance is μ^2. Construct two 95% confidence intervals for μ. First use (2.8) and for the second interval mimic the development in Example 2.8.

2.13 A sample of 1,000 health insurance contracts on adults produced a sample mean of 1,300 for the annual benefits paid with a standard deviation of 400. It is expected that 2,500 contracts will be issued next year. Use the Central Limit Theorem to estimate the probability that benefit payments will be more than 101% of the expected amount.

2.14 The data in the following table are from Herzog and Laverty [60]. A certain class of fifteen year mortgages was followed from issue until December 31, 1993. The issues were split into those that were refinances of existing mortgages and those that were original issues. Each entry in the table provides the number of issues and the percentage of them that were still in effect after the indicated number of years.

	Refinances		Original	
Years	No. issued	Survived	No. issued	Survived
1.5	42,300	99.97	12,813	99.88
2.5	9,756	99.82	18,787	99.43
3.5	1,550	99.03	22,513	98.81
4.5	1,256	98.41	21,420	98.26
5.5	1,619	97.78	26,790	97.45

Draw as much of the two ogives (on the same graph) as is possible from the data. Does it appear from the ogives that the lifetime variable (time to mortgage termination) has a different distribution for refinanced versus original issues?

2.15 The data in Table 2.25 were collected (units are millions of dollars).

(a) Construct the histogram (M95-1).

(b) Estimate the mean.

2.16 Determine a general formula for a non-parametric $(1-\alpha)\%$ confidence interval for π_p when the sample size n is large.

2.17 A sample of size 5 has been taken from a continuous distribution. A confidence interval for the median is formed by using the second and fourth smallest values. Determine the level of confidence. (M94-27).

Table 2.25 Data for Exercise 2.15

Loss	No. of observations
0–2	25
2–10	10
10–100	10
100–1,000	5

2.18 A sample of size 500 has been taken from a continuous distribution. A confidence interval for the median is formed by using the order statistics $X_{(240)}$ and $X_{(260)}$. Use the normal approximation to determine the level of confidence. (M95-18).

Section 2.3

2.19 Six independent observations, $x_1, \ldots, x_6$, were obtained. The variance was estimated as $\hat{\sigma}^2 = \sum_{j=1}^{6}(x_j - \bar{x})^2/6$ with $\bar{x} = \sum_{j=1}^{6} x_j/6$. The true population variance is $\sigma^2 = 2$. Determine the bias of $\hat{\sigma}^2$. (89-47)

2.20 A population of losses has the Pareto distribution (see Appendix A) with $\theta = 6{,}000$ and α unknown. Simulation of the results from maximum likelihood estimation based on samples of size 10 has indicated that $E(\hat{\alpha}) = 2.2$ and $\text{MSE}(\hat{\alpha}) = 1$. Determine $Var(\hat{\alpha})$ if it is known that $\alpha = 2$. (M92-17)

2.21 Two instruments are available for measuring a particular nonzero distance. X is the random variable representing a measurement with the first instrument, and Y the random variable for the second instrument. X and Y are independent with $E(X) = .8m$, $E(Y) = m$, $Var(X) = m^2$, and $Var(Y) = 1.5m^2$ where m is the true distance. Consider estimators of m that are of the form $Z = \alpha X + \beta Y$. Determine the values of α and β that make Z a UMVUE estimator within the class of estimators of this form. (N93-13)

2.22 A population contains six members, with values 1, 1, 2, 3, 5, and 10. A random sample of size three is drawn without replacement. In each case the objective is to estimate the population mean. Note: A spreadsheet with an optimization routine may be the best way to solve this problem.

(a) Determine the bias, variance, and MSE of the sample mean.

(b) Determine the bias, variance, and MSE of the sample median.

(c) Determine the bias, variance, and MSE of the sample midrange (the average of the largest and smallest observations).

(d) Consider an arbitrary estimator of the form $aX_{(1)} + bX_{(2)} + cX_{(3)}$ where $X_{(1)} \leq X_{(2)} \leq X_{(3)}$ are the sample order statistics.

i. Determine a restriction on the values of a, b, and c that will assure that the estimator is unbiased.

ii. Determine the values of a, b, and c that will produce the unbiased estimator with the smallest variance.

iii. Determine the values of a, b, and c that will produce the (possibly biased) estimator with the smallest MSE.

2.23 Two different estimators, $\hat{\theta}_1$ and $\hat{\theta}_2$ are being considered. To test their performance, 75 trials have been simulated, each with the true value set at $\theta = 2$. The following totals were obtained:

$$\sum_{j=1}^{75} \hat{\theta}_{1j} = 165, \sum_{j=1}^{75} \hat{\theta}_{1j}^2 = 375, \sum_{j=1}^{75} \hat{\theta}_{2j} = 147, \sum_{j=1}^{75} \hat{\theta}_{2j}^2 = 312$$

where $\hat{\theta}_{ij}$ is the estimate based on the jth simulation using estimator $\hat{\theta}_i$. Estimate the MSE for each estimator and determine the **relative efficiency** (the ratio of the MSEs). (M95-27)

Section 2.4

2.24 (a) In the definition of minimum layer average severity estimation, verify that the function $G_j(\theta)$ as defined relies only upon the quantities specified in Definition 2.26.

(b) The term **layer average severity** means the expected payment when an insurance policy provides the following benefit when the loss is x

$$\begin{array}{ll} 0, & x \leq c_{j-1} \\ x - c_{j-1}, & c_{j-1} < x < c_j \\ c_j - c_{j-1}, & x \geq c_j. \end{array}$$

Show that the expected payment is given by $E(X \wedge c_j) - E(X \wedge c_{j-1})$.

2.25 A sample of n independent observations, $x_1, \ldots, x_n$ came from a distribution with a pdf of $f(x) = 2\theta x \exp(-\theta x^2)$, $x > 0$. Determine the mle of θ. (89-48)

2.26 The observations 1,000, 850, 750, 1,100, 1,250, and 900 were obtained as a random sample from a gamma distribution with unknown parameters α and θ. Estimate these parameters by the method of moments. (90-34)

2.27 A random sample of claims has been drawn from a loglogistic distribution. In the sample, 80% of the claims exceed 100 and 20% exceed 400. Estimate the loglogistic parameters by percentile matching. (90-44)

2.28 Two states provide workers compensation indemnity benefits. One state caps payments at 50,000 and the other at 100,000. The 5,000 losses in the first state were grouped with 2,900 between 0 and 25,000, 1,200 between 25,000 and 50,000, and 900 above 50,000. In the second state the 2,000 losses were grouped as 1,000 between 0 and 25,000, 300 between 25,000 and 50,000, 200 between 50,000 and 60,000, 300 between 60,000 and 100,000, and 200 over 100,000. Minimum modified chi-square estimation is to be used to estimate the parameters of a Pareto model. When computing the objective function, use only six chi-square terms. That is, combine the two states for the two intervals that have the same boundaries.

(a) Determine the value of the modified chi-square criterion for $\alpha = 3$ and $\theta = 75{,}000$. (90-55)

(b) Determine the minimum modified chi-square parameter estimates.

(c) Determine the maximum likelihood estimates.

2.29 Let $x_1, \ldots, x_n$ be a random sample from a population with cdf $F(x) = x^p,\ 0 < x < 1$.

(a) Determine the mle of p. (91-36)

(b) Determine the method of moments estimate of p. (M95-5)

2.30 A random sample of 10 claims obtained from a gamma distribution is given below:

1,500 6,000 3,500 3,800 1,800 5,500 4,800 4,200 3,900 3,000.

(a) Estimate α and θ by the method of moments. (91-46)

(b) Suppose it is known that $\alpha = 12$. Determine the mle of θ. (91-47)

(c) Determine the mle of α and θ.

2.31 A random sample of 5 claims from a lognormal distribution is given below:

500 1,000 1,500 2,500 4,500.

(a) Estimate μ and σ by the method of moments. Estimate the probability that a loss will exceed 4,500. (M92-10)

(b) Estimate μ and σ by maximum likelihood. Estimate the probability that a loss will exceed 4,500.

2.32 Let $x_1, \ldots, x_n$ be a random sample from a random variable with pdf $f(x) = \theta^{-1}e^{-x/\theta}$, $x > 0$. Determine the mle of θ. (M92-21)

2.33 The random variable X has pdf $f(x) = \beta^{-2}x\exp(-.5x^2/\beta^2)$, $x, \beta > 0$. For this random variable, $E(X) = (\beta/2)\sqrt{2\pi}$ and $Var(X) = 2\beta^2 - \pi\beta^2/2$. You are given the following five observations.

$$4.9 \quad 1.8 \quad 3.4 \quad 6.9 \quad 4.0.$$

(a) Determine the method of moments estimate of β. (M92-26)

(b) Determine the maximum likelihood estimate of β. (M92-27)

2.34 Let $x_1, \ldots, x_n$ be a random sample from a random variable with cdf $F(x) = 1 - x^{-\alpha}$, $x > 1$, $\alpha > 0$. Determine the mle of α. (N92-4)

2.35 The random variable X has pdf $f(x) = \alpha\lambda^\alpha(\lambda + x)^{-\alpha-1}$, $x, \alpha, \lambda > 0$. It is known that $\lambda = 1{,}000$. You are given the following five observations.

$$43 \quad 145 \quad 233 \quad 396 \quad 775.$$

(a) Determine the method of moments estimate of α.

(b) Determine the maximum likelihood estimate of α. (N93-8)

2.36 Demonstrate that (2.14) is the correct formula for the scoring method when there are multiple groups.

2.37 A random sample produced 2 observations between 0 and 2, 4 observations between 2 and 5, and 3 observations between 5 and 8. Estimate the parameter of the exponential distribution by using the Cramér–von Mises estimator with weights $w_j = 1$. (N94-13)

2.38 The following 20 observations were collected. It is desired to estimate $\Pr(X > 200)$. When a parametric model is called for, use the single-parameter Pareto distribution for which $F(x) = 1 - (100/x)^\alpha$, $x > 100$, $\alpha > 0$.

132	149	476	147	135	110	176	107	147	165
135	117	110	111	226	108	102	108	227	102

(a) Determine the empirical estimate of $\Pr(X > 200)$.

(b) Determine the method of moments estimate of the single-parameter Pareto parameter α and use it to estimate $\Pr(X > 200)$.

(c) Determine the maximum likelihood estimate of the single-parameter Pareto parameter α and use it to estimate $\Pr(X > 200)$.

(d) Obtain the variance of the empirical estimator as a function of α.

(e) Use the bootstrap technique to compare the empirical, method of moments, and maximum likelihood estimators.

Table 2.26 Data for Exercise 2.39

Loss	No. of observations	Loss	No. of observations
0–25	5	350–500	17
25–50	37	500–750	13
50–75	28	750–1000	12
75–100	31	1,000–1,500	3
100–125	23	1,500–2,500	5
125–150	9	2,500–5,000	5
150–200	22	5,000–10,000	3
200–250	17	10,000–25,000	3
250–350	15	25,000–	2

2.39 The data in Table 2.26 presents the results of a sample of 250 losses. Consider the inverse exponential distribution with cdf $F(x) = e^{-\theta/x}$, $x > 0$, $\theta > 0$.

(a) Determine the minimum modified chi-square estimate of θ.

(b) Determine the maximum likelihood estimate of θ.

(c) Perform one iteration of the scoring method using your answer to part (b) as the starting value. Verify that your answer was the mle and record the value of the denominator of (2.12).

Section 2.5

2.40 This is a continuation of Exercise 2.29. Let $x_1, \ldots, x_n$ be a random sample from a population with cdf $F(x) = x^p$, $0 < x < 1$.

(a) Determine the asymptotic variance of the maximum likelihood estimator of p.

(b) Use your answer to obtain a general formula for a 95% confidence interval for p.

(c) Determine the maximum likelihood estimator of $E(X)$ and obtain its asymptotic variance and a formula for a 95% confidence interval.

2.41 This is a continuation of Exercise 2.32. Let $x_1, \ldots, x_n$ be a random sample from a population with pdf $f(x) = \theta^{-1}e^{-x/\theta}$, $x > 0$.

(a) Determine the asymptotic variance of the maximum likelihood estimator of θ.

(b) Use your answer to obtain a general formula for a 95% confidence interval for θ. (M94-28)

(c) Determine the maximum likelihood estimator of $Var(X)$ and obtain its asymptotic variance and a formula for a 95% confidence interval.

2.42 A sample of size 40 has been taken from a population with pdf $f(x) = (2\pi\theta)^{-1/2}e^{-x^2/(2\theta)}$, $-\infty < x < \infty$, $\theta > 0$. The mle of θ is $\hat{\theta} = 2$. Approximate the MSE of $\hat{\theta}$. (N94-9)

2.43 Four observations were made from a random variable having the density function $f(x) = 2\lambda x e^{-\lambda x^2}$, $x, \lambda > 0$. Exactly one of the four observations was less than 2.

(a) Determine the mle of λ. (M94-12)

(b) Use the scoring method to determine the estimated variance of the mle of λ.

2.44 Use one iteration of the scoring method to verify that the mle obtained in Exercise 2.28 ($\hat{\alpha} = 5.14825$, $\hat{\theta} = 147{,}130$) is indeed the maximum and then obtain the estimated covariance matrix. Keep the two samples separate so that (Then construct a 95% confidence interval for the mean.

2.45 Estimate the covariance matrix of the mle for Exercise 2.30. Do this by computing approximate derivatives of the loglikelihood. Then construct a 95% confidence interval for the mean. Finally, construct a 95% prediction interval for the next loss.

2.46 For the random variable and data in Exercise 2.31 determine a 95% confidence interval for the probability that a loss will exceed 4,500.

2.47 For the random variable and data in Exercise 2.33 determine the variance of both the method of moments and maximum likelihood estimators of β as a function of β. Use the asymptotic approximation for the mle. Then estimate and compare these variances, in both cases replacing β with its mle. Finally, construct 95% confidence intervals for the mean using both the method of moments and maximum likelihood estimates.

2.48 Estimate the variance of the mle for Exercise 2.35 and use it to construct a 95% confidence interval for $E(X \wedge 500)$.

2.49 Estimate the variance of the mle of α in Exercise 2.38 by determining the information. Then estimate the variance of the mle of $\Pr(X > 200)$ and compare this to the bootstrap variances obtained in that Exercise.

2.50 Use the answer to part (c) of Exercise 2.39 to obtain a 95% confidence interval for $\Pr(X > 10{,}000)$.

Table 2.27 General liability payments, two limits

100,000 Limit		1,000,000 Limit	
Payment	Number	Payment	Number
0–2,500	21	0–2,500	58
2,500–7,500	18	2,500–7,500	61
7,500–12,500	13	7,500–12,500	37
12,500–17,500	4	12,500–17,500	36
17,500–22,500	7	17,500–22,500	22
22,500–32,500	14	22,500–32,500	30
32,500–47,500	9	32,500–47,500	19
47,500–67,500	3	47,500–67,500	15
67,500–100,000	3	67,500–87,500	11
at 100,000	10	87,500–125,000	18
		125,000–175,000	7
		175,000–225,000	7
		225,000–325,000	6
		325,000–475,000	2
		475,000–675,000	2
		675,000–1,000,000	2
		at 1,000,000	3

2.51 Show that whenever a lognormal model is used, the information matrix obtained by inserting the parameter estimates in the true information matrix must be the same as the information matrix obtained by inserting the parameter estimates in the second partial derivatives of the loglikelihood function.

2.52 Show that the equality obtained in the previous exercise does not hold for the Pareto distribution.

Section 2.6

2.53 Complete Example 2.30 assuming that the true density function is given by the corresponding histogram (or equivalently that the true distribution function is given by the ogive). Note that this is a mixed distribution because there is probability $3/217$ at the discrete value 300,000.

2.54 The data in Table 2.27 provide information from two other policy limits for the general liability coverage. Conduct a hypothesis test to determine if it is possible that all four populations (these two limits plus 300,000 and 500,000 as given in Example 2.36) have the same lognormal distribution.

Table 2.28 Data for Exercise 2.58

Loss (x)	No. of losses $\geq x$	Sum of losses $\geq x$
1,000	180	990,000
3,000	118	882,000
5,000	75	713,000
7,000	50	576,000
9,000	34	459,000

2.55 Show that the inverse Gaussian distribution is a scale family but does not have a scale parameter.

2.56 Show that the gamma distribution has a scale parameter.

2.57 Prove Theorem 2.9.

2.58 The entries in Table 2.28 were derived from a random sample of losses. Compute the empirical mean excess loss at each of the five points and state whether or not this sample is consistent with the Pareto distribution. (M94-4)

2.59 The amount of a single claim has a Pareto distribution with $\alpha = 2$ and $\theta = 2{,}000$. Determine the loss elimination ratio for a deductible of 500. (M94-10)

2.60 The distribution of losses for claims in 1993 has a discrete distribution with pf $\Pr(X = 1{,}000k) = 1/6,\ k = 1, \ldots, 6$. The insurance contract calls for a deductible of 1,500 on each loss. Inflation of 5% impacts all claims uniformly from 1993 to 1994. The deductible remains at 1,500 in 1994. Determine the percentage increase in expected payments per loss from 1993 to 1994. (M94-21)

2.61 The distribution of losses for claims in 1993 is lognormal with $\mu = 10$ and $\sigma^2 = 5$. From 1993 to 1994, an inflation rate of 10% impacts all claims uniformly. The insurer has purchased reinsurance that pays the excess over 2,000,000 on any claim. Determine the insurer's expected payment per loss for 1994. (N94-8)

2.62 A random sample of auto glass claims has yielded the following five observed claim amounts: 100, 125, 200, 250, and 300. Determine the value of the empirical mean excess loss function at $x = 250$. (N94-16)

2.63 Claim amounts in 1993 have the normal distribution (the cdf is $F(x) = \Phi\left[(x - \mu)/\sigma\right]$) with $\mu = 1{,}000$ and $\sigma^2 = 10{,}000$. Inflation of 5% impacted all claims uniformly from 1993 to 1994. Determine the distribution for claim amounts in 1994. (N94-28)

2.64 For 1994, loss sizes follow a uniform distribution with pdf $f(x) = 1/2{,}500$, $0 < x < 2{,}500$. Inflation of 3% impacts all losses uniformly from 1994 to 1995. In 1995 a deductible of 100 is applied to all losses. Determine the loss elimination ratio for 1995. (M95-6)

2.65 Losses follow a Pareto distribution with $\alpha > 1$ and θ unspecified. Determine the ratio of the mean excess loss function at $x = 2\theta$ to the mean excess loss function at $x = \theta$. (M95-21)

2.66 In Exercise 2.38 a single-parameter Pareto model was fitted by maximum likelihood. Use this model to estimate the following quantities

(a) The expected payment for a single loss.

(b) The loss elimination ratio if a deductible of 200 is imposed.

(c) A reinsurer is considering providing coverage that will pay 80% of the excess over 500. Compare the expected reinsurance payment per loss as given by the single parameter Pareto model versus the empirical estimate. Do the same for the expected reinsurance payment per payment.

(d) The expected payment per loss if 5% inflation is applied uniformly to all losses.

(e) The loss elimination ratio if a deductible of 200 is imposed on the inflated losses.

2.67 Use the data from Exercise 2.39 to determine the mle for the gamma distribution. A reinsurer is considering covering 80% of the excess over 5,000 of any loss with a maximum payment of 4,000. Determine the expected reinsurance payment on a per loss and per payment basis.

2.68 In *Loss Distributions* [66, p. 161] the data sets in Table 2.29 are presented (but several groups have been combined here in order to reduce the size of the table), representing various limits of automobile bodily injury. The objective is to develop an **increased limits factor**—that is, the ratio of the expected payment per loss with a 25,000 limit to the expected payment per loss with a 10,000 limit. In all cases use a Weibull distribution and maximum likelihood estimates.

(a) Estimate the increased limits factor by obtaining the expected value for the 25,000 limit using only data from policies with that limit and obtaining the expected value for the 10,000 limit using only data from policies with that limit.

(b) Estimate the parameters for a common Weibull distribution using the data from all five limits and then estimate the expected values from that model.

Table 2.29 Automobile bodily injury losses by limit

	Limit				
Range	5,000	10,000	25,000	50,000	100,000
0–250	1,230	415	660	950	4,119
250–500	612	187	268	424	2,047
500–1,000	546	136	311	377	2,031
1,000–3,000	756	256	452	605	3,440
3,000–5,000	226				
5,000–	491				
3,000–6,000		142	214	337	1,905
6,000–8,000		38	77	103	602
8,000–10,000		25			
10,000–		77			
8,000–11,000			59	99	507
11,000–14,000			33	28	255
14,000–19,000			28	61	298
19,000–23,000			19	30	139
23,000–25,000			6		
25,000–			63		
23,000–30,000				27	240
30,000–40,000				14	131
40,000–50,000				16	
50,000–				45	
40,000–55,000					166
55,000–70,000					74
70,000–100,000					119
100,000–					139
Total	3,861	1,276	2,190	3,116	16,212

(c) Assume that the Weibull distributions applicable to each limit have the same value for the parameter τ but have different values for θ. Estimate the single value of τ and the individual values of θ and then estimate the increased limits factor using the value of θ that is appropriate for each of the two limits.

(d) Use the likelihood ratio test to determine which of the three approaches is most appropriate for this problem. For the approach in part (a) assume that unique parameters are fit to each of the five groups.

Section 2.7

2.69 The arguments in Example 2.39 provided mathematical justification of the proper ordering of the three distributions by tail weight. To reinforce this conclusion consider a gamma distribution with parameters $\alpha = 0.2$, $\theta = 500$, a lognormal distribution with parameters $\mu = 3.709290$, $\sigma = 1.338566$, and a Pareto distribution with parameters $\alpha = 2.5$, $\theta = 150$. First demonstrate that all three distributions have the same mean and variance. Then numerically demonstrate that there is a value such that the gamma pdf is smaller than the lognormal and Pareto pdfs for all arguments above that value and that there is another value such that the lognormal pdf is smaller than the Pareto pdf for all arguments above that value.

2.70 Let X have cdf $F_X(x) = 1 - (1+x)^{-\alpha}$, $x, \alpha > 0$. Determine the pdf and cdf of $Y = \theta X$.

2.71 Let X have the Pareto distribution. Determine the cdf of the transformed, inverse, and inverse transformed distributions. Check Appendix A to determine if any of these distributions have special names.

2.72 Let X have the loglogistic distribution. Demonstrate that the inverse distribution also has the loglogistic distribution. Therefore there is no need to identify a separate inverse loglogistic distribution.

2.73 Let Y have the lognormal distribution with parameters μ and σ. Let $Z = \theta Y$. Show that Z also has the lognormal distribution and therefore the addition of a third parameter has not created a new distribution.

2.74 Let X have pdf $f(x) = \exp(-|x/\theta|)/2\theta$ for $-\infty < x < \infty$. Let $Y = e^X$. Determine the pdf and cdf of Y.

2.75 In [122], Venter noted that if X has the transformed gamma distribution and its scale parameter θ has an inverse transformed gamma distribution (where the parameter τ is the same in both distributions) the resulting mixture has the transformed beta distribution. Demonstrate that this is true.

2.76 Determine the limiting distribution of the generalized Pareto distribution as α and θ both go to infinity.

2.77 For the data in Table 2.10 test the null hypothesis of an exponential distribution against the alternative of a Weibull distribution.

2.78 Verify that regardless of the distribution of the X_i, (2.26) must produce a legitimate cdf. That is, the function must be nondecreasing and right continuous and must have a limit of zero as y goes to negative infinity and a limit of one as y goes to positive infinity. Also, obtain the matching formula for the pdf.

2.79 For the data in Table 2.10 test the null hypothesis of a single exponential population against the alternative of a mixture of two exponential distributions.

2.80 In Exercise 2.39 an inverse exponential distribution was fit to a data set and in Exercise 2.67 an inverse Weibull distribution was fit to the same data set. Conduct a likelihood ratio test to determine which of these models is preferred.

2.81 Losses in 1993 follow the density function $f(x) = 3x^{-4}$, $x \geq 1$, where x is losses in millions of dollars. Inflation of 10% impacts all claims uniformly from 1993 to 1994. Determine the cdf of losses for 1994 and use it to determine the probability that a 1994 loss exceeds 2,200,000. (M94-16)

2.82 A common distribution for modeling time to death of humans is the Gompertz distribution. It is usually defined by its force of mortality function $\lambda(x) = Bc^x$, $x > 0$.

(a) Determine the cdf and pdf for the Gompertz distribution.

(b) Is the Gompertz distribution a scale family? If not, create a three parameter version of the Gompertz distribution by adding a scale parameter.

(c) Compare the tail behavior of the gamma, Weibull, and Gompertz (using the two parameters) distributions by seeing which are IFR and which are DFR.

(d) Two thousand disabled individuals aged 30–39 at the time of disablement were followed until they either recovered or died. Table 2.30 gives the number terminated as of various times (in months). Estimate the (two-parameter) Gompertz parameters and compare the model by plotting the Gompertz pdf against a histogram of the data. There were 508 individuals still disabled after 69 months.

2.83 Show that as $\tau \to \infty$ in the transformed beta distribution the result is the inverse transformed gamma distribution.

2.84 Determine the parameters of the limiting gamma distribution used in Example 2.52. Recall that the mode is to be 100.

2.85 Determine the parameters of the limiting inverse gamma distribution used in Example 2.53. Recall that the mean is to be 100.

2.86 Prove that

$$e(x) = \int_x^\infty e^{-\int_x^t \lambda(y)dy} dt, \qquad x \geq 0.$$

Table 2.30 Data for Exercise 2.82

Duration	Number terminated	Duration	Number terminated	Duration	Number terminated
1	21	10	203	19	405
2	33	11	221	20	411
3	63	12	238	21	454
4	81	13	275	33	765
5	113	14	281	45	1,022
6	122	15	319	57	1,252
7	142	16	327	69	1,492
8	176	17	349		
9	201	18	397		

Use this result to show that if $\lambda(x) \geq \lambda$, then $e(x) \leq 1/\lambda$ and that if $\lambda(x) \leq \lambda$, then $e(x) \geq 1/\lambda$.

Use the above to show that if $\lambda(x) \geq \lambda$, then $S(x) \leq e^{-\lambda x}$ and $S_e(x) \leq e^{-\lambda x}$ and if $\lambda(x) \leq \lambda$, then $S(x) \geq e^{-\lambda x}$ and $S_e(x) \geq e^{-\lambda x}$; for all $x \geq 0$.

2.87 Prove that

$$\int_x^\infty y dF(y) = xS(x) + E(X)S_e(x), \quad x \geq 0. \tag{2.34}$$

Use (2.34) to show that

$$S(x) = \frac{\int_x^\infty y dF(y)}{x + e(x)}, \quad x \geq 0. \tag{2.35}$$

Use (2.35) to show that

$$S(x) \leq \frac{E(X)}{x + e(x)}, \quad x \geq 0.$$

Prove that

$$S_e(x) = \frac{e(x)\int_x^\infty y dF(y)}{E(X)[x + e(x)]} \tag{2.36}$$

Prove using (2.36) that

$$S_e(x) \leq \frac{e(x)}{x + e(x)}$$

2.88 A distribution is said to be new worse than used (NWU) if $S(x)S(y) \leq S(x+y)$ for all $x \geq 0$ and $y \geq 0$, and new better than used (NBU) if $S(x)S(y) \geq S(x+y)$ for all $x \geq 0$, $y \geq 0$. Prove that DFR $\Rightarrow$ NWU and that IFR $\Rightarrow$ NBU.

2.89 A distribution is said to be new worse than used in convex ordering (NWUC) if $S_e(x+y) \geq S(x)S_e(y)$ for all $x \geq 0$ and $y \geq 0$, and new better than used in convex ordering (NBUC) if $S_e(x+y) \leq S(x)S_e(y)$ for all $x \geq 0$ and $y \geq 0$.

(a) Prove that NWU $\Rightarrow$ NWUC and NBU $\Rightarrow$ NBUC, where NWU and NBU are defined in Exercise 2.88.

(b) Prove that IMRL $\Rightarrow$ NWUC and DMRL $\Rightarrow$ NBUC. (*Hint*: Recall that $\lambda_e(x) = [e(x)]^{-1}$ and use Exercise 2.88 to show that IMRL implies that $S_e(x)S_e(y) \leq S_e(x+y)$.

2.90 A distribution is said to be new worse than used in expectation (NWUE) if $e(x) \geq e(0)$ or equivalently $S_e(x) \geq S(x)$ for $x \geq 0$, and new better than used in expectation (NBUE) if $e(x) \leq e(0)$ or equivalently $S_e(x) \leq S(x)$ for $x \geq 0$.

(a) Show that NWUC $\Rightarrow$ NWUE and NBUC $\Rightarrow$ NBUE, where NWUC and NBUC are defined in Exercise 2.89.

(b) Use Exercise 2.86 to show that NWUE implies that $S_e(x) \geq e^{-x/E(X)}$ for $x \geq 0$ and NBUE implies that $S_e(x) \leq e^{-x/E(X)}$ for $x \geq 0$.

(c) Use (b) to prove that NWUE implies that $Var(X) \geq [E(X)]^2$ and NBUE implies that $Var(X) \leq [E(X)]^2$.

(d) Show, using Exercise 2.87, that NWUE implies that

$$S(x) \leq \frac{E(X)}{x+E(X)}, \quad x \geq 0.$$

(e) Use (d) to show that NWUE implies that $\Pr[X > E(X)] \leq 1/2$, that is, the mean is at least as large as the (smallest) median.

2.91 Suppose that $1-S_e(x)$ is NWU (see Exercise 2.88), that is, $S_e(x+y) \geq S_e(x)S_e(y)$ for all $x \geq 0$ and $y \geq 0$.

(a) Prove that this may be restated as

$$\frac{1-e^{-\int_0^x [e(t+y)]^{-1}dt}}{x} \leq \frac{1-e^{-\int_0^x [e(t)]^{-1}dt}}{x}.$$

(b) Show that if $1-S_e(x)$ is NWU, then $F(x)$ is NWUE (Exercise 2.90). *Hint*: Let $x \to 0$ in (a) and use L'Hôpital's rule.

(c) Prove that if $1-S_e(x)$ is NWU, then $F(x)$ is NWUC (Exercises 2.88 and 2.89).

(d) Prove that if $1-S_e(x)$ is NBU, then $F(x)$ is NBUC (Exercises 2.88 and 2.89).

Section 2.8

2.92 A classic probability puzzler goes as follows (this version is from Robertson [108]): "You are asked to select one of two envelopes and told that one of the envelopes contains twice as much money as the other. You find 100 in the envelope you select. You are offered the opportunity to select the other envelope, but then you must keep only the money in the second envelope. Is it better to switch or stand?" Construct a Bayesian analysis as follows: Let Θ be the smaller of the two amounts and have prior density $\pi(\theta)$. The other envelope contains the amount 2Θ. The amount in the envelope you select is $X = \Theta$ with probability 0.5 and $X = 2\Theta$ with probability 0.5.[24]

(a) Determine the posterior expected value of each of the two strategies.

(b) Show that if the prior density is $\pi(\theta) = 1/\theta$ the two strategies have the same conditional (on the value of X) expected value.[25]

(c) Show that if the prior density is $\pi(\theta) = \exp(-\theta)$, there will be some values of X for which switching produces a higher conditional expected value and some for which it is lower. Determine the set of X values for which switching is superior.

2.93 Show that if Y is the predictive distribution in Example 2.55, then $\log Y - \log 100$ has the Pareto distribution.

2.94 Determine the posterior distribution of α in Example 2.55 if the prior distribution is an arbitrary gamma distribution. To avoid confusion, denote the first parameter of this gamma distribution by γ. Next determine a particular combination of gamma parameters so that the posterior mean is the maximum likelihood estimate of α regardless of the specific values of $x_1, \ldots, x_n$. Is this prior improper?

2.95 For Example 2.60 demonstrate that the maximum likelihood estimate of α is 1.75.

2.96 Let $x_1, \ldots, x_n$ be a random sample from a lognormal distribution with unknown parameters μ and σ. Let the prior density be $\pi(\mu, \sigma) = \sigma^{-1}$.

[24] This is one of many possible models that could describe how the money gets into the envelopes. It was suggested by James Broffitt. Without postulating a model, it is impossible to answer the question. This is no different from attempting to answer an actuarial question.

[25] Observe that this is the standard noninformative prior for a scale parameter and indeed this is a scale problem due to the doubling/halving of the amounts. The result is then that if you have no useful information about how the amounts were selected, it makes no difference what you do.

(a) Write the posterior pdf of μ and σ up to a constant of proportionality.

(b) Determine Bayesian estimators of μ and σ by using the posterior mode.

(c) Fix σ at the posterior mode as determined in part (b) and then determine the exact (conditional) pdf of μ. Then use it to determine a 95% HPD credibility interval for μ.

2.97 A random sample of size 100 has been taken from a gamma distribution with α known to be 2, but θ unknown. For this sample, $\sum_{j=1}^{100} x_j = 30{,}000$. The prior distribution for θ is inverse gamma with β taking the role of α and λ taking the role of θ.

(a) Determine the exact posterior distribution of θ. At this point the values of β and λ have yet to be specified.

(b) The population mean is 2θ. Determine the posterior mean of 2θ first using the prior distribution with $\beta = \lambda = 0$ (this is equivalent to $\pi(\theta) = \theta^{-1}$) and then with $\beta = 2$ and $\lambda = 250$ (which is a prior mean of 250). Then, in each case, determine a 95% credibility interval with 2.5% probability on each side.

(c) Determine the posterior variance of 2θ and use the Bayesian Central Limit Theorem to construct a 95% credibility interval for 2θ using each of the two prior distributions given in part (b).

(d) Determine the maximum likelihood estimate of θ and then use the estimated variance to construct a 95% confidence interval for 2θ.

Section 2.9

2.98 In Example 2.68 the choice of a lognormal versus a Pareto distribution did not make much difference when the goal was to obtain the expected payment for a layer from 100,000 to 250,000. Compare the two models with regard to pricing a layer from 250,000 to 500,000.

2.99 It has been conjectured that losses have a Pareto distribution with parameters $\alpha = 2$ and $\theta = 1{,}000$. A random sample of size 10 produced 3 losses in the range 0–250, 2 in the range 250–500, 3 in the range 500–1,000, and 2 above 1,000. Perform the chi-square goodness-of-fit test with a significance level of 0.10. Do not require that expected counts be at least 5. (M94-14)

2.100 A random sample of 1,000 observations produced the results in the following table. The parameters of the Pareto distribution as estimated by the minimum modified chi-square technique are $\hat{\alpha} = 3.5$ and $\hat{\theta} = 50$. Of 0.05,

Interval	Number of losses
$[0, 3)$	180
$[3, 7.5)$	180
$[7.5, 15)$	235
$[15, 40)$	255
$[40, \infty)$	150

0.025, 0.01, and 0.005, which is the highest significance level at which you would accept this fitted model? (N94-4)

2.101 Given sample values of 0.1, 0.4, 0.8, 0.8, and 0.9 you wish to test the goodness of fit of the distribution with pdf $f(x) = (1 + 2x)/2, \ 0 \le x \le 1$. Determine the Kolmogorov–Smirnov test statistic (M95-11) and conduct the test at a 5% significance level.

2.102 Determine an appropriate model for the data in Exercise 2.2. Consider a wide variety of possibilities and employ tests as needed.

2.103 Determine an appropriate model for each of the two data sets in Exercise 2.14. If external considerations favor using the same distribution for the two data sets (but possibly with different parameters), would it be reasonable to use a common model?

2.104 Determine an appropriate model for the combined data in Exercise 2.28. Consider a wide variety of possibilities and employ tests as needed.

2.105 In Exercise 2.30 a gamma distribution with maximum likelihood estimates $\hat{\alpha} = 6.341$ and $\hat{\theta} = 599.3$ was fitted to the data. Use the Kolmogorov–Smirnov test to determine if the gamma model is reasonable.

2.106 In Exercise 2.38 a single-parameter Pareto distribution with maximum likelihood estimate $\hat{\alpha} = 2.848$ was fitted to the data. Use the Kolmogorov–Smirnov test to determine if the single-parameter Pareto model is reasonable.

2.107 Determine an appropriate model for the data in Exercise 2.39. Consider a wide variety of possibilities and employ tests as needed.

2.108 Determine an appropriate model for each of the two data sets in Exercise 2.54. If external considerations favor using the same distribution for the two data sets (but possibly with different parameters), would it be reasonable to use a common model with different parameters? Would it be reasonable to use a common model with the same parameters?

2.109 Determine an appropriate model for the two data sets in Exercise 2.68 with limits at 10,000 and 25,000. If external considerations favor using the same distribution for the two data sets (but possibly with different parameters), would it be reasonable to use a common model?

2.110 Determine an appropriate model for the data in Exercise 2.82. Consider a wide variety of possibilities and employ tests as needed.

2.111 Use the Schwartz Bayesian Criterion to answer part (d) of Exercise 2.68. There were a total of 26,655 observations. When a model with common values for θ and τ, the NLL was 54,260.96. When a model with a common value for τ but individual values for θ (6 parameters) was used, the NLL was 54,079.38. When individual estimates were used for both θ and τ (10 parameters), the NLL was 54,047.52.

Section 2.10

2.112 Using the data in Example 2.74, estimate the parameters for the inverse Weibull distribution and then use your model to form a 95% confidence interval for the number of unreported claims.

2.113 Estimate the Weibull parameters by both maximum likelihood and percentile matching for the problem in Example 2.73 for a model shifted at 75.

2.114 Show that (2.32) is true.

2.115 You are given that losses follow a Weibull distribution with parameters $\tau = 1$ and $\theta = 20$. A random sample of losses is collected, but the sample data are truncated from below by a deductible of 10.

(a) Determine the probability that an observed loss is at least 25. (M94-6)

(b) Given that a payment is made, what is the probability that the payment is at most 25? (N94-17)

2.116 X is a random variable for 1993 losses, with pdf $f(x) = 0.1e^{-0.1x}$, $x > 0$. Inflation of 10% impacts all losses uniformly from 1993 to 1994. For 1994 a deductible of d is applied to all losses. Let P be a random variable representing 1994 payments per payment. Determine the value of the cdf of P at the value 5 in 1994. That is, obtain $F_P(5)$. (M94-24)

2.117 A random sample produced two losses that were less than 2,000 and four that were between 2,000 and 5,000. Losses above 5,000 may have occurred, but were not recorded. The model has pdf $f(x) = \lambda e^{-\lambda x}$, $x, \lambda > 0$. Write the likelihood function for this problem. (M95-15)

2.118 Losses follow a Pareto distribution with $\alpha = 2$ and $\theta = 1{,}000$. There are ten losses expected each year. For each loss, the insurer's payment is equal to the entire amount of the loss if the loss is greater than 100. The insurer makes no payment if the loss is less than or equal to 100.

(a) Determine the insurer's expected annual payment. (M95-22)

(b) Determine the insurer's expected annual payment if loss amounts increased uniformly by 10%. (M95-23)

2.119 The data in Exercise 2.1 were truncated from below at 5,000,000. Determine an appropriate model for ground-up losses.

2.120 A model for the data in Exercise 2.2 was obtained in Exercise 2.102. Suppose the data had been truncated from below at 3,200. This would eliminate the first eleven rows of Table 2.24. Determine an appropriate ground-up model from this reduced data set and compare your answer to that for Exercise 2.102. Based on this model determine the expected payment per payment if there is a deductible of 5,000 and a limit of 1,000,000.

2.121 Repeat Exercise 2.119 except this time determine a model for losses shifted at 5,000,000. For each model, determine the expected payment per payment with a deductible of 25,000,000.

2.122 A model for the data in Exercise 2.28 was obtained in Exercise 2.104. Suppose the data had been truncated from below at 25,000. Determine an appropriate ground-up model from this reduced data set and compare your answer to that for Exercise 2.104.

2.123 A single-parameter Pareto model for the data in Exercise 2.38 was obtained in that exercise. Suppose the data had been truncated from below at 125 and censored from above at 200. Determine the maximum likelihood estimate of the single parameter Pareto parameter α from this modified data (the "parameter" θ remains at 100) set. Compare your answer to that from Exercise 2.38.

2.124 A model for the data in Exercise 2.39 was obtained in Exercise 2.107. Suppose the data had been truncated from below at 250. Determine an appropriate ground-up model from this reduced data set and compare your answer to that for Exercise 2.107. Also compare the value of $E(X \wedge 2{,}500)$ to the value obtained in Exercise 2.107.

2.125 Suppose the data in Exercise 2.82 had been truncated from below at 6 months (this would be the case if, for example, there was a 6-month waiting period associated with a long term disability coverage). Determine an appropriate ground-up model from this truncated data and compare it to the answer to Exercise 2.110. Suppose the coverage was retroactive. That is, once disability lasted six months the benefit would be for the entire period. Use the model obtained in this exercise to estimate the expected number of months of disability per benefit.

Section 2.11

2.126 Determine $Var(X_K)$ when using a scale based kernel density estimator that uses a Pareto distribution as the kernel. Assume $\alpha > 2$.

2.127 Assume that the conditional distribution of Y given $X = x$ does not depend on the parameters which describe the marginal distribution of X. Prove that the maximum likelihood estimates of the parameters of the marginal distribution of X are the same when estimated from the marginal distribution and data as they are when estimated from the joint distribution and bivariate data.

2.128 Construct a scale-based kernel density estimator based on the Pareto distribution for the data in Exercise 2.1.

2.129 Let X have the inverse exponential distribution with parameter θ. Let the conditional distribution of Y given $X = x$ have the exponential distribution with scale parameter $\gamma + \beta \log x$.

(a) Determine the joint density function of X and Y.

(b) Simplify expression (1) in Example 2.81 to at least a one-dimensional integral (which could easily be numerically evaluated).

(c) Determine the maximum likelihood estimates of the parameters for the data in Table 2.22. Let X be the loss and Y be the ALAE.

2.130 Consider the data set in Table 2.22. Fit a bivariate distribution using Frank's copula where each marginal distribution has the inverse exponential distribution. Use the likelihood ratio test to determine if the association is significant.

2.13 CASE STUDY

2.13.1 The case study continued

Our analysis of the case study in this section will be confined to issues of claim severity. Because of the numerous different deductibles and limits as well as its excellent statistical properties, we selected maximum likelihood as the estimation method. Writing the likelihood function is a relatively simple matter. If the deductible is d, the limit is u (here u is the maximum *loss*, that

Table 2.31 Fitted distributions, combined losses

Name	No. of parameters	−Loglikelihood
Inverse exponential	1	1,858.56
Lognormal	2	1,781.17
Weibull	2	1,786.02
Loglogistic	2	1,787.71
Paralogistic	2	1,787.90
Inverse paralogistic	2	1,788.43
Pareto	2	1,788.47
Inverse Pareto	2	1,791.96
Inverse Gaussian	2	1,792.51
Inverse Weibull	2	1,793.06
Inverse Burr	3	1,787.18
Generalized Pareto	3	1,788.42

is, the sum of the deductible and the maximum payment as given in Tables 1.1–1.4), and the loss is x, the contribution to the likelihood function is

$$\frac{f(x)}{1-F(d)}$$

when $x < u$ and is

$$\frac{1-F(u)}{1-F(d)}$$

when $x = u$, where $f(x)$ and $F(x)$ are the pdf and cdf of the model, respectively.

The simplex method was used to maximize the likelihood function. Maximization was successful for twelve models. The results are given in Table 2.31. The lognormal distribution has the smallest value of the negative loglikelihood for all twelve models. On the basis of this measure, there is no reason to consider the other models. The parameter estimates are $\hat{\mu} = 10.5430$ and $\hat{\sigma} = 2.31315$.

Although the lognormal distribution "scored" better than any of the others, it would be much more comforting to know if the fit is sufficiently good for us to use this model. If not, we would have to expand our search to more complex models—for example, two-point mixtures. The variety of deductibles makes the chi-square goodness-of-fit test unworkable, and there is no obvious empirical distribution function for use with the Kolmogorov–Smirnov test. However, there is a way to construct an approximate empirical cdf. It turns out that the Kaplan–Meier product-limit estimator developed for survival analysis can be applied. See, for example, London [82, pp. 164–170]. The method works as follows. First, for each loss, $j = 1, \ldots, n$, record d_j, the deductible, u_j, the limit, and x_j, the loss. Note that under our usual definitions, the maximum

Table 2.32 Liability payments—property and liability actuaries

j	d_j	u_j	x_j
1	0	7,500,000	60,664
2	5,000,000	10,000,000	5,116,134
3	0	7,500,000	576,857
4	0	1,000,000	31,698
5	50,000	10,050,000	96,427
6	0	7,500,000	119,206
7	450,000	1,450,000	855,796
8	0	5,000,000	1,519,846
9	500,000	2,500,000	2,500,000
10	0	1,000,000	505
11	250,000	5,250,000	267,833
12	0	3,000,000	20,546
13	22,000,000	32,000,000	32,000,000
14	20,000	5,020,000	24,699
15	150,000	10,150,000	156,055
16	0	1,000,000	10,950
17	25,000,000	35,000,000	25,244,023
18	100,000	1,100,000	355,892
19	0	5,000,000	384,222

payment is $u_j - d_j$. In a general setting, it is possible for d_j to be zero and u_j to be infinity. Now take the set of $2n$ numbers $\{d_1, \ldots, d_n, x_1, \ldots, x_n\}$ and place them in nondecreasing order. If any x_j has the same value as some d_k, then the x_j is placed before the d_k. Also, if two or more x_js have the same value, those for which $x_j = u_j$ are placed after those for which $x_j < u_j$. Label these ordered numbers $t_1, \ldots, t_{2n}$. Next create a second series of numbers $\delta_1, \ldots, \delta_{2n}$, where $\delta_j = 0$ if t_j was equal to some d_k, $\delta_j = 1$ if t_j was equal to some x_k and $x_k = u_k$, and $\delta_j = 2$ if t_j was equal to some x_k and $x_k < u_k$. Before this gets any more complicated, we present an example.

Example 2.87 *Construct the Kaplan–Meier estimate of the empirical distribution function for the data in Table 1.4.*

There were $n = 19$ observations. The values in that table are reproduced in Table 2.32 with the limits and payments as given being converted to limits and losses as required for this analysis. The 38 ordered values along with their origin are given in Table 2.33.

□

Table 2.33 Ordered values for Kaplan–Meier estimation

j	t_j	δ_j	j	t_j	δ_j
1	0	0	20	119,206	2
2	0	0	21	150,000	0
3	0	0	22	156,055	2
4	0	0	23	250,000	0
5	0	0	24	267,833	2
6	0	0	25	355,892	2
7	0	0	26	384,222	2
8	0	0	27	450,000	0
9	0	0	28	500,000	0
10	505	2	29	576,857	2
11	10,950	2	30	855,796	2
12	20,000	0	31	1,519,846	2
13	20,546	2	32	2,500,000	1
14	24,699	2	33	5,000,000	0
15	31,698	2	34	5,116,134	2
16	50,000	0	35	22,000,000	0
17	60,664	2	36	25,000,000	0
18	96,427	2	37	25,244,023	2
19	100,000	0	38	32,000,000	1

We can now use this information to construct the empirical distribution function. Begin with $F_n(0) = 0$. Set counters $r = s = 0$ and argument $y_0 = 0$. For $j = 1, \ldots, 2n$, do the following.

- If $\delta_j = 0$, let $r = r + 1$.
- If $\delta_j = 1$, let $r = r - 1$.
- If $\delta_j = 2$, let $c = r - d$ and $s = s + 1$, set $y_s = t_j$ and $F_n(y_s) = 1 - [1 - F_n(y_{s-1})]c/r$, where d is the number of consecutive js with the same value of t_j and $\delta_j = 2$. Then skip the next $d - 1$ terms.

When the above process is completed, the empirical distribution is obtained by connecting consecutive values of $F_n(y_s)$ with straight lines.

Example 2.88 (Example 2.87 continued) *Construct the empirical distribution function.*

The various required values appear in Table 2.34. Because r reaches zero before the end (it is always zero at the end), the empirical distribution is available only up to 1,519,846. The empirical distribution will reach one only

Table 2.34 Kaplan–Meier estimates

j	r	s	y_s	$F_n(y_s)$	j	r	s	y_s	$F_n(y_s)$
1	1				20	4	8	119,206	0.7407
2	2				21	5			
3	3				22	4	9	156,055	0.7926
4	4				23	5			
5	5				24	4	10	267,833	0.8341
6	6				25	3	11	355,892	0.8756
7	7				26	2	12	384,222	0.9170
8	8				27	3			
9	9				28	4			
10	8	1	505	0.1111	29	3	13	576,857	0.9378
11	7	2	10,950	0.2222	30	2	14	855,796	0.9585
12	8				31	1	15	1,519,846	0.9793
13	7	3	20,546	0.3194	32	0			
14	6	4	24,699	0.4167	33	1			
15	5	5	31,698	0.5139	34	0	16	5,116,134	
16	6				35	1			
17	5	6	60,664	0.5949	36	2			
18	4	7	96,427	0.6759	37	1	17	25,244,023	
19	5				38	0			

if the last observation is a loss ($\delta = 2$) and r is never zero prior to that point. The empirical distribution function is plotted in Figure 2.21. □

Returning to the case study, in Figure 2.22 the empirical distribution function for all payments combined is plotted along with the fitted lognormal distribution function. While the fit is not perfect, it does quite well, particularly at the large losses (those above $\exp(12.5) \doteq 268{,}000$).

The final task for the loss analysis is to check whether or not we should be treating the three samples separately. This can be done using a likelihood ratio test. The key numbers appear in Table 2.35. The likelihood ratio test statistic is $2(1{,}781.17 - 1{,}779.47) = 3.4$. With 4 degrees of freedom, the p-value is 0.4932 and the hypothesis that all three groups have the same severity distribution is accepted.

2.13.2 Exercises

All of the tables referred to in the following exercises appear at the end of this section.

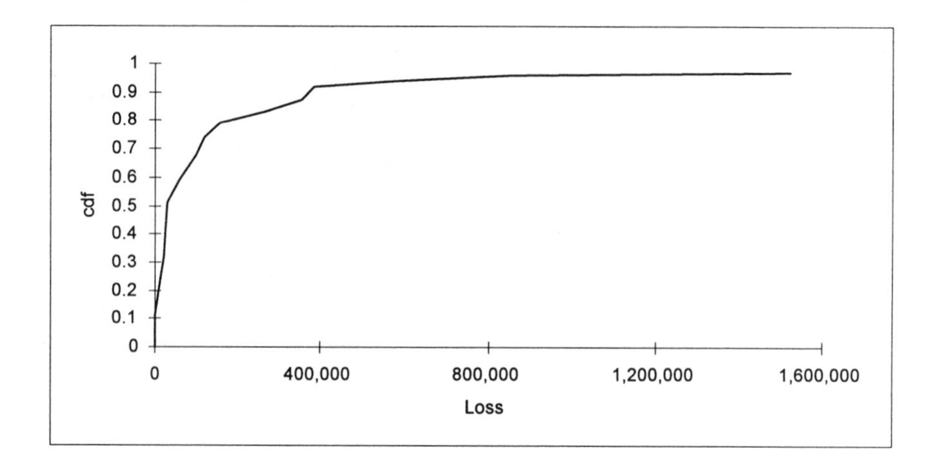

Fig. 2.21 Empirical distribution function—property and liability only

Table 2.35 Likelihood ratio test for separate models

Sample	Parameter		−Loglikelihood
	μ	σ	
All	10.5430	2.31315	1,781.17
Separate			
Life/health	10.8124	2.09550	533.64
Pension	10.3872	2.50254	1,015.37
Property/liability	10.5573	1.90253	230.46
Total			1,779.47

2.131 In 1920, Outwater [97] studied the time to recovery or death for disability cases in several states. Data were obtained from California (1918) and Oregon (1915) on cases of permanent partial disability. Only cases that were terminated (by recovery or death) within one year were included. The data appear in Table 2.36. Determine an appropriate model.

2.132 Outwater [97] also collected data on the duration of cases of temporary total disability in Washington (1913–17, 68,780 observations), Ohio (1915–16, 96,891), and Oregon (1915–17, 11,689). The three data sets are given in Tables 2.37–2.39. Is it reasonable to use the same family to model each population? Is it reasonable to use the same family with the same parameters to model each population? Base your opinions on the likelihood ratio test, the Schwartz Bayesian criterion, and/or tables and graphs. Which method seems to be most appropriate given the large number of observations in each sample?

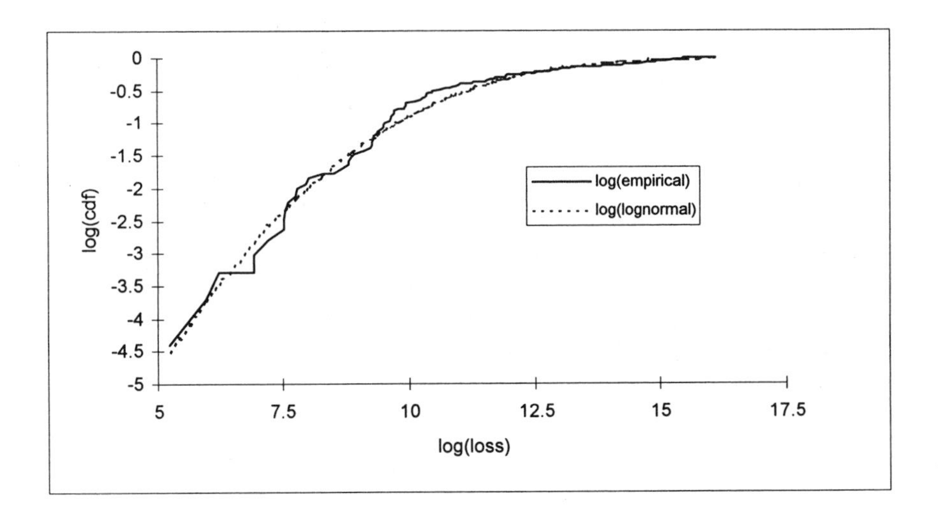

Fig. 2.22 Empirical and fitted lognormal model, all losses

2.133 A particular workers compensation benefit was set at two-thirds of the weekly wage, but must be at least 4 but no more than 10. Mowbray [94] obtained data on the wages relating to 3,092 claims filed in Massachusetts in 1919. The data appear in Table 2.40. Determine a parametric model for the wage level and then determine the expected value and standard deviation of the benefit. (Mowbray obtained an empirical estimate of 9.67). Observations below 4 were possible, there just were not any.

2.134 In New York there were special funds for some infrequent occurrences under workers compensation insurance. One was the event of a case being reopened. Hipp [64] collected data on the time from an accident to when the case was reopened. These covered cases reopened between April 24, 1933 and December 31, 1936. The data appear in Table 2.41. Determine a parametric model for the time from accident to reopening. By definition, at least seven years must have elapsed before a claim can qualify as a reopening, so the model should be conditioned on the time being at least seven years.

2.135 In the first of two papers by Arthur Bailey [5, p. 51], written in 1942 and 1943, he observed that "Another field where a knowledge of sampling distributions could be used to advantage is that of rating procedures for deductibles and excess coverages." In the second paper [6] he presented some data (Table 2.42) on the distribution of loss ratios. In that paper he made the statement that the popular lognormal model provided a good fit and passed the chi-square test. Does it? Is there a better model?

2.136 Longley-Cook [83] was interested in pricing layers of coverage for individual large fire losses. He obtained data on losses during 1946–1949 from the

National Fire Protection Association. A loss had to be for at least 250,000 to qualify. Longley-Cook thought that a function of the following form would closely fit the data.

$$\log[E(X) - E(X \wedge x)] = \log\left[\int_x^{\infty} 1 - F(t)dt\right] = ab^x + c + dx$$

Determine the form of the cdf for the model that satisfies the above relationship. Longley-Cook estimated the parameters as $\hat{a} = 1.47365$, $\hat{b} = 0.0625$, $\hat{c} = 4.81439$, and $\hat{d} = -0.942218$, where x is in millions of dollars. Compare these to the maximum likelihood estimates obtained from the data given in Table 2.43. Can you obtain a better model? Longley-Cook was interested in estimating quantities of the form

$$E[X \wedge 250{,}000(k+1)] - E[X \wedge 250{,}000k],\ k = 1, 2, \ldots$$

which may affect what it means to obtain the best model.

2.137 Comprehensive medical claims were studied by Bevan [14] in 1963. Male (955 payments) and female (1,291 payments) claims were studied separately. The data appear in Table 2.44 where there was a deductible of 25. Can a common model be used? Using an appropriate model, determine the percentage by which the premium should be reduced if the deductible is raised from 25 to 100.

2.138 In 1958, Harwayne [54] investigated the time to pay claims in automobile liability insurance. He fit the model

$$\log_{10} y = -2.0674t^{-0.80599}10^{-0.24841t}$$

where t is time in years and y is the fraction of claims paid by that time (the distribution function). Using the data in Table 2.45, determine a better model of the same form and then see if you can obtain a better model of some other form. The sample size was not indicated. What difficulties does that add to the modeling process?

2.139 In 1961 Kormes [78] studied claim amounts for serious illness insurance in Massachusetts. He collected data (Table 2.46) for both 1958 and 1959. Is the same distribution type appropriate for both years? If so, is the same distribution type with all nonscale parameters equal appropriate? If so, what was the inflation rate from 1958 to 1959?

2.140 In 1964 Dropkin [34] investigated the use of the lognormal distribution for workers compensation losses. One of his data sets was for losses due to major permanent partial disabilities. Data were available for 1960 and 1961 in California. He concluded that the lognormal distribution was not appropriate for these two data sets. Do you agree? Determine appropriate distributions for these two data sets. Is it appropriate to use the same distribution with the same nonscale parameter values? If so, what was the inflation rate from 1960 to 1961? The data in Table 2.47 were condensed by providing wider groupings than in the original.

2.141 In a 1966, paper Hewitt [61] recommended the loggamma distribution. If X has a gamma distribution, then $Y = e^X$ has the loggamma distribution. Note that probability for Y begins at one, not zero. Hewitt's data set was the distribution of workers compensation plans by size of annual premium (Table 2.48). The counts have been rounded to the nearest hundred. Does the loggamma distribution provide an acceptable model? Can you find a better one? Does the large sample size cause any particular problems?

2.142 In 1967 Hewitt [62] was still studying distributions of workers compensation plans by size of premium, but this time he also wanted to relate the experience of the plan to its size. For each plan he obtained the loss ratio, the ratio of the loss to the premium. A regression model and a distributional assumption produced the following results. Given the premium level P, the loss ratio R has a gamma distribution with parameters $\alpha = r$ and $\theta = \epsilon/r$ where ϵ is a constant that is externally available (but not given in the paper). The value of r is based on the regression equation

$$\log_{10} r = -3.264 + 0.773 \log_{10} P.$$

Determine $E[R|P]$ and $E[L|P]$, where $L = RP$. Also determine $Var(L|P)$, $E[L|2P]$, and $Var(L|2P)$. It was Hewitt's contention that insuring one group with premium $2P$ is not the same as insuring two groups each with premium P. Does his model support that claim?

Use the data set in Table 2.49 to determine a distributional model for P. Then use Hewitt's model for L given P to determine the joint density function for (L, P).

2.143 In 1974, Walters [124] studied fire losses under homeowner's insurance coverage. The data are given in Table 2.50. Determine an appropriate parametric model and then obtain the loss elimination ratio for a 100 deductible. In the paper, Walters obtained an empirical value of .111. What was the reason for the unusual class boundaries and therefore what distribution did Walters have in mind?

2.144 In 1976, Finger [40] recommended the lognormal distribution as a model when the goal is to obtain the LER for a policy limit. His argument was that for the lognormal distribution, LER $-[E(X) - E(X \wedge x)]/E(X)$ depends only on $k = x/E(X)$ and the coefficient of variation $\sqrt{Var(X)}/E(X)$. Demonstrate that this is true. Is it true for the gamma distribution? He then looked at the data in Table 2.51 and concluded that the lognormal distribution does not fit the entire range from zero to infinity, but if all observations below 10,000 were deleted, the lognormal distribution provided a good fit. Do you agree?

2.145 In 1977, Miccolis [90] studied increased limits factors. The definition is that for a basic limit of b the increased limits factor for a limit of k is

$$I(k) = \frac{E(X \wedge k)}{E(X \wedge b)}.$$

Thus the increased limits factor is the amount by which the basic limits net premium should be multiplied in order to obtain the net premium for a higher limit. Demonstrate that

$$I'(k) = \frac{1 - F_X(k)}{E(X \wedge b)}$$

and therefore $I(k)$ is increasing. Is $I'(k)$ increasing, or decreasing?

Miccolis presents the following example (Table 2.52) of a table of increased limits factors. Calculate approximate derivatives and state whether or not the table is consistent with some underlying distribution. Give some reasons why the correct increased limits factors may not correspond to an underlying distribution. Finally, assume that losses are lognormally distributed. Determine parameters that provide increased limits factors that are close to those given here but are consistent with the lognormal distribution.

2.146 In 1979, Hewitt and Lefkowitz [63] looked at automobile bodily injury liability data (Table 2.53) and concluded that a two-point mixture of the gamma and loggamma distributions was superior to the lognormal. Do you agree? Can you find a single distribution that is as good as their two-point mixture?

2.147 A 1980 paper by Patrik [98] contained many of the ideas recommended in Chapter 2. One of his examples was data supplied by the Insurance Services Office on Owners, Landlords, and Tenants bodily injury liability. Policies at two different limits were studied. Both were for policy year 1976 with losses developed to the end of 1978. Can the same model (with or without identical parameters) be used for the two limits? The groupings in Table 2.54 have been condensed from those in the paper.

2.148 In 1989, Hayne [56] studied a problem in predicting the number of claims yet to be reported. He begins with a problem and solution similar to that in Example 2.74. Let N be the unknown total number of claims and let the random variable X be the time to reporting of a claim. Also, let n be the number of claims reported as of time t. The estimate of N was $n/F_X(t)$. The error in the estimate due to the fact that the parameters of X were estimated from data was obtained in that example. Hayne was also interested in the variance due to the randomness in the problem itself. His development was as follows.

Given N, n (as a random variable) has a binomial distribution with parameters N and $q = F_X(t)$. Then n is approximately normal with mean Nq and variance $Nq(1-q)$. He then turns this around to claim that N (given n) is approximately normal with mean n/q and variance $n(1-q)/q^2$. The number of outstanding claims (again, given n) is $N-n$ which is approximately normal with mean $n(q^{-1}-1)$ and variance $n(1-q)/q^2$. However, all of this is dependent on θ, the parameter(s) of X and the numerical values are dependent upon $\hat{\theta}$. So finally

$$E(N-n) = E[E(N-n|\hat{\theta})] = E[n(\hat{q}^{-1}-1)] \doteq n(\hat{q}^{-1}-1)$$

and

$$\begin{aligned} Var(N-n) &= E[Var(N-n|\hat{\theta})] + Var[E(N-n|\hat{\theta})] \\ &= E[n(1-\hat{q})/\hat{q}^2] + Var[n(\hat{q}^{-1}-1)] \\ &\doteq n(1-\hat{q})/\hat{q}^2 + n^2 Var(\hat{q})/\hat{q}^4. \end{aligned}$$

Verify each of the algebraic steps and indicate if some are also approximations. Continue Example 2.74 by obtaining a 95% confidence interval that captures the uncertainty in both the claims development and estimation processes.

Table 2.36 California and Oregon permanent partial disability data

Weeks	No.	Weeks	No.	Weeks	No.
0–1	81	11–12	52	22–23	17
1–2	67	12–13	53	23–24	12
2–3	106	13–14	36	24–25	13
3–4	142	14–15	21	25–26	12
4–5	149	15–16	26	26–30	31
5–6	127	16–17	20	30–34	17
6–7	107	17–18	24	34–38	20
7–8	71	18–19	14	38–42	7
8–9	90	19–20	14	42–46	9
9–10	56	20–21	13	46–52	10
10–11	55	21–22	14		
				Total	1,486

Table 2.37 Washington temporary total disability data

Weeks	No.	Weeks	No.	Weeks	No.
0–1	12,313	9–10	780	18–19	137
1–2	17,489	10–11	786	19–20	163
2–3	11,169	11–12	485	20–21	100
3–4	6,658	12–13	899	21–22	378
4–5	5,499	13–14	280	22–23	83
5–6	3,006	14–15	246	23–24	87
6–7	3,425	15–16	250	24–25	72
7–8	1,505	16–17	127	25+	1,507
8–9	1,810	17–18	526		
				Total	69,780

Table 2.38 Ohio temporary total disability data

Weeks	No.
0–1	52,359
1–2	15,591
2–3	10,378
3–4	6,053
5–13	11,124
13+	1,386
Total	96,891

Table 2.39 Oregon temporary total disability data

Weeks	No.	Weeks	No.
0–1	4,136	7–8	148
1–2	3,099	8–9	143
2–3	1,537	9–10	86
3–4	854	10–11	67
4–5	563	11–12	35
5–6	389	12–13	56
6–7	302	13+	274
		Total	11,689

Table 2.40 Massachusetts wage data

Wages	No. of claims
4–9	68
9–14	287
14–19	721
19–24	906
24–29	630
29–34	373
34–39	120
39–44	52
44–49	16
49–54	13
54+	7
Total	3,193

Table 2.41 Time to reopening of a workers compensation claim

Years	No. reopened	Years	No. reopened
7–8	27	15–16	13
8–9	43	16–17	9
9–10	42	17–18	7
10–11	37	18–19	4
11–12	25	19–20	4
12–13	19	20–21	1
13–14	23	21+	0
14–15	10		
		Total	264

Table 2.42 Loss ratio data

Loss ratio	Number
0.0–0.2	16
0.2–0.4	27
0.4–0.6	22
0.6–0.8	29
0.8–1.0	19
1.0–1.5	32
1.5–2.0	10
2.0–3.0	13
3.0+	5
Total	173

Table 2.43 Large fire losses

Loss (10^3)	No. of fires	Loss (10^3)	No. of fires
250–297	248	1,420–1,680	10
297–354	144	1,680–2,000	5
354–420	77	2,000–2,380	9
420–500	57	2,380–2,830	3
500–595	85	2,830–3,360	3
595–707	22	3,360–4,000	2
707–841	33	4,000–4,760	1
841–1,000	19	4,760–5,660	1
1,000–1,190	28	5,660–6,730	1
1,190–1,420	11	6,730–	0
		Total	759

Table 2.44 Comprehensive medical losses

Loss	Male (%)	Female (%)
25–50	19.3	15.4
50–100	28.3	24.0
100–200	16.7	20.3
200–300	9.2	12.6
300–400	6.6	8.0
400–500	4.9	5.3
500–1,000	6.4	9.6
1,000–2,000	3.7	3.1
2,000–3,000	1.9	0.9
3,000–4,000	1.4	0.3
4,000–5,000	0.4	0.1
5,000–6,667	0.5	0.2
6,667–7,500	0.3	0.1
7,500–10,000	0.6	0.1

Table 2.45 Proportion of claims paid by a given time

Years	Proportion paid
1	0.0699
2	0.4237
3	0.7050
4	0.8333
5	0.9090
6	0.9602
7	0.9808

Table 2.46 Serious illness losses

Loss	No. of losses 1958	No. of losses 1959	Loss	No. of losses 1958	No. of losses 1959
0–25	72	44	500–600	11	9
25–50	48	33	600–800	20	10
50–75	52	27	800–1,000	41	156
75–100	46	28	1,000–1,500	41	78
100–150	43	39	1,500–2,000	10	9
150–200	47	23	2,000–3,000	11	15
200–250	32	28	3,000–5,000	12	5
250–300	53	103	5,000–7,500	1	0
300–400	19	10	7,500–10,000	1	0
400–500	26	37			
			Totals	586	654

Table 2.47 Major permanent partial disability losses

Loss	No. of losses 1960	No. of losses 1961	Loss	No. of losses 1960	No. of losses 1961
0–100	6	3	13,000–14,000	233	326
100–500	1	2	14,000–15,000	152	276
500–1,000	5	2	15,000–17,500	310	486
1,000–3,000	11	8	17,500–20,000	178	307
3,000–5,000	33	78	20,000–25,000	183	284
5,000–6,000	68	108	25,000–35,000	94	145
6,000–7,000	159	236	35,000–50,000	34	70
7,000–8,000	253	328	50,000–75,000	26	35
8,000–9,000	310	402	75,000–100,000	7	13
9,000–10,000	355	433	100,000–150,000	1	2
10,000–11,000	346	432	150,000–200,000	1	2
11,000–12,000	286	369	200,000–	0	0
12,000–13,000	219	374			
			Totals	3,271	4,721

Table 2.48 Workers' compensation plans by size of premium

Amount of premium	No. of plans
0–100	397,200
100–500	348,600
500–750	55,600
750–1,000	28,000
1,000–5,000	71,500
5,000–25,000	13,900
25,000–50,000	1,500
50,000–100,000	700
100,000–250,000	400
250,000–	100
Total	917,500

Table 2.49 Workers' compensation plans by size of premium

Amount of premium	No. of plans
0–500	2,946
500–750	3,126
750–1,000	2,889
1,000–1,500	3,718
1,500–2,500	3,576
2,500–5,000	2,891
5,000–7,500	939
7,500–10,000	465
10,000–15,000	454
15,000–25,000	316
25,000–50,000	256
50,000–100,000	91
100,000–	55
Total	21,722

Table 2.50 Homeowner's fire losses

Fire loss	No. of claims	Fire loss	No. of claims
0–1.78	151	562.34–1,000.00	257
1.78–3.16	14	1,000.00–1,778.28	157
3.16–5.62	93	1,778.28–3,162.29	96
5.62–10.00	228	3,162.29–5,623.38	71
10.00–17.78	736	5,623.38–10,000.00	75
17.78–31.62	1,159	10,000.00–17,782.80	100
31.62–56.23	1,225	17,782.80–31,622.85	22
56.23–100.00	1,120	31,622.85–56,233.75	1
100.00–177.83	821	56,233.75–100,000.00	1
177.83–316.23	636	100,000.00–	0
316.23–562.34	396		
		Total	7,359

Table 2.51 Loss data

Loss (10^3)	Number
0–10	1,496
10–25	365
25–100	267
100–300	99
300–1000	15
1000–	1
Total	2,243

Table 2.52 Increased limits factors

Limit, k (10^3)	$I(k)$	Limit, k (10^3)	$I(k)$
25	1.000	1,250	2.575
50	1.250	1,500	2.700
100	1.425	1,750	2.825
200	1.625	2,000	2.950
250	1.705	2,500	3.100
300	1.775	3,000	3.300
350	1.865	4,000	3.600
400	1.915	5,000	3.800
500	1.975	7,500	4.300
750	2.175	10,000	4.800
1,000	2.400		

Table 2.53 Automobile bodily injury liability losses

Loss	Number	Loss	Number
0–50	27	750–1,000	8
50–100	4	1,000–1,500	16
100–150	1	1,500–2 ,000	8
150–200	2	2,000–2 ,500	11
200–250	3	2,500–3 ,000	6
250–300	4	3,000–4 ,000	12
300–400	5	4,000–5 ,000	9
400–500	6	5,000–7 ,500	14
500–750	13	7,500–	40
		Total	189

Table 2.54 OLT bodily injury liability losses

Loss (10^3)	300 Limit	500 Limit	Loss (10^3)	300 Limit	500 Limit
0–0.2	10,075	3,977	11–12	56	22
0.2–0.5	3,049	1,095	12–13	47	23
0.5–1	3,263	1,152	13–14	20	6
1–2	2,690	991	14–15	151	51
2–3	1,498	594	15–20	151	54
3–4	964	339	20–25	109	44
4–5	794	307	25–50	154	53
5–6	261	103	50–75	24	14
6–7	191	79	75–100	19	5
7–8	406	141	100–200	22	6
8–9	114	52	200–300	6	9
9–10	279	89	300–500	10[a]	3
10–11	58	23	500–		0
			Totals	24,411	9,232

[a]losses for 300+

3

Frequency Distributions—Models for the Number of Payments

3.1 INTRODUCTION

The purpose of this chapter is to introduce a large class of counting distributions. Counting distributions are discrete distributions with probabilities only on the non-negative integers; that is, probabilities are defined only at the points $0, 1, 2, 3, 4, \ldots$. The purpose of studying counting distributions in an insurance context is simple. Counting distributions describe the number of events such as losses to the insured or claims to the insurance company. With an understanding of both the number of claims and the size of claims, one can have a deeper understanding of a variety of issues surrounding insurance than if one has only information about total losses. The description of total losses in terms of numbers and amounts separately also allows one to address issues of modification of an insurance contract. Another reason for separating numbers and amounts of claims is that models for the number of claims are fairly easy to obtain and experience has shown that the commonly used distributions really do model the propensity to generate losses. Some additional perspectives on the historical development of these models is provided by the exercises in the last section of this chapter. This chapter, then, is devoted to the study of the claim (or loss) number process and the interaction with the claim size distribution which was studied in Chapter 2.

As in Chapter 2 we will focus on parametric models of claim numbers rather than the empirical distribution based on a set of data. The most important reason for using a parametric model is that it summarizes the information about a distribution in terms of the form of the distribution (e.g. Poisson,

Table 3.1 Number of hospital liability claims by year

Year	Number of claims
1985	6
1986	2
1987	3
1988	0
1989	2
1990	1
1991	2
1992	5
1993	1
1994	3

Table 3.2 Hospital liability claims by frequency

Frequency	Number of observations
0	1
1	2
2	3
3	2
4	0
5	1
6	1
7+	0

binomial, etc.) and the parameter values of the distribution. Also, as in Chapter 2, we find that parametric models serve to smooth the empirical data. The empirical distribution has each point in the distribution as an independent parameter. Hence, the dimensionality of the information is greatly reduced when using a parametric distribution. Consider the following example.

Example 3.1 *A hospital liability policy has experienced the number of claims over the past ten years as given in Table 3.1.*

These data can be summarized in a different way. We can count the number of years in which exactly zero claims occurred, one claim occurred, etc. as in Table 3.2.

This now constitutes an empirical distribution of the number of claims per year. It provides information about the variability of the number of claims per year and is useful for predicting the number of claims in future years and understanding the variability that might be expected in future years. □

Although in practice the risk exposure may have changed over the years because of the increase in hospital size and other factors, we will assume that this is not the case or, equivalently, that numbers have been adjusted accordingly. We will discuss this adjustment in a later section of this chapter.

For the purpose of developing a distribution to describe the number of claims in future years, the information in our table is insufficient to tell us directly about the probability of occurrence of a number of claims per year larger than those observed. For example, in the past ten years, there were no years in which there were 8 claims. However, we know that 8 claims could occur in some year in the future and we would like to assign a probability to that occurrence.

Parametric models can be fitted to this data. For example, a Poisson distribution has positive probabilities for all non-negative integers. Fitting a Poisson distribution to the data would automatically assign probabilities at all integers, including those greater than the maximum number of observed occurrences. Hence, the fitting of a parametric distribution then provides automatic extrapolation beyond the data. By extending the distribution beyond the maximum observation we can study the shape of the right tail of the distribution.

Parametric models also allow us to develop frequency models for insurance situations for which no data exist—for example, a policy with a different level of deductible. Consider an automobile insurance policy with a 100 deductible. Suppose the insurer is interested in designing a policy with an increased deductible. The number of claims arising from the policy with the increased deductible will, of course, be different from the policy with the lower deductible. We will show how to change the parameters of the frequency distribution to reflect the change in deductible. That change is based on the severity distribution, specifically the probability that a loss will lie between the two deductibles under consideration.

We now formalize some of the notation that will be used in this and subsequent chapters. Let the **probability function** (pf) p_k denote the probability that exactly k events (such as claims or losses) occur. Let N be a random variable representing the number of such events. Then

$$p_k = \Pr(N = k), \qquad k = 0, 1, 2, \ldots.$$

Definition 3.1 *The* ***probability generating function*** *(pgf) of a discrete random variable N with pf p_k is*

$$P(z) = P_N(z) = E\left(z^N\right) = \sum_{k=0}^{\infty} p_k z^k. \tag{3.1}$$

As is true with the moment generating function, the pgf can be used to generate moments. In particular, $P'(1) = E(N)$ and $P''(1) = E[N(N-1)]$ (see Exercise 3.2). The remainder of this chapter is organized by class of distribution. Since it is statistical in nature, much of the material may have been previously studied by the reader. Consequently, we will pass through the basic material rather quickly.

3.2 THE POISSON DISTRIBUTION

3.2.1 Some properties

The Poisson distribution has pf

$$p_k = \frac{e^{-\lambda}\lambda^k}{k!}, \qquad k = 0, 1, 2, \ldots.$$

The probability generating function from (3.1) is

$$P(z) = e^{\lambda(z-1)}, \qquad \lambda > 0.$$

The mean and variance can be computed from the probability generating function as follows:

$$\begin{aligned} E(N) &= P'(1) = \lambda \\ E[N(N-1)] &= P''(1) = \lambda^2 \\ Var(N) &= E[N(N-1)] + E(N) - [E(N)]^2 \\ &= \lambda^2 + \lambda - \lambda^2 \\ &= \lambda. \end{aligned}$$

It can be seen that for the Poisson distribution the variance is equal to the mean. The Poisson distribution can arise from a Poisson process (to be discussed in Chapter 6). The Poisson distribution and Poisson processes are also discussed in many textbooks in statistics and actuarial science including Panjer and Willmot [102].

The Poisson distribution has at least two additional useful properties. The first is given in the following theorem and the second is discussed in the next subsection.

Theorem 3.1 *Let $N_1, \ldots, N_n$ be independent Poisson variables with parameters $\lambda_1, \ldots, \lambda_n$. Then $N = N_1 + \cdots + N_n$ has a Poisson distribution with parameter $\lambda_1 + \cdots + \lambda_n$.*

Proof: The pgf of the sum of independent random variables is the product of the individual pgfs. For the sum of Poisson random variables we have

$$\begin{aligned} P_N(z) &= \prod_{j=1}^{n} P_{N_j}(z) = \prod_{j=1}^{n} \exp[\lambda_j(z-1)] \\ &= \exp\left[\sum_{j=1}^{n} \lambda_j(z-1)\right] \\ &= e^{\lambda(z-1)}. \end{aligned}$$

where $\lambda = \lambda_1 + \cdots + \lambda_n$. Just as is true with moment generating functions, the pgf is unique and therefore N must have a Poisson distribution with parameter λ. □

3.2.2 Decomposing Poisson frequencies

The Poisson distribution has a property that is particularly useful in modeling insurance risks. Suppose that the number of claims in a fixed time period, such as one year, follows a Poisson distribution. Further suppose that the claims can be classified into m distinct types. For example, claims could be classified by size, such as those below a fixed limit and those above the limit. It turns out that if one is interested in studying the number of claims above the limit, that distribution is also Poisson but with a new Poisson parameter.

This is also useful when considering removing or adding a part of an insurance coverage. Suppose that the number of claims for a complicated medical benefit coverage follows a Poisson distribution. Consider the "types" of claims to be the different medical procedures or medical benefits under the plan. If one of the benefits is removed from the plan, again it turns out that the distribution of the number of claims under the revised plan will still have a Poisson distribution but with a new parameter.

It is also interesting to note that in each of the cases mentioned in the previous paragraph, the number of claims of the different types will not only be Poisson distributed, but also be independent of each other; that is, the distributions of the number of claims above the limit and the number below the limit will be independent. This is a somewhat surprising result. For example, suppose we currently sell a policy with a deductible of 50 and experience has indicated that a Poisson distribution with a certain parameter is a valid model for the number of payments. Further suppose we are also comfortable with the assumption that the number of losses in a period also has the Poisson distribution, but we do not know the parameter. Without additional information, it is impossible to infer the value of the Poisson parameter should the deductible be lowered or removed entirely. We now formalize these ideas in the following theorem.

Theorem 3.2 *Suppose that the number of events N is a Poisson random variable with mean λ. Further suppose that each event can be classified into one of m types with probabilities $p_1, \ldots, p_m$ independent of all other events. Then the number of events $N_1, \ldots, N_m$ corresponding to event types $1, \ldots, m$ respectively, are mutually independent Poisson random variables with means $\lambda p_1, \ldots, \lambda p_m$ respectively.*

Proof: For fixed $N = n$, the conditional joint distribution of $(N_1, \ldots, N_m)$ is multinomial with parameters $(n, p_1, \ldots, p_m)$. Also for fixed $N = n$, the conditional marginal distribution of N_j is binomial with parameters (n, p_j).

The joint pf of $(N_1, ..., N_m)$ is given by

$$\begin{aligned}
\Pr(N_1 = n_1, ..., N_m = n_m) &= Pr(N_1 = n_1, ..., N_m = n_m | N = n) \\
&\quad \times Pr(N = n) \\
&= \frac{n!}{n_1! n_2! \cdots n_m!} p_1^{n_1} \cdots p_m^{n_m} \frac{e^{-\lambda}\lambda^n}{n!} \\
&= \prod_{j=1}^{m} e^{-\lambda p_j} \frac{(\lambda p_j)^{n_j}}{n_j!}.
\end{aligned}$$

where $n = n_1 + n_2 + \cdots + n_m$. Similarly, the marginal pf of N_j is given by

$$\begin{aligned}
\Pr(N_j = n_j) &= \sum_{n=n_j}^{\infty} \Pr(N_j = n_j | N = n) \Pr(N = n) \\
&= \sum_{n=n_j}^{\infty} \binom{n}{n_j} p_j^{n_j} (1 - p_j)^{n-n_j} \frac{e^{-\lambda}\lambda^n}{n!} \\
&= e^{-\lambda} \frac{(\lambda p_j)^{n_j}}{n_j!} \sum_{n=n_j}^{\infty} \frac{[\lambda(1 - p_j)]^{n-n_j}}{(n - n_j)!} \\
&= e^{-\lambda} \frac{(\lambda p_j)^{n_j}}{n_j!} e^{\lambda(1-p_j)} \\
&= e^{-\lambda p_j} \frac{(\lambda p_j)^{n_j}}{n_j!}.
\end{aligned}$$

Hence the joint pf is the product of the marginal pfs, establishing mutual independence. □

Example 3.2 *In a study of medical insurance the expected number of claims per individual policy is 2.3 and the number of claims is Poisson distributed. You are considering removing one medical procedure from the coverage under this policy. Based on historical studies, this procedure accounts for approximately 10% of the claims. Determine the new frequency distribution.*

From Theorem 3.2, we know that the distribution of the number of claims expected under the revised insurance policy after removing the procedure from coverage is Poisson with mean $0.9(2.3) = 2.07$. In carrying out studies of the distribution of total claims, and hence the appropriate premium under the new policy, one also needs to study the change in the amounts of losses, the severity distribution, since the distribution of amounts of losses for the procedure which was removed may be different from distribution of amounts when all procedures are covered. This will be discussed in the next chapter. □

3.2.3 Estimation

The principles of estimation discussed in Chapter 2 can be applied equally to frequency distributions. We will now illustrate the methods of estimation by fitting a Poisson model.

Example 3.3 (Example 3.1 continued) *Estimate the Poisson parameter using the method of moments and the method of maximum likelihood.*

The total number of claims for the period 1985 through 1994 is 25. Hence, the average number of claims per year is 2.5. The average can also be computed from Table 3.2. Let n_k denote the number of years in which a frequency of exactly k claims occurred. The expected frequency (sample mean) is

$$\hat{\lambda} = \frac{\sum_{k=0}^{\infty} k n_k}{\sum_{k=0}^{\infty} n_k}$$

where n_k represents the number of observed values at frequency k. Hence the method of moments estimate of the Poisson parameter is 2.5.

Maximum likelihood estimation can easily be carried out on these data. The likelihood contribution of an observation of k is p_k. Then the likelihood for the entire set of observations is

$$L = \prod_{k=0}^{\infty} p_k^{n_k}$$

and the loglikelihood is

$$l = \sum_{k=0}^{\infty} n_k \log p_k.$$

The likelihood and loglikelihood functions are considered to be functions of the unknown parameters. In the case of the Poisson distribution, there is only one parameter, making the maximization easy.

For the Poisson distribution we obtain

$$p_k = \frac{e^{-\lambda}\lambda^k}{k!}$$

and we have

$$\log p_k = -\lambda + k \log \lambda - \log k!.$$

The loglikelihood is

$$l = -\lambda n + \sum_{k=0}^{\infty} k\ n_k \log \lambda - \sum_{k=0}^{\infty} n_k \log k!.$$

Differentiating the loglikelihood with respect to λ, we obtain

$$\frac{dl}{d\lambda} = -n + \sum_{k=0}^{\infty} k\ n_k \frac{1}{\lambda}.$$

By setting the derivative of the loglikelihood to zero, the maximum likelihood estimate is obtained as the solution of the resulting equation. The estimator is then

$$\hat{\lambda} = \frac{\sum_{k=0}^{\infty} k n_k}{n}.$$

From this it can be seen that for the Poisson distribution the maximum likelihood and the method of moments estimators are identical. This estimator has mean

$$E(\hat{\lambda}) = E(N) = \lambda$$

and variance

$$Var(\hat{\lambda}) = Var(N)/n = \lambda/n.$$

See Examples 2.12 and 2.13. Hence, $\hat{\lambda}$ is unbiased. From Theorem 2.2, the maximum likelihood estimator is asymptotically normally distributed with mean λ and variance

$$\begin{aligned} Var(\hat{\lambda}) &= \left\{-nE\left[\frac{d^2}{d\lambda^2}\log p_N\right]\right\}^{-1} \\ &= \left\{-nE\left[\frac{d^2}{d\lambda^2}(-\lambda + N\log\lambda - \log N!)\right]\right\}^{-1} \\ &= \left[nE(N/\lambda^2)\right]^{-1} \\ &= \left(n\lambda^{-1}\right)^{-1} = \lambda/n. \end{aligned}$$

In this case the asymptotic approximation to the variance is equal to its true value. From this information, we can construct an approximate 95% confidence interval for the true value of the parameter. The interval is $\hat{\lambda} \pm 1.96(\hat{\lambda}/n)^{1/2}$. In the case of the data in Example 3.1, the interval becomes (1.52, 3.48). This confidence interval is only an approximation because it relies on large sample theory. The data set in Example 3.1 is very small and such a confidence interval should be used with caution. □

Example 3.4 *The following example is taken from Douglas [31, p. 253]. An insurance company's records for one year show the number of accidents per day which resulted in a claim to the insurance company for a particular insurance coverage. The results are in Table 3.3. Determine if a Poisson model is appropriate.*

A Poisson model is fitted to these data. The method of moments and the maximum likelihood method both lead to the estimate of the mean

$$\hat{\lambda} = \frac{742}{365} = 2.0329.$$

The resulting Poisson model using this parameter value yields the distribution and expected numbers of claims per day as given in Table 3.4.

Table 3.3 Data for Example 3.4

No. of claims/day	Observed no. of days
0	47
1	97
2	109
3	62
4	25
5	16
6	4
7	3
8	2
9+	0

Table 3.4 Observed and expected frequencies

Number of claims/day, k	Poisson probability, $\hat{p}_k$	Expected number, $365\hat{p}_k$	Observed number, n_k
0	0.1310	47.8	47
1	0.2662	97.2	97
2	0.2706	98.8	109
3	0.1834	66.9	62
4	0.0932	34.0	25
5	0.0379	13.8	16
6	0.0128	4.7	4
7	0.0037	1.4	3
8	0.0009	0.3	2
9+	0.0003	0.1	0

The results in Table 3.4 show that the Poisson distribution fits the data pretty well. We can test formally by using a chi-square test statistic. This was done in Subsection 2.9.1 in connection with fitting models to grouped data from continuous distributions. The test statistic is

$$Q = \sum_{k=0}^{\infty} \frac{(n_k - E_k)^2}{E_k}$$

where E_k, the expected number of counts is given by

$$E_k = n\hat{p}_k = n \Pr(N = k; \hat{\boldsymbol{\theta}})$$

where $\hat{\boldsymbol{\theta}}$ indicates that all parameters have been replaced by estimates.

Table 3.5 Chi-square goodness-of-fit test

Claims/day	Observed	Expected	Chi-square
0	47	47.8	0.01
1	97	97.2	0.00
2	109	98.8	1.06
3	62	66.9	0.36
4	25	34.0	2.39
5	16	13.8	0.34
6+	9	6.5	0.97
Totals	365	365	5.14

For the test to be valid (at least approximately), each cell should have at least 5 expected observations. In the above example, this suggests grouping the cells at the upper end so that all observations of 6 or above are combined. This results in Table 3.5. Any time such a table is made, the expected count for the last group is

$$E_{k+} = n\hat{p}_{k+} = n(1 - \hat{p}_0 - \cdots - \hat{p}_{k-1}).$$

The number of degrees of freedom for this test statistic is given by

$$d = (\text{\# of cells}) - (\text{\# of estimated parameters}) - 1.$$

Hence, $d = 7 - 1 - 1 = 5$.

The null hypothesis is that the underlying distribution is the Poisson distribution. The null hypothesis is rejected if Q exceeds $\chi^2_{d,\alpha}$, where α is the significance level of the test.

The p-value for the test is the probability that a random value from the chi-square distribution (with 5 degrees of freedom) exceeds 5.14. For this data set, the p-value is 0.3994. This can be obtained from tables of the χ^2 distribution. With typical values of α such as 0.01, 0.05 and 0.10, the null hypothesis cannot be rejected. We then conclude that the Poisson distribution is an adequate fit.

For this example, we also have

$$Var(\hat{\lambda}) \doteq \frac{\hat{\lambda}}{n} = \frac{2.0329}{365} = 0.0056$$

resulting in an approximate 95% confidence interval for λ as $\hat{\lambda} \pm 1.96(\hat{\lambda}/n)^{1/2} = (1.89, 2.18)$. This gives a qualitative statement about the estimate of λ. □

In Chapter 2, the Kolmogorov–Smirnov test was presented as an alternative method of testing the goodness-of-fit. That test is appropriate only for continuous models and so cannot be used here.

All of the examples presented so far have assumed that the counts at each observed frequency are known. Occasionally, data are collected so that this is not given. The most common example is to have a final entry given as $k+$ where the count is the number of times k or more claims were observed. If n_{k+} is the number of times this was observed, the contribution to the likelihood function is

$$(p_k + p_{k+1} + \cdots)^{n_{k+}} = (1 - p_0 - \cdots - p_{k-1})^{n_{k+}}.$$

The same adjustments apply to grouped frequency data of any kind. Suppose there were 5 observations at frequencies 3–5. The contribution to the likelihood function is

$$(p_3 + p_4 + p_5)^5.$$

Example 3.5 (Example 3.4 continued) *Assume the data were given as in Table 3.5, that is, it was only known that there were 9 observations of 6 or more. Determine the mle for the Poisson distribution.*

The likelihood function is

$$L = p_0^{47} p_1^{97} p_2^{109} p_3^{62} p_4^{25} p_5^{16} (1 - p_0 - p_1 - p_2 - p_3 - p_4 - p_5)^9$$

and when written as a function of λ, it becomes somewhat complicated. While the derivative can be taken, solving the equation when it is set equal to zero will require numerical methods such as Newton–Raphson or the solver routine which accompanies many spreadsheet programs. It may be just as easy to use a numerical method to directly maximize the function. The simplex method as outlined in Appendix C is a good choice. A reasonable starting value can be obtained by assuming that all 9 observations were exactly at 6 and then using the sample mean. Of course, this will understate the true mle, but should be a good place to start. For this particular example, the mle is 2.0226, which is very close to the value obtained when all the counts were recorded. The test statistic for the chi-square goodness-of-fit test is 5.17, and with 5 degrees of freedom the model still passes (the p-value is .0.395). □

3.3 THE NEGATIVE BINOMIAL DISTRIBUTION

3.3.1 Some properties

The negative binomial distribution has been used extensively as an alternative to the Poisson distribution. Like the Poisson distribution, it has positive probabilities on the non-negative integers. Because it has two parameters, it has more flexibility in shape than the Poisson.

Definition 3.2 *The probability function of the* ***negative binomial distribution*** *is given by*

$$\Pr(N=k) \quad = \quad p_k = \binom{k+r-1}{k}\left(\frac{1}{1+\beta}\right)^r\left(\frac{\beta}{1+\beta}\right)^k,$$

$$k=0,1,2,..., \quad r>0,\ \beta>0. \quad (3.2)$$

The binomial coefficient is to be evaluated as

$$\binom{x}{k}=\frac{x(x-1)\cdots(x-k+1)}{k!}.$$

While k must be an integer, x may be any real number. It can also be written as

$$\binom{x}{k}=\frac{\Gamma(x+1)}{\Gamma(k+1)\Gamma(x-k+1)}$$

which may be useful because $\log\Gamma(x)$ is available in many spreadsheets, programming languages, and mathematics packages.

It is not difficult to show that the probability generating function for the negative binomial distribution is

$$P(z)=[1-\beta(z-1)]^{-r}.$$

From this it follows that the mean and variance of the negative binomial distribution are

$$E(N)=r\beta \quad \text{and} \quad Var(N)=r\beta(1+\beta).$$

Because β is positive, it can be seen that the variance of the negative binomial distribution exceeds the mean. This is in contrast to the Poisson distribution for which the variance is equal to the mean. This suggests that for a particular set of data, if the observed variance is larger than the observed mean, the negative binomial might be a better candidate than the Poisson distribution as a model to be fitted.

The negative binomial distribution is a generalization of the Poisson in at least two different ways, namely as a mixed Poisson distribution with a gamma mixing distribution (see Subsection 3.3.2) and as a compound Poisson distribution with a logarithmic secondary distribution (see Section 3.7). Another view of the Poisson distribution is presented in Chapter 6. There, among other assumptions, the rate at which claims occur is assumed constant over time. If the rate is linearly increasing with regard to the number of past claims, then the number of claims in any period will have the negative binomial distribution. See *Insurance Risk Models* [102, Theorem 3.6.1] for this derivation of the negative binomial distribution.

The **geometric distribution** is the special case of the negative binomial distribution when $r = 1$. The geometric distribution is, in some senses, the discrete analogue of the continuous exponential distribution. Both the geometric and exponential distributions have an exponentially decaying probability function and hence the memoryless property. The memoryless property can be interpreted in various contexts as follows. If the exponential distribution is a distribution of lifetimes, then the expected future lifetime is constant for any age. If the exponential distribution describes the size of insurance claims, then the memoryless property can be interpreted as follows: *Given that a claim exceeds a certain level d, the expected amount of the claim in excess of d is constant and so does not depend on d.* That is, if a deductible of d is imposed, the expected payment per claim will be unchanged, but of course the expected number of payments will decrease. If the geometric distribution describes the number of claims, then the memoryless property can be interpreted as follows: *Given that there are at least m claims, the probability distribution of the number of claims in excess of m does not depend on m.* Among continuous distributions, the exponential distribution is used to distinguish between "subexponential" distributions with heavy (or fat) tails, and distributions with light (or thin) tails. Similarly for frequency distributions, distributions that decay in the tail slower than the geometric distribution are often considered to have long tails, whereas distributions that decay more rapidly than the geometric have short tails. The negative binomial distribution has a long tail, that is, decays more slowly than the geometric distribution when $r < 1$ and decays more rapidly than the geometric distribution when $r > 1$.

3.3.2 As a mixture of Poissons

The negative binomial can be generated in several ways. One very natural way in studying claim numbers is as a mixture of Poissons. Suppose that we know that a risk has a Poisson number of claims distribution when the risk parameter λ is known. This is essentially what was done in Section 3.2.

In this section, we will treat λ as being the outcome of a random variable Λ. We will denote the pf of Λ by $u(\lambda)$, where Λ may be continuous or discrete, and denote the cdf by $U(\lambda)$. The idea that λ is the outcome of a random variable can be justified in several ways. First, we can think of the population of risks as being heterogeneous with respect to the risk parameter Λ. In practice this makes sense. Consider a block of insurance policies with the same premium, such as a group of automobile drivers in the same rating category. Such categories are usually broad ranges such as 0–7,500 miles per year, garaged in a rural area, commuting less than 50 miles per week, and so on. We know that not all drivers in the same rating category are the same even though they may "appear" to be the same from the point of view of the insurer and are charged the same premium. The parameter λ measures the expected number of accidents. If λ varies across the population of drivers, then we can think of the insured individual as a sample value drawn from the population of possible drivers. This means implicitly that λ is unknown to the insurer but follows some distribution, in this case $u(\lambda)$, over the population of drivers. The true value of λ is unobservable. All we observe are the number of accidents coming from the driver. There is now an additional degree of uncertainty, that is, uncertainty about the parameter.

This is the same mixing process that was discussed with regard to continuous distributions in Subsection 2.7.3. In some contexts this is referred to as "parameter uncertainty." In the Bayesian context, the distribution of Λ is called a "prior distribution" and the parameters of its distribution are sometimes called "hyper-parameters." The role of the distribution $u(\cdot)$ is very important in credibility theory, the subject of Chapter 5. When the parameter λ is unknown, the probability that exactly k claims will arise can be written as the expected value of the same probability but conditional on $\Lambda = \lambda$ where the expectation is taken with respect to the distribution of Λ. From the law of total probability, we can write

$$\begin{aligned} p_k &= \Pr(N=k) \\ &= E[\Pr(N=k|\Lambda)] \\ &= \int_0^\infty \Pr(N=k|\Lambda=\lambda)u(\lambda)d\lambda \\ &= \int_0^\infty \frac{e^{-\lambda}\lambda^k}{k!}u(\lambda)d\lambda. \end{aligned}$$

Now suppose Λ has a gamma distribution. Then

$$p_k = \int_0^\infty \frac{e^{-\lambda}\lambda^k}{k!}\frac{\lambda^{\alpha-1}e^{-\frac{\lambda}{\theta}}}{\theta^\alpha\Gamma(\alpha)}d\lambda = \frac{1}{k!}\frac{1}{\theta^\alpha\Gamma(\alpha)}\int_0^\infty e^{-\lambda(1+\frac{1}{\theta})}\lambda^{k+\alpha-1}d\lambda.$$

From the definition of the gamma distribution in Appendix A, this expression can be evaluated as

$$\begin{aligned} p_k &= \frac{\Gamma(k+\alpha)}{k!\Gamma(\alpha)}\frac{\theta^k}{(1+\theta)^{k+\alpha}} \\ &= \binom{k+\alpha-1}{k}\left(\frac{\theta}{1+\theta}\right)^k\left(\frac{1}{1+\theta}\right)^\alpha. \end{aligned}$$

This formula is of the same form as equation (3.2), demonstrating that the mixed Poisson, with a gamma mixing distribution, is the same as a negative binomial distribution.

3.3.3 Estimation

In this subsection, we examine estimation techniques for the negative binomial distribution. As in the last section, which dealt with the Poisson distribution, we only consider the method of moments and the method of maximum likelihood.

The moment equations are

$$r\beta = \frac{\sum_{k=0}^{\infty} k n_k}{n} \tag{3.3}$$

and

$$r\beta(1+\beta) = \frac{\sum_{k=0}^{\infty} k^2 n_k}{n} - \left(\frac{\sum_{k=0}^{\infty} k n_k}{n}\right)^2. \tag{3.4}$$

Note that this variance estimate is obtained by dividing by n, not $n-1$. This is a common, though not required, approach when using the method of moments.

Example 3.6 (Example 3.1 continued) *Estimate the negative binomial parameters by the method of moments.*

The sample mean and the sample variance are 2.5 and 3.05 (verify this) respectively. Hence, the moment equations for the data in this example are

$$r\beta = 2.5 \quad \text{and} \quad r\beta(1+\beta) = 3.05.$$

From these two equations, the moment-based estimators of the parameters are $\hat{r} = 11.364$ and $\hat{\beta} = 0.22$. □

It is convenient to reparameterize the distribution by denoting the mean by μ and getting rid of r. For our example the moment equations can then be rewritten as $\mu = 2.5$ and $\mu(1+\beta) = 3.05$. Then the moment-based estimators for μ and β are $\hat{\mu} = 2.5$ and $\hat{\beta} = 0.22$.

When compared to the Poisson distribution with the same mean μ, it can be seen that β is a measure of "extra-Poisson" variation. A value of $\beta = 0$ means no extra-Poisson variation, while a value of $\beta = 0.22$ implies a 22% increase in the variance when compared to the Poisson distribution with the same mean.

We now examine maximum likelihood estimation. The loglikelihood for the negative binomial distribution is

$$\begin{aligned} l &= \sum_{k=0}^{\infty} n_k \log p_k \\ &= \sum_{k=0}^{\infty} n_k \left[\log \binom{r+k-1}{k} - r\log(1+\beta) + k\log\beta - k\log(1+\beta)\right]. \end{aligned}$$

The loglikelihood is a function of the two parameters β and r. In order to find the maximum of the loglikelihood, we differentiate with respect to each of the parameters, set the derivatives equal to zero, and solve for the parameters. The derivatives of the loglikelihood are

$$\frac{\partial l}{\partial \beta} = \sum_{k=0}^{\infty} n_k \left(\frac{k}{\beta} - \frac{r+k}{1+\beta} \right) \tag{3.5}$$

and

$$\begin{aligned}
\frac{\partial l}{\partial r} &= -\sum_{k=0}^{\infty} n_k \log(1+\beta) + \sum_{k=0}^{\infty} n_k \frac{\partial}{\partial r} \log \frac{(r+k-1)\cdots r}{k!} \\
&= -n \log(1+\beta) + \sum_{k=0}^{\infty} n_k \frac{\partial}{\partial r} \log \prod_{m=0}^{k-1} (r+m) \\
&= -n \log(1+\beta) + \sum_{k=0}^{\infty} n_k \frac{\partial}{\partial r} \sum_{m=0}^{k-1} \log(r+m) \\
&= -n \log(1+\beta) + \sum_{k=1}^{\infty} n_k \sum_{m=0}^{k-1} \frac{1}{r+m}
\end{aligned} \tag{3.6}$$

Setting these equations to zero yields

$$\hat{\mu} = \hat{r}\hat{\beta} = \frac{\sum_{k=0}^{\infty} k n_k}{n} = \bar{x} \tag{3.7}$$

and

$$n \log(1+\hat{\beta}) = \sum_{k=1}^{\infty} n_k \left(\sum_{m=0}^{k-1} \frac{1}{\hat{r}+m} \right). \tag{3.8}$$

Note that the maximum likelihood estimator of the mean is the sample mean (as, by definition, in the method of moments). Equations (3.7) and (3.8) can be solved numerically. Replacing $\hat{\beta}$ in (3.8) by $\hat{\mu}/\hat{r}$ yields the equation

$$H(\hat{r}) = n \log(1 + \frac{\bar{x}}{\hat{r}}) - \sum_{k=1}^{\infty} n_k \left(\sum_{m=0}^{k-1} \frac{1}{\hat{r}+m} \right) = 0. \tag{3.9}$$

If the right-hand side of (3.4) is greater than the right-hand side of (3.3), it can be shown that there is a unique solution of equation (3.9). If not, then the negative binomial model is probably not a good model to use since the sample variance does not exceed the sample mean.

Equation (3.9) can be solved numerically for $\hat{r}$ using the Newton–Raphson method. The required equation for the kth iteration is

$$r_k = r_{k-1} - \frac{H(r_{k-1})}{H'(r_{k-1})}.$$

A useful starting value for r_0 is the moment-based estimator of r. Of course, any numerical root-finding method (e.g., bisection, secant) may be used.

The loglikelihood is a function of two variables. It can be maximized directly using a method such as the simplex method described in Appendix C. For the case of the negative binomial distribution, since we know the estimator of the mean must be the sample mean, by reparameterizing from (r, β) to (μ, β) the maximization problem is reduced to one dimension. There are also good theoretical grounds for such a parameterization for a larger class of distributions. This will be discussed later in a more formal way. Essentially, the reparameterization will result in estimators of parameters other than the mean that are independent, at least for large samples, of the sample mean. This will allow better separate qualitative statements about the parameters.

Example 3.7 *Determine the maximum likelihood estimates of the negative binomial parameters for the data in Example 3.1.*

The scoring method (see the end of this section) was used to obtain $\hat{r} = 10.9650$ and $\hat{\beta} = 0.227998$. □

Example 3.8 *Tröbliger [120] studied the driving habits of 23,589 automobile drivers in a class of automobile insurance by counting the number of accidents per driver in a one-year time period. The data as well as fitted Poisson and negative binomial distributions are given in Table 3.6. Determine which model is preferred.*

From Table 3.6, it can be seen that the chi-square statistic is much lower (3.60) for the fitted negative binomial than for the fitted Poisson distribution (203.87). Notice that these numbers are directly comparable since they are based on the same number of degrees of freedom. This is because the grouping for the negative binomial distribution is different due to the heavier tail and fewer cells required to get at least 5 expected claims in the right tail.

The p-values in the above table are taken from standard χ^2 tables. They show that at typical levels of significance (1%–10%), the Poisson distribution is rejected while the negative binomial distribution cannot be rejected. □

The loglikelihood values can be used to support the hypothesis that the negative binomial is the better of the two models. Clearly, the likelihood for the negative binomial is larger than the likelihood for the Poisson. This should be expected since the negative binomial distribution has two parameters and the Poisson has one and since the Poisson distribution is a special limiting case ($\lambda = r\beta$, $\beta \to 0$) of the negative binomial distribution. Hence, the relevant question is: *Is the improvement in the likelihood sufficient to justify the more complicated model?* This is the question of parsimony, which suggests that, all things being equal, a simpler model should be selected.

The likelihood ratio test discussed in Subsections 2.6.4 and 2.9.2 can be used. The hypotheses are:

$$H_0: \text{ Poisson}$$
$$H_1: \text{ Negative binomial.}$$

Table 3.6 Two models for automobile claims frequency

No. of claims/year	No. of drivers	Fitted Poisson expected	χ^2	Fitted negative binomial expected	χ^2
0	20,592	20,420.9	1.43	20,596.8	0.00
1	2,651	2,945.1	29.37	2,631.0	0.15
2	297	212.4	33.72	318.4	1.43
3	41	10.2		37.8	0.27
4	7	0.4		4.4	
5	0	0.0	139.35	0.5	1.74
6	1	0.0		0.1	
7+	0	0.0		0.0	
Totals	23,589	23,589.0	203.87	23,589.0	3.60
Parameters		$\lambda = 0.144220$		$r = 1.11790$ $\beta = 0.129010$	
Degrees of freedom		$4 - 1 - 1 = 2$		$5 - 2 - 1 = 2$	
p-Value		$< 1\%$		16.53%	
At 1% level		Reject		Accept	
At 10% level		Reject		Accept	
At 20% level		Reject		Reject	
−Loglikelihood		10,297.84		10,223.42	

If l_0 and l_1 are the corresponding loglikelihood values, the test statistic is

$$Q = 2\,(l_1 - l_0)$$

which has a χ^2 distribution with 1 degree of freedom, since the negative binomial distribution has one more parameter. At a 5% significance level, the null hypothesis is rejected in favor of the alternative hypothesis if $Q > 3.84$.

Example 3.9 *Conduct the likelihood ratio test for the models in Example 3.8.*

In the example $2(l_1 - l_0) = 2(-10{,}223.42 + 10{,}297.84) = 148.84 > 3.84$. Hence, based on this test, the negative binomial distribution has a significantly better fit than the Poisson distribution. □

This test complements the chi-square test of fit that was done earlier. However, the likelihood ratio test does not address the issue of whether the negative binomial distribution is a good fit. It only addresses the question: *Is the negative binomial distribution a significantly better fit than the Poisson?* The chi-square test should also be done to answer the question: *Is the negative binomial a good fit?*

3.3.4 The scoring method for discrete models

The scoring method was introduced in Subsection 2.4.4 for use with grouped data and continuous distributions. Here we are considering discrete distributions, but in some sense grouping continuous data creates a discrete distribution, where p_k is the probability of an observation falling in the kth class. All of the formulas are the same and are repeated here for convenience.

Let an arbitrary parametric discrete distribution have probability function $p_k(\boldsymbol{\theta})$ where $\boldsymbol{\theta}$ may be a vector. For a given parameter value, the vector of scores is $\mathbf{S}(\boldsymbol{\theta})$ with ith element

$$S_i = \frac{\partial \log L}{\partial \theta_i} = \sum_{k=0}^{\infty} n_k \frac{\partial p_k(\boldsymbol{\theta})}{\partial \theta_i} \frac{1}{p_k(\boldsymbol{\theta})} \tag{3.10}$$

and the estimated information matrix is $\mathbf{I}(\boldsymbol{\theta})$ with ijth element

$$I_{ij} = n \sum_{k=0}^{\infty} \frac{\partial p_k(\boldsymbol{\theta})}{\partial \theta_i} \frac{\partial p_k(\boldsymbol{\theta})}{\partial \theta_j} \frac{1}{p_k(\boldsymbol{\theta})}.$$

The iteration for the maximum likelihood estimate is then

$$\boldsymbol{\theta}_m = \boldsymbol{\theta}_{m-1} + [\mathbf{I}(\boldsymbol{\theta}_{m-1})]^{-1}\mathbf{S}(\boldsymbol{\theta}_{m-1}). \tag{3.11}$$

The estimated covariance matrix of the maximum likelihood estimator is $[\mathbf{I}(\hat{\boldsymbol{\theta}})]^{-1}$.

For the negative binomial distribution, as parameterized in this section, the required derivatives can be obtained from (3.5) and (3.6) by noting that $\partial p_k(\boldsymbol{\theta})/\partial \theta_i = p_k(\boldsymbol{\theta}) \partial \log p_k(\boldsymbol{\theta})/\partial \theta_i$. We then have

$$\frac{\partial p_k(\boldsymbol{\theta})}{\partial \beta} = p_k(\boldsymbol{\theta}) \left(\frac{k}{\beta} - \frac{r+k}{1+\beta} \right)$$

and

$$\frac{\partial p_k(\boldsymbol{\theta})}{\partial r} = p_k(\boldsymbol{\theta}) \left[-\log(1+\beta) + \sum_{m=0}^{k-1} \frac{1}{r+m} \right]$$

where the sum is zero when $k = 0$.

3.4 THE BINOMIAL DISTRIBUTION

3.4.1 Some properties

The binomial distribution is another counting distribution that arises naturally and frequently in claim number modeling. It possesses some properties different from the Poisson and the negative binomial that make it particularly useful. First, its variance is smaller than its mean. This makes it useful for fitting to data sets in which the observed sample variance is less than the sample mean. This contrasts with the negative binomial where the variance exceeds the mean, and it also contrasts with the Poisson distribution where the variance is equal to the mean.

Secondly, it is useful because it describes a physical situation in which a collection of m risks are each subject to claim or loss. We can formalize this as follows. Consider m independent and identical risks each with probability q of making a claim. This might apply to a life insurance situation in which all the individuals under consideration are in the same mortality class; that is, they may all be male smokers at age 35 and duration 5 of an insurance policy. In that case, q is the probability that a person with those attributes will die in the next year. Then the number of claims for a single person follows a Bernoulli distribution, a distribution with probability $1-q$ at 0 and probability q at 1. The probability generating function of the number of claims per individual is then given by

$$P(z) = (1-q)z^0 + qz^1 = 1 + q(z-1).$$

Now if there are m such independent individuals, then the probability generating functions can be multiplied together to give the probability generating function of the total number of claims arising from the group of m individuals. That probability generating function is

$$P(z) = [1 + q(z-1)]^m, \qquad 0 < q < 1.$$

Then from this, it is easy to show that the probability of exactly k claims from the group is

$$p_k = \Pr(N = k) = \binom{m}{k} q^k (1-q)^{m-k}, \qquad k = 0, 1, \ldots, m$$

the pf for a binomial distribution with parameters m and q. From this Bernoulli trial framework, it is clear that at most m events (claims) can occur. Hence, the distribution only has positive probabilities on the non-negative integers up to and including m.

Consequently, an additional attribute of the binomial distribution that is sometimes useful is that it has finite support; that is, the range of values for which there exist positive probabilities has finite length. This may be useful, for instance, in modeling the number of individuals injured in an automobile accident or the number of family members covered under a health insurance policy. In each case it is reasonable to have an upper limit on the range of possible values. It is useful also in connection with situations where it is believed that it is unreasonable to assign positive probabilities beyond some point. For example, if one is modeling the number of accidents per automobile during a one-year period, it is probably physically impossible for there to be more than some number, say 12, of claims during the year given the time it would take to repair the automobile between accidents. If a model with probabilities that extend beyond 12 were used, those probabilities should be very small so that they have little impact on any decisions that are made. The mean and variance of the binomial distribution are given by

$$E(N) = mq, \quad Var(N) = mq(1-q).$$

3.4.2 Estimation

The binomial distribution has two parameters, m and q. Frequently, the value of m is known and fixed. In this case, only one parameter, q, needs to be estimated. In many insurance situations, q is interpreted as the probability of some event such as death or disability. In such cases the value of q is usually estimated as

$$\hat{q} = \frac{\text{Number of observed events}}{\text{Maximum number of possible events}}$$

which is the method of moments estimator when m is known.

In situations where frequency data are in the form of the previous examples in this chapter, the value of the parameter m, the largest possible observation, may be known and fixed or unknown. In any case, m must be no smaller than the largest observation. The loglikelihood is

$$\begin{aligned} l &= \sum_{k=0}^{m} n_k \log p_k \\ &= \sum_{k=0}^{m} n_k \left[\log \binom{m}{k} + k \log q + (m-k) \log(1-q) \right]. \end{aligned}$$

When m is known and fixed, one needs only maximize l with respect to q.

$$\frac{\partial l}{\partial q} = \frac{1}{q} \sum_{k=0}^{m} k n_k - \frac{1}{1-q} \sum_{k=0}^{m} (m-k) n_k.$$

Setting this equal to zero yields

$$\hat{q} = \frac{1}{m} \frac{\sum_{k=0}^{m} k n_k}{\sum_{k=0}^{m} n_k}$$

which is the sample proportion of observed events. For the method of moments, with m fixed, the estimator of q is the same as the maximum likelihood estimator since the moment equation is

$$mq = \frac{\sum_{k=0}^{m} k n_k}{\sum_{k=0}^{m} n_k}.$$

Table 3.7 Number of claims per policy

No. of claims/policy	No. of policies
0	5,367
1	5,893
2	2,870
3	842
4	163
5	23
6	1
7	1
8+	0

When m is unknown, the mle of q is

$$\hat{q} = \frac{1}{\hat{m}} \frac{\sum_{k=0}^{\infty} kn_k}{\sum_{k=0}^{\infty} n_k} \tag{3.12}$$

where $\hat{m}$ is the mle of m. An easy way to approach the maximum likelihood estimation of m and q is to create a "likelihood profile" for various possible values of m as follows:

Step 1: Start with $\hat{m}$ equal to the largest observation.
Step 2: Obtain $\hat{q}$ using (3.12).
Step 3: Calculate the loglikelihood at these values.
Step 4: Increase $\hat{m}$ by 1.
Step 5: Repeat Steps 2–4 until a maximum is found.

Example 3.10 *The number of claims per policy during a one-year period for a block of 15,160 insurance policies are given in Table 3.7. Obtain moment-based and maximum likelihood estimators.*

The sample mean and variance are 0.985422 and 0.890355, respectively. The variance is smaller than the mean, suggesting the binomial as a reasonable distribution to try. The method of moments leads to

$$mq = 0.985422$$

and

$$mq(1-q) = 0.890355.$$

Hence, $\hat{q} = 0.096474$ and $\hat{m} = 10.21440$. However, m can only take on integer values. We choose $\hat{m} = 10$ by rounding. Then we adjust the estimate of $\hat{q}$ to 0.0985422 from the first moment equation. Doing this will result in a

Table 3.8 Binomial likelihood profile

$\hat{m}$	$\hat{q}$	$-$Loglikelihood
7	0.140775	19,273.56
8	0.123178	19,265.37
9	0.109491	19,262.02
10	0.098542	19,260.98
11	0.089584	19,261.11
12	0.082119	19,261.84

model variance which differs from the sample variance since $10(0.0985422)(1-0.0985422) = 0.888316$. This shows one of the pitfalls of using the method of moments with integer-valued parameters.

We now turn to maximum likelihood estimation. From the data $m \geq 7$. If m is known, then only q needs to be estimated. If m is unknown, then we can produce a likelihood profile by maximizing the likelihood for fixed values of m starting at 7 and increasing until a maximum is found. The results are in Table 3.8.

The largest loglikelihood value occurs at $m = 10$. If, a priori, the value of m is unknown, then the mle of the parameters are $\hat{m} = 10$ and $\hat{q} = 0.0985422$. This is the same as the adjusted moment estimates. This is not necessarily the case for all data sets. With these parameters, the chi-square goodness-of-fit test statistic is 0.36. With three degrees of freedom, the p-value is 0.9484 and so the binomial model is clearly acceptable. □

3.5 THE $(a, b, 0)$ CLASS

3.5.1 Some properties

The following definition characterizes the members of this class of distributions.

Definition 3.3 *Let p_k be the pf of a discrete random variable. It is a member of the* ***(a,b,0) class of distributions****, provided that there exists constants a and b such that*

$$\frac{p_k}{p_{k-1}} = a + \frac{b}{k}, \qquad k = 1, 2, 3, \ldots.$$

This recursion describes the relative size of successive probabilities in the counting distribution. The probability at zero, p_0, can be obtained from the recursive formula since the probabilities must add up to 1. This provides a boundary condition. The $(a, b, 0)$ class of distributions is a two-parameter class, the two parameters being a and b. By substituting in the probability

Table 3.9 Members of the $(a, b, 0)$ class

Distribution	a	b	p_0
Poisson	0	λ	$e^{-\lambda}$
Binomial	$-\dfrac{q}{1-q}$	$(m+1)\dfrac{q}{1-q}$	$(1-q)^m$
Negative binomial	$\dfrac{\beta}{1+\beta}$	$(r-1)\dfrac{\beta}{1+\beta}$	$(1+\beta)^{-r}$
Geometric	$\dfrac{\beta}{1+\beta}$	0	$(1+\beta)^{-1}$

function for each of the Poisson, binomial, and negative binomial distributions on the left-hand side of the recursion, it can be seen that each of these three distributions satisfies the recursion and that the values of a and b are as given in Table 3.9. In addition the table gives the value of p_0, the starting value for the recursion. Also in the table is the geometric distribution, the one parameter special case ($r = 1$) of the negative binomial distribution.

It can be shown (see Panjer and Willmot [102, Chapter 6]) that these are the only possible distributions satisfying this recursive formula.

The recursive formula can be rewritten as

$$k\frac{p_k}{p_{k-1}} = ak + b, \qquad k = 1, 2, 3, \ldots.$$

The expression on the left-hand side is a linear function in k. Note from Table 3.9 that the slope a of the straight line is 0 for the Poisson distribution, is negative for the binomial distribution, and is positive for the negative binomial distribution, including the geometric. This suggests a graphical way of indicating which of the three distributions should be selected for fitting. First, one can plot

$$k\frac{\hat{p}_k}{\hat{p}_{k-1}} = k\frac{n_k}{n_{k-1}}$$

against k. The observed values should form approximately a straight line if one of these models is to be selected, and the value of the slope should be an indication of which of the models should be selected. Note that this cannot be done if any of the observations are 0. Hence this procedure is less useful for a small number of observations.

Example 3.11 *Consider the accident data in Table 3.10 which is taken from Thyrion [119]. For the 9,461 automobile insurance policies studied, the number of accidents under the policy is recorded in the table. Also recorded in the table is the observed value of the quantity that should be linear.*

Figure 3.1 plots the value of the quantity of interest against k, the number of accidents. It can be seen from the graph that the quantity of interest

Table 3.10 Accident profile

Number of accidents, k	Number of policies, n_k	$k\dfrac{n_k}{n_{k-1}}$
0	7,840	
1	1,317	0.17
2	239	0.36
3	42	0.53
4	14	1.33
5	4	1.43
6	4	6.00
7	1	1.75
8+	0	
Total	9,461	

looks approximately linear except for the point at $k = 6$. The reliability of the quantities as k increases diminishes because the number of observations becomes small and the variability of the results grows. This illustrates the weakness of this *ad hoc* procedure. Visually, all the points appear to have equal value. However, the points on the left are more reliable than the points on the right. From the graph, it can be seen that the slope is positive and the data appear approximately linear. This suggests the negative binomial distribution is an appropriate model. Whether or not the slope is significantly different from 0 is also not easily judged from the graph. By rescaling the vertical axis of the graph, the slope can be made to look steeper and hence the slope could be made to appear to be significantly different from 0. Graphically, it is difficult to distinguish between the Poisson and the negative binomial distribution, since the Poisson requires a slope of 0. However, we can say that the binomial distribution is probably not a good choice since there is no evidence of a negative slope. In this case it is advisable to fit both the Poisson and negative binomial distributions and compare their relative likelihoods.

It is also possible to compare the appropriateness of the distributions by looking at the relationship of the variance to the mean. For this data set, the mean number of claims per policy is 0.2144. The variance is 0.2889. Because the variance exceeds the mean, the negative binomial should be considered as an alternative to the Poisson. Again this is a qualitative comment because we have no formal way of determining whether the variance is sufficiently larger than the mean to warrant use of the negative binomial. In order to do some formal analysis, Table 3.11 gives the results of maximum likelihood estimation of the parameters of the Poisson and negative binomial distributions and the negative loglikelihood in each case. The Poisson distribution can

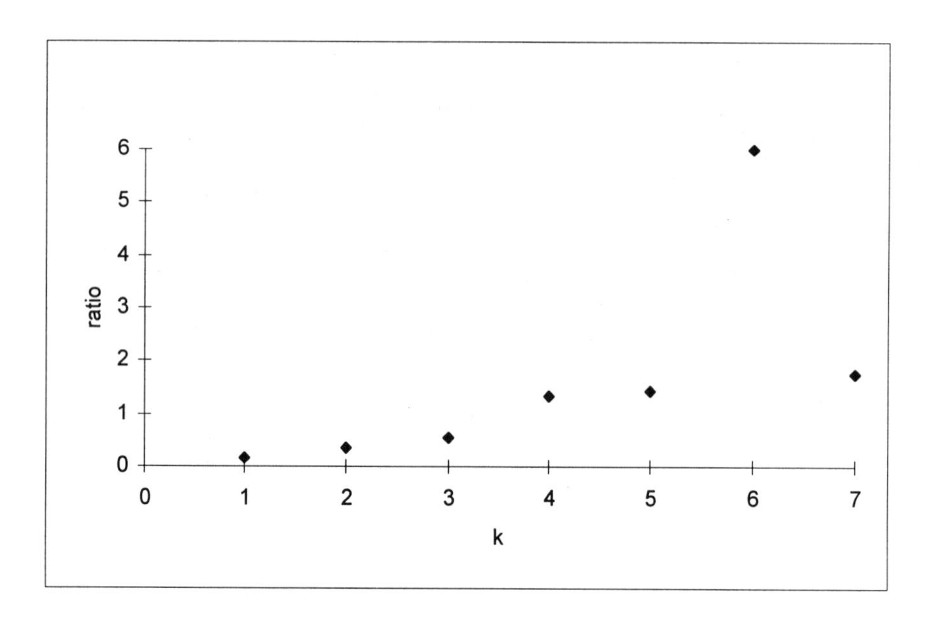

Fig. 3.1 Plot of the ratio kn_k/n_{k-1} against k

Table 3.11 Poisson - negative binomial comparison

Distribution	Parameter estimates	−Loglikelihood
Poisson	$\hat{\lambda} = 0.2143537$	5,490.78
Negative binomial	$\hat{\beta} = 0.3055594$ $\hat{r} = 0.7015122$	5,348.04

be considered a special case of the negative binomial distribution by setting $r\beta = \lambda$ and letting β approach 0. Since the Poisson is a special case of the negative binomial, we can compare loglikelihoods because the likelihood ratio test requires that the loglikelihood increase by at least 1.92 at the 5% significance level. From Table 3.11, it can be seen that the loglikelihood changes by 142.74, a quantity much greater than 1.92. This suggests that the negative binomial is a significantly better choice than the Poisson. However, the chi-square goodness-of-fit test statistic is 8.77 with two degrees of freedom. The p-value is 0.0125 and it is clear that a better model is needed. We will consider alternatives to the negative binomial for this data set in later sections.

In subsequent sections of this chapter we will expand the class of the distributions beyond the three discussed in this section by constructing more general models related to the Poisson, binomial, and negative binomial distributions.

3.6 TRUNCATION AND MODIFICATION AT ZERO

3.6.1 The $(a,b,1)$ class

At times, the distributions discussed previously do not adequately describe the characteristics of some data sets encountered in practice. This may be because the tail of the negative binomial is not heavy enough or because the distributions in the $(a, b, 0)$ class cannot capture the shape of the data set in some other part of the distribution.

In this section, we address the problem of a poor fit at the left-hand end of the distribution, in particular—the probability at zero.

For insurance count data, the probability at zero is the probability that no claims occur during the period under study. For applications in insurance where the probability of occurrence of a loss is low, the probability at zero has the largest value. Thus, it is important to pay special attention to the fit at this point.

There are also situations that naturally occur which generate unusually large probabilities at zero. Consider the case of group dental insurance. If, in a family, both husband and wife have coverage with their employer-sponsored plans and both group insurance contracts provide coverage for all family members, the claims will be made to the insurer of the plan which provides the better benefits; and, no claims may be made under the other contract. Then, in conducting studies for a specific insurer, one may find a higher than expected number of individuals who made no claim.

Similarly, it is possible to have situations in which there is less than the expected number, or even zero, occurrences at zero. For example, if one is counting the number of claims from accidents resulting in a claim, the minimum observed value is 1.

An adjustment of the probability at zero is easily handled for the Poisson, binomial and negative binomial distributions.

The $(a, b, 0)$ class is easily generalized to allow for an adjustment of the probability at zero.

Definition 3.4 *Let p_k be the pf of a discrete random variable. It is a member of the* ***(a,b,1) class of distributions****, provided that there exists constants a and b such that*

$$\frac{p_k}{p_{k-1}} = a + \frac{b}{k}, \qquad k = 2, 3, 4, \ldots.$$

Note that the only difference from the $(a, b, 0)$ class is that the recursion begins at p_1, rather than p_0. This forces the distribution from $k = 1$ to $k = \infty$ to have the same shape as the $(a, b, 0)$ class in the sense that the probabilities are the same up to a constant of proportionality since $\sum_{k=1}^{\infty} p_k$ can be set to any number in the interval $(0, 1]$. The remaining probability is at $k = 0$.

We will distinguish between the situations in which $p_0 = 0$ and those where $p_0 > 0$. The first subclass is called the **truncated** (more specifically,

zero-truncated) distributions. The members are the zero-truncated Poisson, zero-truncated binomial, and zero-truncated negative binomial (and its special case, the zero-truncated geometric) distributions.

The second subclass will be referred to as the **zero-modified** distributions because the probability is modified from that for the $(a, b, 0)$ class. These distributions can be viewed as a mixture of an $(a, b, 0)$ distribution and a degenerate distribution with all the probability at zero. Alternatively, they can be called "truncated with zeros" distributions since the distribution can be viewed as a mixture of a truncated distribution and a degenerate distribution with all the probability at zero. We now show this more formally. Note that all zero-truncated distributions can be considered as zero-modified distributions, with the particular modification being to set $p_0 = 0$.

With three types of distributions, notation can become confusing. When writing about discrete distributions in general we will continue to let $p_k = \Pr(N = k)$. When referring to a zero-truncated distribution, we will use p_k^T and when referring to a zero-modified distribution we will use p_k^M. Once again, it is possible for a zero-modified distribution to be a zero-truncated distribution.

Let $P(z) = \sum_{k=0}^{\infty} p_k z^k$ denote the pgf of a member of the $(a, b, 0)$ class. Let $P^M(z) = \sum_{k=0}^{\infty} p_k^M z^k$ denote the pgf of the corresponding member of the $(a, b, 1)$ class; that is,

$$p_k^M = c \cdot p_k, \qquad k = 1, 2, 3, \ldots.$$

and p_0^M is an arbitrary number. Then

$$\begin{aligned} P^M(z) &= p_0^M + \sum_{k=1}^{\infty} p_k^M z^k \\ &= p_0^M + c \sum_{k=1}^{\infty} p_k z^k \\ &= p_0^M + c[P(z) - p_0]. \end{aligned}$$

Since $P^M(1) = P(1) = 1$,

$$1 = p_0^M + c(1 - p_0)$$

resulting in

$$c = \frac{1 - p_0^M}{1 - p_0} \text{ or } p_0^M = 1 - c(1 - p_0).$$

This relationship is necessary to ensure that the p_k^M sum to 1. We then have

$$\begin{aligned} P^M(z) &= p_0^M + \frac{1 - p_0^M}{1 - p_0}[P(z) - p_0] \\ &= \left(1 - \frac{1 - p_0^M}{1 - p_0}\right) 1 + \frac{1 - p_0^M}{1 - p_0} P(z). \end{aligned} \tag{3.13}$$

This is a weighted average of the pgfs of the degenerate distribution and the corresponding $(a, b, 0)$ member. Furthermore,

$$p_k^M = \frac{1 - p_0^M}{1 - p_0} p_k, \qquad k = 1, 2, ... \tag{3.14}$$

Let $P^T(z)$ denote the pgf of the zero-truncated distribution corresponding to an $(a, b, 0)$ pgf $P(z)$. Then, by setting $p_0^M = 0$ in (3.13) and (3.14),

$$P^T(z) = \frac{P(z) - p_0}{1 - p_0}$$

and

$$p_k^T = \frac{p_k}{1 - p_0}, \qquad k = 1, 2, \ldots. \tag{3.15}$$

Then from (3.14)

$$p_k^M = (1 - p_0^M) p_k^T, \qquad k = 1, 2, ... \tag{3.16}$$

and

$$P^M(z) = p_0^M \cdot 1 + (1 - p_0^M) P^T(z). \tag{3.17}$$

Then the zero-modified distribution is also the weighted average of a degenerate distribution and the zero-truncated member of the $(a, b, 0)$ class.

Although we have only discussed the zero-modified distributions of the $(a, b, 0)$ class, the $(a, b, 1)$ class admits additional distributions. The (a, b) parameter space can be expanded to admit an extension of the negative binomial distribution to include cases where $-1 < r \leq 0$. For the $(a, b, 0)$ class, $r > 0$ is required. By adding the additional region to the sample space, the "extended" truncated negative binomial (ETNB) distribution has parameter restrictions, $r > -1$ and $\beta > 0$.

To show that the recursive equation

$$p_k = p_{k-1}(a + \frac{b}{k}), \qquad k = 2, 3, ... \tag{3.18}$$

with $p_0 = 0$ defines a proper distribution, it is sufficient to show that for any value of p_1, the successive values of p_k obtained recursively are each positive and that $\sum_{k=1}^{\infty} p_k < \infty$. For the ETNB, this must be done for the parameter space

$$a = \frac{\beta}{1 + \beta}, \qquad \beta > 0$$

$$b = (r - 1) \frac{\beta}{1 + \beta}, \qquad r > -1$$

(see Exercise 3.20).

When $r = 0$, the special case of the ETNB is the logarithmic distribution with

$$p_k^T = \frac{\left(\frac{\beta}{1+\beta}\right)^k}{k\log(1+\beta)}, \qquad k = 1, 2, 3, \ldots \tag{3.19}$$

(see Exercise 3.21). The pgf of the logarithmic distribution is

$$P^T(z) = 1 - \frac{\log[1-\beta(z-1)]}{\log(1+\beta)} \tag{3.20}$$

(see Exercise 3.22). The zero-modified logarithmic distribution is created by assigning an arbitrary probability at zero and reducing the remaining probabilities.

It is also interesting that the special extreme case with $-1 < r < 0$ and $\beta \to \infty$ is a proper distribution. However, no moments exist (see Exercise 3.23). Distributions with no moments are not particularly interesting for modeling claim numbers (unless the right tail is subsequently modified) since an infinite number of claims are expected. This might be difficult to price! We will ignore the case where $\beta \to \infty$ from here on.

There are no other members of the $(a, b, 1)$ class except the zero-modified versions of the ones discussed above. A summary is given in Table 3.12. Because the algebraic form of the ETNB is the same as for the zero-truncated negative binomial (only the parameter space is modified), no additional mathematics is required to obtain the maximum likelihood estimators of the extended version of the distribution.

Example 3.12 *Consider a negative binomial random variable with parameters $\beta = 0.5$ and $r = 2.5$. Determine the first four probabilities for this random variable. Then determine the corresponding probabilities for the zero-truncated and zero-modified (with $p_0^M = 0.6$) versions.*

From Table 3.12 we have, for the negative binomial distribution,

$$\begin{aligned} p_0 &= (1+0.5)^{-2.5} = 0.362887 \\ a &= 0.5/1.5 = 1/3 \\ b &= (2.5-1)(0.5)/1.5 = 1/2. \end{aligned}$$

The first three recursions are

$$\begin{aligned} p_1 &= 0.362887\left(\frac{1}{3} + \frac{1}{2}\frac{1}{1}\right) = 0.302406 \\ p_2 &= 0.302406\left(\frac{1}{3} + \frac{1}{2}\frac{1}{2}\right) = 0.176404 \\ p_3 &= 0.176404\left(\frac{1}{3} + \frac{1}{2}\frac{1}{3}\right) = 0.088202. \end{aligned}$$

Table 3.12 Members of the $(a, b, 1)$ class

Distribution[a]	p_0	a	b	Parameter space
Poisson	$e^{-\lambda}$	0	λ	$\lambda > 0$
ZT Poisson	0	0	λ	$\lambda > 0$
ZM Poisson	arbitrary	0	λ	$\lambda > 0$
Binomial	$(1-q)^m$	$-\frac{q}{1-q}$	$(m+1)\frac{q}{1-q}$	$0 < q < 1$
ZT binomial	0	$-\frac{q}{1-q}$	$(m+1)\frac{q}{1-q}$	$0 < q < 1$
ZM binomial	arbitrary	$-\frac{q}{1-q}$	$(m+1)\frac{q}{1-q}$	$0 < q < 1$
Negative binomial	$(1+\beta)^{-r}$	$\frac{\beta}{1+\beta}$	$(r-1)\frac{\beta}{1+\beta}$	$r > 0,\ \beta > 0$
ETNB	0	$\frac{\beta}{1+\beta}$	$(r-1)\frac{\beta}{1+\beta}$	$r > -1,\ \beta > 0$
ZM ETNB	arbitrary	$\frac{\beta}{1+\beta}$	$(r-1)\frac{\beta}{1+\beta}$	$r > -1,\ \beta > 0$
Geometric	$(1+\beta)^{-1}$	$\frac{\beta}{1+\beta}$	0	$\beta > 0$
ZT geometric	0	$\frac{\beta}{1+\beta}$	0	$\beta > 0$
ZM geometric	arbitrary	$\frac{\beta}{1+\beta}$	0	$\beta > 0$
Logarithmic	0	$\frac{\beta}{1+\beta}$	$-\frac{\beta}{1+\beta}$	$\beta > 0$
ZM logarithmic	arbitrary	$\frac{\beta}{1+\beta}$	$-\frac{\beta}{1+\beta}$	$\beta > 0$

[a]ZT = "zero truncated" ZM = "zero modified."

For the zero-truncated random variable, $p_0^T = 0$ by definition. The recursions start with (from (3.15)) $p_1^T = 0.302406/(1 - 0.362887) = 0.474651$. Then

$$
\begin{aligned}
p_2^T &= 0.474651\left(\frac{1}{3} + \frac{1}{2}\frac{1}{2}\right) = 0.276880 \\
p_3^T &= 0.276880\left(\frac{1}{3} + \frac{1}{2}\frac{1}{3}\right) = 0.138440.
\end{aligned}
$$

If the original values were all available, then the zero-truncated probabilities could have all been obtained by multiplying them by $1/(1 - 0.362887) = 1.569580$.

For the zero-modified random variable, $p_0^M = 0.6$ arbitrarily. From (3.14), $p_1^M = (1-0.6)(0.302406)/(1-0.362887) = 0.189860$. Then

$$\begin{aligned} p_2^M &= 0.189860\left(\frac{1}{3}+\frac{1}{2}\frac{1}{2}\right) = 0.110752 \\ p_3^M &= 0.110752\left(\frac{1}{3}+\frac{1}{2}\frac{1}{3}\right) = 0.055376. \end{aligned}$$

□

3.6.2 Estimation

Estimation of the parameters for the $(a, b, 1)$ class follows the same general principles that were used in connection with the $(a, b, 0)$ class.

Assuming that the data are in the same form as the previous examples in this chapter, the likelihood is

$$L = \left(p_0^M\right)^{n_0} \prod_{k=1}^{\infty} (p_k^M)^{n_k} = \left(p_0^M\right)^{n_0} \prod_{k=1}^{\infty} \left[(1-p_0^M)p_k^T\right]^{n_k}$$

The loglikelihood is, using (3.16)

$$\begin{aligned} l &= n_0 \log p_0^M + \sum_{k=1}^{\infty} n_k [\log(1-p_0^M) + \log p_k^T] \\ &= n_0 \log p_0^M + \sum_{k=1}^{\infty} n_k \log(1-p_0^M) + \sum_{k=1}^{\infty} n_k [\log p_k - \log(1-p_0)]. \end{aligned}$$

The three parameters of the $(a, b, 1)$ class are p_0^M, a, and b, where a and b determine $p_1, p_2, \ldots$.

Then it can be seen that

$$l = l_0 + l_1$$

where l_0 depends only on the parameter p_0^M and l_1 is independent of p_0^M, depending only on a and b. This simplifies the maximization since

$$\frac{\partial l}{\partial p_0^M} = \frac{\partial l_0}{\partial p_0^M} = \frac{n_0}{p_0^M} - \sum_{k=1}^{\infty} \frac{n_k}{1-p_0^M} = \frac{n_0}{p_0^M} - \frac{n-n_0}{1-p_0^M}$$

resulting in

$$\hat{p}_0^M = \frac{n_0}{n}$$

the proportion of observations at zero. This is the natural estimator since p_0^M represents the probability of an observation of zero.

Similarly, because the likelihood factors conveniently, the estimation of a and b is independent of p_0^M. Note that although a and b are parameters, maximization should not be done with respect to them. That is because not all values of a and b produce admissible probability distributions.[1] For the zero-modified Poisson distribution, the relevant part of the loglikelihood is

$$\begin{aligned} l_1 &= \sum_{k=1}^{\infty} n_k \left[\log \frac{e^{-\lambda}\lambda^k}{k!} - \log(1-e^{-\lambda}) \right] \\ &= -(n-n_0)\lambda + \left(\sum_{k=1}^{\infty} k\, n_k \right) \log \lambda - (n-n_0)\log(1-e^{-\lambda}) + c \\ &= -(n-n_0)[\lambda + \log(1-e^{-\lambda})] + n\bar{x}\log\lambda + c \end{aligned}$$

where $\bar{x}$ is the sample mean and c is independent of λ,

$$\bar{x} = \frac{1}{n}\sum_{k=0}^{\infty} k n_k$$

and

$$n = \sum_{k=0}^{\infty} n_k.$$

Hence,

$$\begin{aligned} \frac{\partial l_1}{\partial \lambda} &= -(n-n_0) - (n-n_0)\frac{e^{-\lambda}}{1-e^{-\lambda}} + n\frac{\bar{x}}{\lambda} \\ &= -\frac{n-n_0}{1-e^{-\lambda}} + \frac{n\bar{x}}{\lambda}. \end{aligned}$$

Setting this to zero yields

$$\bar{x}(1-e^{-\lambda}) = \frac{n-n_0}{n}\lambda. \tag{3.21}$$

By graphing each side as a function of λ, it is clear that if $n_0 > 0$, there exist exactly two roots: one is $\lambda = 0$, the other is $\lambda > 0$. Equation (3.21) can be solved numerically to obtain $\hat{\lambda}$. Note that since $\hat{p}_0^M = n_0/n$ and $p_0 = e^{-\lambda}$, equation (3.21) can be rewritten as

$$\bar{x} = \frac{1-\hat{p}_0^M}{1-p_0}\lambda. \tag{3.22}$$

Because the right-hand side of (3.22) is the theoretical mean of the zero-modified Poisson distribution (when $\hat{p}_0^M$ is replaced with p_0^M), equation (3.22)

[1] Maximization can be done with respect to any parameterization since maximum likelihood estimation is invariant under parameter transformations. However, it is more difficult to maximize over bounded regions because numerical methods are difficult to constrain and analytic methods will fail due to a lack of differentiability. Therefore, estimation is usually done with respect to particular class members, such as the Poisson.

is a moment equation. Hence, an alternative estimation method yielding the same results as the maximum likelihood method is to equate the zero probability to the sample proportion at zero and the theoretical mean to the sample mean. This suggests that whenever a distribution under consideration is zero-modified, a modified moment method, by fixing the zero-probability to the observed proportion at zero and equating the low order moments, can be used to get starting values for a procedure of numerically maximizing the likelihood function. Since the maximum likelihood method has better asymptotic properties, it is preferable to use the modified moment method only to obtain starting values.

For the purpose of obtaining estimates of the asymptotic variance of the mle of λ, it is easy to obtain

$$\frac{\partial^2 l}{\partial \lambda^2} = (n - n_0)\frac{e^{-\lambda}}{(1 - e^{-\lambda})^2} - \frac{n\bar{x}}{\lambda^2}$$

and the expected value is obtained by observing that $E(\bar{x}) = (1 - p_0^M)\lambda/(1 - e^{-\lambda})$. Finally, p_0^M may be replaced by its estimator, n_0/n. The variance of $\hat{p}_0^M$ is obtained by observing that the numerator, n_0 has a binomial distribution and therefore the variance is $p_0^M(1 - p_0^M)/n$.

For the zero-modified binomial distribution, we obtain

$$\begin{aligned}
l_1 &= \sum_{k=1}^{m} n_k \left\{ \log\left[\binom{m}{k} q^k (1-q)^{m-k}\right] - \log[1 - (1-q)^m] \right\} \\
&= \left(\sum_{k=1}^{m} k n_k\right) \log q + \sum_{k=1}^{m} (m-k) n_k \log(1-q) \\
&\quad - \sum_{k=1}^{m} n_k \log[1 - (1-q)^m] + c \\
&= n\bar{x}\log q + m(n - n_0)\log(1-q) - n\bar{x}\log(1-q) \\
&\quad -(n - n_0)\log[1 - (1-q)^m] + c
\end{aligned}$$

and

$$\frac{\partial l_1}{\partial q} = \frac{n\bar{x}}{q} - \frac{m(n-n_0)}{1-q} + \frac{n\bar{x}}{1-q} - \frac{(n-n_0)m(1-q)^{m-1}}{1-(1-q)^m}.$$

Setting this to zero yields

$$\bar{x} = \frac{1 - \hat{p}_0^M}{1 - p_0} mq \tag{3.23}$$

where we recall that $p_0 = (1-q)^m$. This equation matches the theoretical mean with the sample mean.

If m is known and fixed, the mle of p_0^M is still

$$\hat{p}_0^M = \frac{n_0}{n}.$$

However, even with m known, (3.23) must be solved numerically for q. When m is unknown and also needs to be estimated, the above procedure can be followed for different values of m until the maximum of the likelihood function is obtained.

The zero-modified negative binomial (or extended truncated negative binomial) distribution is a bit more complicated since three parameters need to be estimated. Of course, the mle of p_0^M is again $\hat{p}_0^M = n_0/n$ as before, reducing the problem to the estimation of r and β. The part of the loglikelihood relevant to r and β is

$$l_1 = \sum_{k=1}^{\infty} n_k \log p_k - (n - n_0)\log(1 - p_0). \tag{3.24}$$

Hence

$$\begin{aligned} l_1 &= \sum_{k=1}^{\infty} n_k \log \left[\binom{k+r-1}{k} \left(\frac{1}{1+\beta}\right)^r \left(\frac{\beta}{1+\beta}\right)^k \right] \\ &\quad -(n-n_0)\log\left[1 - \left(\frac{1}{1+\beta}\right)^r\right]. \end{aligned} \tag{3.25}$$

This function needs to be maximized over the (r, β) plane to obtain the mles. This can be done numerically using any maximization procedure such as the simplex method (described in Appendix C). Starting values can be obtained by the modified moment method by setting $\hat{p}_0^M = n_0/n$ and equating the first two moments of the distribution to the first two sample moments. It is generally easier to use raw moments (moments about the origin) than central moments for this purpose. In practice, it may be more convenient to maximize (3.24) rather than (3.25) since one can take advantage of the recursive scheme

$$p_k = p_{k-1}(a + \frac{b}{k})$$

in evaluating (3.24). This makes computer programming a bit easier.

For zero-truncated distributions there is no need to estimate the probability at zero because it is known to be zero. The remaining parameters are estimated using the same formulas developed for the zero-modified distributions.

Example 3.13 *The data set in Table 3.13 come from Beard et al. [10]. Determine a model that adequately describes the data.*

When a Poisson distribution is fitted to it, the resulting fit is very poor. The geometric distribution is tried as a one-parameter alternative. It has loglikelihood

$$\begin{aligned} l &= -n\log(1+\beta)+\sum_{k=1}^{\infty} n_k \log[\beta/(1+\beta)]^k \\ &= -n\log(1+\beta)+\sum_{k=1}^{\infty} kn_k[\log\beta-\log(1+\beta)] \\ &= -n\log(1+\beta)+n\bar{x}[\log\beta-\log(1+\beta)] \\ &= -(n+n\bar{x})\log(1+\beta)+n\bar{x}\log\beta \end{aligned}$$

where $\bar{x}=\sum_{k=1}^{\infty} k\, n_k/n$ and $n=\sum_{k=0}^{\infty} n_k$.

Differentiation reveals that the loglikelihood has a maximum at

$$\hat{\beta}=\bar{x}.$$

The mle of the geometric mean is the same as the moment estimator. From Table 3.13, it is clear that the geometric distribution is a poor fit. Corresponding zero-modified distributions might be reasonable alternatives. The results are given in the table. Since the mle of the probability at zero is the observed proportion at zero, there is a perfect fit at zero for zero-modified distributions. The mle of the remaining parameter is obtained by numerically maximizing the loglikelihood. The results show that the zero-modified Poisson distribution is still a poor fit. Even though the expected number of observations appear to compare favorably to the observed numbers, the high sample size demands a better fit. The zero-modified geometric provides a much better fit in the tail region and provides an almost perfect fit overall (with the same number of parameters as the zero-modified Poisson). □

3.6.3 The scoring method with the $(a,b,0)$ and $(a,b,1)$ classes

It has been seen that maximum likelihood estimation requires numerical maximization. As long as iterations are to be used, it makes sense to consider the scoring method because then the estimated information and covariance matrices are produced automatically. Recall from Subsection 3.3.4 that the required quantity is

$$\frac{\partial p_k(\boldsymbol{\theta})}{\partial \theta_i}$$

where $\boldsymbol{\theta}$ is the vector of parameters. From the recursions for members of the two classes,

$$\begin{aligned} \frac{\partial p_k(\boldsymbol{\theta})}{\partial \theta_i} &= \frac{\partial (a+b/k)p_{k-1}(\boldsymbol{\theta})}{\partial \theta_i} \\ &= \left(\frac{\partial a}{\partial \theta_i}+\frac{1}{k}\frac{\partial b}{\partial \theta_i}\right)p_{k-1}(\boldsymbol{\theta})+(a+b/k)\frac{\partial p_{k-1}(\boldsymbol{\theta})}{\partial \theta_i}. \end{aligned}$$

Table 3.13 Fitted distributions to Beard data

Accidents	Observed	Poisson	Geometric	ZM Poisson	ZM geometric
0	370,412	369,246.9	372,206.5	370,412.0	370,412.0
1	46,545	48,643.6	43,325.8	46,432.1	46,555.2
2	3,935	3,204.1	5,043.2	4,138.6	3,913.6
3	317	140.7	587.0	245.9	329.0
4	28	4.6	68.3	11.0	27.7
5	3	0.1	8.0	0.4	2.3
6+	0	0.0	1.0	0.0	0.2
Parameters		λ: 0.13174	β: 0.13174	p_0^M: 0.87934	p_0^M: 0.87934
				λ: 0.17827	β: 0.091780
Chi-square		543.0	643.4	64.8	0.58
Degrees of Freedom		2	4	2	2
p-Value		$< 1\%$	$< 1\%$	$< 1\%$	74.9%
$-$Loglikelihood		171,373	171,479	171,160	171,133

The above holds for $k = 1, 2, \ldots$ for members of the $(a, b, 0)$ class and for $k = 2, 3, \ldots$ for members of the $(a, b, 1)$ class. Thus there is also a recursion for the derivatives, although this one involves not only prior derivatives, but prior values of the pf itself. The derivatives of a and b with respect to the parameters are easy to obtain for the various class members.

With regard to starting the recursions, the required derivatives are taken with regard to the value of p_0 or p_1 as appropriate. These values can be found in Appendix B. Note that here p_k refers to p_k, p_k^M, or p_k^T as required.

Example 3.14 (Example 3.13 continued) *Determine the iterative formula for the ZM geometric distribution and perform one iteration from a starting value of $\beta = 0.13174$ (which is the sample mean).*

The required derivatives with respect to the single parameter β are

$$\begin{aligned}\frac{\partial a}{\partial \beta} &= \frac{\partial \beta(1+\beta)^{-1}}{\partial \beta} = (1+\beta)^{-2}\\ \frac{\partial b}{\partial \beta} &= \frac{\partial 0}{\partial \beta} = 0\\ \frac{\partial p_1^T}{\partial \beta} &= \frac{\partial (1+\beta)^{-1}}{\partial \beta} = -(1+\beta)^{-2}\end{aligned}$$

and the recursion for subsequent derivatives is

$$\frac{\partial p_k^T}{\partial \beta} = (1+\beta)^{-2} p_{k-1}^T + \frac{\beta}{1+\beta}\frac{\partial p_{k-1}^T}{\partial \beta}, \qquad k = 2, 3, \ldots.$$

When using the scoring method for zero-modified distributions, the appropriate p_1 to use is the one for the zero-truncated version. All sums (for

Table 3.14 One iteration of the scoring method

k	n_k	p_k^T	$p_k^{T\prime}$	$n_k p_k^{T\prime}/p_k^T$	$(p_k^{T\prime})^2/p_k^T$
1	46,545	0.883595	−0.780740	−41,126.9	0.689858
2	3,935	0.102855	0.598977	22,915.6	3.488151
3	317	0.011973	0.150027	3,972.2	1.879928
4	28	0.001394	0.026811	538.7	0.515792
5	3	0.000162	0.004209	77.8	0.109204
6+	0	0.000021	0.000717	0.0	0.024036
Totals	50,828	1.000000	0.000000	−13,622.7	6.706970

the likelihood function as well as for the components of the scoring method) should ignore the zero term. The idea is to pretend that the nonzero portion of the sample comes from a zero-truncated distribution regardless of the actual situation. The mle of the probability at zero is already known to be n_0/n and because it is independent of the mles of the other parameters, it need not be included in the estimated covariance matrix.

Using $\beta = 0.13174$ the calculations for the first iteration are displayed in Table 3.14. The final line ($k = 6+$) is obtained by noting that $\sum p_k^T = 1$ and $\sum p_k^{T\prime} = 0$ where $p_k^{T\prime} = \partial p_k^T/\partial\beta$. The new value of $\hat{\beta}$ is, using (3.10)–(3.11),

$$0.13174 + (-13{,}622.7)/[50{,}828(6.706970)] = 0.091779.$$

The estimate of the variance of $\hat{\beta}$ is $1/[50{,}828(6.706970)] = 2.9334 \times 10^{-6}$. After just a few more iterations the process converges at $\hat{\beta} = 0.0917801$ with estimated variance 1.97144×10^{-6}. □

3.7 COMPOUND FREQUENCY MODELS

A larger class of distributions can be created by the processes of compounding any two discrete distributions. The term "compounding" reflects the idea that the pgf of the new distribution $P(z)$ is written as

$$P(z) = P_1[P_2(z)] \tag{3.26}$$

where $P_1(z)$ and $P_2(z)$ are called the "primary" and "secondary" distributions, respectively.

The compound distributions arise naturally as follows. Let N be a counting random variable with pgf $P_1(z)$. Let $M_1, M_2, \ldots$ be identically and independently distributed random variables with pgf $P_2(z)$. Assuming that the M_js do not depend on N, the pgf of the random sum

$$S = M_1 + M_2 + \cdots + M_N$$

is $P(z) = P_1[P_2(z)]$. This is shown as follows

$$\begin{aligned} P(z) &= \sum_{k=0}^{\infty} \Pr(S = k) z^k \\ &= \sum_{k=0}^{\infty}\sum_{n=0}^{\infty} \Pr(S = k|N = n)\Pr(N = n) z^k \\ &= \sum_{n=0}^{\infty} \Pr(N = n) \sum_{k=0}^{\infty} \Pr(M_1 + \cdots + M_n = k|N = n) z^k \\ &= \sum_{n=0}^{\infty} \Pr(N = n)[P_2(z)]^n \\ &= P_1[P_2(z)]. \end{aligned}$$

In insurance contexts, this distribution can arise naturally. If N represents the number of accidents arising in a portfolio of risks and $\{M_k;\ k = 1, 2, ..., N\}$ represents the number of claims (injuries, number of cars, etc.) from the accidents, then S represents the total number of claims from the portfolio. This kind of interpretation is not necessary to justify the use of a compound distribution. If a compound distribution fits data well, that may be enough justification itself.

Example 3.15 *Demonstrate that any of the zero-modified (or "with zeros") distributions are compound distributions.*

Consider a primary Bernoulli distribution. It has pgf $P_1(z) = 1 - q + qz$. Then consider an arbitrary secondary distribution with pgf $P_2(z)$. Then, from (3.26) we obtain

$$P(z) = P_1[P_2(z)] = 1 - q + qP_2(z).$$

From (3.13) this is the pgf of a ZM distribution with

$$q = \frac{1 - p_0^M}{1 - p_0}.$$

That is, the ZM distribution has assigned arbitrary probability p_0^M at zero, while p_0 is the probability assigned at zero by the secondary distribution. □

Example 3.16 *Consider the case where both M and N have the Poisson distribution. Determine the pgf of this distribution.*

This distribution is called the Poisson–Poisson or Neyman Type A distribution. Let $P_1(z) = e^{\lambda_1(z-1)}$ and $P_2(z) = e^{\lambda_2(z-1)}$. Then

$$P(z) = e^{\lambda_1[e^{\lambda_2(z-1)} - 1]}.$$

When λ_2 is a lot larger than λ_1—for example, $\lambda_1 = 0.1$ and $\lambda_2 = 10$—then the resulting distribution will have two local modes. □

The probability of exactly k claims can be written as

$$\begin{aligned}\Pr(S=k) &= \sum_{n=0}^{\infty}\Pr(S=k|N=n)\Pr(N=n)\\ &= \sum_{n=0}^{\infty}\Pr(M_1+\cdots+M_N=k|N=n)\Pr(N=n)\\ &= \sum_{n=0}^{\infty}\Pr(M_1+\cdots+M_n=k)\Pr(N=n). \qquad (3.27)\end{aligned}$$

Letting $g_n = \Pr(S=n)$, $p_n = \Pr(N=n)$ and $f_n = \Pr(M=n)$, this is rewritten as

$$g_k = \sum_{n=0}^{\infty} p_n f_k^{*n} \qquad (3.28)$$

where f_k^{*n}, $k = 0, 1, \ldots$ is the "n-fold convolution" of the function f_k, $k = 0, 1, \ldots$, that is the probability that the sum of n random variables, which are each i.i.d. with probability function f_k, will take on value k.

When $P_1(z)$ is chosen to be a member of the $(a, b, 0)$ class, then

$$p_k = \left(a + \frac{b}{k}\right) p_{k-1}, \qquad k = 1, 2, \ldots \qquad (3.29)$$

and a simple recursive formula can be used. This formula avoids the use of convolutions and thus reduces the computations considerably.

Theorem 3.3 *For the model described in this section, if (3.29) is satisfied for $P_1(z)$, then*

$$g_k = \frac{1}{1 - af_0}\sum_{j=1}^{k}\left(a + \frac{bj}{k}\right) f_j g_{k-j}, \qquad k = 1, 2, 3, \ldots. \qquad (3.30)$$

Proof: From (3.29),

$$np_n = a(n-1)p_{n-1} + (a+b)p_{n-1}.$$

Multiplying each side by $[P_2(z)]^{n-1}P_2'(z)$ and summing over n yields

$$\begin{aligned}\sum_{n=1}^{\infty} np_n[P_2(z)]^{n-1}P_2'(z) &= a\sum_{n=1}^{\infty}(n-1)p_{n-1}[P_2(z)]^{n-1}P_2'(z)\\ &\quad +(a+b)\sum_{n=1}^{\infty}p_{n-1}[P_2(z)]^{n-1}P_2'(z).\end{aligned}$$

Because $P(z) = \sum_{n=0}^{\infty} p_n [P_2(z)]^n$, the previous equation is

$$P'(z) = a \sum_{n=0}^{\infty} n p_n [P_2(z)]^n P_2'(z) + (a+b) \sum_{n=0}^{\infty} p_n [P_2(z)]^n P_2'(z).$$

Therefore

$$P'(z) = aP'(z)P_2(z) + (a+b)P(z)P_2'(z).$$

Each side can be expanded in powers of z. The coefficients of z^{k-1} in such an expansion must be the same on both sides of the equation. Hence, for $k = 1, 2, \ldots$ we have

$$\begin{aligned} kg_k &= a\sum_{j=0}^{k}(k-j)f_j g_{k-j} + (a+b)\sum_{j=0}^{k} j f_j g_{k-j}, \\ &= akf_0 g_k + a\sum_{j=1}^{k}(k-j)f_j g_{k-j} + (a+b)\sum_{j=1}^{k} j f_j g_{k-j} \\ &= akf_0 g_k + ak\sum_{j=1}^{k} f_j g_{k-j} + b\sum_{j=1}^{k} j f_j g_{k-j}. \end{aligned}$$

Therefore,

$$g_k = af_0 g_k + \sum_{j=1}^{k}\left(a + \frac{bj}{k}\right) f_j g_{k-j}.$$

Rearrangement yields (3.30). □

In order to use (3.30) the starting value g_0 is required. Its value will depend on the particular primary distribution. These will be developed in the examples. If the primary distribution is a member of the $(a, b, 1)$ class, the proof must be modified to reflect the fact that the recursion for the primary distribution begins at $k = 2$. The result is the following.

Theorem 3.4 *If the primary distribution is a member of the* $(a, b, 1)$ *class the recursive formula is*

$$g_k = \frac{[p_1 - (a+b)p_0]f_k + \sum_{j=1}^{k}\left(a + \frac{bj}{k}\right) f_j g_{k-j}}{1 - af_0}, \qquad k = 1, 2, 3, \ldots.$$

Proof: It is similar to the proof of Theorem 3.3 and is left to the reader. □

Example 3.17 *Develop the recursive formula for the case where the primary distribution is Poisson.*

In this case $a = 0$ and $b = \lambda$, yielding the recursive form

$$g_k = \frac{\lambda}{k} \sum_{j=1}^{k} j f_j g_{k-j}.$$

The starting value is, from (3.27)

$$\begin{aligned} g_0 &= \sum_{n=0}^{\infty} \Pr(M_1 + \cdots + M_n = 0) \Pr(N = n) \\ &= \sum_{n=0}^{\infty} (f_0)^n e^{-\lambda} \lambda^n / n! \\ &= e^{-\lambda(1-f_0)}. \end{aligned} \tag{3.31}$$

Distributions of this type are called compound Poisson distributions. When the secondary distribution is specified, the compound distribution is called Poisson–X, where X is the name of the secondary distribution. □

There is an easier way to obtain g_0 and it applies to any compound distribution.

Theorem 3.5 *For any compound distribution, $g_0 = P_1(f_0)$ where $P_1(z)$ is the pgf of the primary distribution and f_0 is the probability that the secondary distribution takes on the value zero.*

Proof: Arguing as in (3.31), we obtain

$$\begin{aligned} g_0 &= \sum_{n=0}^{\infty} \Pr(M_1 + \cdots + M_n = 0) \Pr(N = n) \\ &= \sum_{n=0}^{\infty} (f_0)^n \Pr(N = n) \\ &= P_1(f_0). \end{aligned}$$

□

Note that the secondary distribution is not required to be in any special form. However, for purposes in the next chapter, it will also be convenient to choose it from the $(a, b, 0)$ or the $(a, b, 1)$ class. The choice of $P_1(z)$ and $P_2(z)$ from the $(a, b, 0)$ or $(a, b, 1)$ classes provides a rich variety of shapes of the compound distribution with pgf $P(z) = P_1[P_2(z)]$.

Example 3.18 *Demonstrate that the negative binomial distribution is a Poisson–logarithmic distribution.*

The negative binomial distribution has pgf

$$P(z) = [1 - \beta(z-1)]^{-r}.$$

Suppose $P_1(z)$ is Poisson(λ) and $P_2(z)$ is logarithmic(β), then

$$\begin{aligned} P_1[P_2(z)] &= \exp\{\lambda[P_2(z)-1]\} \\ &= \exp\left\{\lambda\left[1 - \frac{\log[1-\beta(z-1)]}{\log(1+\beta)} - 1\right]\right\} \\ &= \exp\left\{\frac{-\lambda}{\log(1+\beta)}\log[1-\beta(z-1)]\right\} \\ &= [1-\beta(z-1)]^{-\lambda/[\log(1+\beta)]} \\ &= [1-\beta(z-1)]^{-r} \end{aligned}$$

where $r = \lambda/\log(1+\beta)$. This shows that the negative binomial distribution can be written as a compound Poisson distribution with a logarithmic secondary distribution. □

The above example shows that the "Poisson–logarithmic" distribution does not create a new distribution beyond the $(a, b, 0)$ and $(a, b, 1)$ classes. As a result, this combination of distributions is not useful to us. The following theorem shows that certain other combinations are also of no use in expanding the class of distributions through compounding. Suppose $P(z) = P_1[P_2(z); \theta]$ as before. Now, $P_2(z)$ can always be written as

$$P_2(z) = f_0 + (1 - f_0)P_2^*(z) \tag{3.32}$$

where $P_2^*(z)$ is the pgf of the conditional distribution over the positive range (in other words, the zero-truncated version).

Theorem 3.6 *Suppose the pgf $P_1(z; \theta)$ satisfies*

$$P_1(z; \theta) = B[\theta(z-1)]$$

for some parameter θ and some function $B(z)$ which is independent of θ. That is, the parameter θ and the argument z only appear in the pgf as $\theta(z-1)$. There may be other parameters as well, but they may appear anywhere in the pgf. Then $P(z) = P_1[P_2(z); \theta]$ can be rewritten as

$$P(z) = P_1[P_2^T(z); \theta(1-f_0)].$$

Proof:

$$\begin{aligned} P(z) &= P_1[P_2(z); \theta] \\ &= P_1[f_0 + (1-f_0)P_2^T(z); \theta] \\ &= B\{\theta[f_0 + (1-f_0)P_2^T(z) - 1]\} \\ &= B\{\theta(1-f_0)[P_2^T(z) - 1]\} \\ &= P_1[P^T(z); \theta(1-f_0)]. \end{aligned}$$

□

This shows that adding, deleting, or modifying the probability at zero in the secondary distribution does not add a new distribution since it is equivalent to modifying the parameter θ of the primary distribution. This means that, for example, a Poisson primary distribution with either a Poisson, zero-truncated Poisson, or zero-modified Poisson secondary distribution will still lead to a Neyman Type A (Poisson–Poisson) distribution.

3.7.1 Estimation

For the method of moments, the first few moments can be matched with the sample moments. The system of equations can be solved to obtain the moment based estimators. Note that the number of parameters in the compound model is the sum of the number of parameters in the primary and secondary distributions. The first two theoretical moments for compound distributions are

$$\begin{aligned} E(S) &= E(N)E(M) \\ Var(S) &= E(N)Var(M) + E(M)^2 Var(N). \end{aligned}$$

These results are developed in the next chapter. The first three moments for the compound Poisson distribution are developed in the next section.

Maximum likelihood estimation is also carried out as before. The loglikelihood to be maximized is

$$l = \sum_{k=0}^{\infty} n_k \log g_k.$$

When g_k is the probability of a compound distribution, the loglikelihood can be maximized numerically using a procedure, such as the simplex method (described in Appendix C). The first and second derivatives of the loglikelihood can be obtained by using approximate differentiation methods as applied directly to the loglikelihood function at the maximum value. Implementation of the scoring method will be discussed at the end of this section.

Example 3.19 *Determine various properties of the Poisson–zero-truncated geometric distribution. This distribution is also called the Polya–Aeppli distribution.*

For the zero-truncated geometric the pgf is

$$P_2(z) = \frac{[1-\beta(z-1)]^{-1} - (1+\beta)^{-1}}{1-(1+\beta)^{-1}}$$

and therefore the pgf of the Polya–Aeppli distribution is

$$\begin{aligned} P(z) &= P_1[P_2(z)] = \exp\left(\lambda\left\{\frac{[1-\beta(z-1)]^{-1} - (1+\beta)^{-1}}{1-(1+\beta)^{-1}} - 1\right\}\right) \\ &= \exp\left\{\lambda\frac{[1-\beta(z-1)]^{-1} - 1}{1-(1+\beta)^{-1}}\right\}. \end{aligned}$$

The mean is

$$P'(1) = \lambda(1+\beta)$$

and the variance is

$$P''(1) + P'(1) - [P'(1)]^2 = \lambda(1+\beta)(1+2\beta).$$

From Theorem 3.5, the probability at zero is

$$g_0 = P_1(0) = e^{-\lambda}.$$

The successive values of g_k are computed easily using the compound Poisson recursion

$$g_k = \frac{\lambda}{k}\sum_{j=1}^{k} j f_j g_{k-j}, \qquad k = 1,2,3,... \tag{3.33}$$

where $f_j = \beta^{j-1}/(1+\beta)^j$, $j = 1,2,....$ For any values of λ and β, the loglikelihood function can be easily evaluated. □

Example 3.20 *The data in Table 3.15, from Simon [112], represent the observed number of claims per contract for 298 contracts. Determine an appropriate model.*

The Poisson, negative binomial and Polya–Aeppli distributions are fitted to the data. The Polya–Aeppli and the negative binomial are both plausible distributions. The p-value of the chi-square statistic and the loglikelihood both indicate that the Polya–Aeppli is slightly better than the negative binomial. □

Another useful compound Poisson distribution is the Poisson–extended truncated negative binomial (Poisson–ETNB) distribution. Although it does not matter if the secondary distribution is modified or truncated, we prefer the truncated version here so that the parameter r may be extended.[2] Special cases are: $r = 1$, which is the Poisson–geometric (also called Polya–Aeppli); $r = 0$, which is the Poisson–logarithmic (negative binomial); and $r = -0.5$, which is called the Poisson–inverse Gaussian. This name is not consistent with the others. Here the inverse Gaussian distribution is a mixing distribution (see Section 3.9).

Example 3.21 *Consider the data in Table 3.16 on automobile liability policies in Switzerland taken from Bühlmann [19]. The Poisson, negative binomial, and Poisson–inverse Gaussian distributions are fitted to these data.*

[2] This does not contradict Theorem 3.6. When $-1 < r \le 0$, it is still the case that changing the probability at zero will not produce new distributions. What is true is that there is no probability at zero which will lead to an ordinary $(a, b, 0)$ negative binomial secondary distribution.

Table 3.15 Fit of Simon data

Number of claims/contract	Number of contracts	Fitted distributions		
		Poisson	Negative binomial	Polya–Aeppli
0	99	54.0	95.9	98.7
1	65	92.2	75.8	70.6
2	57	78.8	50.4	50.2
3	35	44.9	31.3	32.6
4	20	19.2	18.8	20.0
5	10	6.5	11.0	11.7
6	4	1.9	6.4	6.6
7	0	0.5	3.7	3.6
8	3	0.1	2.1	2.0
9	4	0.0	1.2	1.0
10	0	0.0	0.7	0.5
11	1	0.0	0.4	0.3
12+	0	0.0	0.5	0.3
Parameters		$\lambda = 1.70805$	$\beta = 1.15907$	$\lambda = 1.10551$
			$r = 1.47364$	$\beta = 0.545039$
Chi-square		72.64	4.06	2.84
Degrees of freedom		4	5	5
p-Value		<1%	54.05%	72.39%
−Loglikelihood		577.0	528.8	528.5

The Poisson distribution is a very bad fit. Its tail is far too light compared with the actual experience. The negative binomial distribution appears to be much better, but cannot be accepted since the p-value of the chi-square statistic is very small. The large sample size requires a better fit.

The Poisson–inverse Gaussian (P.–I.G.) distribution provides an almost perfect fit (p-value is large). Note that the P.–I.G. has two parameters, like the negative binomial. This example shows that the P.–I.G. can have a much heavier right hand tail than the negative binomial. □

Example 3.22 (Example 3.11 continued) *Members of the* $(a, b, 0)$ *class were not sufficient to describe these data. Determine a suitable model.*

Thirteen different distributions were fit to the data. The results of that process revealed six models with p-values above 0.01 for the chi-square goodness-of-fit test. Information about those models is given in Table 3.17. The likelihood ratio test indicates that the three parameter model with the smallest negative loglikelihood (Poisson–ETNB) is not significantly better than the

Table 3.16 Fit of Bühlmann data

No. of accidents	Observed frequency	Fitted Distributions: Poisson	Negative. binomial	P.–I.G.[a]
0	103,704	102,629.6	103,723.6	103,710.0
1	14,075	15,922.0	13,989.9	14,054.7
2	1,766	1,235.1	1,857.1	1,784.9
3	255	63.9	245.2	254.5
4	45	2.5	32.3	40.4
5	6	0.1	4.2	6.9
6	2	0.0	0.6	1.3
7+	0	0.0	0.1	0.3
Parameters		$\lambda = 0.155140$	$\beta = 0.150232$	$\lambda = 0.144667$
			$r = 1.03267$	$\beta = 0.310536$
Chi-square		1,332.3	12.12	0.78
Degrees of freedom		2	2	3
p-Values		<1%	<1%	85.5%
−Loglikelihood		55,108.5	54,615.3	54,609.8

[a]P.–I.G. stands for Poisson–inverse Gaussian

Table 3.17 Six useful models for Example 3.22

Model	Number of parameters	Negative loglikelihood	χ^2	p-value
Negative binomial	2	5,348.04	8.77	0.0125
ZM logarithmic	2	5,343.79	4.92	0.1779
Poisson–inverse Gaussian	2	5,343.51	4.54	0.2091
ZM negative binomial	3	5,343.62	4.65	0.0979
Geometric–negative binomial	3	5,342.70	1.96	0.3754
Poisson–ETNB	3	5,342.51	2.75	0.2525

two-parameter Poisson–inverse Gaussian model. The latter appears to be an excellent choice. □

3.7.2 The scoring method for compound distributions

As in the previous sections, the key element for the scoring method is the derivative of the pf with respect to a parameter. Using the notation of this section we are interested in obtaining

$$\frac{\partial g_k}{\partial \theta_i}.$$

From (3.30) we have, for primary distributions in the $(a, b, 0)$ class,

$$\frac{\partial g_k}{\partial \theta_i} = \frac{\partial}{\partial \theta_i} \frac{1}{1 - af_0} \sum_{j=1}^{k} \left(a + \frac{bj}{k}\right) f_j g_{k-j}.$$

The parameters of the compound distribution come separately from the primary and secondary distributions. If θ_i is a parameter from the secondary distribution (that is, it determines the values $f_0, f_1, \ldots$), the only place it appears is in the f_js (and, of course, it also appears in earlier values of g_k). For such parameters

$$\begin{aligned}
\frac{\partial g_k}{\partial \theta_i} &= \frac{1}{1 - af_0} \sum_{j=1}^{k} \left(a + \frac{bj}{k}\right) \frac{\partial f_j}{\partial \theta_i} g_{k-j} \\
&+ \frac{a}{(1 - af_0)^2} \frac{\partial f_0}{\partial \theta_i} \sum_{j=1}^{k} \left(a + \frac{bj}{k}\right) f_j g_{k-j} \\
&+ \frac{1}{1 - af_0} \sum_{j=1}^{k} \left(a + \frac{bj}{k}\right) f_j \frac{\partial g_{k-j}}{\partial \theta_i}.
\end{aligned}$$

If the secondary distribution is a member of the $(a, b, 0)$ or $(a, b, 1)$ class, then the derivatives of f_j can be obtained directly, or, if necessary, by recursion as in the previous section.

If θ_i is a parameter from the primary distribution, then it appears in a, b, and previous values of g_k. For such parameters

$$\begin{aligned}
\frac{\partial g_k}{\partial \theta_i} &= \frac{1}{1 - af_0} \sum_{j=1}^{k} \left(\frac{\partial a}{\partial \theta_i} + \frac{j}{k} \frac{\partial b}{\partial \theta_i}\right) f_j g_{k-j} \\
&+ \frac{f_0}{(1 - af_0)^2} \frac{\partial a}{\partial \theta_i} \sum_{j=1}^{k} \left(a + \frac{bj}{k}\right) f_j g_{k-j} \\
&+ \frac{1}{1 - af_0} \sum_{j=1}^{k} \left(a + \frac{bj}{k}\right) f_j \frac{\partial g_{k-j}}{\partial \theta_i}.
\end{aligned}$$

Example 3.23 (Example 3.21 continued) *For the Poisson–inverse Gaussian model, perform one iteration of the scoring method using $\lambda = 0.144667$ and $\beta = 0.310536$ as starting values. Demonstrate that these are the maximum likelihood estimates and obtain the estimated covariance matrix.*

The numerical values are given in Table 3.18. Although the results are presented to six (and later four) decimal places, more accuracy was used and kept in the intermediate calculations. The development follows.

For the ETNB secondary distribution (recalling that for the Poisson–inverse Gaussian distribution, $r = -0.5$), $a = \beta/(1+\beta) = 0.236953$ and $b = -1.5a = -0.355430$. The first probability is

$$f_1 = -0.5\beta[(1+\beta)^{.5} - (1+\beta)]^{-1} = 0.936763.$$

Recursions provide

$$\begin{aligned} f_2 &= (0.236953 - 0.355430/2)0.936763 = 0.055492 \\ f_3 &= (0.236953 - 0.355430/3)0.055492 = 0.006575 \\ f_4 &= (0.236953 - 0.355430/4)0.006575 = 0.000974 \\ f_5 &= (0.236953 - 0.355430/5)0.000974 = 0.000161 \\ f_6 &= (0.236953 - 0.355430/6)0.000161 = 0.000029. \end{aligned}$$

The Poisson–inverse Gaussian probabilities can be obtained from (3.30). The starting value is $g_0 = e^{-0.144667} = 0.865310$ and the recursive formula is

$$g_k = \sum_{j=1}^{k} \frac{0.144667j}{k} f_j g_{k-j}.$$

The probabilities are

$$\begin{aligned} g_1 &= \frac{0.144667}{1} 1(9.936763)(0.865310) = 0.117266 \\ g_2 &= \frac{0.144667}{2} [1(0.936763)(0.117266) + 2(0.055492)(0.865310)] \\ &= 0.014892 \\ g_3 &= \frac{0.144667}{3} [1(0.936763)(0.014892) + 2(0.055492)(0.117266) \\ &\quad +3(0.006575)(0.865310)] = 0.002123 \\ g_4 &= \frac{0.144667}{4} [1(0.936763)(0.002123) + 2(0.055492)(0.014892) \\ &\quad +3(0.006575)(0.117266) + 4(0.000974)(0.865310)] = 0.000337 \\ g_5 &= \frac{0.144667}{5} [1(0.936763)(0.000337) + 2(0.055492)(0.002123) \\ &\quad +3(0.006575)(0.014892) + 4(0.000974)(0.117266) \\ &\quad +5(0.000161)(0.865310)] = 0.000058 \\ g_6 &= \frac{0.144667}{6} [1(0.936763)(0.000058) + 2(0.055492)(0.000337) \\ &\quad +3(0.006575)(0.002123) + 4(0.000974)(0.014892) \\ &\quad +5(0.000161)(0.117266) + 6(0.000029)(0.865310)] = 0.000010 \end{aligned}$$

and by subtraction, $g_{7+} = 0.000002$.

For $\partial g_k/\partial\beta$ we have

$$\begin{aligned}
\frac{\partial g_0}{\partial\beta} &= 0\\
\frac{\partial g_k}{\partial\beta} &= \sum_{j=1}^{k}\frac{\lambda j}{k}\left(\frac{\partial f_j}{\partial\beta}g_{k-j}+f_j\frac{\partial g_{k-j}}{\partial\beta}\right), \quad k=1,2,\ldots
\end{aligned}$$

and for the partial derivative of f_j with respect to β we have

$$\begin{aligned}
\frac{\partial f_1}{\partial\beta} &= -\frac{\partial}{\partial\beta}(0.5\beta)[(1+\beta)^{.5}-(1+\beta)]^{-1}\\
&= -\frac{[(1+\beta)^{.5}-(1+\beta)](0.5)-0.5\beta[0.5(1+\beta)^{-.5}-1]}{[(1+\beta)^{0.5}-(1+\beta)]^2}\\
&= -0.166635\\
\frac{\partial f_j}{\partial\beta} &= \left(\frac{\partial}{\partial\beta}\frac{\beta}{1+\beta}+\frac{1}{j}\frac{\partial}{\partial\beta}\frac{-1.5\beta}{1+\beta}\right)f_{j-1}+\left(\frac{\beta}{1+\beta}-\frac{1}{j}\frac{1.5\beta}{1+\beta}\right)\frac{\partial f_{j-1}}{\partial\beta}\\
&= \left(1-\frac{1.5}{j}\right)\left[\frac{1}{(1+\beta)^2}f_{j-1}+\frac{\beta}{1+\beta}\frac{\partial f_{j-1}}{\partial\beta}\right], \quad j=2,3,\ldots.
\end{aligned}$$

For $\partial g_k/\partial\lambda$ we have

$$\begin{aligned}
\frac{\partial g_0}{\partial\lambda} &= \frac{\partial e^{-\lambda}}{\partial\lambda}=-e^{-\lambda}\\
\frac{\partial g_k}{\partial\lambda} &= \sum_{j=1}^{k}\frac{j}{k}f_j\left(g_{k-j}+\lambda\frac{\partial g_{k-j}}{\partial\lambda}\right).
\end{aligned}$$

The elements of the **S** vector are $(0.003, 0.098)$, which are close to zero, the value that will indicate that the maximum has been achieved. The last three totals are multiplied by the sample size (119,853) to give the elements of the information matrix:

$$\begin{bmatrix} 5{,}079.96 & 11{,}079.4\\ 11{,}079.4 & 799{,}411 \end{bmatrix}$$

and its inverse:

$$\begin{bmatrix} 2.02988\times10^{-4} & -2.81331\times10^{-6}\\ -2.81331\times10^{-6} & 1.28991\times10^{-6} \end{bmatrix}.$$

The standard deviations of the estimators are the square root of the diagonal elements, 0.014247 and 0.00113574. The correlation is the off-diagonal element divided by the product of the standard deviations, −0.173861. □

Table 3.18 Scoring method for the Poisson–inverse Gaussian distribution

k	n_k	$\frac{\partial f_k}{\partial \beta}$	$\frac{\partial g_k}{\partial \beta}$	$\frac{\partial g_k}{\partial \lambda}$	$\frac{n_k}{g_k}\frac{\partial g_k}{\partial \beta}$	$\frac{n_k}{g_k}\frac{\partial g_k}{\partial \lambda}$	$\frac{\left(\frac{\partial g_k}{\partial \beta}\right)^2}{g_k}$	$\frac{\frac{\partial g_k}{\partial \beta}\frac{\partial g_k}{\partial \lambda}}{g_k}$	$\frac{\left(\frac{\partial g_k}{\partial \lambda}\right)^2}{g_k}$
0	103,704	0	0	−.8653	.0000	−103,704	0	0	.8653
1	14,075	−.1666	−.0209	.6933	−2,504	83,217	.0037	−.1233	4.0992
2	1,766	.1265	.0130	.1430	1,542	16,955	.0114	.1249	1.3726
3	255	.0311	.0057	.0240	683	2,885	.0152	.0643	.2718
4	45	.0070	.0016	.0041	217	546	.0078	.0197	.0496
5	6	.0016	.0004	.0007	43	75	.0029	.0052	.0091
6	2	.0003	.0001	.0001	19	26	.0009	.0013	.0017
7+	0	.0001	.0000	.0000	0	0	.0004	.0004	.0004
Total	119,853				0.003	0.098	.0424	.0924	6.6699

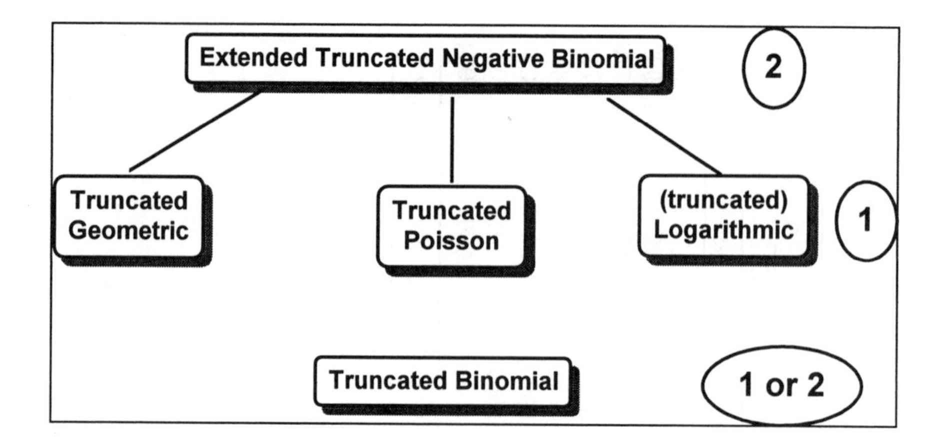

Fig. 3.2 Discrete distributions I: Zero-truncated distributions

3.8 AN INVENTORY OF DISCRETE DISTRIBUTIONS

In previous sections, we introduced the simple $(a, b, 0)$ class, generalized to the $(a, b, 1)$ class, and then used compounding to create a larger class of distributions. The probabilities of these distributions can all be carried out by using simple recursive procedures. In this section we classify these distributions into a logical family as was done with the continuous distributions in Section 2.7.

When a distribution is identified as a special case of a more general distribution, we can use a formal likelihood ratio test to determine if the more general (and complex) distribution is justified on the basis of the improvement in fit.

Figures 3.2–3.5 indicate relationships between four subsets of the distributions that can be constructed by compounding distributions from the $(a, b, 0)$ and $(a, b, 1)$ classes. The number of parameters is given in the circle accompanying each name. Note that whenever the binomial distribution appears, it is labeled as having 1 or 2 parameters depending on whether m is known or unknown in advance. Figure 3.2 gives the zero-truncated versions of the basic distributions. Figure 3.3 gives the set of compound Poisson distributions resulting from using the distributions in Figure 3.2 as secondary distributions. Some distributions appear in more than one figure in order to indicate additional interconnections.

The generalized Poisson–Pascal distribution is a compound Poisson with a negative binomial or extended truncated negative binomial (ETNB) as a secondary distribution. Using the ETNB allows a larger range for the parameter r, since for the ETNB we have $r > -1$. Figure 3.4 gives the class of $(a, b, 1)$ distributions that include a modification at zero. Figure 3.5 gives a few other distributions that are generalizations of the geometric distribution and the binomial distribution. Of course, these families can be expanded by

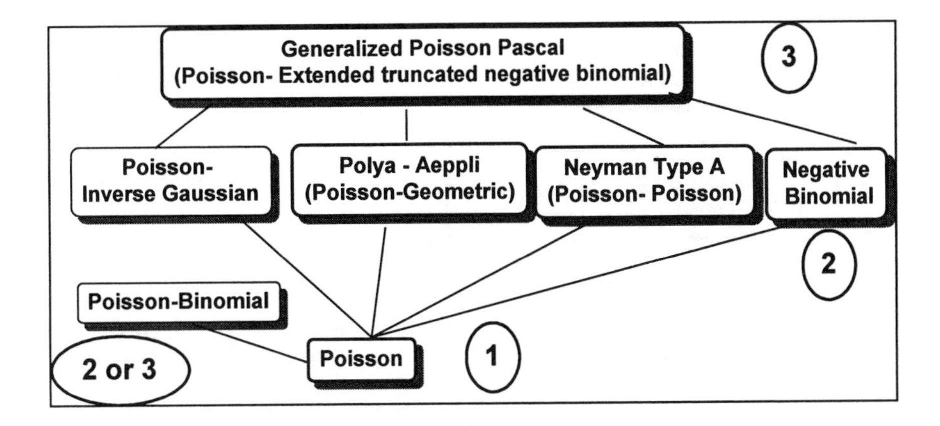

Fig. 3.3 Discrete distributions II: Compound Poisson distributions

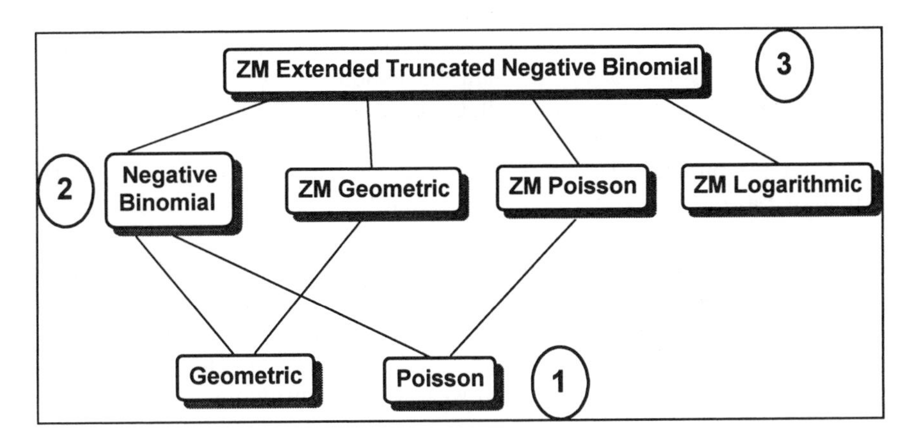

Fig. 3.4 Discrete distributions III: Zero-modified distributions

further compounding. For example, the zero-probabilities in the distributions in Figure 3.3 could be modified. Of course, when this is done, an additional parameter is introduced. This should be justified by an increase in the log-likelihood of 1.92 (at the 5% significance level) as was shown in Subsection 3.3.3.

We have not listed compound distributions where the primary distribution is one of the two parameter models such as the negative binomial or Poisson–inverse Gaussian. This was done because these distributions are often themselves compound Poisson distributions and, as such, are generalizations of the distributions in Figure 3.3.

For a given data set, the recommended strategy is to begin by fitting the simple one-parameter models and examining the quality of fit through a test-

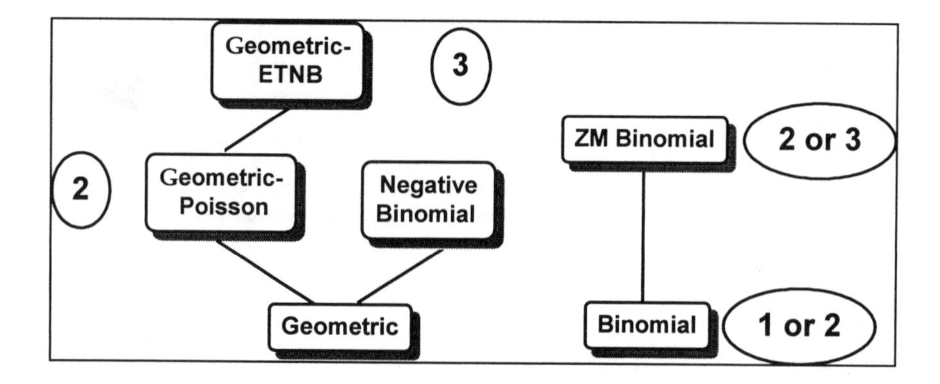

Fig. 3.5 Discrete distributions IV: Miscellaneous distributions

of-fit statistic such as the chi-square statistic. If the fit is deemed unacceptable, the two-parameter models should be fitted and tested. This process can be repeated as necessary until suitable candidates for a model are found. The likelihood ratio test is also available to indicate if there is value in extra parameters.

From a computational point of view, this procedure allows one to use the estimated parameter(s) of the simpler model to be starting values for the maximization of the loglikelihood of the more complex model. When doing this the initial value of the additional parameter can be arbitrarily selected. This procedure allows one to avoid using the method of moments or other estimation methods as initial values for the maximization procedure. All other considerations in fitting and testing distributions are the same as those discussed in Chapter 2 in connection with fitting continuous distributions to grouped data.

The distributions in Figures 3.2–3.5 form a particularly rich set of distributions in terms of shape. However, many other distributions are also possible. Many others are discussed in Johnson, Kotz, and Kemp [73], Douglas [31], and Panjer and Willmot [102].

3.8.1 The compound Poisson class

The compound Poisson distributions in Figure 3.3 form an interesting class. In this section we compare the skewness (3rd moment) of these distributions to develop an appreciation of the amount by which the skewness, and hence the tails of these distributions, can vary even when the mean and variance are fixed. Consider a compound Poisson distribution with pgf

$$P(z) = \exp\{\lambda[P_2(z) - 1]\}$$

where $P_2(z)$ is the pgf of the secondary distribution. From this (see Exercise 3.37) and Definition 2.11, the mean and second and third central moments of the compound Poisson distribution are

$$\begin{aligned}\mu_1' &= \mu = \lambda m_1', \\ \mu_2 &= \sigma^2 = \lambda m_2', \text{ and} \\ \mu_3 &= \lambda m_3'\end{aligned} \tag{3.34}$$

where m_j' is the jth raw moment of the secondary distribution. The coefficient of skewness is

$$\gamma_1 = \frac{\mu_3}{\sigma^3} = \frac{m_3'}{\lambda^{1/2}(m_2')^{3/2}}.$$

For the Poisson–binomial distribution, with a bit of algebra (see Exercise 3.38) we obtain

$$\begin{aligned}\mu &= \lambda mq, \\ \sigma^2 &= \mu[1+(m-1)q], \text{ and} \\ \mu_3 &= 3\sigma^2 - 2\mu + \frac{m-2}{m-1}\frac{(\sigma^2-\mu)^2}{\mu}.\end{aligned} \tag{3.35}$$

Carrying out similar exercises for the Neyman Type A, negative binomial, Polya–Aeppli and generalized Poisson–Pascal distributions yields

$$\begin{aligned}
\text{Negative binomial:} &\quad \mu_3 = 3\sigma^2 - 2\mu + 2\frac{(\sigma^2-\mu)^2}{\mu} \\
\text{Polya–Aeppli:} &\quad \mu_3 = 3\sigma^2 - 2\mu + \frac{3}{2}\frac{(\sigma^2-\mu)^2}{\mu} \\
\text{Neyman Type A:} &\quad \mu_3 = 3\sigma^2 - 2\mu + \frac{(\sigma^2-\mu)^2}{\mu} \\
\text{Generalized Poisson–Pascal:} &\quad \mu_3 = 3\sigma^2 - 2\mu + \frac{r+2}{r+1}\frac{(\sigma^2-\mu)^2}{\mu}.
\end{aligned}$$

The range of r is $-1 < r < \infty$ since the Generalized Poisson–Pascal is a compound Poisson distribution with an extended truncated negative binomial secondary distribution.

Note that for fixed mean and variance, the skewness only changes through the coefficient in the last term for each of the five distributions. For the Poisson distribution, $\mu_3 = \lambda = 3\sigma^2 - 2\mu$, and so the third term for each expression for μ_3 represents the change from the Poisson distribution. For the Poisson–binomial distribution, if $m = 1$ the distribution is Poisson since it is equivalent to a Poisson–zero-truncated binomial because truncation at zero leaves only probability at 1. Another view is that from (3.35) we have

$$\begin{aligned}\mu_3 &= 3\sigma^2 - 2\mu + \frac{m-2}{m-1}\frac{(m-1)^2q^4\lambda^2m^2}{\lambda mq} \\ &= 3\sigma^2 - 2\mu + (m-2)(m-1)q^3\lambda m\end{aligned}$$

which reduces to the Poisson value for μ_3 when $m = 1$. Hence, it is necessary that $m \geq 2$ for non-Poisson distributions to be created. Then the coefficient satisfies

$$0 \leq \frac{m-2}{m-1} < 1.$$

For the Generalized Poisson–Pascal, since $r > -1$, the coefficient satisfies

$$1 < \frac{r+2}{r+1} < \infty.$$

For the Neyman Type A distribution, the coefficient is exactly 1. Hence, these three distributions provide any desired degree of skewness greater than that of the Poisson distribution. Note that the Polya–Aeppli and the negative binomial distributions are special cases of the generalized Poisson Pascal with $r = 1$ and $r = 0$, respectively.

Example 3.24 *The data in Table 3.19 are taken from Hossack et al. [68] and give the distribution of the number of claims on automobile insurance policies in Australia. Determine an appropriate frequency model.*

The data set is very large and, as a result, requires a very close correspondence of the model to the data. The results are given in Table 3.19.

From Table 3.19, it is seen that the negative binomial distribution does not fit well while the fit of the Poisson–inverse Gaussian is marginal at best ($p = 2.88\%$). However, by adding the parameter $r > 0$, the fit was improved dramatically. The P.–i.G. is a special case ($r = -0.5$) of the Generalized Poisson Pascal (G.P.P.). Hence, a likelihood ratio test can be formally applied to determine if the additional parameter r is justified. Since the loglikelihood increases by 5, which is more than 1.92, the G.P.P. is a significantly better fit. The chi-square test shows that the G.P.P. provides an adequate fit. □

3.9 MIXED FREQUENCY MODELS

Many of the compound distributions discussed in previous sections can arise in a way that is very different from compounding. In this section, we examine mixing distributions by treating one or more parameters as being "random" in some sense. This section expands on the ideas discussed in Subsection 3.3.2 in connection with the gamma mixture of the Poisson distribution being negative binomial.

We assume that the parameter is distributed over the population under consideration (the "collective") and that the sampling scheme that generates our data has two stages. First, a value of the parameter is selected. Then, given that parameter value, an observation is generated using that parameter value.

In automobile insurance, for example, classification schemes attempt to put individuals into (relatively) homogeneous groups for the purpose of pricing.

Table 3.19 Fit of Hossack et al. data

No. of claims	Observed frequency	Fitted distributions: Negative binomial	P.–i.G.	G.P.P.
0	565,664	565,708.1	565,712.4	565,661.2
1	68,714	68,570.0	68,575.6	68,721.2
2	5,177	5,317.2	5,295.9	5,171.7
3	365	334.9	344.0	362.9
4	24	18.7	20.8	29.6
5	6	1.0	1.2	3.0
6+	0	0.0	0.1	0.4
Parameters		$\beta = 0.0350662$	$\lambda = 0.123304$	$\lambda = 0.123395$
		$r = 3.57784$	$\beta = 0.0712027$	$\beta = 0.233862$
				$r = -0.846872$
Chi-square		12.13	7.09	0.29
Degrees of freedom		2	2	1
p-value		<1%	2.88%	58.9%
−Loglikelihood		251,117	251,114	251,109

Variables used to develop the classification scheme might include age, experience, a history of violations, accident history, and other variables. Since there will always be some residual variation in accident risk within each class, mixed distributions provide a framework for modeling this heterogeneity.

Let $P(z|\theta)$ denote the pgf of the number of events (e.g., claims) if the risk parameter is known to be θ. The parameter, θ, might be the Poisson mean, for example, in which case the measurement of risk is the expected number of events in a fixed time period.

Let $U(\theta) = Pr(\Theta \leq \theta)$ be the cdf of Θ, where Θ is the risk parameter, which is viewed as a random variable. Then $U(\theta)$ represents the probability that, when a value of Θ is selected (e.g. a driver is included in our sample), the value of the risk parameter does not exceed θ. Let $u(\theta)$ be the pf or pdf of Θ. Then

$$P(z) = \int P(z|\theta)u(\theta)d\theta \text{ or } P(z) = \sum P(z|\theta_i)u(\theta_i) \tag{3.36}$$

is the unconditional pgf of the number of events (where the formula selected depends on whether Θ is discrete or continuous[3]). The corresponding proba-

[3]We could have written the more general $P(z) = \int P(z|\theta)dU(\theta)$, which would include situations where Θ has a distribution that is partly continuous and partly discrete.

bilities are denoted by

$$p_k = \int p_k(\theta)u(\theta)d\theta \text{ or } p_k = \sum p_k(\theta_i)u(\theta_i). \tag{3.37}$$

The "mixing distribution" denoted by $U(\theta)$ may be of the discrete or continuous type or even a combination of discrete and continuous types. "Discrete mixtures" are mixtures of distributions when the mixing function is of the discrete type. Similarly for "continuous mixtures." This phenomenon was introduced for continuous mixtures of severity distributions in Subsection 2.7.3 and for finite discrete mixtures in Subsection 2.7.6.

It should be noted that the mixing distribution is unobservable, since the data are drawn from the mixed distribution.

Example 3.25 *Demonstrate that the zero-modified distributions may be created by using two-point mixtures.*

Suppose

$$P(z) = p \cdot 1 + (1-p)P_2(z|\theta).$$

This is a (discrete) two-point mixture of a degenerate distribution and the distribution with pgf $P_2(z|\theta)$. From (3.32) this is also a compound Bernoulli distribution. □

Example 3.26 *Suppose drivers can be classified as "good drivers" and "bad drivers," each group with its own Poisson distribution. Determine the pf for this model and fit it to the data from Example 3.8. This model and its application to the data set are from Tröbliger [120].*

From (3.37) the pf is

$$p_k = p\frac{e^{-\lambda_1}\lambda_1^k}{k!} + (1-p)\frac{e^{-\lambda_2}\lambda_2^k}{k!}.$$

The maximum likelihood estimates were calculated by Tröbliger to be $\hat{p} = 0.94, \hat{\lambda}_1 = 0.11$ and $\hat{\lambda}_2 = 0.70$. This means that about 6% of drivers were "bad" with a risk of $\lambda_1 = 0.70$ expected accidents per year and 94% were "good" with a risk of $\lambda_2 = 0.11$ expected accidents per year. Note that it is not possible to return to the data set and identify which were the "bad" drivers. □

This example illustrates two important points about finite mixtures. First, the model is probably oversimplified in the sense that risks (e.g., drivers) probably exhibit a continuum of risk levels rather than just two. The second point is that finite mixture models have a lot of parameters to be estimated. The simple two-point Poisson mixture above has 3 parameters. Increasing the number of distributions in the mixture to r will then involve $r-1$ mixing parameters in addition to the total number of parameters in the r component distributions. As a result of this, continuous mixtures are frequently preferred.

3.9.1 Poisson mixtures

The class of mixed Poisson distributions has played an important role in actuarial mathematics.

From (3.36) the pgf of the mixed Poisson distribution is

$$P(z) = \int e^{\lambda\theta(z-1)} u(\theta) d\theta \quad \text{or} \quad P(z) = \sum e^{\lambda\theta_i(z-1)} u(\theta_i)$$

by introducing a scale parameter λ, for convenience.

Douglas [31] proves that for any mixed Poisson distribution, the mixing distribution is unique. This means that two different mixing distributions cannot lead to the same mixed Poisson distribution. This allows us to identify the mixing distribution in some cases.

There is also an important connection between mixed Poisson distributions and compound Poisson distributions which were discussed in the last section.

Definition 3.5 *A distribution is said to be* ***infinitely divisible*** *if for all values of* $n = 1, 2, 3, ...,$ *its characteristic function* $\varphi(z)$ *can be written as*

$$\varphi(z) = [\varphi_n(z)]^n.$$

where $\varphi_n(z)$ *is the characteristic function of some random variable.*

In other words, taking the $(1/n)$th power of the characteristic function still results in a characteristic function. The various generating functions are defined as follows.

Definition 3.6 *The* ***characteristic function*** *of a random variable* X *is* $\varphi_X(z) = E(e^{izX})$. *The* ***moment generating function*** *of* X *is* $M_X(z) = E(e^{zX})$. *The* ***Laplace transform*** *of* X *is* $L_X(z) = E(e^{-zX})$.

In Definition 3.5, "characteristic function" could have been replaced by "moment generating function," "probability generating function," "Laplace transform," or some other transform. That is, if the definition is satisfied for one of these transforms, it will be satisfied for all others which exist for the particular random variable. We choose the characteristic function since it exists for all distributions while the moment generating function does not exist for some distributions with heavy tails. The Laplace transform exists for all random variables which do not take on negative values. Because many earlier results involved probability generating functions, it is useful to note the relationship between it and the characteristic function.

Theorem 3.7 *If the probability generating function exists for a random variable* X, *then* $P_X(z) = \varphi(-i \log z)$ *and* $\varphi_X(z) = P(e^{iz})$.

Proof:

$$P_X(z) = E(z^X) = E(e^{X \log z}) = E[e^{-i(i \log z)X}] = \varphi_X(-i \log z)$$

and

$$\varphi_X(z) = E(e^{izX}) = E[(e^{iz})^X] = P_X(e^{iz}).$$

□

The following distributions, among others, are infinitely divisible: normal, gamma, Poisson, negative binomial. The binomial distribution is not infinitely divisible, since the exponent m in its pgf must take on integer values. Dividing m by $n = 1, 2, 3, ...$ will result in nonintegral values. In fact, no distributions with a finite range of support (the range over which positive probabilities exist) can be infinitely divisible. Now to the important result.

Theorem 3.8 *Suppose $P(z)$ is a mixed Poisson pgf with an infinitely divisible mixing distribution. Then $P(z)$ is also a compound Poisson pgf and may be expressed as*

$$P(z) = e^{\lambda[P_2(z)-1]}$$

where $P_2(z)$ is a pgf. If one adopts the convention that $P_2(0) = 0$, then $P_2(z)$ is unique. □

A proof can be found in Feller [38, Ch. 12]. As a result of this theorem, if one chooses any infinitely divisible mixing distribution, the corresponding mixed Poisson distribution can be equivalently described as a compound Poisson distribution. For some distributions, this is a distinct advantage in carrying out numerical work since the recursive formula (3.33) can be used in evaluating the probabilities, once the secondary distribution is identified. For most cases, this identification is easily carried out.

Let $P(z)$ be the pgf of a mixed Poisson distribution with arbitrary mixing distribution $U(\theta)$. Then (with formulas given for the continuous case)

$$\begin{aligned} P(z) &= \int e^{\lambda\theta(z-1)}u(\theta)d\theta = \int \left[e^{\lambda(z-1)}\right]^\theta u(\theta)d\theta \\ &= E\left\{\left[e^{\lambda(z-1)}\right]^\theta\right\} = P_\Theta\left[e^{\lambda(z-1)}\right] \end{aligned} \quad (3.38)$$

where $P_\Theta(z)$ is the pgf of the mixing distribution.

Example 3.27 *Use the above results to demonstrate that a gamma mixture of Poisson variables is negative binomial.*

If the mixing distribution is gamma, it has the following probability generating function where β plays the role of $1/\theta$ in Appendix A.

$$P_\Theta(z) = \left(\frac{\beta}{\beta - \log z}\right)^\alpha, \qquad \beta > 0,\ \alpha > 0.$$

It is clearly infinitely divisible because $[P_\Theta(z)]^{1/n}$ is the pgf of a gamma distribution with parameters α/n and β. Then the pgf of the mixed Poisson distribution is

$$\begin{aligned} P(z) &= \left\{\frac{\beta}{\beta - \log\left[e^{\lambda(z-1\}}\right]}\right\}^{\alpha} \\ &= \left[\frac{\beta}{\beta - \lambda(z-1)}\right]^{\alpha} \\ &= \left[1 - \frac{\lambda}{\beta}(z-1)\right]^{-\alpha} \end{aligned}$$

which is the form of the pgf of the negative binomial distribution. This important idea will be revisited in connection with credibility theory in Chapter 5. □

It was shown in Example 3.18 that a compound Poisson distribution with a logarithmic secondary distribution is a negative binomial distribution. Therefore the theorem holds true for this case. Many other similar relationships can be identified for both continuous and discrete mixing distributions.

Further examination of (3.38) reveals that if $u(\theta)$ is the pf for any discrete random variable with pgf $P_\Theta(z)$, then the pgf of the mixed Poisson distribution is $P_\Theta\left[e^{\lambda(z-1)}\right]$, a compound distribution with a Poisson secondary distribution.

Example 3.28 *Demonstrate that the Neyman Type A distribution can be obtained by mixing.*

If in (3.38) the mixing distribution has pgf

$$P_\Theta(z) = e^{\mu(z-1)}$$

then the mixed Poisson distribution has pgf

$$P(z) = \exp\{\mu[e^{\lambda(z-1)} - 1]\}$$

the pgf of a compound Poisson with a Poisson secondary distribution, that is, the Neyman Type A distribution. □

A further interesting result obtained by Holgate [67] is that if a mixing distribution is absolutely continuous and unimodal, then the resulting mixed Poisson distribution is also unimodal. Multimodality can occur when discrete mixing functions are used. For example, the Neyman Type A distribution can have more than one mode. The reader should try this calculation for various combinations of the two parameters.

Most continuous distributions in this book involve a scale parameter. This means that scale changes to distributions do not cause a change in the form of the distribution, only in the value of its scale parameter. For the mixed Poisson distribution, with pgf (3.38), any change in λ is equivalent to a change in the scale parameter of the mixing distribution. Hence, it may be convenient to simply set $\lambda = 1$ where a mixing distribution with a scale parameter is used.

Example 3.29 *Show that a mixed Poisson with an inverse Gaussian mixing distribution is the same as a Poisson–ETNB distribution with $r = -0.5$.*

The inverse Gaussian distribution is described in Appendix A. It has pdf

$$f(x) = \left(\frac{\theta}{2\pi x^3}\right)^{1/2} \exp\left[-\frac{\theta}{2x}\left(\frac{x-\mu}{\mu}\right)^2\right], \qquad x > 0$$

which is conveniently rewritten as

$$f(x) = \frac{\mu}{(2\pi\beta x^3)^{1/2}} \exp\left[-\frac{(x-\mu)^2}{2\beta x}\right], \qquad x > 0$$

where $\beta = \mu^2/\theta$. The pgf of this distribution is (see Exercise 3.40)

$$P(z) = \exp\left\{-\frac{\mu}{\beta}[(1 - 2\beta \log z)^{1/2} - 1]\right\}. \tag{3.39}$$

Hence, the inverse Gaussian distribution is infinitely divisible ($[P(z)]^{1/n}$ is also inverse Gaussian, but with μ replaced by μ/n). From (3.38) with $\lambda = 1$, the pgf of the mixed distribution is

$$P(z) = \exp\left\{-\frac{\mu}{\beta}\{[1 + 2\beta(1-z)]^{1/2} - 1\}\right\}.$$

By setting

$$\lambda = \frac{\mu}{\beta}[(1 + 2\beta)^{1/2} - 1]$$

and

$$P_2(z) = \frac{[1 - 2\beta(z-1)]^{1/2} - (1 + 2\beta)^{1/2}}{1 - (1 + 2\beta)^{1/2}}$$

we see that

$$P(z) = \exp\{\lambda[P_2(z) - 1]\}$$

where $P_2(z)$ is the pgf of the extended truncated negative binomial distribution with $r = -1/2$.

Hence, the Poisson–inverse Gaussian distribution is a compound Poisson distribution with an ETNB ($r = -1/2$) secondary distribution. □

Example 3.30 *Show that the Poisson–ETNB (also called the generalized Poisson Pascal distribution) can be written as a mixed distribution.*

In the previous section, it was shown that the skewness of this distribution can be made arbitrarily large by taking r towards -1. This suggest a potentially very heavy tail for the compound distribution for values of r that are close to -1. When $r = 0$, the secondary distribution is the logarithmic and the corresponding compound Poisson distribution is a negative binomial distribution. When $r > 0$, the pgf is

$$P(z) = \exp\left\{\lambda\{\frac{[1-\beta(z-1)]^{-r}-(1+\beta)^{-r}}{1-(1+\beta)^{-r}} - 1\}\right\} \tag{3.40}$$

which can be rewritten as

$$P(z) = P_\Theta[e^{\lambda(z-1)}]$$

where

$$P_\Theta(z) = e^{\lambda_1[(1-\beta_1 \log z)^{-r}-1]} \tag{3.41}$$

and where $\lambda_1 > 0$ and $\beta_1 > 0$ are appropriately chosen parameters.

The pgf (3.41) is the pgf of a compound Poisson distribution with a gamma secondary distribution. This type of distribution (compound Poisson with continuous severity) has a mass point at zero and is of the continuous type over the positive real axis.

When $-1 < r < 0$, (3.40) can be reparameterized as

$$P(z) = \exp\{-\lambda\{[1-\mu(z-1)]^\alpha - 1\}\}$$

where $\lambda > 0$, $\mu > 0$ and $0 < \alpha < 1$. This can be rewritten as

$$P(z) = P_\Theta[e^{(1-z)}]$$

where

$$P_\Theta(z) = \exp\{-\lambda[(1-\mu\log z)^\alpha - 1]\}. \tag{3.42}$$

Feller [39, pp. 448, 581] shows that

$$P_\alpha(z) = e^{-(-\log z)^\alpha}$$

is the pgf of the stable distribution with pdf

$$f_\alpha(x) = \frac{1}{\pi}\sum_{k=1}^{\infty}\frac{\Gamma(k\alpha+1)}{k!}(-1)^{k-1}x^{-(k\alpha+1)}\sin(\alpha k\pi), \quad x > 0.$$

Then (see Exercise 3.41), it can be shown that the mixing distribution with pgf (3.42) has pdf

$$f(x) = \frac{e^\lambda}{\mu\lambda^{1/\alpha}}e^{-x/\mu}f_\alpha\left(\frac{x}{\mu\lambda^{1/\alpha}}\right), \quad x > 0. \tag{3.43}$$

Although this is a complicated mixing distribution, it is infinitely divisible. The corresponding secondary distribution in the compound Poisson formulation of this distribution is the ETNB. Consequently, in numerical work, one needs to only work with the simple recursive formula (3.30).

The stable distribution can have arbitrarily large moments and is then very heavy-tailed. This might be the case for a very diffuse mixing distribution. In the case of Example 3.24, the value of the parameter was -0.85, indicating a high degree of heterogeneity in the risk characteristics of the insurance policies being studied. □

3.9.2 Other mixed models

Many other mixed models can be constructed beginning with a simple distribution.

Example 3.31 *Determine the pf for a mixed binomial with a beta mixing distribution. This distribution is called binomial-beta, negative hypergeometric, or Polya–Eggenberger.*

The beta distribution has pdf

$$u(q) = \frac{\Gamma(a+b)}{\Gamma(a)\Gamma(b)} q^{a-1}(1-q)^{b-1}, \qquad a>0, b>0 \ .$$

Then the mixed distribution has probabilities

$$\begin{aligned} p_k &= \int_0^1 \binom{m}{k} q^k (1-q)^{m-k} \frac{\Gamma(a+b)}{\Gamma(a)\Gamma(b)} q^{a-1}(1-q)^{b-1} dq \\ &= \frac{\Gamma(a+b)\Gamma(m+1)\Gamma(a+k)\Gamma(b+m-k)}{\Gamma(a)\Gamma(b)\Gamma(k+1)\Gamma(m-k+1)\Gamma(a+b+m)} \\ &= \frac{\binom{-a}{k}\binom{-b}{m-k}}{\binom{-a-b}{m}} \qquad k=0,1,2,\ldots. \end{aligned}$$

□

Example 3.32 *Determine the pf for a mixed negative binomial distribution with mixing on the parameter $p=(1+\beta)^{-1}$. Let p have a beta distribution. The mixed distribution is called the generalized Waring.*

Arguing as in Example 3.31 we have

$$\begin{aligned} p_k &= \frac{\Gamma(r+k)}{\Gamma(r)\Gamma(k+1)} \frac{\Gamma(a+b)}{\Gamma(a)\Gamma(b)} \int_0^1 p^{a+r-1}(1-p)^{b+k-1} dp \\ &= \frac{\Gamma(r+k)}{\Gamma(r)\Gamma(k+1)} \frac{\Gamma(a+b)}{\Gamma(a)\Gamma(b)} \frac{\Gamma(a+r)\Gamma(b+k)}{\Gamma(a+r+b+k)}, \qquad k=0,1,2,\ldots. \end{aligned}$$

When $b = 1$, this distribution is called the Waring distribution. When $r = b = 1$, it is termed the Yule distribution. □

Examples 3.31 and 3.32 are mixtures. However, the mixing distributions are not infinitely divisible, because the beta distribution has finite support. Hence, we cannot write the distributions as compound Poisson distributions to take advantage of the compound Poisson structure used in the previous section.

In this chapter, we focused on distributions that are easily handled computationally. Although many other discrete distributions are available, we believe that those discussed form a sufficiently rich class for most problems.

3.10 THE INTERACTION OF FREQUENCY WITH SEVERITY AND EXPOSURE

In this and the previous chapter we have developed models for two of the key elements of an insurance process. The frequency, say N, represents the number of claims generated by a portfolio of policies. Aside from the factors directly related to the insurance process (such as underwriting characteristics and the legal environment) there are two key factors that impact the distribution of N. One is the size of the exposure base. Clearly the larger the portfolio (in terms of number of lives, square feet of building, miles driven, etc.) the larger we expect N to be. Furthermore, the number of payments is affected by the nature of the coverage. If a deductible is imposed or raised, fewer payments will result.

3.10.1 Effect of exposure

Assume that the current portfolio consists of n entities, each of which could produce claims. Let N_j be the number of claims produced by the jth entity. Then $N = N_1 + \cdots + N_n$. If we assume that the N_j are independent and identically distributed, then

$$P_N(z) = [P_{N_1}(z)]^n.$$

Now suppose the portfolio is expected to expand to n^* entities, with frequency N^*. Then

$$P_{N^*}(z) = [P_{N_1}(z)]^{n^*} = [P_N(z)]^{n^*/n}.$$

Thus, if N is infinitely divisible, the distribution of N^* will have the same form as that of N, but with modified parameters.

Example 3.33 *It has been determined from past studies that the number of workers compensation claims for a group of 300 employees in a certain occu-*

Table 3.20 Automobile claims by year

Year	Exposure	Claims
1986	2,145	207
1987	2,452	227
1988	3,112	341
1989	3,458	335
1990	3,698	362
1991	3,872	359

pation class has the negative binomial distribution with $\beta = 0.3$ and $r = 10$. Determine the frequency distribution for a group of 500 such individuals.

The pgf of N^* is

$$\begin{aligned} P_{N^*}(z) &= [P_N(z)]^{500/300} = \{[1 - 0.3(z-1)]^{-10}\}^{500/300} \\ &= [1 - 0.3(z-1)]^{-16.67} \end{aligned}$$

which is negative binomial with $\beta = 0.3$ and $r = 16.67$. □

For the $(a, b, 0)$ class, all members except the binomial have this property. For the $(a, b, 1)$ class, none of the members do. For compound distributions, it is the primary distribution that must be infinitely divisible. In particular, compound Poisson and compound negative binomial (including the geometric) distributions will be preserved under an increase in exposure. Earlier, some reasons were given to support the use of zero-modified distributions. If exposure adjustments are anticipated, it may be better to choose a compound model, even if the fit is not quite as good. It should be noted that compound models have the ability to place large amounts of probability at zero.

A second use for exposure adjustments is when data have been collected from portfolios with different exposures. Consider the following example.

Example 3.34 *Determine the maximum likelihood estimate of the Poisson parameter for the data in Table 3.20.*

Let λ be the Poisson parameter for a single exposure. If year k has e_k exposures, then the number of claims has a Poisson distribution with parameter λe_k. If n_k is the number of claims in year k, the likelihood function is

$$L = \prod_{k=1}^{6} \frac{e^{-\lambda e_k}(\lambda e_k)^{n_k}}{n_k!}.$$

The mle is found by

$$l = \log L = \sum_{k=1}^{6} [-\lambda e_k + n_k \log(\lambda e_k) - \log(n_k!)]$$

$$\partial l/\partial\lambda = \sum_{k=1}^{6}[-e_k + n_k\lambda^{-1}] = 0$$

$$\hat{\lambda} = \frac{\sum_{k=1}^{6} n_k}{\sum_{k=1}^{6} e_k} = \frac{1{,}831}{18{,}737} = 0.09772.$$

□

In this example the answer is what we expected it to be, the average number of claims per exposure. This technique will work for any distribution in the $(a, b, 0)$[4] and compound classes. But care must be taken in the interpretation of the model. For example, if we use a negative binomial distribution, we are assuming that each exposure unit produces claims according to a negative binomial distribution. This is different from assuming that total claims have a negative binomial distribution because they arise from individuals who each have a Poisson distribution, but with different parameters.

We can check the Poisson model in the previous example by creating an analog of the chi-square goodness-of-fit test. For each year we are assuming that the number of claims is the result of the sum of a number (given by the exposure) of independent and identical random variables. In that case the Central Limit Theorem indicates that a normal approximation may be appropriate. The expected count (E_k) is the exposure times the estimated expected value for one exposure unit while the variance (V_k) is the exposure times the estimated variance for one exposure unit. The test statistic is then

$$Q = \sum_k \frac{(n_k - E_k)^2}{V_k}$$

and has an approximate chi-square distribution with degrees of freedom equal to the number of data points less the number of estimated parameters.

Example 3.35 (Example 3.34 continued) *Conduct the approximate goodness-of-fit test for the Poisson distribution.*

The expected count is $E_k = \lambda e_k$ and the variance is $V_k = \lambda e_k$ also. The test statistic is

$$\begin{aligned} Q &= \frac{(207-209.61)^2}{209.61} + \frac{(227-239.61)^2}{239.61} + \frac{(341-304.11)^2}{304.11} \\ &\quad + \frac{(335-337.92)^2}{337.92} + \frac{(362-361.37)^2}{361.37} + \frac{(359-378.38)^2}{378.38} \\ &= 6.19. \end{aligned}$$

With five degrees of freedom, the 5% critical value is 11.07 and the Poisson hypothesis is accepted. □

[4] For the binomial distribution, the usual problem that m must be an integer remains.

3.10.2 Effect of severity modifications

To begin, suppose X, the severity, represents ground-up losses and there are no coverage modifications. Let N denote the number of losses. Now consider a coverage modification such that v is the probability that a loss will result in a payment. For example, if there is a deductible of d, $v = \Pr(X > d)$. Let X^* denote the conditional payment amount, given that a payment is made. There are two ways to proceed.

3.10.2.1 Unmodified frequency We could leave the frequency distribution unchanged. The severity distribution becomes a two-point mixture with a probability of $1 - v$ being degenerate at zero and a probability v of being X^*.

Example 3.36 *Suppose N has a negative binomial distribution with $\beta = 0.3$ and $r = 10$. Further suppose there are only six possible ground-up loss values:* 25, 50, 75, 100, 200, *and* 500. *The probabilities are* 0.2, 0.3, 0.2, 0.15, 0.1, *and* 0.05, *respectively. Provide a model for this portfolio if a deductible of* 50 *is imposed.*

If the negative binomial parameters are to be maintained, the payment distribution must be shifted to take on the values 0, 25, 50, 150, and 450 with probabilities 0.5, 0.2, 0.15, 0.1, and 0.05. □

For the above alteration, as well as the next one, to hold, it must be assumed that the imposition of coverage modifications does not affect the process that produces losses, or the type of individual who will purchase insurance. For example, those who buy a 250 deductible on an automobile property damage coverage may (correctly) view themselves as less likely to be involved in an accident than those who buy full coverage. Similarly, an employer may find that the rate of permanent disability declines when reduced benefits are provided to employees in the first few years of employment.

3.10.2.2 Modified frequency As before, let N be the number of losses. Let $I_j = 1$ if the jth loss results in a payment and $I_j = 0$ otherwise. Each I_j has a Bernoulli distribution with parameter v. Then $N^* = I_1 + \cdots + I_N$ represents the number of payments. If $I_1, I_2, \ldots$ are mutually independent and are also independent of N, then N^* has a compound distribution with N as the primary distribution and a Bernoulli secondary distribution. Thus

$$P_{N^*}(z) = P_N[1 + v(z - 1)].$$

Example 3.37 (Example 3.36 continued) *Provide an alternative model.*

We have $v = \Pr(X > 50) = 0.2 + 0.15 + 0.1 + 0.05 = 0.5$ and so

$$\begin{aligned} P_{N^*}(z) &= \{1 - 0.3[1 + 0.5(z - 1) - 1]\}^{-10} \\ &= [1 - 0.15(z - 1)]^{-10}. \end{aligned}$$

Table 3.21 Frequency adjustments

N	Parameters for N^*
Poisson	$\lambda^* = v\lambda$
ZM Poisson	$p_0^{M*} = \dfrac{p_0^M - e^{-\lambda} + e^{-\nu\lambda} - p_0^M e^{-v\lambda}}{1 - e^{-\lambda}}, \ \lambda^* = v\lambda$
Binomial	$q^* = vq$
ZM binomial	$p_0^{M*} = \dfrac{p_0^M - (1-q)^m + (1-vq)^m - p_0^M(1-vq)^m}{1-(1-q)^m}$ $q^* = vq$
Negative binomial	$\beta^* = v\beta, \ r^* = r$
ZM negative binomial	$p_0^{M*} = \dfrac{p_0^M - (1+\beta)^{-r} + (1+v\beta)^{-r} - p_0^M(1+v\beta)^{-r}}{1-(1+\beta)^{-r}}$ $\beta^* = v\beta, \ r^* = r$
ZM logarithmic	$p_0^{M*} = 1 - (1 - p_0^M)\log(1+v\beta)/\log(1+\beta)$ $\beta^* = v\beta$

Then we see that N^* is negative binomial with $\beta^* = .15$ and $r^* = 10$. The distribution for the amount per payment is conditional on $X > 50$ and thus places probabilities 0.4, 0.3, 0.2, and 0.1 on the values 25, 50, 150, and 450.□

As indicated in Table 3.21, all members of the $(a, b, 0)$ and $(a, b, 1)$ classes retain their type after the change.

Example 3.38 *Verify the entries in Table 3.21 for the zero-modified negative binomial distribution.*

$$\begin{aligned} P_{N^*}(z) &= P_N[1 + v(z-1)] \\ &= P_0^M + (1 - p_0^M)P_N^T[1 + v(z-1)] \end{aligned}$$

using (3.17) and where $P_N^T(z)$ is the corresponding zero-truncated negative binomial pgf. Inserting the pgf of the zero-truncated negative binomial distribution (see Appendix B) produces

$$\begin{aligned} P_{N^*}(z) &= p_0^M + (1 - p_0^M)\frac{\{1 - \beta[v(z-1)]\}^{-r} - (1+\beta)^{-r}}{1 - (1+\beta)^{-r}} \\ &= p_0^M + (1 - p_0^M) \\ &\quad \times \frac{\{1 - \beta[v(z-1)]\}^{-r} - (1+v\beta)^{-r} + (1+v\beta)^{-r} - (1+\beta)^{-r}}{1 - (1+\beta)^{-r}} \end{aligned}$$

$$\begin{aligned} &= p_0^M + (1 - p_0^M)\frac{(1+v\beta)^{-r} - (1+\beta)^{-r}}{1-(1+\beta)^{-r}} \\ &\quad +(1-p_0^M)\frac{1-(1+v\beta)^{-r}}{1-(1+\beta)^{-r}}\frac{\{1-\beta[v(z-1)]\}^{-r} - (1+v\beta)^{-r}}{1-(1+v\beta)^{-r}}. \end{aligned}$$

Now let

$$1 - p_0^{M*} = (1 - p_0^M)\frac{1-(1+v\beta)^{-r}}{1-(1+\beta)^{-r}}$$

in which case

$$\begin{aligned} p_0^{M*} &= p_0^M + (1-p_0^M)\frac{(1+v\beta)^{-r} - (1+\beta)^{-r}}{1-(1+\beta)^{-r}} \\ &= \frac{p_0^M - p_0^M(1+v\beta)^{-r} + (1+v\beta)^{-r} - (1+\beta)^{-r}}{1-(1+\beta)^{-r}} \end{aligned}$$

and so

$$P_{N^*}(z) = p_0^{M*} + (1-p_0^{M*})\frac{\{1-\beta[v(z-1)]\}^{-r} - (1+v\beta)^{-r}}{1-(1+v\beta)^{-r}}$$

which is the pgf of a zero-modified negative binomial distribution with parameters as given in Table 3.21. □

Example 3.39 *Suppose N is zero-modified negative binomial with $\beta = 0.3$, $r = 4$, and $p_0^M = 0.4$. If $v = 0.5$, determine the distribution of N^*.*

The value of r is unchanged and $\beta^* = 0.5(0.3) = 0.15$.

$$p_0^{M*} = \frac{0.4 - 0.4(1.15)^{-4} + (1.15)^{-4} - (1.3)^{-4}}{1-(1.3)^{-4}} = 0.60462.$$ □

Now consider the case where N has a compound distribution. We have

$$P_{N^*}(z) = P_1\{P_2[1 + v(z-1)]\}$$

and so N^* is also a compound distribution with the same primary distribution as N but a different secondary distribution. If the secondary distribution is a member of the $(a, b, 0)$ class, it will remain so, and N^* will have the same type of compound distribution as N. The only case that requires a special adjustment is developed in the following example.

Example 3.40 *Suppose N is Poisson–ETNB with $\lambda = 5$, $\beta = 0.3$, and $r = 4$. If $v = 0.5$, determine the distribution of N^*.*

From the discussion above, N^* is compound Poisson with $\lambda^* = 5$, but the secondary distribution is a zero-modified negative binomial with (from Table 3.21) $\beta^* = 0.5(0.3) = 0.15$,

$$p_0^{M*} = \frac{0 - 1.3^{-4} + 1.15^{-4} - 0(1.15)^{-4}}{1 - 1.3^{-4}} = 0.34103$$

and $r^* = 4$. This would be sufficient, except we have acquired the habit of using the ETNB as the secondary distribution. From Theorem 3.6 a compound Poisson distribution with a zero-modified secondary distribution is equivalent to a compound Poisson distribution with a zero-truncated secondary distribution. The Poisson parameter must be changed to $\lambda^* = (1-p_0^{M*})\lambda$. Therefore, N^* has a Poisson–ETNB distribution with $\lambda^* = (1 - 0.34103)5 = 3.29485$, $\beta^* = 0.15$, and $r^* = 4$. □

All of the above analyses apply equally well for an increase in the deductible. If N is the frequency distribution for a deductible of d and N^* is for a deductible of $d^* > d$, then we have $v = [1 - F_X(d^*)]/[1 - F_X(d)]$, the proportion of claims that continue to produce a claim after the deductible is raised. Note that if we begin with a model for a shifted (at d) distribution, the new value is $v = 1 - F_X(d^* - d)$.

The process can also be reversed. Suppose N is again the frequency when the deductible is d. Now contemplate a new deductible of $d^* < d$. Provided we have information about the severity distribution for losses down to at least d^*, we can obtain the distribution for N^*. In fact, there is no change. We again use $v = [1 - F_X(d^*)]/[1 - F_X(d)]$ only now $v > 1$. This means that it is possible for the parameters of N^* to be outside the acceptable range, in which case we cannot generalize the model with regard to the extra claims.

Example 3.41 (Example 3.40 continued) *For what values of v will N^* still have a Poisson–ETNB distribution?*

We need $0 < p_0^{M*} < 1$, that is

$$
\begin{aligned}
0 &< \frac{-1.3^{-4} + (1+0.3v)^{-4}}{1-1.3^{-4}} < 1 \\
0 &< -1.3^{-4} + (1+0.3v)^{-4} < 0.64987 \\
0.35013 &< (1+0.3v)^{-4} < 1 \\
1.3 &> 1+0.3v > 1 \\
1 &> v > 0.
\end{aligned}
$$

Therefore, there is no situation in which lowering the deductible will preserve the Poisson–ETNB distribution. □

Finally, another way to reduce frequency is to apply truncation from above (recall that a policy limit will not do this, all claims above the limit are still paid). Then with a new deductible of d^* and truncating limit u^* we have $v = [F_X(u^*) - F_X(d^*)]/[F_X(u) - F_X(d)]$. The results with regard to changing the frequency distribution due to truncation of the severity are summarized in Appendix D.

3.11 EXERCISES

Section 3.1

3.1 The probability generating function has that name because

$$p_k = \Pr(N = k) = P_N^{(k)}(0)/k!.$$

That is, the probability of k observations can be obtained by evaluating the kth derivative of the pgf at zero. Demonstrate that this is true.

3.2 The **moment generating function** (mgf) is defined as

$$M_N(z) = E\left(e^{zN}\right) = \sum_{k=0}^{\infty} p_k e^{zk}.$$

Demonstrate that $P_N(z) = M_N(\log z)$. Use the fact that $E(N^k) = M_N^{(k)}(0)$ to show that $P'(1) = E(N)$ and $P''(1) = E[N(N-1)]$.

Section 3.2

3.3 A portfolio of 10,000 risks produced the following numbers of claims (data from M94-7).

No. of claims	No. of policies
0	9,048
1	905
2	45
3	2
4+	0

(a) Determine the mle of λ for a Poisson model.

(b) Determine a 95% confidence interval for λ.

(c) Conduct the chi-square goodness-of-fit test at $\alpha = 0.05$.

3.4 An alternative confidence interval was given in Example 2.8. Use this formula to construct a 95% confidence interval for the Poisson parameter in Exercise 3.3.

No. of claims	Underinsured	Uninsured
0	901	947
1	92	50
2	5	2
3	1	1
4	1	0
5+	0	0

3.5 An automobile insurance policy provides benefits for accidents caused by both underinsured and uninsured motorists. Data on 1,000 policies revealed the following information.

(a) Determine the mle of λ for a Poisson model for each of the variables N_1 = number of underinsured claims and N_2 = number of uninsured claims.

(b) Conduct the chi-square goodness-of-fit test at $\alpha = 0.05$ for each model in part (a).

(c) Assume that N_1 and N_2 are independent. Use Theorem 3.1 to determine a model for $N = N_1 + N_2$.

3.6 An alternative method of obtaining a model for N in Exercise 3.5 would be to record the total number of underinsured and uninsured claims for each of the 1,000 policies. Suppose this was done and the results were as follows.

No.of claims	No. of policies
0	861
1	121
2	13
3	3
4	1
5	0
6	1
7+	0

(a) Determine the mle of λ for a Poisson model.

(b) The answer to part (a) matched the answer to part (c) of the previous exercise. Demonstrate that this must always be so.

(c) Conduct the chi-square goodness-of-fit test at $\alpha = 0.05$.

(d) Is there reason to believe (from these data) that N_1 and N_2 are independent? Does your answer agree with your intuition concerning this coverage?

3.7 The following data represent the number of prescriptions filled in one year for a group of elderly members of a group insurance plan.

No. of prescriptions	Frequency	No. of prescriptions	Frequency
0	82	16–20	40
1–3	49	21–25	38
4–6	47	26–35	52
7–10	47	36–	91
11–15	57		

(a) Determine the mle of λ for a Poisson model.

(b) Determine a 95% confidence interval for λ.

(c) Conduct the chi-square goodness-of-fit test at $\alpha = 0.05$.

Section 3.3

3.8 Use the data from Exercise 3.3 (repeated here for convenience).

No. of claims	No. of policies
0	9,048
1	905
2	45
3	2
4+	0

(a) Determine the method of moments estimates of the parameters of the negative binomial model.

(b) Conduct the chi-square goodness-of-fit test, using the method of moments estimates.

(c) Write (3.9) as it applies to this problem. Determine the maximum likelihood estimates of the parameters of the negative binomial distribution.

(d) Conduct the likelihood ratio test to determine if a negative binomial model is more appropriate than a Poisson model.

No. of claims	No. of policies
0	861
1	121
2	13
3	3
4	1
5	0
6	1
7+	0

3.9 Use the data from Exercise 3.6 (repeated above for convenience).

(a) Determine the method of moments estimates of the parameters of the negative binomial model.

(b) Conduct the chi-square goodness-of-fit test, using the method of moments estimates.

(c) Determine the maximum likelihood estimates of the parameters of the negative binomial distribution.

(d) Conduct the likelihood ratio test to determine if a negative binomial model is more appropriate than a Poisson model.

(e) Based on these results, is there any reason to change your answer to part (d) of Exercise 3.6?

(f) Using the maximum likelihood estimates from part (c), perform one iteration of the scoring method to verify that the likelihood has been maximized and to obtain an estimate of the asymptotic covariance matrix of $\hat{\beta}$ and $\hat{r}$.

(g) Use the covariance matrix obtained in part (f) to construct 95% confidence intervals for each parameter. Is there reason to believe that the two intervals are independent?

3.10 From the development in Subsection 3.3.2 it is clear that if the negative binomial distribution arises as a mixture of Poissons, the parameters of the gamma distribution can be obtained from the parameters of the negative binomial distribution. Assume that the negative binomial distribution obtained in Exercise 3.9 did arise from a mixture of Poissons. Use the method of moments estimates to estimate the parameters of the gamma mixing distribution. (Note that being convinced that the negative binomial distribution applies does not assure us that it arises as a gamma mixture of Poissons. However, if, through evidence obtained elsewhere, we are convinced that the observations are some kind of Poisson mixture and are further convinced that the distribution is negative binomial, then the mixing distribution must be gamma.)

3.11 Use (3.7) to determine the mle of β for the geometric distribution. In addition, determine the variance of the mle and verify that it matches the asymptotic variance as given in Theorem 2.2.

3.12 Consider a geometric model for each of the data sets in Exercises 3.8 and 3.9. For each data set do the following.

(a) Determine the mle of β.

(b) Estimate the variance of $\hat{\beta}$ and use it to construct an approximate 95% confidence interval for β.

(c) Conduct the chi-square goodness-of-fit test.

(d) Conduct a likelihood ratio test to determine if a negative binomial model is preferable to a geometric model.

(e) Of the Poisson, geometric, and negative binomial models, which is the best?

3.13 Use the data from Exercise 3.7 (repeated here for convenience).

No. of prescriptions	Frequency	No. of prescriptions	Frequency
0	82	16–20	40
1–3	49	21–25	38
4–6	47	26–35	52
7–10	47	36–	91
11–15	57		

(a) Determine the mles for the geometric and negative binomial distributions.

(b) Conduct the chi-square goodness-of-fit test for each model.

(c) Use the likelihood ratio test to determine which model (of Poisson, geometric, and negative binomial) is best.

(d) Construct a 95% confidence interval for the parameter of the geometric distribution.

(e) Suppose the group insurance provides 10 for each prescription with a maximum annual benefit of 200. For the geometric distribution, express the expected benefit as a function of the parameter β. Determine the mle of the expected benefit and construct an approximate 95% confidence interval.

Section 3.4

3.14 Assume that the binomial parameter m is known. Consider the maximum likelihood estimator of q.

(a) Show that the mle is unbiased.

(b) Determine the variance of the maximum likelihood estimator.

(c) Show that the asymptotic variance as given in Theorem 2.2 is the same as that developed in part (b).

(d) Determine a simple formula for a confidence interval using (2.8) that is based on replacing q with $\hat{q}$ in the variance term.

(e) Determine a more complicated formula for a confidence interval using (2.7) that is not based on such a replacement. This should be done in a manner similar to that used in Example 2.8.

3.15 Use the data from Exercises 3.3 and 3.8 (repeated here for convenience). For parts (a)–(c) assume that the value of m is known to be 4.

No. of claims	No. of policies
0	9,048
1	905
2	45
3	2
4+	0

(a) Determine the mle of q of the binomial model.

(b) Conduct the chi-square goodness-of-fit test, using the mle.

(c) Construct 95% confidence intervals for q using the methods developed in parts (d) and (e) of Exercise 3.14.

(d) Determine the mles of m and q by constructing a likelihood profile.

(e) Conduct the likelihood ratio test to determine if a binomial model is more appropriate than a Poisson model.

(f) Of Poisson, negative binomial, geometric, and binomial, which model best fits the data? Explain.

3.16 Use the data from Exercises 3.6 and 3.9 (repeated here for convenience). For parts (a)–(c) assume that the value of m is known to be 7.

No. of claims	No. of policies
0	861
1	121
2	13
3	3
4	1
5	0
6	1
7+	0

(a) Determine the mle of q of the binomial model.

(b) Conduct the chi-square goodness-of-fit test, using the mle.

(c) Construct 95% confidence intervals for q using the methods developed in parts (d) and (e) of Exercise 3.14.

(d) Determine the mles of m and q by constructing a likelihood profile.

(e) Conduct the likelihood ratio test to determine if a binomial model is more appropriate than a Poisson model.

(f) Of Poisson, negative binomial, geometric, and binomial, which model best fits the data? Explain.

Section 3.5

3.17 For each of the data sets in Exercises 3.15 and 3.16 construct a plot similar to that in Figure 3.1. From each graph, determine the most appropriate model from the $(a, b, 0)$ class. Compare your response to your answer to part (f) of each of those exercises.

3.18 Use your knowledge of the permissible ranges for the parameters of the Poisson, negative binomial, and binomial to determine all possible values of a and b for these members of the $(a, b, 0)$ class. Because these are the only members of the class, all other pairs must not lead to a legitimate probability distribution (non-negative values that sum to one). Show that the pair $a = -1$ and $b = 1.5$ (which is not on the list of possible values) does not lead to a legitimate distribution.

3.19 The following table gives the number of medical claims per reported automobile accident.

(a) Construct a plot similar to Figure 3.1. Does it appear that a member of the $(a, b, 0)$ class will provide a good model? If so, which one?

No. of medical claims	No. of accidents
0	529
1	146
2	169
3	137
4	99
5	87
6	41
7	25
8+	0

(b) Determine the mles of the parameters for each member of the $(a, b, 0)$ class.

(c) Based on the chi-square goodness-of-fit test and the likelihood ratio test, which member of the $(a, b, 0)$ class provides the best fit? Is this model acceptable?

Section 3.6

3.20 Show that for the negative binomial distribution with any $\beta > 0$ and $r > -1$, the successive values of p_k given by (3.18) are, for any p_1, positive and $\sum_{k=1}^{\infty} p_k < \infty$.

3.21 Show that when, in the zero-truncated negative binomial distribution, $r = 0$, the pf is as given in (3.19).

3.22 Show that the pgf of the logarithmic distribution is as given in (3.20).

3.23 Show that for the ETNB distribution with $-1 < r < 0$ and $\beta \to \infty$, the mean does not exist (that is, the sum which defines the mean does not converge). Because this random variable takes on non-negative values, this also shows that no other positive moments exist.

3.24 Determine the iterative formula for the scoring method for the zero-modified Poisson distribution and perform one iteration for the data in Example 3.13 using starting values $p_0^M = 0.879337$ and $\lambda = 0.178267$. Confirm that these starting values are the maximum likelihood estimates and obtain the estimated covariance matrix.

3.25 Repeat Exercise 3.24 for the zero-modified logarithmic distribution with starting values $p_0^M = 0.879337$ and $\beta = 0.189011$. Conduct the chi-square goodness-of-fit test. Is this model superior to the zero-modified geometric distribution?

3.26 For the four data sets introduced in earlier exercises (3.3, 3.6, 3.7, and 3.19) you have determined the best model from among members of the $(a, b, 0)$ class. For each data set determine the mles of the zero-modified Poisson, geometric, logarithmic, and negative binomial distributions. Use the chi-square goodness-of-fit test and likelihood ratio tests to determine the best of the eight models considered and state whether or not the selected model is acceptable.

3.27 Use the data from Exercise 3.7 to obtain the estimated covariance matrix for the mles of the zero-modified geometric distribution. Use them to obtain the following.

(a) Construct 95% confidence intervals for p_0^M and β.

(b) Determine the probability that the procedure used in part (a) produced intervals that both contain the true parameter value.

(c) As in part (e) of Exercise 3.13 determine a 95% confidence interval for the expected payment.

3.28 A frequency model that has not been mentioned to this point is the **zeta distribution**. It is a zero-truncated distribution with $p_k^T = k^{-(\rho+1)}/\zeta(\rho + 1)$, $k = 1, 2, \ldots, \rho > 0$. The denominator is the zeta function, which must be evaluated numerically as $\zeta(\rho + 1) = \sum_{k=1}^{\infty} k^{-(\rho+1)}$. The zero-modified zeta distribution can be formed in the usual way. More information can be found in Luong and Doray [86].

(a) Verify that the zeta distribution is not a member of the $(a, b, 1)$ class.

(b) Determine the maximum likelihood estimates of the parameters of the zero-modified zeta distribution for the data in Example 3.13.

(c) Is the zero-modified zeta distribution superior to the zero-modified geometric distribution?

Section 3.7

3.29 For the data in Table 3.16 determine the method of moments estimates of the parameters of the Poisson–Poisson distribution where the secondary distribution is the ordinary (not zero-truncated) Poisson distribution. Perform the chi-square goodness-of-fit test using this model.

3.30 According to Theorem 3.6 there is a Poisson–zero-truncated Poisson distribution that has exactly the same pf as the Poisson–Poisson distribution obtained in Exercise 3.29. Determine the parameters of this equivalent model. Working directly with this model, verify that it has the same mean, variance, and probabilities at zero and one as the original Poisson–Poisson model.

3.31 Do all the members of the $(a, b, 0)$ class satisfy the condition of Theorem 3.6? For those that do, identify the parameter (or function of its parameters) that plays the role of θ in the theorem.

3.32 The maximum likelihood estimates of the parameters of the Poisson–Poisson distribution for the data in Table 3.16 are $\lambda_1 = 1.101229$ for the primary distribution and $\lambda_2 = 0.140879$ for the secondary distribution. Verify that these are the maximum likelihood estimates by performing one iteration of the scoring method. Also determine the estimated covariance matrix of the mles.

3.33 In Exercise 3.26 the best model from among the members of the $(a, b, 0)$ and $(a, b, 1)$ classes was selected for the data sets in Exercises 3.3, 3.6, 3.7,and 3.19. Fit the Poisson–Poisson, Polya–Aeppli, Poisson–inverse Gaussian, and Poisson–ETNB (generalized Poisson–Pascal) distributions to these data and use the chi-square goodness-of-fit test as well as the likelihood ratio test to determine if any of these distributions should replace the one selected in Exercise 3.26. Is the current best model acceptable?

3.34 For $i = 1, \ldots, n$ let S_i have a compound Poisson frequency distribution with Poisson parameter λ_i and a secondary distribution with pgf $P_2(z)$. Note that all n of the variables have the same secondary distribution. Determine the distribution of $S = S_1 + \cdots + S_n$.

3.35 Show that the following three distributions are identical: (1) geometric–geometric, (2) Bernoulli–geometric, (3) zero–modified geometric. That is, for any one of the distributions with arbitrary parameters, show that there is a member of the other two distribution types that has the same pf or pgf.

3.36 Show that the binomial–geometric and negative binomial–geometric distributions are identical.

Section 3.8

3.37 Show that for any pgf, $P^{(k)}(1) = E[N(N-1)\cdots(N-k+1)]$, provided the expectation exists. Here $P^{(k)}(z)$ indicates the kth derivative. Use this result to confirm the three moments as given in (3.34).

3.38 Verify the three moments as given in (3.35).

3.39 The five data sets presented in this problem are all taken from Lemaire [80]. For each data set compute the first three moments and then use them to make a guess at an appropriate model from among our compound Poisson collection (Poisson, geometric, negative binomial, Poisson–binomial (with $m = 2$ and $m = 3$), Polya–Aeppli, Neyman Type A, Poisson–inverse Gaussian, and

Generalized Poisson–Pascal). Then determine maximum likelihood estimates for the parameters of each of these models, conduct the chi-square goodness-of-fit test, and then use this information to select the best model from these choices.

(a) The data in Table 3.22 represent counts from third-party automobile liability coverage in Belgium.

Table 3.22 Data for Excercise 3.39(a)

No. of claims	No. of policies
0	96,978
1	9,240
2	704
3	43
4	9
5+	0

(b) The data in Table 3.23 represent the number of deaths due to horse kicks in the Prussian army between 1875 and 1894. The counts are the number of deaths in a corps (there were 10 of them) in a given year, and thus there are 200 observations. This data set is often cited as the inspiration for the Poisson distribution. For using any of our models what additional assumption about the data must be made?

Table 3.23 Data for Excercise 3.39(b)

No. of deaths	No. of corps
0	109
1	65
2	22
3	3
4	1
5+	0

(c) The data in Table 3.24 represent the number of major international wars per year from 1500 through 1931.

(d) The data in Table 3.25 represent the number of runs scored in each half-inning of World Series baseball games played from 1947 through 1960.

Table 3.24 Data for Excercise 3.39(c)

No. of wars	No. of years
0	223
1	142
2	48
3	15
4	4
5+	0

Table 3.25 Data for Excercise 3.39(d)

No. of runs	No. of half innings
0	1,023
1	222
2	87
3	32
4	18
5	11
6	6
7+	3

(e) The data in Table 3.26 represent the number of goals per game per team in the 1966–1967 season of the National Hockey League.

Section 3.9

3.40 Show that the pgf for the inverse Gaussian distribution is as given in (3.39).

3.41 Derive the expression in (3.43).

3.42 Show that the negative binomial–Poisson compound distribution is the same as a mixed Poisson distribution with a negative binomial mixing distribution.

3.43 For $i = 1, \ldots, n$ let N_i have a mixed Poisson distribution with parameter λ. Let the mixing distribution for N_i have pgf $P_i(z)$. Show that $N = N_1 + \cdots + N_n$ has a mixed Poisson distribution and determine the pgf of the mixing distribution.

Table 3.26 Data for Excercise 3.39(e)

No. of goals	No. of games
0	29
1	71
2	82
3	89
4	65
5	45
6	24
7	7
8	4
9	1
10+	3

3.44 Let N have a Poisson distribution with (given that $\Theta = \theta$) parameter $\lambda\theta$. Let the distribution of the random variable Θ have a scale parameter. Show that the mixed distribution does not depend on the value of λ.

3.45 Let N have a Poisson distribution with (given that $\Theta = \theta$) parameter θ. Let the distribution of the random variable Θ have pdf $u(\theta) = \alpha^2(\alpha + 1)^{-1}(\theta + 1)e^{-\alpha\theta}$, $\theta > 0$. Determine the pf of the mixed distribution. Also, show that the mixed distribution is also a compound distribution.

3.46 Verify that the estimates presented in Example 3.26 are the maximum likelihood estimates. (Because only two decimals are presented, it is probably sufficient to observe that the likelihood function takes on smaller values at each of the nearby points.) The negative binomial distribution was fit to these data in Example 3.8. Which of these two models is preferable?

Section 3.10

3.47 Redo Example 3.34 assuming that each exposure unit has a geometric distribution. Conduct the approximate chi-square goodness-of-fit test. Is the geometric preferable to the Poisson model?

3.48 A group life insurance policy has an accidental death rider. For ordinary deaths, the benefit is 10,000, however, for accidental deaths the benefit is 20,000. The insureds are approximately the same age, so it is reasonable to assume they all have the same claim probabilities. Let them be 0.97 for no claim, 0.01 for an ordinary death claim, and 0.02 for an accidental death claim. A reinsurer has been asked to bid on providing an excess reinsurance that will pay 10,000 for each accidental death.

(a) The claim process can be modeled with a frequency component that has the Bernoulli distribution (the event is claim/no claim) and a two-point severity component (the probabilities are associated with the two claim levels, given that a claim occurred). Specify the probability distributions for the frequency and severity random variables.

(b) Suppose the reinsurer wants to retain the same frequency distribution. Determine the modified severity distribution that will reflect the reinsurer's payments.

(c) Determine the reinsurer's frequency and severity distributions when the severity distribution is to be conditional on a reinsurance payment being made.

3.49 Individual losses have a Pareto distribution with $\alpha = 2$ and $\theta = 1{,}000$. With a deductible of 500 the frequency distribution for the number of payments is Poisson–inverse Gaussian with $\lambda = 3$ and $\beta = 2$. If the deductible is raised to 1,000, determine the distribution for the number of payments. Also, determine the pdf of the severity distribution (per payment) when the new deductible is in place.

3.50 Losses have a Pareto distribution with $\alpha = 2$ and $\theta = 1{,}000$. The frequency distribution for a deductible of 500 is zero-truncated logarithmic with $\beta = 4$. Determine a model for the number of payments when the deductible is reduced to 0.

3.51 Determine a model for the annual number of hurricanes causing in excess of five million dollars in damage, using the data in Exercise 2.1. A ground-up inverse Gaussian ($\mu = 203{,}500$ and $\theta = 38{,}060.3$) severity model (for losses in thousands of dollars) was determined in Exercise 2.119. Determine the distribution for the annual number of hurricanes causing in excess of five hundred million dollars in damage.

3.12 CASE STUDY

3.12.1 The case study continued

The frequency data, originally presented in Table 1.5, are reproduced here in Table 3.27. There are two ways to handle data presented in this format. The first is to use the ideas presented in Subsection 3.10.1. The combined data set is considered as a sample of size 12 from an infinitely divisible population. The parameters for each observation differ because they must reflect the sample size. This is done by multiplying the appropriate parameter for a single observation by the sample size. For the Poisson distribution the parameter to multiply is λ, for the negative binomial it is r, and for compound distributions

Table 3.27 Exposures and loss counts

	Life/health		Pension		Property/liability	
Year	Exposure	Claims	Exposure	Claims	Exposure	Claims
1990	853	20	1,446	27	639	5
1991	1,105	14	1,780	35	725	8
1992	1,148	16	1,717	36	685	4
1993	1,270	21	2,065	24	864	11

Table 3.28 Fitted models: combined frequency data, method one

Model	Parameter(s)		−Loglikelihood
Poisson	$\lambda = 0.0154578$		38.2111
Negative binomial	$\beta = 0.865869$	$r = 0.0178523$	36.9086
Poisson–Poisson	$\lambda = 0.0192067$	$\lambda = 0.804811$	36.9284
Polya–Aeppli	$\lambda = 0.0108939$	$\beta = 0.418946$	36.9148
Poisson–inverse Gaussian	$\lambda = 0.0115555$	$\beta = 1.80695$	36.9163
Poisson–ETNB	$\lambda = 0.0112188$	$\beta = 1.03748$	36.9081
	$r = -0.156906$		

it is the parameter of the primary distribution. For the Poisson distribution the maximum likelihood estimate is the total number of losses divided by the total exposure. The results from fitting six distributions are given in Table 3.28. The likelihood ratio test indicates that the Poisson distribution ($\lambda = 0.0154578$) is the best model choice.

A second approach is to assume that no policyholder had more than one loss in a given year. Due to the low frequency of losses, this may be a reasonable assumption. (With the Poisson model above, the chance of one policyholder having more than one loss is about one in ten thousand.) In that case, the standard estimation methods of this chapter may be used. However, with counts of only zero and one, one-parameter distributions are all that can be used. That is, none of the two-parameter models had maximum likelihood estimates. For the two one-parameter distributions the results are $\lambda = 0.0154578$ and a negative loglikelihood of 1,142.49 for the Poisson distribution and $\beta = 0.0154578$ and a negative loglikelihood of 1,144.19 for the geometric distribution. The Poisson distribution is again selected and the parameter estimate is the sample mean. Note that because the mle is the sample mean, it was not actually necessary to assume that no policyholder had more than one claim. However, that assumption was used in the calculation of the negative loglikelihood, and therefore was used to arrive at the Poisson distribution in the first place.

Table 3.29 Likelihood ratio test for separate models

Sample		λ	−Loglikelihood
All		0.0154578	38.2111
Separate	Life/health	0.0162249	11.2885
	Pension	0.0174087	13.6449
	Property/liability	0.0096121	8.6411
	Total		33.5745

Our final task is to determine if the same Poisson parameter applies to all three categories of actuary. The results are summarized in Table 3.29 using the first estimation method. The test statistic is $2(38.2111 - 33.5745) = 9.2732$. With two degrees of freedom the p-value is 0.0097 and so there is a significant difference in the frequency distributions for the three groups. If estimation is done using the at most one claim per policyholder assumption, the test statistic turns out to be the exactly the same.

3.12.2 Exercises

3.52 The data in Table 3.30 were collected by Fisher [41] on coal mining disasters in the United States over 25 years ending about 1910. This particular compilation counted the number of disasters per year that claimed the lives of 5 to 9 miners. The data appear in Table 3.30. In the article Fisher claimed that a Poisson distribution was a good model. Do you agree? Can you find a better model?

3.53 The data in Table 3.31 were collected by Downey and Kelly [32] in a study of workers compensation. There were 1,203 occupation classifications, and the number of deaths in Pennsylvania in 1916 were recorded by classification. The data were grouped, but the total number of deaths in each range were also recorded. Determine an appropriate frequency model for the number of deaths per classification. Is your result meaningful, or was some important information missing?

3.54 One possible basis for reinsurance (or in this case for insurance on top of self-insurance) is to have the insurance coverage become effective when more than a specified number of individuals are involved in a claim. Cahill [22] studied this for workers compensation insurance. Three different hazard levels (by occupation) were studied. In each case the number of persons involved in an accident was recorded. The data appear in Table 3.32. Is it reasonable to use the same distribution family for all three hazard groups? If so, is it reasonable to use the same parameters? If the models differ, either

Table 3.30 Mining disasters per year

No. of disasters	No. of years
0	1
1	1
2	3
3	4
4	5
5	2
6	2
7	3
8	1
9	0
10	1
11	1
12	1
13+	0

Table 3.31 Workers' compensation deaths

No. of deaths	No. of classes	Total deaths
0	934	0
1	141	141
2–4	82	225
5–9	29	195
10–99	15	225
100+	2	490

by family or parameter, is it possible to rank the three groups with regard to the number of persons per accident?

3.55 In 1958 Longley-Cook [84] examined employment patterns of casualty actuaries. One of his tables listed the number of CAS members employed by casualty companies in 1949 (55 actuaries) and 1957 (78 actuaries). Using the data in Table 3.33 determine a model for the number of actuaries per company which employs at least one actuary and find out whether the distribution has changed over the eight-year period.

3.56 An early proponent of the negative binomial distribution was Dropkin [33], who studied driving habits in California. His data set was on the number of accidents in one year. The data were collected from 1956–1958 and appear

Table 3.32 Number of injuries per accident, by hazard level

No. of persons injured	No. of accidents		
	Group A	Group B	Group C
1	95	541	1,278
2	5	16	15
3	4	3	4
4	0	1	0
5	0	0	0
6	0	0	1
7	0	1	1
8	0	0	0
9	0	0	0
10	0	0	0
11	0	0	0
12	0	0	0
13	0	0	1
Totals	104	562	1,300

Table 3.33 Number of actuaries per company

No. of actuaries	No. of companies—1949	No. of companies—1957
1	17	23
2	7	7
3–4	3	3
5–9	2	3
10+	0	1

in Table 3.34. He claimed that the negative binomial distribution was superior to the Poisson. Do you agree? Can you find a better model? His conclusion was that drivers are not homogeneous with regard to accident propensity and therefore experience rating of individual drivers was appropriate.

3.57 Harwayne [55] was curious as to the relationship between driving record and number of accidents. His data on California drivers included the number of violations. For the six data sets represented by each column in Table 3.35, is the same distributional family appropriate? If so, are the same parameters appropriate? Is it reasonable to conclude that the expected number of accidents increases with the number of violations?

Table 3.34 Distribution of accidents in California

No. of accidents	No. of drivers
0	81,714
1	11,306
2	1,618
3	250
4	40
5+	7

Table 3.35 Number of accidents by number of violations

	No. of violations					
No. of accidents	0	1	2	3	4	5+
0	51,365	17,081	6,729	3,098	1,548	1,893
1	3,997	3,131	1,711	963	570	934
2	357	353	266	221	138	287
3	34	41	44	31	34	66
4	4	6	6	6	4	14
5+	0	1	1	1	3	1

3.58 An automobile claims study was conducted by Roberts [106] using data collected in 1957. Of interest was the frequency of claims producing excess losses. The data were the number of claims per accident which produced excess losses given that there were excess losses. Determine an appropriate model using the data in Table 3.36. The counts in that table are the number of losses in excess of 10,000 for an accident given that the accident produced at least one such loss. Suppose it has been determined that the probability that a loss is above 20,000 given that it is above 10,000 is 0.75. As well, it has been determined that the probability that an accident produces no claims in excess of 10,000 is 0.85. Determine a frequency model for the number of loss per accident that are in excess of 10,000. Also determine models for the number of ground-up losses and the number of losses in excess of 20,000.

3.59 In 1961, Simon [112] proposed using the zero-modified negative binomial distribution. His data set was the number of accidents in one year along various one-mile stretches of Oregon highway. The data appear in Table 3.37. Simon claimed that the zero-modified negative binomial distribution was superior to the negative binomial. Do you agree? Can you obtain a better model?

Table 3.36 Distribution of the number of excess losses

No. of claims	No. of accidents	No. of claims	No. of accidents
1	2,072	8	8
2	301	9	6
3	165	10	4
4	120	24	1
5	69	28	1
6	50	39	1
7	15		

Table 3.37 Number of accidents per year

No. of accidents	No. of stretches	No. of accidents	No. of stretches
0	99	6	4
1	65	7	0
2	57	8	3
3	35	9	4
4	20	10	0
5	10	11	1

3.60 In 1970, Weber [126] studied accident records for California drivers. He was able to follow the same drivers over a 34.5-month period. For each driver he recorded the number of accidents in 1961, 1962, and the first 10.5 months of 1963. He also recorded the number of accidents for each driver over the entire 34.5-month period. The results are given in Table 3.38. He used the negative binomial distribution throughout. Is it a reasonable model for the four data sets? If it is a reasonable model, keep using it, else determine an infinitely divisible model that is satisfactory for all four data sets. For the model you have selected, is it reasonable to use the same value for the parameter(s) that do not relate to infinite divisibility for all four cases, and let the remaining parameter be proportional to the number of months? If this is so, what can you conclude?

Table 3.38 Number of accidents for various time periods

No. of accidents	Year			
	1961	1962	1963	1961–1963
0	138,343	138,087	139,326	122,593
1	9,072	9,211	8,140	21,350
2	547	650	505	3,425
3	44[a]	58[a]	35[a]	530
4				89
5+				19

[a]Entry is for 3+.

4

Aggregate Loss Models

4.1 INTRODUCTION

The purpose of this chapter is to develop models of aggregate losses, the total amount paid on all claims occurring in a fixed time period on a defined set of insurance contracts (see Definition 1.13). There are two ways to go about adding the claims in order to obtain the total for the period.

One method is to record the payments as they are made and then add them up. In that case we can represent the aggregate losses as a sum, S, of a random number, N, of individual payment amounts $(X_1, X_2, \cdots, X_N)$. Hence,

$$S = X_1 + X_2 + \cdots + X_N, \qquad N = 0, 1, 2, ... \tag{4.1}$$

where $S = 0$ when $N = 0$.

Following Definition 1.14, the **collective risk model** has this representation with the X_js being independent and identically distributed (iid) random variables, unless otherwise specified. More formally, the independence assumptions are:

1. Conditional on $N = n$, the random variables $X_1, X_2, \cdots, X_n$ are iid random variables

2. Conditional on $N = n$, the common distribution of the random variables $X_1, X_2, \cdots, X_n$ does not depend on n

3. The distribution of N does not depend in any way on the values of $X_1, X_2, \ldots$.

Following Definition 1.15, the **individual risk model** represents the aggregate loss as a sum, S, of a fixed number, n, of insurance contracts. The loss amounts for the n contracts are $(X_1, X_2, \cdots, X_n)$ where the X_js are assumed to be independent but are not assumed to be identically distributed. The distribution of the X_js usually has a probability mass at zero, corresponding to the probability of no loss or payment.

The individual risk model is used to "add together" the losses or payments from a fixed number of insurance contracts or sets of insurance contracts. It is used in modeling the losses of a group life or health insurance policy that covers a group of n employees. Each employee can have different coverage (life insurance benefit as a multiple of salary) and different levels of loss probabilities (different ages and health status).

In the special case where the X_js are identically distributed, the individual risk model becomes the special case of the collective risk model, with the distribution of N being the degenerate distribution with all of the probability at $N = n$; that is, $\Pr(N = n) = 1$.

The distribution of S in (4.1) is obtained from the distribution of N and the distribution of the X_js. Using this approach, the frequency and the severity of claims are modelled separately. The information about these distributions is used to obtain information about S. An alternative to this approach is to simply gather information about S (e.g., total losses each month for a period of months) and to use some model from Chapter 2 to model the distribution of S. Modeling the distribution of N and the distribution of the X_js separately has some distinct advantages:

1. The expected number of claims changes as the number of insured policies changes. Growth in the volume of business needs to be accounted for in forecasting the number of claims in future years based on past years' data.

2. The effects of general economic inflation and additional claims inflation are reflected in the losses incurred by insured parties and the claims paid by insurance companies. Such effects are often masked when insurance policies have deductibles and policy limits which do not depend on inflation and aggregate results are used.

3. The impact of changing individual deductibles and policy limits is more easily studied. This is done by changing the specification of the severity distribution.

4. The impact on claims frequencies of changing deductibles is better understood.

5. Data that are heterogeneous in terms of deductibles and limits can be combined to obtain the hypothetical loss size distribution. This is useful when data from several years in which policy provisions were changing are combined.

6. Models developed for noncovered losses to insureds, claim costs to insurers, and claim costs to reinsurers can be mutually consistent. This is useful for a direct insurer which is studying the consequence of shifting losses to a reinsurer.

7. The shape of the distribution of S depends on the shapes of both distributions of N and X. The understanding of the relative shapes is useful when modifying policy details. For example, if the severity distribution has a much heavier tail than the frequency distribution, the shape of the tail of the distribution of aggregate claims or losses will be determined by the severity distribution and will be insensitive to the choice of frequency distribution.

In summary, a more accurate and flexible model can be constructed by examining frequency and severity separately.

In constructing the model (4.1) for S, if N represents the actual number of losses to the insured, then the X_js can represent (i) the losses to the insured, (ii) the claim payments of the insurer, (iii) the claim payments of a reinsurer, or (iv) the deductibles (self-insurance) paid by the insured. In each case, the interpretation of S is different and the severity distribution can be constructed in a consistent manner.

Because the random variables N, $X_1, X_2, \ldots$, and S provide much of the focus for this chapter and the two that follow, we want to be especially careful when referring to them. We will refer to N as the **claim count random variable** and will refer to its distribution as the **claim count distribution**. The expression **number of claims** will also be used, and, occasionally, just **claims** will be used. Another term that will commonly be used is **frequency distribution**. The X_js are the **individual** or **single loss random variables**. The modifier **individual** or **single** will be dropped when the reference is clear. In Chapter 2 a distinction was made between losses and payments. Strictly speaking, the X_js are payments because they represent a real cash transaction. However, the term **loss** is more customary, and we will continue with it. Another common term for the X_js is **severity**. Finally, S is the **aggregate loss random variable** or the **total loss random variable**.

Example 4.1 *An insurer estimates that individual losses to an insured follow a distribution with pdf $f_X(x)$. The insurer pays 80% of individual losses in excess of 1,000 with a maximum payment of 100,000. It reinsures that portion of any payments in excess of 50,000. Develop models for each of the following: (a) the total loss, preinsurance, to the policyholder, (b) the aggregate loss to the insurer, prior to the reinsurance payment, (c) the aggregate loss to the reinsurer, (d) the aggregate loss to the insurer, after the reinsurance payment, and (e) the aggregate loss to the insured.*

(a) The aggregate losses with no insurance are $S = X_1 + X_2 + \cdots + X_N$, where the X_js have the distribution with pdf $f_X(x)$.

(b) The aggregate payments by the insurer (before recovery of reinsurance) are $S = Y_1 + Y_2 + \cdots + Y_N$, where

$$Y_j = \begin{cases} 0, & X_j \leq 1{,}000 \\ 0.80(X_j - 1{,}000), & 1{,}000 < X_j \leq 126{,}000 \\ 100{,}000, & X_j > 126{,}000. \end{cases}$$

(c) The aggregate payments by the reinsurer are $S = Y_1 + Y_2 + \cdots + Y_N$, where

$$Y_j = \begin{cases} 0, & X_j \leq 63{,}500 \\ 0.80(X_j - 63{,}500), & 63{,}500 < X_j \leq 126{,}000 \\ 50{,}000, & X_j > 126{,}000. \end{cases}$$

(d) The aggregate costs to the insurer, after recovery of reinsurance payments, are $S = Y_1 + Y_2 + \cdots + Y_N$, where

$$Y_j = \begin{cases} 0, & X_j \leq 1{,}000 \\ 0.80(X_j - 1{,}000), & 1{,}000 < X_j \leq 63{,}500 \\ 50{,}000, & X_j > 63{,}500. \end{cases}$$

(e) The aggregate costs to the insured; that is, the uninsured costs are $S = Y_1 + Y_2 + \cdots + Y_N$, where

$$Y_j = \begin{cases} X_j, & X_j \leq 1{,}000 \\ 800 + 0.20X_j, & 1{,}000 < X_j \leq 126{,}000 \\ X_j - 100{,}000, & X_j > 126{,}000. \end{cases}$$

□

4.2 MODEL CHOICES

In many cases of fitting frequency or severity distributions to data, several distributions may be good candidates for models. However, some distributions may be preferable for a variety of practical reasons.

In general, it is useful for the severity distribution to be from a scale family (see Definition 2.31) since the choice of currency (e.g., U.S. dollars or British pounds) should not affect the result. Also, scale families are easy to adjust for inflationary effects over time (this is, in effect, a change in currency; e.g.,

1994 U.S. dollars to 1995 U.S. dollars). When forecasting the costs for a future year, the anticipated rate of inflation can be factored in easily by adjusting the parameters.

A similar consideration applies to frequency distributions. As a block of an insurance company's business grows, the number of claims can be expected to grow, all other things being equal. If one chooses models that have probability generating functions of the form

$$P_N(z;\alpha) = Q(z)^\alpha \tag{4.2}$$

for some parameter α, then the expected number of claims is proportional to α. Increasing the volume of business by $100r\%$ results in expected claims being proportional to $\alpha^* = (1+r)\alpha$. Since r is any value satisfying $r > -1$, the distributions satisfying (4.2) should allow α to take on any positive values. Such distribution are said to infinitely divisible (see Definition 3.5).

A related consideration also suggests frequency distributions that are infinitely divisible. This relates to the concept of invariance over the time period of study. Ideally the model selected should not depend on the length of the time period used in the study of claims frequency. The expected frequency should be proportional to the length of the time period, after any adjustment for growth in business. This means that a study conducted over a period of 10 years can be used to develop claims frequency distributions for periods of a month, a year, or any other period. Furthermore, the form of the distribution for a one-year period is the same as for a one-month period with a change of parameter. The parameter α corresponds to the length of a time period. For example, if $\alpha = 1.7$ in (4.2) for a one-month period, then the identical model with $\alpha = 20.4$ is an appropriate model for a one-year period.

Distributions that have a modification at zero are not of the form (4.2). However, it may be still desirable to use a zero-modified distribution if the physical situation suggests it. For example, if a certain proportion of policies never make a claim, due to duplication of coverage or other reason, it may be appropriate to use this same proportion in future periods for a policy selected at random.

4.3 THE COMPOUND MODEL FOR AGGREGATE CLAIMS

Let S denote aggregate losses associated with a set of N observed claims $X_1, X_2, \cdots, X_N$ satisfying the independence assumptions following (4.1). The approach in this chapter is to

1. Develop a model for the distribution of N based on data.

2. Develop a model for the common distribution of the X_js based on data.

3. Using these two models, carry out necessary calculations to obtain the distribution of S.

Completion of the first two steps follows the ideas developed in Chapters 2 and 3. We now presume that these two models are developed and that we only need to carry out numerical work in obtaining solutions to problems associated with the distribution of S. These might involve pricing a stop-loss reinsurance contract, and they require analyzing the impact of changes in deductibles, coinsurance levels, and maximum payments on individual losses.

The random sum

$$S = X_1 + X_2 + \cdots + X_N$$

(where N has a counting distribution) has a distribution function

$$\begin{aligned} F_S(x) &= \Pr(S \leq x) \\ &= \sum_{n=0}^{\infty} p_n \Pr(S \leq x | N = n) \\ &= \sum_{n=0}^{\infty} p_n F_X^{*n}(x) \end{aligned} \tag{4.3}$$

where $F_X(x) = \Pr(X \leq x)$ is the common distribution function of the X_js and $p_n = \Pr(N = n)$. In (4.3), $F_X^{*n}(x)$ is the "n-fold convolution" of the cdf of X. It can be obtained as

$$F_X^{*0}(x) = \begin{cases} 0, & x < 0 \\ 1, & x \geq 0 \end{cases}$$

and

$$F_X^{*k}(x) = \int_{-\infty}^{\infty} F_X^{*(k-1)}(x-y)\,dF_X(y). \tag{4.4}$$

If X is a continuous random variable with no probability on negative values, (4.4) reduces to

$$F_X^{*k}(x) = \int_0^x F_X^{*(k-1)}(x-y) f_X(y)\,dy$$

and by differentiating the pdf is

$$f_X^{*k}(x) = \int_0^x f_X^{*(k-1)}(x-y) f_X(y)\,dy.$$

In the case of discrete random variables with probabilities at $0, 1, 2, \ldots$, equation (4.4) reduces to

$$F_X^{*k}(x) = \sum_{y=0}^{x} F_X^{*(k-1)}(x-y) f_X(y), \qquad x = 0, 1, \ldots.$$

The corresponding pf is

$$f_X^{*k}(x) = \sum_{y=0}^{x} f_X^{*(k-1)}(x-y) f_X(y), \qquad x = 0, 1, \ldots.$$

The distribution (4.3) is called a **compound distribution** and the pf for the distribution of aggregate losses is

$$f_S(x) = \sum_{n=0}^{\infty} p_n f_X^{*n}(x).$$

Arguing as in Section 3.7, the pgf of S is

$$\begin{aligned}
P_S(z) &= E[z^S] \\
&= \sum_{n=0}^{\infty} E[z^{X_1+X_2+\cdots+X_n}|N=n]\Pr(N=n) \\
&= \sum_{n=0}^{\infty} E\left[\prod_{j=1}^{n} z^{X_j}\right]\Pr(N=n) \\
&= \sum_{n=0}^{\infty} \Pr(N=n)[P_X(z)]^n \\
&= E[P_X(z)^N] = P_N[P_X(z)] \qquad (4.5)
\end{aligned}$$

due to the independence of $X_1, \ldots, X_n$ for fixed n.

A similar relationship exists for the other generating functions. It is sometimes more convenient to use the characteristic function

$$\varphi_S(z) = E(e^{izS}) = P_N[\varphi_X(z)]$$

which always exists. Panjer and Willmot [102] use the Laplace transform

$$L_S(z) = E(e^{-zS}) = P_N[L_X(z)]$$

which always exists for random variables defined on non-negative values. With regard to the moment generating function, we have

$$M_S(z) = P_N[M_X(z)].$$

The pgf of compound distributions was discussed in Section 3.7 where the "secondary" distribution plays the role of the claim size distribution in this chapter.

In the case where $P_N(z) = P_1[P_2(z)]$—that is, N is itself a compound distribution—then $P_S(z) = P_1\{P_2[P_X(z)]\}$, which in itself produces no additional difficulties.

From (4.5), the moments of S can be obtained in terms of the moments of N and the X_js. The first three moments are

$$\begin{aligned} E(S) &= \mu'_{S1} = \mu'_{N1}\mu'_{X1} = E(N)E(X) \\ Var(S) &= \mu_{S2} = \mu'_{N1}\mu_{X2} + \mu_{N2}(\mu'_{X1})^2 \\ E\{[S - E(S)]^3\} &= \mu_{S3} = \mu'_{N1}\mu_{X3} + 3\mu_{N2}\mu'_{X1}\mu_{X2} + \mu_{N3}(\mu'_{X1})^3. \end{aligned} \tag{4.6}$$

Here, the first subscript indicates the appropriate random variable, the second subscript indicates the "order" of the moment, and the superscript is a "prime" (′) for raw moments (moments about the origin) and is unprimed for central moments (moments about the mean). The moments can be used on their own to provide approximations for probabilities of aggregate claims by matching the first few model and sample moments.

Example 4.2 *The observed mean (and standard deviation) of the number of claims and the individual losses over the past 10 months are 6.7 (2.3) and 179,247 (52,141), respectively. Determine the mean and variance of aggregate claims per month.*

$$\begin{aligned} E(S) &= 6.7(179{,}247) = 1{,}200{,}955. \\ Var(S) &= 6.7(52{,}141)^2 + (2.3)^2(179{,}247)^2 \\ &= 1.88180 \times 10^{11}. \end{aligned}$$

Hence, the mean and standard deviation of aggregate claims are 1,200,955 and 433,797, respectively. □

Example 4.3 (Example 4.2 continued) *Using normal and lognormal distributions as approximating distributions for aggregate claims, calculate the probability that claims will exceed 140% of expected costs. That is,*

$$\Pr(S > 1.40 \times 1{,}200{,}955) = \Pr(S > 1{,}681{,}337).$$

For the normal distribution

$$\begin{aligned} \Pr(S > 1,681,337) &= \Pr\left(\frac{S - E(S)}{\sqrt{Var(S)}} > \frac{1,681,337 - 1,200,955}{433,797}\right) \\ &= \Pr(Z > 1.107) = 1 - \Phi(1.107) = 0.134. \end{aligned}$$

For the lognormal distribution, from Appendix A, the mean and second raw moment of the lognormal distribution are

$$E(S) = \exp(\mu + \sigma^2/2) \quad \text{and} \quad E(S^2) = \exp(2\mu + 2\sigma^2).$$

Equating these to 1.200955×10^6 and $1.88180 \times 10^{11} + (1.200955 \times 10^6)^2 = 1.63047 \times 10^{12}$ and taking logarithms results in the following two equations in two unknowns:

$$\mu + \sigma^2/2 = 13.99863$$

$$2\mu + 2\sigma^2 = 28.11989.$$

From this $\mu = 13.93731$ and $\sigma^2 = .1226361$. Then

$$\begin{aligned} \Pr(S > 1{,}681{,}337) &= 1 - \Phi\left[\frac{\log 1,681,337 - 13.93731}{(0.1226361)^{0.5}}\right] \\ &= 1 - \Phi(1.135913) = 0.128. \end{aligned}$$

The normal distribution provides a good approximation when $E(N)$ is large. In particular, if N has the Poisson, binomial, or negative binomial distributions, a version of the central limit theorem indicates that as λ, m, or r respectively goes to infinity, the distribution of S becomes normal. In this example, $E(N)$ is small so the distribution of S is likely to be skewed. In this case the lognormal distribution may provide a good approximation, although there is no theory to support this choice. □

Example 4.4 (Group dental insurance) *Under a group dental insurance plan covering employees and their families, the premium for each married employee is the same regardless of the number of family members. The insurance company has compiled statistics showing that the annual cost (adjusted to current dollars) of dental care per person for the benefits provided by the plan has the distribution in Table 4.1 (given in units of 25 dollars).*

Table 4.1 Loss distribution for Example 4.4

x	$f_X(x)$
1	0.150
2	0.200
3	0.250
4	0.125
5	0.075
6	0.050
7	0.050
8	0.050
9	0.025
10	0.025

Furthermore, the distribution of the number of persons per insurance certificate (that is, per employee) receiving dental care in any year has the distribution given in Table 4.2.

Table 4.2 Frequency distribution for Example 4.4.

n	p_n
0	0.05
1	0.10
2	0.15
3	0.20
4	0.25
5	0.15
6	0.06
7	0.03
8	0.01

The insurer is now in a position to calculate the distribution of the cost per year per married employee in the group. The cost per married employee is

$$f_S(x) = \sum_{n=0}^{8} p_n f_X^{*n}(x).$$

Determine the pf of S up to 525. Determine the mean and standard deviation of total payments per employee.

The distribution up to amounts of 525 is given in Table 4.3. To obtain $f_S(x)$, each row of the matrix of convolutions of $f_X(x)$ is multiplied by the probabilities from the row below the table and the products are summed.

The reader may wish to verify using (4.6) that the first two moments of the distribution $f_S(x)$ are

$$E(S) = 12.58$$

$$Var(S) = 58.7464.$$

Hence the annual cost of the dental plan has mean $12.58 \times 25 = 314.50$ dollars and standard deviation 191.6155 dollars. (Why can't the calculations be done from Table 4.3?) □

It is common for insurance to be offered in which a deductible is applied to the aggregate losses for the period. It is an insurance coverage when the losses occur to a policyholder and is reinsurance when the losses occur to an insurance company. The latter version is a common method for an insurance company to protect itself against an adverse year (as opposed to protecting against a single, very large claim). More formally, we present the following definition.

Table 4.3 Aggregate probabilities for Example 4.4

x	$f_X^{*0}(x)$	f_X^{*1}	f_X^{*2}	f_X^{*3}	f_X^{*4}	f_X^{*5}	f_X^{*6}	f_X^{*7}	f_X^{*8}	$f_S(x)$
0	1	0	0	0	0	0	0	0	0	.05000
1	0	.150	0	0	0	0	0	0	0	.01500
2	0	.200	.02250	0	0	0	0	0	0	.02338
3	0	.250	.06000	.00338	0	0	0	0	0	.03468
4	0	.125	.11500	.01350	.00051	0	0	0	0	.03258
5	0	.075	.13750	.03488	.00270	.00008	0	0	0	.03579
6	0	.050	.13500	.06144	.00878	.00051	.00001	0	0	.03981
7	0	.050	.10750	.08569	.01999	.00198	.00009	.00000	0	.04356
8	0	.050	.08813	.09750	.03580	.00549	.00042	.00002	.00000	.04752
9	0	.025	.07875	.09841	.05266	.01194	.00136	.00008	.00000	.04903
10	0	.025	.07063	.09338	.06682	.02138	.00345	.00031	.00002	.05190
11	0	0	.06250	.08813	.07597	.03282	.00726	.00091	.00007	.05138
12	0	0	.04500	.08370	.08068	.04450	.01305	.00218	.00022	.05119
13	0	0	.03125	.07673	.08266	.05486	.02062	.00448	.00060	.05030
14	0	0	.01750	.06689	.08278	.06314	.02930	.00808	.00138	.04818
15	0	0	.01125	.05377	.08081	.06934	.03826	.01304	.00279	.04576
16	0	0	.00750	.04125	.07584	.07361	.04677	.01919	.00505	.04281
17	0	0	.00500	.03052	.06811	.07578	.05438	.02616	.00829	.03938
18	0	0	.00313	.02267	.05854	.07552	.06080	.03352	.01254	.03575
19	0	0	.00125	.01673	.04878	.07263	.06573	.04083	.01768	.03197
20	0	0	.00063	.01186	.03977	.06747	.06882	.04775	.02351	.02832
21	0	0	0	.00800	.03187	.06079	.06982	.05389	.02977	.02479
$p_n =$	.05	.10	.15	.20	.25	.15	.06	.03	.01	

Definition 4.1 *Insurance on the aggregate losses, subject to a deductible, is called* **stop-loss insurance**. *The expected cost of this insurance is called the* **net stop-loss premium** *and can be computed as* $E[(S-d)_+]$, *where* d *is the deductible and the notation* $(\cdot)_+$ *means to use the value in parentheses if it is positive, but to use zero otherwise.*

For any aggregate distribution,

$$E[(S-d)_+] = \int_d^\infty [1 - F_S(x)]dx.$$

If the distribution is continuous, the net stop-loss premium can be computed directly from the definition as

$$E[(S-d)_+] = \int_d^\infty (x-d)f_S(x)dx.$$

Similarly, for discrete random variables,

$$E[(S-d)_+] = \sum_{x>d}(x-d)f_S(x).$$

Any time there is an interval with no aggregate probability, the following result may simplify calculations.

Theorem 4.1 *Suppose* $\Pr(a < S < b) = 0$. *Then, for* $a \leq d \leq b$,

$$E[(S-d)_+] = \frac{b-d}{b-a}E[(S-a)_+] + \frac{d-a}{b-a}E[(S-b)_+].$$

That is, the net stop-loss premium can be calculated via linear interpolation.

Proof: From the assumption, $F_S(x) = F_S(a)$, $a \leq x < b$. Then,

$$\begin{aligned} E[(S-d)_+] &= \int_d^\infty [1 - F_S(x)]dx \\ &= \int_a^\infty [1 - F_S(x)]dx - \int_a^d [1 - F_S(x)]dx \\ &= E[(S-a)_+] - \int_a^d [1 - F_S(a)]dx \\ &= E[(S-a)_+] - (d-a)[1 - F_S(a)]. \end{aligned} \tag{4.7}$$

Then, by setting $d = b$ in (4.7),

$$E[(S-b)_+] = E[(S-a)_+] - (b-a)[1 - F_S(a)]$$

and therefore

$$1 - F_S(a) = \frac{E[(S-a)_+] - E[(S-b)_+]}{b-a}.$$

Substituting this in (4.7) produces the desired result. □

Further simplification is available in the discrete case, provided S places probability at equally spaced values.

Theorem 4.2 *Assume* $\Pr(S = kh) = f_k \geq 0$ *for some fixed* $h > 0$ *and* $k = 0, 1, \ldots$ *and* $\Pr(S = x) = 0$ *for all other* x. *Then, for* $d = jh$,

$$E[(S-d)_+] = h \sum_{m=0}^{\infty} \{1 - F_S[(m+j)h]\}.$$

Proof:

$$\begin{aligned} E[(S-d)_+] &= \sum_{x>d} (x-d) f_S(x) dx \\ &= \sum_{k=j}^{\infty} (kh - jh) f_k \\ &= h \sum_{k=j}^{\infty} \sum_{m=0}^{k-j-1} f_k \\ &= h \sum_{m=0}^{\infty} \sum_{k=m+j+1}^{\infty} f_k \\ &= h \sum_{m=0}^{\infty} \{1 - F_S[(m+j)h]\}. \end{aligned}$$

□

In the discrete case with probability at equally spaced values, a simple recursion holds.

Corollary 4.3 *Under the conditions of Theorem 4.2,*

$$E\{[S-(j+1)h]_+\} = E[(S-jh)_+] - h[1 - F_S(jh)].$$

□

This result is easy to use because when $d = 0$, $E[(S-0)_+] = E(S) = E(N)E(X)$, which can be obtained directly from the frequency and severity distributions.

Example 4.5 (Example 4.4 continued) *The insurer is examining the effect of imposing an aggregate deductible per employee. Determine the reduction in the net premium as a result of imposing deductibles of 25, 30, 50, and 100 dollars.*

From Table 4.3, the cdf at 0, 25, 50, and 75 dollars has values 0.05, 0.065, 0.08838, and 0.12306. With $E(S) = 25(12.58) = 314.5$ we have

$$\begin{aligned}
E[(S-25)_+] &= 314.5 - 25(1-0.05) = 290.75 \\
E[(S-50)_+] &= 290.75 - 25(1-0.065) = 267.375 \\
E[(S-75)_+] &= 267.375 - 25(1-0.08838) = 244.5845 \\
E[(S-100)_+] &= 244.5845 - 25(1-0.12306) = 222.661.
\end{aligned}$$

From Theorem 4.1, $E[(S-30)_+] = \frac{20}{25}290.75 + \frac{5}{25}267.375 = 286.07$. When compared to the original premium of 314.5, the reductions are 23.75, 28.43, 47.125, and 91.839 for the four deductibles. □

4.4 ANALYTIC RESULTS

For most choices of distributions of N and the X_js, the compound distributional values can only be obtained numerically. Subsequent sections in this chapter are devoted to such numerical procedures.

However, for certain combinations of choices, simple analytic results are available, thus reducing the computational problems considerably.

Example 4.6 (Compound geometric–exponential) *Suppose $X_1, X_2, \cdots$ are iid with common exponential distribution with mean θ and mgf $M_X(z) = (1-\theta z)^{-1}$ (see Appendix A). Suppose that N has a geometric distribution with parameter β and pgf $P_N(z) = [1-\beta(z-1)]^{-1}$ (see Appendix B). Determine the distribution of S.*

The mgf of S is

$$\begin{aligned}
M_S(z) &= P_N[M_X(z)] \\
&= \{1-\beta[(1-\theta z)^{-1}-1]\}^{-1} \\
&= \frac{1}{1+\beta}1 + \frac{\beta}{1+\beta}[1-\theta(1+\beta)z]^{-1}
\end{aligned}$$

with a bit of algebra.

This is a two-point mixture of a degenerate distribution with probability 1 at zero and an exponential distribution with mean $\theta(1+\beta)$. Hence, the pf of S can be written as

$$f_S(x) = \begin{cases} \dfrac{1}{1+\beta}, & x = 0 \\ \dfrac{\beta}{\theta(1+\beta)^2}\exp\left[-\dfrac{x}{\theta(1+\beta)}\right], & x > 0. \end{cases}$$

It has a point mass of $(1+\beta)^{-1}$ at zero, and an exponentially decaying density over the positive axis. Its cdf can be written as

$$F_S(x) = 1 - \frac{\beta}{1+\beta}\exp\left[-\frac{x}{\theta(1+\beta)}\right], \qquad x \geq 0.$$

It has a jump at zero and is continuous otherwise. This example will arise again in Chapter 6 in connection with ruin theory. □

Example 4.7 (Exponential severities) *Determine the cdf of S for any compound distribution with exponential severities.*

The mgf of the sum of n independent exponential random variables, each with mean θ is

$$M_{X_1+X_2+\cdots+X_n}(z) = (1-\theta z)^{-n}$$

which is the mgf of the gamma distribution with cdf

$$F_X^{*n}(x) = \Gamma(n; x/\theta)$$

(see Appendix A). For integer values of α, the values of $\Gamma(\alpha; x)$ can be calculated exactly (see Appendix A for the derivation) as

$$\Gamma(n; x) = 1 - \sum_{j=0}^{n-1} \frac{x^j e^{-x}}{j!}, \qquad n = 1, 2, 3, \ldots. \tag{4.8}$$

From (4.3)

$$F_S(x) = p_0 + \sum_{n=1}^{\infty} p_n \Gamma(n; x/\theta).$$

Substituting in (4.8) yields

$$F_S(x) = 1 - \sum_{n=1}^{\infty} p_n \sum_{j=0}^{n-1} \frac{(x/\theta)^j e^{-x/\theta}}{j!}, \qquad x \geq 0. \tag{4.9}$$

Interchanging the order of summation yields

$$\begin{aligned} F_S(x) &= 1 - e^{-x/\theta} \sum_{j=0}^{\infty} \frac{(x/\theta)^j}{j!} \sum_{n=j+1}^{\infty} p_n \\ &= 1 - e^{-x/\theta} \sum_{j=0}^{\infty} \bar{P}_j \frac{(x/\theta)^j}{j!}, \qquad x \geq 0 \end{aligned}$$

where $\bar{P}_j = \sum_{n=j+1}^{\infty} p_n$ for $j = 0, 1, \ldots$. □

For frequency distributions which assign probability to all non-negative integers, (4.9) can be evaluated by taking sufficient terms in the first summation. For distributions for which $\Pr(N > n^*) = 0$, the first summation becomes finite. For example, for the binomial frequency distribution, (4.9) becomes

$$F_S(x) = 1 - \sum_{n=1}^{m} \binom{m}{n} q^n (1-q)^{m-n} \sum_{j=0}^{n-1} \frac{(x/\theta)^j e^{-x/\theta}}{j!}. \tag{4.10}$$

Example 4.8 (Compound negative binomial–exponential) *Determine the distribution of S when the frequency distribution is negative binomial with an integer value for the parameter r and the severity distribution is exponential.*

The mgf of S is

$$\begin{aligned} M_S(z) &= P_N[M_X(z)] \\ &= P_N[(1-\theta z)^{-1}] \\ &= \{1 - \beta[(1-\theta z)^{-1} - 1]\}^{-r}. \end{aligned}$$

With a bit of algebra, this can be rewritten as

$$M_S(z) = \left\{1 + \frac{\beta}{1+\beta}[\{1 - \theta(1+\beta)z\}^{-1} - 1]\right\}^r$$

which is of the form

$$M_S(z) = P_N^*(M_X^*(z))$$

where

$$P_N^*(z) = \left\{1 + \frac{\beta}{1+\beta}(z-1)\right\}^r$$

the pgf of the binomial distribution with parameters r and $\beta/(1+\beta)$, and $M_X^*(z)$ is the mgf of the exponential distribution with mean $\theta(1+\beta)$.

This transformation reduces the computation of the distribution function to the finite sum of the form of (4.10)

$$\begin{aligned} F_S(x) &= 1 - \sum_{n=1}^{r} \binom{r}{n} \left(\frac{\beta}{1+\beta}\right)^n \left(\frac{1}{1+\beta}\right)^{r-n} \\ &\quad \times \sum_{j=0}^{n-1} \frac{\{x\theta^{-1}(1+\beta)^{-1}\}^j e^{-x\theta^{-1}(1+\beta)^{-1}}}{j!}. \end{aligned}$$

□

Example 4.9 (Severity distributions closed under convolution) *A distribution is said to be **closed under convolution** when adding iid members of a family produces another member of that family. Further assume that adding n members of a family produces a member with all but one parameter unchanged and the remaining parameter is multiplied by n. Determine the distribution of S when the severity distribution has this property.*

The condition means that if $f_X(x;a)$ is the pf of each X_j, then the pf of $X_1 + X_2 + \cdots + X_n$ is $f_X(x;na)$. This means that

$$\begin{aligned} f_S(x) &= \sum_{n=1}^{\infty} p_n f_X^{*n}(x;a) \\ &= \sum_{n=1}^{\infty} p_n f_X(x;na) \end{aligned}$$

eliminating the need to carry out evaluation of the convolution. Severity distributions that are closed under convolution include the gamma and inverse Gaussian distributions. See Exercise 4.27. □

4.5 COMPUTING THE AGGREGATE CLAIMS DISTRIBUTION

The computation of the compound distribution function

$$F_S(x) = \sum_{n=0}^{\infty} p_n F_X^{*n}(x) \tag{4.11}$$

or the corresponding probability (density) function is generally not an easy task, even in the simplest of cases. In this section we discuss a number of approaches to numerical evaluation of (4.11) for specific choices of the frequency and severity distributions as well as for arbitrary choices of one or both distributions.

One approach is to use an **approximating distribution** to avoid direct calculation of (4.11). This approach was used in Example 4.3 where the method of moments was used to estimate the parameters of the approximating distribution. The advantage of this method is that it is simple and easy to apply. However, the disadvantages are significant. First, there is no way of knowing how good the approximation is. Choosing different approximating distributions can result in very different results, particularly in the right-hand tail of the distribution. Of course, the approximation should improve as more moments are used; but after four moments, one quickly runs out of distributions!

The approximating distribution may also fail to accommodate special features of the true distribution. For example, when the loss distribution is of the continuous type and there is a maximum possible claim (for example, when there is a policy limit), the severity distribution may have a point-mass ("atom" or "spike") at the maximum. The true aggregate claims distribution is of the mixed type with spikes at integral multiples of the maximum corresponding to $1, 2, 3, \ldots$ claims at the maximum. These spikes, if large, can have a significant effect on the probabilities near such multiples. These jumps in the aggregate claims distribution function cannot be replicated by a smooth approximating distribution.

The second method to evaluate (4.11) or the corresponding pdf is **direct calculation**. The most difficult (or computer intensive) part is the evaluation of the n-fold convolutions of the severity distribution for $n = 2, 3, 4, \ldots$. In some situations, there is an analytic form—for example, when the severity distribution is closed under convolution, as defined in Example 4.9 and illustrated in Examples 4.6–4.8. Otherwise the convolutions need to be evaluated numerically using

$$F_X^{*k}(x) = \int_{-\infty}^{\infty} F_X^{*(k-1)}(x-y)dF_X(y). \tag{4.12}$$

When the losses are limited to non-negative values (as is usually the case) the range of integration becomes finite, reducing (4.12) to

$$F_X^{*k}(x) = \int_{0-}^{x} F_X^{*(k-1)}(x-y)dF_X(y). \tag{4.13}$$

These integrals are written in Lebesgue–Stieltjes form because of possible jumps in the cdf $F_X(x)$ at zero and at other points.[1] Numerical evaluation of (4.13) requires numerical integration methods. Because of the first term inside the integral, (4.13) needs to be evaluated for all possible values of x. This quickly becomes technically overpowering!

A simple way to avoid these technical problems is to replace the severity distribution by a discrete distribution defined at multiples $0, 1, 2 \ldots$ of some convenient monetary unit such as 1,000. This reduces (4.13) to (in terms of the new monetary unit)

$$F_X^{*k}(x) = \sum_{y=0}^{x} F_X^{*(k-1)}(x-y)f_X(y).$$

The corresponding pf is

$$f_X^{*k}(x) = \sum_{y=0}^{x} f_X^{*(k-1)}(x-y)f_X(y).$$

In practice, the monetary unit can be made sufficiently small to accommodate spikes at maximum insurance amounts. One needs only the spike to be a multiple of the monetary unit to have it located at exactly the right point. As the monetary unit of measurement becomes small, the discrete distribution function needs to approach the true distribution function. The

[1]Without going into the formal definition of the Lebesgue–Stieltjes integral, it suffices to interpret $\int g(y)dF_X(y)$ as to be evaluated by integrating $g(y)f_X(y)$ over those y-values for which X has a continuous distribution and then adding $g(y_i)\Pr(X = y_i)$ over those points where $\Pr(X = y_i > 0)$. This allows for a single notation to be used for continuous, discrete, and mixed random variables.

simplest approach is to round all amounts to the nearest multiple of the monetary unit; for example, round all losses or claims to the nearest 1,000. More sophisticated methods will be discussed later in this chapter.

When the severity distribution is defined on non-negative integers $0, 1, 2, \ldots$, calculating $f_X^{*k}(x)$ for integral x requires $x+1$ multiplications. Then carrying out these calculations for all possible values of k and x up to n requires a number of multiplications that are of order n^3, written as $O(n^3)$, to obtain the distribution of (4.11) for $x = 0$ to $x = n$. When the maximum value, n, for which the aggregate claims distribution is calculated is large, the number of computations quickly becomes prohibitive, even for fast computers. For example, in real applications n can easily be as large as 1,000. This requires about 10^9 multiplications. Further, if $\Pr(X = 0) > 0$, an infinite number of calculations are required to obtain any single probability. This is because $F_X^{*n}(x) > 0$ for all n and all x and so the sum in (4.11) contains an infinite number of terms. When $\Pr(X = 0) = 0$ we have $F_X^{*n}(x) = 0$ for $n > x$ and so (4.11) will have no more than $x + 1$ positive terms. Table 4.3 provides an example of this latter case.

Alternative methods to more quickly evaluate the aggregate claims distribution are discussed in the next two sections. The first such method, **the recursive method**, reduces the number of computations discussed above to $O(n^2)$, which is a considerable savings in computer time, a reduction of about 99.9% when n = 1,000 compared to direct calculation. However, the method is limited to certain frequency distributions. Fortunately, it includes all frequency distributions discussed in Chapter 3 and Appendix B.

The second such method, **the inversion method**, numerically inverts a transform, such as the characteristic function, using a general or specialized inversion software package. Two versions of this method are discussed in this chapter.

4.6 THE RECURSIVE METHOD

Suppose that the severity distribution $f_X(x)$ is defined on $0, 1, 2, ..., r$ representing multiples of some convenient monetary unit. The number r represents the largest possible payment and could be infinite. Further, suppose that the frequency distribution, p_k, is a member of the $(a, b, 1)$ class and therefore satisfies

$$p_k = \left(a + \frac{b}{k}\right) p_{k-1}, \qquad k = 2, 3, 4, \ldots.$$

Then the following result holds.

Theorem 4.4 *For the model described above,*

$$f_S(x) = \frac{[p_1 - (a+b)p_0] f_X(x) + \sum_{y=1}^{x \wedge r} \left(a + \frac{by}{x}\right) f_X(y) f_S(x-y)}{1 - a f_X(0)}. \qquad (4.14)$$

Proof: This result is identical to Theorem 3.4 with appropriate substitution of notation and recognition that the argument of $f_X(x)$ cannot exceed r. □

Recall from Chapter 2 that $x \wedge r$ is notation for $\min(x, r)$.

Corollary 4.5 *For the $(a, b, 0)$ class, the result (4.14) reduces to*

$$f_S(x) = \frac{\sum_{y=1}^{x \wedge r} \left(a + \frac{by}{x}\right) f_X(y) f_S(x-y)}{1 - a f_X(0)}. \tag{4.15}$$

Note that when the severity distribution has no probability at zero, the denominator of (4.14) and (4.15) equals one. Further, in the case of the Poisson distribution (4.15) reduces to

$$f_S(x) = \frac{\lambda}{x} \sum_{y=1}^{x \wedge r} y f_X(y) f_S(x-y), \qquad x = 1, 2, \ldots. \tag{4.16}$$

The starting value of the recursive schemes (4.14) and (4.15) is $f_S(0) = P_N[f_X(0)]$ following Theorem 3.5 with an appropriate change of notation. In the case of the Poisson distribution we have

$$f_S(0) = e^{-\lambda[1 - f_X(0)]}.$$

Starting values for other frequency distributions are found in Appendix F.

4.6.1 Applications to compound frequency models

When the frequency distribution can be represented as a compound distribution (e.g., Neyman Type A, Poisson-inverse Gaussian) involving only distributions from the $(a, b, 0)$ or $(a, b, 1)$ classes, the recursive formula (4.14) can be used two or more times to obtain the aggregate claims distribution. If the frequency distribution can be written as

$$P_N(z) = P_1[P_2(z)]$$

then the aggregate claims distribution has pgf

$$\begin{aligned} P_S(z) &= P_N[P_X(z)] \\ &= P_1\{P_2[P_X(z)]\} \end{aligned}$$

which can be rewritten as

$$P_S(z) = P_1[P_{S_1}(z)] \tag{4.17}$$

where

$$P_{S_1}(z) = P_2[P_X(z)]. \tag{4.18}$$

Now (4.18) is the same form as an aggregate claims distribution. Thus, if $P_2(z)$ is in the $(a, b, 0)$ or $(a, b, 1)$ class, the distribution of S_1 can be calculated using (4.14). The resulting distribution is the "severity" distribution in equation (4.18). Thus, a second application of (4.14) to (4.17) results in the distribution of S.

This simple idea can be extended to higher levels of compounding, by repeatedly applying the same concepts. The computer time required to carry out two applications will be about twice that of one application of (4.14). However, the total number of computations is still of order $O(n^2)$ rather than $O(n^3)$ in the direct method.

When the severity distribution has a maximum possible value at r, the computations are speeded up even more, since the sum in (4.14) will be restricted to at most r nonzero terms. In this case, then the computations can be considered to be of order $O(n)$.

4.6.2 Underflow/overflow problems

The recursion (4.14) starts with the calculated value of $P(S = 0) = P_N[f_X(0)]$. For large insurance portfolios, this probability is very small, sometimes smaller than the smallest number that can be represented on the computer. When this occurs, this initial value is represented on the computer as zero and the recursion (4.14) fails. This problem can be overcome in several different ways (see Panjer and Willmot [101]). One of the easiest ways is to start with an arbitrary set of values for $f_S(0), f_S(1), ..., f_S(k)$ such as $(0, 0, 0, \ldots, 0, 1)$, where k is sufficiently far to the left in the distribution so that $F_S(k)$ is still negligible. Setting k to a point 6 standard deviations to the left of the mean is usually sufficient. Recursion (4.14) is used to generate values of the distribution with this set of starting values until the values are consistently less than $f_S(k)$. The "probabilities" are then summed and divided by the sum so that the "true" probabilities add to 1. Trial and error will dictate how small k should be for a particular problem.

Another method to obtain probabilities when the starting value is too small is to carry out the calculations for a subset of the portfolio. For example, for the Poisson distribution with mean λ, find a value of $\lambda^* = \lambda/2^n$ so that the probability at zero is representable on the computer when λ^* is used as the Poisson mean. Equation (4.14) is now used to obtain the aggregate claims distribution when λ^* is used as the Poisson mean. If $P_*(z)$ is the pgf of the aggregate claims using Poisson mean λ^*, then $P_S(z) = [P_*(z)]^{2^n}$. Hence one can obtain successively the distributions with pgfs $[P_*(z)]^2$, $[P_*(z)]^4$, $[P_*(z)]^8, ..., [P_*(z)]^{2^n}$ by convoluting the result at each stage with itself. This requires an additional n convolutions in carrying out the calculations, but involves no approximations. This procedure can be carried out for any frequency distributions that are closed under convolution. For the negative binomial distribution, the analogous procedure starts with $r^* = r/2^n$. For the binomial distribution, the parameter m must be integer valued. A slight modification can be used. Let $m^* = [m/2^n]$ when $[\cdot]$ indicates the "integer part of" function. When the n convolutions are carried out, one still needs to carry out the calculations using (4.14) for parameter $m - m^*2^n$. This result is then convoluted with the result of the n convolutions. For compound frequency distributions, only the primary distribution needs to be closed under convolutions.

4.6.3 Numerical stability

Any recursive formula requires accurate computation of values since each such value will be used in computing subsequent values. Recursive schemes suffer the risk of errors propagating through all subsequent values and potentially "blowing up." In the recursive formula (4.14), errors are introduced through rounding or truncation at each stage since computers represent numbers with a finite number of significant digits. The question about stability is, *"How fast do the errors in the calculations grow as the computed values are used in successive computations?"*

The question of error propagation in recursive formulas has been a subject of study of numerical analysts. This work has been extended by Panjer and Wang [100] to study the recursive formula (4.14). The analysis is quite complicated and well beyond the scope of this book. However, some general conclusions can be made here.

Errors are introduced in subsequent values through the summation

$$\sum_{y=1}^{x} \left(a + \frac{by}{x}\right) f_X(y) f_S(x-y)$$

in recursion (4.14). In the extreme right-hand tail of the distribution of S, this sum is positive (or at least non-negative), and subsequent values of the sum will be decreasing. The sum will stay positive, even with rounding errors, when each of the three factors in each term in the sum is positive. In this case, the recursive formula is stable, producing relative errors that do not grow fast. For the Poisson and negative binomial based distributions, the factors in each term are always positive.

On the other hand, for the binomial distribution, the sum can have negative terms, since a is negative, b is positive, and y/x is a positive function not exceeding 1. In this case, the negative terms can cause the successive values to "blow up" with alternating signs. When this occurs, the nonsensical results are immediately obvious. Although this does not happen frequently in practice, the reader should be aware of this possibility in models based on the binomial distribution.

4.6.4 Continuous severity

The recursive method has been developed for discrete severity distributions, while in Chapter 2, continuous distributions are proposed for severity. In the case of continuous severities, the analog of the recursion (4.14) is an integral equation, the solution of which is the aggregate claims distribution.

Theorem 4.6 *For the $(a, b, 1)$ class of frequency distributions and any continuous severity distribution with probability on the positive real line the following integral equation holds:*

$$f_S(x) = p_1 f_X(x) + \int_0^x \left(a + \frac{by}{x}\right) f_X(y) f_S(x-y)dy \tag{4.19}$$

The proof of this result is beyond the scope of this book. For a detailed proof see Theorems 6.14.1 and 6.16.1 of Panjer and Willmot [102], along with the associated corollaries. They consider the more general (a, b, m) class of distributions, which allow for arbitrary modification of m initial values of the distribution. Note that the initial term is $p_1 f_X(x)$, not $[p_1 - (a+b)p_0]f_X(x)$ as in (4.14). Also, (4.19) holds for members of the $(a, b, 0)$ class as well.

Integral equations of the form (4.19) are Volterra integral equations of the second kind. Numerical solution of this type of integral equation has been studied in the text by Baker [9]. We will develop a method using a discrete approximation of the severity distribution in order to use the recursive method (4.14) and avoid the more complicated methods of Baker [9].

4.6.5 Constructing arithmetic distributions

In order to implement recursive methods, the easiest approach is to construct a discrete severity distribution on multiples of a convenient unit of measurement h, the **span**. Such a distribution is called arithmetic since it is defined on the non-negative integers. In order to "arithmetize" a distribution, it is important to preserve the properties of the original distribution both locally through the range of the distribution and globally—that is, for the entire distribution. This should preserve the general shape of the distribution and at the same time preserve global quantities such as moments.

The methods suggested here apply to the discretization ("arithmetization") of continuous, mixed, and nonarithmetic discrete distributions.

4.6.5.1 Method of rounding (mass dispersal) Let f_j denote the probability placed at jh, $j = 0, 1, 2, \ldots$. Then set[2]

$$f_0 = \Pr(X < h/2) = F_X\left(\frac{h}{2} - 0\right)$$

[2] The notation $F_X(x - 0)$ indicates that discrete probability at x should not be included. For continuous distributions this will make no difference.

$$\begin{aligned} f_j &= \Pr\left(jh - \frac{h}{2} \le X < jh + \frac{h}{2}\right) \\ &= F_X\left(jh + \frac{h}{2} - 0\right) - F_X\left(jh - \frac{h}{2} - 0\right), \qquad j = 1, 2, \ldots . \end{aligned}$$

This method splits the probability between $(j+1)h$ and jh and "assigns" it to $j+1$ and j. This, in effect, rounds all amounts to the nearest convenient monetary unit, h, the span of the distribution.

4.6.5.2 Method of local moment matching In this method we construct an arithmetic distribution that matches p moments of the arithmetic and the true severity distributions. Consider an arbitrary interval of length ph, denoted by $[x_k, x_k + ph)$. We will locate point masses $m_0^k, m_1^k, \cdots, m_p^k$ at points $x_k, x_k + h, \cdots, x_k + ph$ so that the first p moments are preserved. The system of $p+1$ equations reflecting these conditions is

$$\sum_{j=0}^{p} (x_k + jh)^r m_j^k = \int_{x_k - 0}^{x_k + ph - 0} x^r \, dF_X(x), \qquad r = 0, 1, 2, ..., p \tag{4.20}$$

where the notation "-0" at the limits of the integral are to indicate that discrete probability at x_k is to be included, but discrete probability at $x_k + ph$ is to be excluded.

Arrange the intervals so that $x_{k+1} = x_k + ph$, and so the endpoints coincide. Then the point masses at the endpoints are added together. With $x_0 = 0$, the resulting discrete distribution has successive probabilities:

$$\begin{aligned} f_0 &= m_0^0, \; f_1 = m_1^0, \; f_2 = m_2^0, \ldots, \\ f_p &= m_p^0 + m_0^1, \; f_{p+1} = m_1^1, \; f_{p+2} = m_2^1, \ldots . \end{aligned} \tag{4.21}$$

By summing (4.20) for all possible values of k, with $x_0 = 0$, it is clear that p moments are preserved for the entire distribution and that the probabilities add to one exactly. The only remaining point is to solve the system of equations (4.20).

Theorem 4.7 *The solution of (4.20) is*

$$m_j^k = \int_{x_k - 0}^{x_k + ph - 0} \prod_{i \ne j} \frac{x - x_k - ih}{(j - i)h} \, dF_X(x), \qquad j = 0, 1, \ldots, p. \tag{4.22}$$

Proof: The Lagrange formula for collocation of a polynomial $f(y)$ at points $y_0, y_1, ..., y_n$ is

$$f(y) = \sum_{j=0}^{n} f(y_j) \prod_{i \ne j} \frac{y - y_i}{y_j - y_i}.$$

Applying this formula to the polynomial $f(y) = y^r$ over the points x_k, $x_k + h, ..., x_k + ph$ yields

$$x^r = \sum_{j=0}^{p} (x_k + jh)^r \prod_{i \neq j} \frac{(x - x_k - ih)}{(j - i)h}, \qquad r = 0, 1, \ldots p.$$

Integrating over the interval $[x_k, x_k + ph)$ with respect to the severity distribution results in

$$\int_{x_k - 0}^{x_k + ph - 0} x^r dF_X(x) = \sum_{j=0}^{p} (x_k + jh)^r m_j^k$$

where m_j^k is given by (4.22). Hence, the solution (4.22) preserves the first p moments, as required. □

Example 4.10 *Suppose X has the exponential distribution with pdf $f(x) = 0.1e^{-0.1x}$. Use a span of $h = 2$ to discretize this distribution by the method of rounding and by matching the first moment.*

For the method of rounding, the general formulas are

$$\begin{aligned} f_0 &= F(1) = 1 - e^{-0.1(1)} = 0.09516 \\ f_j &= F(2j + 1) - F(2j - 1) = e^{-0.1(2j-1)} - e^{-0.1(2j+1)}. \end{aligned}$$

The first few values are given in Table 4.4.

For matching the first moment we have $p = 1$ and $x_k = 2k$. The key equations become

$$\begin{aligned} m_0^k &= \int_{2k}^{2k+2} \frac{x - 2k - 2}{-2} (0.1)e^{-0.1x} dx = 5e^{-0.1(2k+2)} - 4e^{-0.1(2k)} \\ m_1^k &= \int_{2k}^{2k+2} \frac{x - 2k}{2} (0.1)e^{-0.1x} dx = -6e^{-0.1(2k+2)} + 5e^{-0.1(2k)} \end{aligned}$$

and then

$$\begin{aligned} f_0 &= m_0^0 = 5e^{-0.2} - 4 = 0.09365 \\ f_j &= m_1^{j-1} + m_0^j = 5e^{-0.1(2j-2)} - 10e^{-0.1(2j)} + 5e^{-0.1(2j+2)}. \end{aligned}$$

The first few values also are given in Table 4.4. A more direct solution for matching the first moment is provided in Exercise 4.37. □

This method of local moment matching was introduced by Gerber and Jones [50] and Gerber[49] and studied by Panjer and Lutek [99] for a variety of empirical an analytical severity distributions. In assessing the impact of errors on aggregate stop-loss net premiums (aggregate excess-of-loss pure

Table 4.4 Discretization of the exponential distribution by two methods

j	f_j-rounding	f_j-matching
0	0.09516	0.09365
1	0.16402	0.16429
2	0.13429	0.13451
3	0.10995	0.11013
4	0.09002	0.09017
5	0.07370	0.07382
6	0.06034	0.06044
7	0.04940	0.04948
8	0.04045	0.04051
9	0.03311	0.03317
10	0.02711	0.02716

premiums), Panjer and Lutek [99] found that two moments were usually sufficient and that adding a third moment requirement adds only marginally to the accuracy. Furthermore, the rounding method and the first moment method ($p = 1$) had similar errors while the second moment method ($p = 2$) provided significant improvement. The specific formulas for the method of rounding and the method of matching the first moment are given in Appendix G.

The methods described here are qualitatively similar to numerical methods used to solve Volterra integral equations such as (4.19) developed in numerical analysis (see, for example, Baker [9]).

4.7 INVERSION METHODS

Inversion methods discussed in this section are used to obtain numerically the probability function, or some related function such as a net stop-loss premium (aggregate excess-of-loss pure premium), from a known expression for a transform, such as the pgf, mgf or cf of the desired function.

Compound distributions lend themselves naturally to this approach since their transforms are compound functions and are easily evaluated when both frequency and severity components are known. The pgf and cf of the aggregate loss distribution are

$$P_S(z) = P_N[P_X(z)]$$

and

$$\varphi_S(z) = E[e^{iSz}] = P_N[\varphi_X(z)] \tag{4.23}$$

respectively. The characteristic function always exists and is unique. Conversely, for a given characteristic function, there always exists a unique distribution. The objective of inversion methods is to obtain the distribution numerically from the characteristic function (4.23).

4.7.1 Fast Fourier transform

The fast Fourier transform (FFT) is an algorithm that can be used for inverting characteristic functions to obtain densities of discrete random variables. The FFT comes from the field of signal processing. It was first used for the inversion of characteristic functions of compound distributions by Bertram [13] and is explained in detail with applications to aggregate loss calculation by Robertson [107].

Definition 4.2 *For any continuous function $f(x)$, the* ***Fourier transform*** *is the mapping*

$$\tilde{f}(z) = \int_{-\infty}^{\infty} f(x)e^{izx}dx. \tag{4.24}$$

The original function can be recovered from its Fourier transform as

$$f(x) = \frac{1}{2\pi}\int_{-\infty}^{\infty} \tilde{f}(z)e^{-izx}dz.$$

When $f(x)$ is a probability density function, then $\tilde{f}(z)$ is its characteristic function. For our applications, $f(x)$ will be real-valued. From (4.24), $\tilde{f}(z)$ is complex-valued. When $f(x)$ is a probability function of a discrete (or mixed) distribution, the definitions can be easily generalized (see, for example, Fisz [43]).

Definition 4.3 *Let f_x denote a function defined for all integer values of x that is periodic with period length n (that is, $f_{x+n} = f_x$ for all x). For the vector $(f_0, f_1, ..., f_{n-1})$, the* ***discrete Fourier transform*** *is the mapping $\tilde{f}_x$, $x = \ldots -1, 0, 1, \ldots$, defined by*

$$\tilde{f}_k = \sum_{j=0}^{n-1} f_j \exp\left(\frac{2\pi i}{n}jk\right), \qquad k = \ldots -1, 0, 1, \ldots. \tag{4.25}$$

This mapping is bijective. In addition, $\tilde{f}_k$ is also periodic with period length n. The inverse mapping is

$$f_j = \frac{1}{n}\sum_{k=0}^{n-1} \tilde{f}_k \exp\left(-\frac{2\pi i}{n}kj\right), \qquad j = \ldots -1, 0, 1, \ldots. \tag{4.26}$$

This inverse mapping recovers the values of the original function.

Because of the periodic nature of f and $\tilde{f}$, we can think of the discrete Fourier transform as a bijective mapping of n points into n points. From

(4.25), it is clear that, in order to obtain n values of $\tilde{f}_k$, the number of terms that need to be evaluated is of order n^2, that is, $O(n^2)$.

The **fast Fourier transform (FFT)** is an algorithm that reduces the number of computations required to be of order $O(n \log_2 n)$. This can be a dramatic reduction in computations when n is large. The algorithm exploits the property that a discrete Fourier transform of length n can be rewritten as the sum of two discrete transforms, each of length $n/2$, the first consisting of the even-numbered points and the second consisting of the odd-numbered points.

$$\begin{aligned} \tilde{f}_k &= \sum_{j=0}^{n-1} f_j \exp\left(\frac{2\pi i}{n} jk\right) \\ &= \sum_{j=0}^{n/2-1} f_{2j} \exp\left(\frac{2\pi i}{n} 2jk\right) + \sum_{j=0}^{n/2-1} f_{2j+1} \exp\left[\frac{2\pi i}{n}(2j+1)k\right] \\ &= \sum_{j=0}^{m-1} f_{2j} \exp\left(\frac{2\pi i}{m} jk\right) + \exp\left(\frac{2\pi i}{n} k\right) \sum_{j=0}^{m-1} f_{2j+1} \exp\left(\frac{2\pi i}{m} jk\right) \end{aligned}$$

when $m = n/2$. Hence

$$\tilde{f}_k = \tilde{f}_k^a + \exp\left(\frac{2\pi i}{n} k\right) \tilde{f}_k^b. \tag{4.27}$$

These can, in turn, be written as the sum of two transforms of length $m/2$. This can be continued successively. For the lengths $n/2$, $m/2$, etc., to be integers the FFT algorithm begins with a vector of length $n = 2^r$. The successive writing of the transforms into transforms of half the length will result, after r times, in transforms of length 1. Knowing the transform of length 1 will allow one to successively compose the transforms of length 2, 2^2, $2^3, ..., 2^r$ by simple addition using (4.27). Details of the methodology are found in Press et al. [103].

In our applications, we use the FFT to invert the characteristic function when discretization of the severity distribution is done. This is carried out as follows:

1. Discretize the severity distribution using some methods such as those described in the previous section, obtaining the discretized severity distribution
$$f_X(0), f_X(1), ..., f_X(n-1)$$
where $n = 2^r$ for some integer r and n is the number of points desired in the distribution $f_S(x)$ of aggregate claims.

2. Apply the FFT to this vector of values, obtaining $\varphi_X(z)$, the characteristic function of the **discretized** distribution. The result is also a vector of $n = 2^r$ values.

3. Transform this vector using the pgf transformation of the claim frequency distribution, obtaining $\varphi_S(z) = P_N[\varphi_X(z)]$, which is the characteristic function, that is, the discrete Fourier transform of the aggregate claims distribution, a vector of $n = 2^r$ values.

4. Apply the inverse fast Fourier transform (IFFT), which is identical to the FFT except for a sign change and a division by n (see (4.26)). This gives a vector of length $n = 2^r$ values representing the exact distribution of aggregate claims for the discretized severity model.

The FFT procedure requires a discretization of the severity distribution. When the number of points in the severity distribution is less than $n = 2^r$, the severity distribution vector must be "padded" with zeros until it is of length n.

When the severity distribution places probability on values beyond $x = n$, as is the case with most distributions discussed in Chapter 2, the probability that is missed in the right-hand tail beyond n can introduce some minor error in the final solution since the function and its transform are both assumed to be periodic with period n, when in reality they are not. The authors suggest putting all the remaining probability at the final point at $x = n$ so that the probabilities add up to one exactly. This allows for periodicity to be used for the severity distribution in the FFT algorithm and ensures that the final set of aggregate probabilities will sum to one. However, it is imperative that n be selected to be large enough so that most all the aggregate probability occurs by the nth point. The following example provides an extreme illustration.

Example 4.11 *Suppose the random variable X takes on the values 1, 2, and 3 with probabilities 0.5, 0.4, and 0.1, respectively. Further suppose the number of claims has the Poisson distribution with parameter $\lambda = 3$. Use the FFT to obtain the distribution of S, using $n = 8$ and $n =$ 4,096.*

In either case, the probability distribution of X is completed by adding one zero at the beginning (because S places probability at zero, the initial representation of X must also have the probability at zero given) and either 4 or 4,092 zeros at the end. The results from employing the FFT and IFFT appear in Table 4.5. For the case $n = 8$ the eight probabilities sum to 1. For the case $n =$ 4,096, those probabilities also sum to 1, but there is not room here to show them all. It is easy to apply the recursive formula to this problem, which verifies that all of the entries for $n =$ 4,096 are accurate to the five decimal places presented. On the other hand, with $n = 8$, the FFT gives values that are clearly distorted. If any generalization can be made it is that more of the extra probability has been added to the smaller values of S. □

Since the FFT and IFFT algorithms are available in many computer software packages and because the computer code is short, easy to write, and

Table 4.5 Aggregate probabilities computed by the FFT and IFFT

s	$n = 8$ $f_S(s)$	$n = 4{,}096$ $f_S(s)$
0	0.11227	0.04979
1	0.11821	0.07468
2	0.14470	0.11575
3	0.15100	0.13256
4	0.14727	0.13597
5	0.13194	0.12525
6	0.10941	0.10558
7	0.08518	0.08305

available (e.g., [103, pp. 411–412]), no further technical details about the algorithm are given here. The reader can read any one of numerous books dealing with FFTs for a more detailed understanding of the algorithm. The technical details which allow the speeding up of the calculations from $O(n^2)$ to $O(n \log_2 n)$ relate to the detailed properties of the discrete Fourier transform. Robertson [107] gives a good explanation of the FFT as applied to calculating the distribution of aggregate claims.

4.7.2 Direct numerical inversion

The inversion of the characteristic function (4.23) has been done using approximate integration methods by Heckman and Meyers [58] in the case of Poisson, binomial and negative binomial claim frequencies and continuous severity distributions. The method is easily extended to other frequency distributions.

In this method, the severity distribution function is replaced by a piecewise linear distribution. It further uses a maximum single loss amount so the cdf jumps to one at the maximum possible individual loss. The range of the severity random variable is divided into intervals of possibly unequal length. The remaining steps parallel those of the FFT method. Consider the cdf of the severity distribution $F_X(x)$, $0 \le x < \infty$. Let $0 = x_0 < x_1 < \cdots < x_n$ be arbitrarily selected loss values. Then the probability that losses lie in the interval $(x_{k-1}, x_k]$ is given by $f_k = F_X(x_k) - F_X(x_{k-1})$. Using a uniform density d_k over this interval results in the approximating density function $f^*(x) = d_k = f_k/(x_k - x_{k-1})$ for $x_{k-1} < x \le x_k$. Any remaining probability, $f_{n+1} = 1 - F_X(x_n)$ is placed as a spike at x_n. This approximating pdf is selected to make evaluation of the cf easy. It is not required for direct inversion. The cf of the approximating severity distribution is

$$\begin{aligned}
\varphi_X(z) &= \int_0^\infty e^{izx} dF_X(x) \\
&= \sum_{k=1}^n \int_{x_{k-1}}^{x_k} d_k e^{izx} dx + f_{n+1} e^{izx_n} \\
&= \sum_{k=1}^n d_k \frac{e^{izx_k} - e^{izx_{k-1}}}{iz} + f_{n+1} e^{izx_n}.
\end{aligned}$$

The cf can be separated into real and imaginary parts by using Euler's formula

$$e^{i\theta} = \cos(\theta) + i\sin(\theta).$$

Then the real part of cf is

$$\begin{aligned}
a(z) = \mathrm{Re}[\varphi_X(z)] &= \frac{1}{z}\sum_{k=1}^n d_k[\sin(zx_k) - \sin(zx_{k-1})] \\
&\quad + f_{n+1}\cos(zx_n)
\end{aligned}$$

and the imaginary part is

$$\begin{aligned}
b(z) = \mathrm{Im}[\varphi_X(z)] &= \frac{1}{z}\sum_{k=1}^n d_k[\cos(zx_{k-1}) - \cos(zx_k)] \\
&\quad + f_{n+1}\sin(zx_n).
\end{aligned}$$

The cf of aggregate losses (4.23) is obtained as

$$\varphi_S(z) = P_N[\varphi_X(z)] = P_N[a(z) + ib(z)]$$

which can be rewritten as

$$\varphi_S(z) = r(z)e^{i\theta(z)}$$

since it is complex valued.

The distribution of aggregate claims is obtained as

$$F_S(x) = \frac{1}{2} + \frac{1}{\pi}\int_0^\infty \frac{r(z/\sigma)}{z} \sin\left[\frac{zx}{\sigma} - \theta(z/\sigma)\right] dz \tag{4.28}$$

where σ is the standard deviation of the distribution of aggregate losses. Approximate integration techniques are used to evaluate (4.28) for any value of x. The reader is referred to Heckman and Meyers [58] for details. They also obtain the net stop-loss (excess pure) premium for the aggregate loss distribution as

$$\begin{aligned}
P(d) &= E[(S-d)_+] = \int_d^\infty (s-d)dF_S(s) \\
&= \mu - \frac{d}{2} \\
&\quad + \frac{\sigma}{\pi}\int_0^\infty \frac{r(z/\sigma)}{z^2}\left\{\cos[\theta(z/\sigma)] - \cos\left[\frac{zd}{\sigma} - \theta(z/\sigma)\right]\right\} dz \quad (4.29)
\end{aligned}$$

from (4.28) where μ is the mean of the aggregate loss distribution and d is the deductible.

Equation (4.28) provides only a single value of the distribution, while (4.29) provides only one value of the premium, but it does so quickly. The error of approximation depends on the spacing of the numerical integration method, but is controllable.

4.8 SIMULATION

The analytic methods presented in the previous two sections have two features in common. First, they are exact up to the level of the approximation introduced. For recursion and the FFT, that involves replacing the true severity distribution with an arithmetized approximation. For Heckman–Meyers a histogram approximation is required. Furthermore, Heckman–Meyers requires a numerical integration. In each case, the errors can be reduced to near zero by increasing the number of points used. Second, both recursion and inversion assume that aggregate claims can be written $S = X_1 + \cdots + X_N$ with $N, X_1, X_2, \ldots$ independent and the X_js identically distributed.

There is no need to be concerned about the first feature because the approximation error can be made as small as desired. However, the second restriction may prevent the model from reflecting reality. In this section we indicate some common ways in which the independence or identical distribution assumptions may fail to hold. We then introduce simulation, a method that does not require these assumptions.

Before doing this we must introduce some concepts that will be explained more fully in Chapter 6. When the X_js are iid it does not matter how we go about labeling the losses—that is, which loss is called X_1, which one X_2, and so on. With the assumption removed, the labels become important. Because S is the aggregate loss for one year, time is a factor. One way of identifying the losses is to let X_1 be the first loss, X_2 be the second loss, and so on. Then let T_j be the random variable that records the time of the jth loss. Without going into much detail about the claims paying process, we do want to note that T_j may be the time at which the loss occurred, the time it was reported, or the time payment was made. In the latter two cases it may be that $T_j > 1$, which occurs when the report of the loss or the payment of the claim takes place at a time subsequent to the end of the time period of the coverage, usually one year. If the timing of the losses is important, we will need to know the joint distribution of $(T_1, T_2, \ldots, X_1, X_2, \ldots)$.

4.8.1 Examples of lack of independence or identical distributions

There are two common ways to have the assumption fail to hold. One is through accounting for time (and in particular the time value of money) and the other is through coverage modifications. The latter may have a time factor as well. The following examples provide some illustrations.

Example 4.12 (Time value of loss payments) *Suppose the quantity of interest, S, is the present value of all payments made in respect of a policy issued today and covering loss events that occur in the next year. Develop a model for S and indicate where the assumptions are invalid.*

Let T_j be the time of the payment of the jth loss. While T_j records the time of the payment, the subscripts are selected in order of the loss event. Let $T_j = C_j + L_j$ where C_j is the time of the event and L_j is the time from occurrence to payment. Assume they are independent and the L_js are independent of each other. Let the time between events, $C_j - C_{j-1}$ (where $C_0 = 0$), be iid with an exponential distribution with mean 0.2 years.

Let X_j be the amount paid at time T_j on the loss that occurred at time C_j. Assume that X_j and C_j are independent (the amount of the claim does not depend on when in the year it occurred) but X_j and L_j are positively correlated (a specific distributional model will be specified when the example is continued). This is reasonable because the more expensive losses may take longer to settle.

Finally, let V_t be a random variable that represents the value, which, if invested today, will accumulate to one in t years. It is independent of all X_j, C_j, and L_j. But clearly, for $s \neq t$, V_s and V_t are dependent.

We then have

$$S = \sum_{j=1}^{N} X_j V_{T_j}.$$

The various dependencies were established in the development of the random variables. □

Example 4.13 (Out-of-pocket maximum) *Suppose there is a deductible, d, on individual losses. However, in the course of a year, the policyholder will pay no more than u. Develop a model for the insurer's aggregate payments and indicate where the assumptions fail to hold.*

Let X_j be the amount of the jth loss. Here the assignment of j does not matter. Let $W_j = X_j \wedge d$ be the amount paid by the policyholder due to the deductible and let $Y_j = X_j - W_j$ be the amount paid by the insurer. Then $R = W_1 + \cdots + W_N$ is the total amount paid by the policyholder prior to imposing the out-of-pocket maximum. Then the amount actually paid by the policyholder is $R_u = R \wedge u$. Let $S = X_1 + \cdots + X_N$ be the total losses, and then the aggregate amount paid by the insurer is $T = S - R_u$. Note that the distributions of S and R_u are based on iid severity distributions. The analytic methods described earlier can be used to obtain their distributions. But because they are dependent, their individual distributions cannot be combined to produce the distribution of T. There is also no way to write T as a random sum of iid variables. At the beginning of the year, it appears that T will be the sum of iid Y_js, but at some point the Y_js may be replaced by X_js as the out-of-pocket maximum is reached. □

4.8.2 The simulation approach

To paraphrase a line associated with the home state of one of the authors, "If you build (a model of) it, he (the answer) will come." The beauty of simulation is that once the model is created, the aggregate distribution will follow with no additional creative thought required.[3] The entire process can be summarized in three steps, where the goal is to determine the cdf of some random variable S.

1. Build a model for S which depends on random variables $X, Y, Z, \ldots$, where their distributions and any dependencies are known.

2. For $i = 1, \ldots, n$ generate pseudorandom values $x_i, y_i, z_i, \ldots$ and then compute s_i using the model from step 1.

3. The cdf of S may be approximated by $F_{\hat{S}}(s)$, the empirical cdf based on the pseudorandom sample $s_1, \ldots, s_n$.

Two questions remain. First, what does it mean to generate a pseudorandom variable? Consider a random variable X with cdf $F_X(x)$. This is the real random variable produced by some phenomenon of interest. For example, it may be the result of the experiment "collect one automobile bodily injury medical payment at random and record its value." We assume that the cdf is known. For example, it may be the Pareto cdf, $F_X(x) = 1 - \left(\frac{1{,}000}{1{,}000+x}\right)^3$. Now consider a second random variable, X^*, resulting from some other process, but with the same Pareto distribution. A random sample from X^*, say $x_1^*, \ldots, x_n^*$, would be impossible to distinguish from one taken from X. That is, given the n numbers, we could not tell if they arose from automobile claims or something else. This means that instead of learning about X by observing automobile claims, we could learn about it by observing X^*. Obtaining a random sample from a Pareto distribution is still probably difficult, so we have not yet accomplished much.

We can make some progress by making a concession. Let us accept as a replacement for a random sample from X^*, a sequence of numbers $x_1^{**}, \ldots, x_n^{**}$ which is not a random sample at all, but simply a sequence of numbers which may not be independent, or even random, but were generated by some known process that is related to the random variable X^*. Such a sequence is called

[3] This is not entirely true. A great deal of creativity may be employed in designing an efficient simulation. The brute force approach used here will work, it just may take your computer longer to produce the answer.

a pseudorandom sequence since anyone who did not know their origin could not distinguish them from a random sample from X^* (and therefore from X). Such a sequence will be satisfactory for our purposes.

The field of developing processes for generating pseudorandom sequences of numbers has been well-developed. One fact that makes it easier to do this is that it is sufficient to be able to generate such sequences for the uniform distribution on the interval $(0,1)$. That is because if U has the uniform$(0,1)$ distribution, then $X = F_X^{-1}(U)$ will have $F_X(x)$ as its cdf. Therefore, we simply obtain uniform pseudorandom numbers $u_1^{**}, \ldots, u_n^{**}$ and then let $x_j^{**} = F_X^{-1}(u_j^{**})$. A very brief discussion of a common method of generating pseudouniform numbers is given in Appendix H along with tricks for the particular case of generating Poisson and normal (and therefore lognormal) pseudorandom variables.

Example 4.14 *Generate 10,000 pseudo-Pareto (with $\alpha = 3$, and $\theta = 1{,}000$) variates and verify that they are indistinguishable from real Pareto observations.*

The pseudouniform values were obtained using the built-in generator supplied with a commercial programming language.[4] The pseudo-Pareto values are calculated from

$$u^{**} = 1 - \left(\frac{1{,}000}{1{,}000 + x^{**}}\right)^3.$$

That is,

$$x^{**} = 1{,}000[(1 - u^{**})^{-1/3} - 1].$$

So, for example, if the first value generated is $u_1^{**} = 0.54246$, we have $x_1^{**} = 297.75$. This was repeated 10,000 times. The results are displayed in Table 4.6 where a chi-square goodness-of-fit test is conducted. The expected counts use the Pareto distribution with $\alpha = 3$ and $\theta = 1{,}000$. Because the parameters are known, there are 9 degrees of freedom. At a significance level of 5% the critical value is 16.92 and we conclude that the pseudorandom sample could have been a random sample from this Pareto distribution. □

The second question is: What value of n should be used? We know that any consistent estimator will be arbitrarily close to the true value with high probability as the sample size is increased. In particular, empirical estimators have this attribute. With a little effort we should be able to determine the value of n that will get us as close as we want with a specified probability. Often, the central limit theorem will help, as in the following example.

Example 4.15 (Example 4.14 continued) *Use simulation to estimate the mean, $F_X(1{,}000)$ and $\pi_{0.9}$, the 90th percentile of the Pareto distribution with*

[4] Not all pseudouniform generators are created equal. Verify that yours is a good one before using it.

Table 4.6 Chi-square test of simulated Pareto observations

Interval	Observed	Expected	Chi-square
0–100	2,519	2,486.85	0.42
100–250	2,348	2,393.15	0.85
250–500	2,196	2,157.04	0.70
500–750	1,071	1,097.07	0.62
750–1,000	635	615.89	0.59
1,000–1,500	589	610.00	0.72
1,500–2,500	409	406.76	0.01
2,500–5,000	192	186.94	0.14
5,000–10,000	36	38.78	0.20
10,000–	5	7.51	0.84
Total	10,000	10,000	5.10

$\alpha = 3$ *and* $\theta = 1{,}000$. *In each case, stop the simulations when you are 95% confident that the answer is within* $\pm 1\%$ *of the true value.*

In this example we know the values. Here, $\mu = 500$, $F_X(1{,}000) = 0.875$, and $\pi_{.9} = 1{,}154.43$. For instructional purposes we will behave as if we do not know these values.

Begin by using $\bar{x}$ to estimate μ. The central limit theorem tells us that for a sample of size n,

$$\begin{aligned} 0.95 &= \Pr(0.99\mu \le \bar{X}_n \le 1.01\mu) \\ &= \Pr\left(-\frac{0.01\mu}{\sigma/\sqrt{n}} \le \frac{\bar{X}_n - \mu}{\sigma/\sqrt{n}} \le \frac{0.01\mu}{\sigma/\sqrt{n}}\right) \\ &\doteq \Pr\left(-\frac{0.01\mu}{\sigma/\sqrt{n}} \le Z \le \frac{0.01\mu}{\sigma/\sqrt{n}}\right) \end{aligned}$$

where Z has the standard normal distribution. Our goal is achieved when

$$\frac{0.01\mu}{\sigma/\sqrt{n}} = 1.96$$

which means $n = 38{,}416(\sigma/\mu)^2$. Because we do not know the values of σ and μ, we estimate them with the sample standard deviation and mean. The estimates improve with n, so our stopping rule is to cease simulating when

$$n \ge \frac{38{,}416 s^2}{\bar{x}^2}.$$

For a particular simulation conducted by the authors, the criterion was met when $n = 106{,}934$ at which point $\bar{x} = 501.15$, a relative error of 0.23%, well within our goal.

We now turn to the estimation of $F_X(1{,}000)$. The empirical estimator is the sample proportion below 1,000, say P_n/n where P_n is the number below 1,000 after n simulations. The central limit theorem tells us that P_n/n is approximately normal with mean $F_X(1{,}000)$ and variance $F_X(1{,}000)[1 - F_X(1{,}000)]/n$. Arguing as above, the requirement will be met when

$$n \geq 38{,}416\frac{n - P_n}{P_n}.$$

For our simulation, the criterion was met at $n = 5{,}548$,at which point the estimate was $4{,}848/5{,}548 = 0.87383$, which has a relative error of 0.13%.

Finally, for $\pi_{0.9}$, we can use the nonparametric confidence interval developed in Subsection 2.2.3. Here the 95% confidence interval is determined from

$$0.95 = \Pr(Y_a \leq \pi_{0.9} \leq Y_b)$$

where $Y_1 \leq Y_2 \leq \cdots \leq Y_n$ are the order statistics from the simulated sample, $a = [0.9n - 1.96\sqrt{0.9(0.1)n}]$, $b = [0.9n + 1.96\sqrt{0.9(0.1)n}] + 1$ (where $[\cdot]$ is the greatest integer function and the one is added for conservatism), and the process terminates when both

$$\hat{\pi}_{0.9} - Y_a \leq 0.01\hat{\pi}_{0.9}$$

and

$$Y_b - \hat{\pi}_{0.9} \leq 0.01\hat{\pi}_{0.9}.$$

For the example, this occurred when $n = 126{,}364$, and the estimated 90th percentile is 1,153.97, a relative error of 0.04%. □

4.8.3 Two examples

We close by completing the two examples using the simulation approach. The models have been selected arbitrarily, but we should assume they were determined by a careful estimation process using the techniques of Chapters 2 and 3.

Example 4.16 (Example 4.12 continued) *The model is completed with the following specifications. The amount of a payment has the Pareto distribution with parameters $\alpha = 3$ and $\theta = 1{,}000$. The time from the occurrence of a claim to its payment (L_j) has a Weibull distribution with $\tau = 1.5$ and $\theta = \log(X_j)/6$. This models the dependence by having the scale parameter depend on the size of the loss. The discount factor will be modeled by assuming that for $t > s$, $[\log(V_s/V_t)]/(t - s)$ has a normal distribution with mean 0.06 and variance $0.0004(t - s)$. We do not need to specify a model for the number of losses. Instead, we use the model given earlier for the time between losses. Use simulation to determine the expected present value of aggregate payments.*

The mechanics of a single simulation will be done in detail, and that should indicate how the process is to be done. Begin by generating iid exponential interloss times until their sum exceeds one (in order to obtain one year's worth of claims). The individual variates are generated from pseudouniform numbers using

$$u = 1 - e^{-5x}$$

which yields

$$x = -0.2\log(1-u).$$

For the first simulation, the uniform pseudorandom numbers and the corresponding x-values are (0.25373, 0.0585), (0.46750, 0.1260), (0.23709, 0.0541), (0.75780, 0.2836), and (0.96642, 0.6788). At this point the simulated xs total 1.2010 and therefore there were four loss events, occurring at times $c_1 = 0.0585$, $c_2 = 0.1845$, $c_3 = 0.2386$, and $c_4 = 0.5222$.

The four loss amounts are found from inverting the Pareto cdf. That is,

$$x = 1{,}000[(1-u)^{-1/3} - 1].$$

The four pseudouniform numbers are 0.71786, 0.47779, 0.61084, and 0.68579. This produces the four losses $x_1 = 524.68$, $x_2 = 241.80$, $x_3 = 369.70$, and $x_4 = 470.93$.

The times from occurrence to payment have a Weibull distribution. The equation to solve is

$$u = 1 - e^{-[6l/\log(x)]^{1.5}}$$

where x is the loss. Solving for the lag time l yields

$$l = \log(x)[-\log(1-u)]^{2/3}/6.$$

For the first lag we have $u = 0.23376$ and so

$$l_1 = \log(524.68)[-\log 0.76624]^{2/3}/6 = 0.4320.$$

Similarly, with the next three values of u being 0.85799, 0.12951, and 0.72085 we have $l_2 = 1.4286$, $l_3 = 0.2640$, and $l_4 = 1.2068$. The payment times of the four losses are the sum of c_j and l_j, namely $t_1 = 0.4905$, $t_2 = 1.6131$, $t_3 = 0.5026$, and $t_4 = 1.7290$.

Finally, we generate the discount rate. They must be generated in order of increasing t_j so we first obtain $v_{0.4905}$. We begin with a normal variate with mean 0.06 and variance $0.0004(0.4905) = 0.0001962$. Using inversion, the simulated value is $0.0592 = [\log(1/v_{0.4905})]/0.4905$ and so $v_{0.4905} = 0.9714$. Note that for the first value we have $s = 0$, and $v_0 = 1$. For the second value we require a normal variate with mean 0.06 and variance $(0.5026 - 0.4905)(0.0004) = 0.00000484$. The simulated value is $0.0604 = [\log(0.9714/v_{0.5026})]/0.0121$ for $v_{0.5026} = 0.9707$. We next have $0.0768 = [\log(0.9707/v_{1.6131})]/1.1105$ and so $v_{1.6131} = 0.8913$, and lastly, $0.0628 = [\log(0.8913/v_{1.7290})]/0.1159$ for $v_{1.7290} = 0.8848$.

Table 4.7 Negative binomial cumulative probabilities

n	$F_N(n)$	n	$F_N(n)$
0	0.03704	8	0.76589
1	0.11111	9	0.81888
2	0.20988	10	0.86127
3	0.31962	11	0.89467
4	0.42936	12	0.92064
5	0.53178	13	0.94062
6	0.62282	14	0.95585
7	0.70086	15	0.96735

We are now ready to determine the first simulated value of the aggregate present value. It is

$$\begin{aligned} s_1 &= 524.68(0.9714) + 241.80(0.8913) + 369.70(0.9707) + 470.93(0.8848) \\ &= 1{,}500.74. \end{aligned}$$

The process was then repeated until there was 95% confidence that the estimated mean was within 1% of the true mean. This took 26,944 simulations, producing a sample mean of 2,299.16. □

Example 4.17 (Example 4.13 continued) *For this example, set the deductible d at 250 and the out-of-pocket maximum at $u = 1{,}000$. Assume that the number of losses has the negative binomial distribution with $r = 3$ and $\beta = 2$. Further assume that individual losses have the Weibull distribution with $\tau = 2$ and $\theta = 600$. Determine the 95th percentile of the insurer's losses.*

In order to simulate the negative binomial claim counts, we require the cdf of the negative binomial distribution. There is no closed form, but a table can be constructed, and one appears here as Table 4.7. The number of losses for the year is generated by obtaining one pseudouniform value—for example, $u = 0.47515$—and then determining the smallest entry in the table that is larger than 0.47515. The simulated value appears to its left. In this case our first simulation produced $n = 5$ losses.

The amounts of the 5 losses are obtained from the Weibull distribution. Inversion of the cdf produces

$$x = 600[-\log(1-u)]^{1/2}.$$

The five simulated values are 544.04, 453.67, 217.87, 681.98 and 449.83. The total loss is 2,347.39. The policyholder pays 250.00+250.00+217.87+250.00+250.00 = 1,217.87, but the out-of-pocket maximum limits this to 1,000. Thus our first simulated value has the insurer paying 1,347.39.

The goal was set to be 95% confident that the estimated 95th percentile would be within 2% of the true value. This required 11,476 simulations, producing an estimated 95th percentile of 6,668.18. □

4.9 CALCULATIONS WITH APPROXIMATE DISTRIBUTIONS

Whenever the severity distribution is calculated using an approximate method, the result is, of course, an approximation to the true aggregate distribution. In particular, the true aggregate distribution is often continuous (except, perhaps, with discrete probability at zero or at an aggregate censoring limit) while the approximate distribution either is discrete with probability at equally spaced values (recursion and FFT), is discrete with probability $1/n$ at arbitrary values (simulation), or has a piecewise linear distribution function (Heckman–Meyers). In this section we introduce reasonable ways to obtain values of $F_S(x)$ and $E[(S \wedge x)^k]$ from those approximating distributions. In all cases we assume that the true distribution of aggregate payments is continuous, except perhaps with discrete probability at $S = 0$.

4.9.1 Arithmetic distributions

For recursion and the FFT, the approximating distribution can be written as $p_0, p_1, \ldots$, where $p_j = \Pr(S^* = jh)$ and S^* refers to the approximating distribution. While several methods of "undiscretizing" this distribution are possible, we will introduce only one. It assumes we can obtain $g_0 = \Pr(S = 0)$, the true probability that aggregate payments are zero. The method is based on constructing a continuous approximation to S^* by assuming the probability p_j is uniformly spread over the interval $(j-1/2)h$ to $(j+1/2)h$, for $j = 1, 2, \ldots$. For the interval from 0 to $h/2$, a discrete probability of g_0 is placed at zero and the remaining probability, $p_0 - g_0$, is spread uniformly over the interval. Let S^{**} be the random variable with this mixed distribution. All quantities of interest are then computed using S^{**}.

Example 4.18 *Let N have the geometric distribution with $\beta = 2$ and let X have the exponential distribution with $\theta = 100$. Use recursion with a span of 2 to approximate the aggregate distribution and then obtain a continuous approximation.*

The exponential distribution was discretized using the method which preserves the first moment. The probabilities appear in Table 4.8. Also presented are the aggregate probabilities computed using the recursive formula. We also note that $g_0 = \Pr(N = 0) = (1 + \beta)^{-1} = 1/3$. For $j = 1, 2, \ldots$ the continuous approximation has pdf $f_{S^{**}}(x) = f_{S^*}(2j)/2$, $2j-1 < x \le 2j+1$. We also have $\Pr(S^{**} = 0) = 1/3$ and $f_{S^{**}}(x) = (0.335556 - 1/3)/1 = 0.002223$, $0 < x \le 1$. This density is plotted in Figure 4.1. □

Table 4.8 Discrete approximation to the aggregate payments distribution

j	x	$f_X(x)$	$p_j = f_{S^*}(x)$
0	0	0.009934	0.335556
1	2	0.019605	0.004415
2	4	0.019216	0.004386
3	6	0.018836	0.004356
4	8	0.018463	0.004327
5	10	0.018097	0.004299
6	12	0.017739	0.004270
7	14	0.017388	0.004242
8	16	0.017043	0.004214
9	18	0.016706	0.004186
10	20	0.016375	0.004158

Returning to the original problem, it is possible to work out the general formulas for the basic quantities. For the cdf,

$$\begin{aligned} F_{S^{**}}(x) &= g_0 + \int_0^x \frac{p_0 - g_0}{h/2} ds \\ &= g_0 + \frac{2x}{h}(p_0 - g_0), \qquad 0 \le x \le h/2 \end{aligned}$$

and

$$\begin{aligned} F_{S^{**}}(x) &= \sum_{i=0}^{j-1} p_i + \int_{(j-1/2)h}^{x} \frac{p_j}{h} ds \\ &= \sum_{i=0}^{j-1} p_i + \frac{x-(j-1/2)h}{h} p_j, \qquad (j-1/2)h < x \le (j+1/2)h. \end{aligned}$$

For the limited expected value,

$$\begin{aligned} E[(S^{**} \wedge x)^k] &= 0^k g_0 + \int_0^x s^k \frac{p_0 - g_0}{h/2} ds + x^k[1 - F_{S^{**}}(x)] \\ &= \frac{2x^{k+1}(p_0 - g_0)}{h(k+1)} + x^k[1 - F_{S^{**}}(x)], \qquad 0 < x \le h/2 \end{aligned}$$

and

$$\begin{aligned} E[(S^{**} \wedge x)^k] &= 0^k g_0 + \int_0^{h/2} s^k \frac{p_0 - g_0}{h/2} ds + \sum_{i=1}^{j-1} \int_{(i-1/2)h}^{(i+1/2)h} s^k \frac{p_i}{h} ds \\ &\quad + \int_{(j-1/2)h}^{x} s^k \frac{p_j}{h} ds + x^k[1 - F_{S^{**}}(x)] \end{aligned}$$

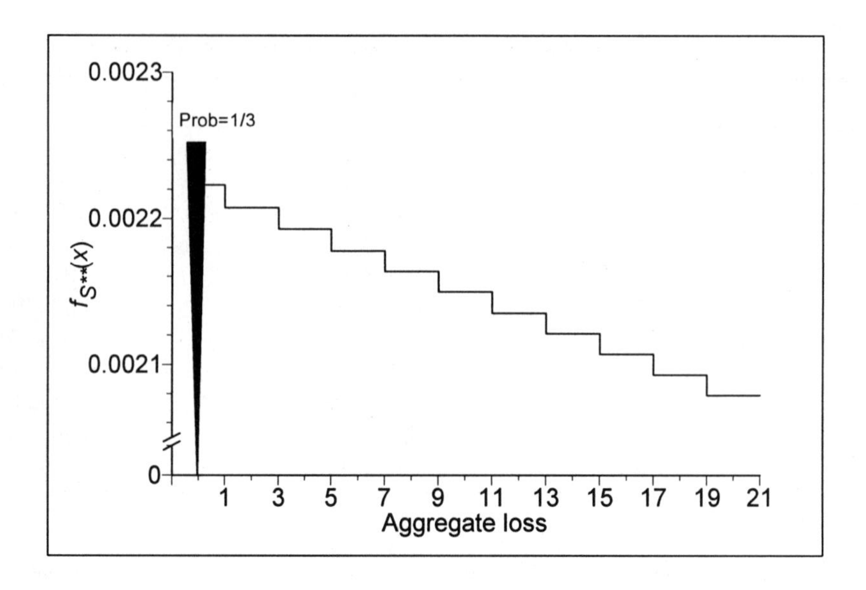

Fig. 4.1 Undiscretized aggregate loss density

$$
\begin{aligned}
&= \frac{(h/2)^k(p_0 - g_0)}{k+1} + \sum_{i=1}^{j-1} \frac{h^k[(i+1/2)^{k+1} - (i-1/2)^{k+1}]}{k+1} p_i \\
&\quad + \frac{x^{k+1} - [(j-1/2)h]^{k+1}}{h(k+1)} p_j \\
&\quad + x^k[1 - F_{S^{**}}(x)], \quad (j-1/2)h < x \le (j+1/2)h.
\end{aligned}
$$

For $k = 1$ this reduces to

$$
\begin{aligned}
E(S^{**} \wedge x) &= x(1-g_0) - \frac{x^2}{h}(p_0 - g_0), \quad 0 < x \le h/2 \\
&= \frac{h}{4}(p_0 - g_0) + \sum_{i=1}^{j-1} ihp_i + \frac{x^2 - [(j-1/2)h]^2}{2h} p_j \\
&\quad + x[1 - F_{S^{**}}(x)], \quad (j-1/2)h < x \le (j+1/2)h. \quad (4.30)
\end{aligned}
$$

These formulas are summarized in Appendix G.

Example 4.19 (Example 4.18 continued) *Compute the cdf and LEV at integral values from 1 to 10 using S^*, S^{**}, and the exact distribution of aggregate losses.*

The exact distribution is available for this example. It was developed in Example 4.6 where it was determined that $\Pr(S = 0) = 1/3$ and the pdf for

Table 4.9 Comparison of true aggregate payment values and two approximations

	cdf			LEV		
x	S	S^*	S^{**}	S	S^*	S^{**}
1	0.335552	0.335556	0.335556	0.66556	0.66444	0.66556
2	0.337763	0.339971	0.337763	1.32890	1.32889	1.32890
3	0.339967	0.339971	0.339970	1.99003	1.98892	1.99003
4	0.342163	0.344357	0.342163	2.64897	2.64895	2.64896
5	0.344352	0.344357	0.344356	3.30571	3.30459	3.30570
6	0.346534	0.348713	0.346534	3.96027	3.96023	3.96025
7	0.348709	0.348713	0.348712	4.61264	4.61152	4.61263
8	0.350876	0.353040	0.350876	5.26285	5.26281	5.26284
9	0.353036	0.353040	0.353039	5.91089	5.90977	5.91088
10	0.355189	0.357339	0.355189	6.55678	6.55673	6.55676

the continuous part is

$$f_S(x) = \frac{\beta}{\theta(1+\beta)^2} \exp\left[-\frac{x}{\theta(1+\beta)}\right] = \frac{2}{900} e^{-x/300}, \quad x > 0.$$

From this we have

$$F_S(x) = \frac{1}{3} + \int_0^x \frac{2}{900} e^{-s/300} ds = 1 - \frac{2}{3} e^{-x/300}$$

and

$$E(S \wedge x) = \int_0^x \frac{2s}{900} e^{-s/300} ds + x\frac{2}{3} e^{-x/300} = 200[1 - e^{-x/300}].$$

The requested values are given in Table 4.9. □

4.9.2 Empirical distributions

When the approximate distribution is obtained by simulation the result is an empirical distribution. Unlike approximations produced by recursion or the FFT, simulation does not place the probabilities at equally spaced values. This makes it less clear how the approximate distribution should be smoothed. On the other hand, simulation usually involves tens or hundreds of thousands of points and therefore the individual points are likely to be close to each other. For these reasons it seems sufficient to simply use the empirical distribution as the answer. That is, all calculations should be done using the approximate empirical random variable, S^*. The formulas for the commonly required quantities are very simple. Let $x_1, x_2, \ldots, x_n$ be the simulated values. Then

$$F_{S^*}(x) = \frac{\text{number of } x_j \leq x}{n}$$

Table 4.10 Simulated values of aggregate losses

j	x_j	j	x_j
1–331	0	346	6.15
332	0.04	347	6.26
333	0.12	348	6.58
334	0.89	349	6.68
335	1.76	350	6.71
336	2.16	351	6.82
337	3.13	352	7.76
338	3.40	353	8.23
339	4.38	354	8.67
340	4.78	355	8.77
341	4.95	356	8.85
342	5.04	357	9.18
343	5.07	358	9.88
344	5.81	359	10.12
345	5.94		

and

$$E[(S^* \wedge x)^k] = \frac{1}{n} \sum_{x_j < x} x_j^k + x^k[1 - F_{S^*}(x)].$$

Example 4.20 (Example 4.18 continued) *Simulate 1,000 observations from the compound model with geometric frequency and exponential severity. Use the results to obtain values of the cdf and LEV for the integers from 1 to 10. The small sample size was selected so that only about thirty values between zero and ten (not including zero) are expected.*

The simulations produced an aggregate payment of zero 331 times. The set of nonzero values that were less than ten plus the first value past ten are presented in Table 4.10. Other than zero, none of the values appeared more than once in the simulation. The requested values from the empirical distribution along with the true values are given in Table 4.11. □

4.9.3 Piece-wise cdf

When using the Heckman–Meyers inversion method, the output is approximate values of the cdf $F_S(x)$ at any set of desired values. The values are approximate because the severity distribution function is required to be piece-wise linear and because approximate integration is used. Let $S^\#$ denote an arbitrary random variable with cdf values as given by the Heckman–Meyers method at arbitrarily selected points $0 = x_1 < x_2 < \cdots < x_n$ and let $F_j = F_{S^\#}(x_j)$. Also, set $F_n = 1$ so that no probability is lost. The easiest

Table 4.11 Empirical and smoothed values from a simulation

x	$F_{S^*}(x)$	$F_S(x)$	$E(S^* \wedge x)$	$E(S \wedge x)$
0	0.331	0.333	0.0000	0.0000
1	0.334	0.336	0.6671	0.6656
2	0.335	0.338	1.3328	1.3289
3	0.336	0.340	1.9970	1.9900
4	0.338	0.342	2.6595	2.6490
5	0.341	0.344	3.3206	3.3057
6	0.345	0.347	3.9775	3.9603
7	0.351	0.349	4.6297	4.6126
8	0.352	0.351	5.2784	5.2629
9	0.356	0.353	5.9250	5.9109
10	0.358	0.355	6.5680	6.5568

way to complete the description of the smoothed distribution is to connect these points with straight lines. Let $S^{\#\#}$ be the random variable with this particular cdf. Intermediate values of the cdf of $S^{\#\#}$ are found by interpolation.

$$F_{S^{\#\#}}(x) = \frac{(x - x_{j-1})F_j + (x_j - x)F_{j-1}}{x_j - x_{j-1}}.$$

The formula for the limited expected value is (for $x_{j-1} < x \leq x_j$)

$$\begin{aligned}
E[(S^{\#\#} \wedge x)^k] &= \sum_{i=2}^{j-1} \int_{x_{i-1}}^{x_i} s^k \frac{F_i - F_{i-1}}{x_i - x_{i-1}} ds \\
&\quad + \int_{x_{j-1}}^{x} s^k \frac{F_j - F_{j-1}}{x_j - x_{j-1}} ds + x^k [1 - F_{S^{\#\#}}(x)] \\
&= \sum_{i=2}^{j-1} \frac{(x_i^{k+1} - x_{i-1}^{k+1})(F_i - F_{i-1})}{(k+1)(x_i - x_{i-1})} \\
&\quad + \frac{(x^{k+1} - x_{j-1}^{k+1})(F_j - F_{j-1})}{(k+1)(x_j - x_{j-1})} \\
&\quad + x^k \left[1 - \frac{(x - x_{j-1})F_j + (x_j - x)F_{j-1}}{x_j - x_{j-1}} \right]
\end{aligned}$$

and when $k = 1$,

$$\begin{aligned}
E(S^{\#\#} \wedge x) &= \sum_{i=2}^{j-1} \frac{(x_i + x_{i-1})(F_i - F_{i-1})}{2} + \frac{(x^2 - x_{j-1}^2)(F_j - F_{j-1})}{2(x_j - x_{j-1})} \\
&\quad + x \left[1 - \frac{(x - x_{j-1})F_j + (x_j - x)F_{j-1}}{x_j - x_{j-1}} \right].
\end{aligned}$$

4.9.4 Effect of coverage modifications

When there are coverage modifications affecting individual payments, all of the undiscretizations are affected in some way. There are two ways to approach incorporating modifications: (1) by discretizing first and then modifying or (2) by doing the modification first. Although the first method is usually easier, the second is normally more accurate. Because a computer will likely be used in either case, we favor accuracy over ease and therefore present only the second approach. One consequence of this choice is that calculations are done with the severity random variable for the amount paid per payment and the frequency random variable for the number of payments. We illustrate the procedure for arithmetized distributions with the following example.

Example 4.21 *Suppose the number of losses has a Poisson distribution with a mean of 10 and the amount of each loss has a Pareto distribution with* $\alpha = 3$ *and* $\theta = 100$. *Obtain the undiscretized distribution of aggregate payments when a per loss deductible of 50 is applied. Use the method of rounding with a span of 5 for the discretization.*

Because the modification is applied first, we work with Y, the random variable which represents the amount paid per payment. For this example the cdf is

$$\begin{aligned}
F_Y(y) &= \Pr(Y \le y) = \Pr(X \le y + 50 | X > 50) \\
&= \frac{F_X(y+50) - F_X(50)}{1 - F_X(50)} \\
&= \frac{1 - \left(\dfrac{100}{150+y}\right)^3 - 1 + \left(\dfrac{100}{150}\right)^3}{1 - 1 + \left(\dfrac{100}{150}\right)^3} \\
&= 1 - \left(\frac{150}{150+y}\right)^3.
\end{aligned}$$

The discretized distribution of this random variable appears in Table 4.12. Because this severity variable represents payments per payment, the frequency variable must be the number of payments, not the number of losses. The formulas developed in Section 3.10 indicate that the distribution remains Poisson and the new parameter is

$$\begin{aligned}
\lambda &= 10 \Pr(X > 50) \\
&= 10 \left(\frac{100}{150}\right)^3 = 2.96296.
\end{aligned}$$

The results of performing the recursions are also presented in Table 4.12. □

The next example illustrates another type of deductible.

Table 4.12 Discretization done after applying the deductible of 50

x	$f_Y(x)$	$f_{S^*}(x)$	x	$f_Y(x)$	$f_{S^*}(x)$
0	0.0484	0.0596	55	0.0287	0.0157
5	0.0878	0.0155	60	0.0260	0.0156
10	0.0773	0.0157	65	0.0237	0.0155
15	0.0684	0.0158	70	0.0216	0.0153
20	0.0607	0.0159	75	0.0198	0.0151
25	0.0540	0.0160	80	0.0181	0.0150
30	0.0483	0.0160	85	0.0166	0.0148
35	0.0432	0.0160	90	0.0153	0.0146
40	0.0389	0.0159	95	0.0141	0.0144
45	0.0350	0.0159	100	0.0130	0.0142
50	0.0317	0.0158			

Example 4.22 (Example 4.21 continued) *Repeat the example, only this time employ a franchise deductible of 50.*

Again, we begin with the cdf of Y, the payment per payment.

$$F_Y(y) = \begin{cases} 0, & y < 50 \\ \Pr(X < y|X > 50) = 1 - \left(\dfrac{150}{100+y}\right)^3, & y \geq 50. \end{cases}$$

The discretization of this distribution appears in Table 4.13. The aggregate distribution has the correct probability at zero. Note that when converting the distribution of S^* to S^{**} the probability at 50 is spread uniformly over the interval from 50 to 52.50, not from 47.5 to 52.5. □

For simulation, it is again best to work with Y. Consider a deductible of d. No simulations will be wasted producing losses below d, which result in no addition to the aggregate payment.

For Heckman–Meyers, it is just as easy to work with Y^*. There will be discrete probability at zero, but this is reflected in the characteristic function of the severity distribution by an additive term of $\Pr(Y^* = 0)$.

4.10 COMPARISON OF METHODS

The recursive method has some significant advantages. The time required to compute an entire distribution of n points is reduced to $O(n^2)$ from $O(n^3)$ for the direct convolution method. Furthermore, it provides exact values when the severity distribution is itself discrete (arithmetic). The only source of error is in the discretization of the severity distribution. Except for binomial

Table 4.13 Discretization done after applying the franchise deductible of 50

x	$f_Y(x)$	$f_{S^*}(x)$	x	$f_Y(x)$	$f_{S^*}(x)$
0	0	0.0517	55	0.0878	0.0134
5	0	0	60	0.0773	0.0118
10	0	0	65	0.0684	0.0105
15	0	0	70	0.0607	0.0093
20	0	0	75	0.0540	0.0083
25	0	0	80	0.0483	0.0074
30	0	0	85	0.0432	0.0066
35	0	0	90	0.0389	0.0060
40	0	0	95	0.0350	0.0054
45	0	0	100	0.0317	0.0054
50	0.0484	0.0074			

models, the calculations are guaranteed to be numerically stable. This method is very easy to program in a few lines of computer code. However, it has a few disadvantages. The recursive method only works for the classes of frequency distributions described in Chapter 3. Using distributions not based on the $(a, b, 0)$ and $(a, b, 1)$ classes requires modification of the formula or developing a new recursion. Numerous other recursions have recently been developed in the actuarial and statistical literature.

The FFT method is easy to use in that it uses standard "canned" routines available with many software packages. It is faster than the recursive method when n is large since it requires calculations of order $n \log_2 n$ rather than n^2. However, if the severity distribution has a fixed (and not too large) number of points, the recursive method will require fewer computations since the sum in (4.14) will have at most r terms, reducing the order of required computations to be of order n, rather than n^2 in the case of no upper limit of the severity. The FFT method can be extended to the case where the severity distribution can take on negative values. Like the recursive method, it produces the entire distribution.

The direct inversion method has been demonstrated to be very fast in calculating a single value of the aggregate distribution or the net stop-loss (excess pure) premium for a single deductible d. However, it requires a major computer programming effort. It has been developed by Heckman and Meyers [58] specifically for $(a, b, 0)$ frequency models. It is possible to generalize the computer code to handle any distribution with a pgf that is a relatively simple function. This method is much faster than the recursive method when the expected number of claims is large. The speed does not depend on the size of λ in the case of the Poisson frequency model. In addition to being complicated to program, the method involves approximate integration whose errors depend on the method and interval size.

Through the use of transforms, both the FFT and inversion methods are able to handle convolutions efficiently. For example, suppose a reinsurance agreement was to cover the aggregate losses of three groups, each with unique frequency and severity distributions. If S_i, $i = 1, 2, 3$, are the aggregate losses for each group, the characteristic function for the total aggregate losses $S = S_1 + S_2 + S_3$ is $\varphi_S(z) = \varphi_{S_1}(z)\varphi_{S_2}(z)\varphi_{S_3}(z)$ and so the only extra work is some multiplications prior to the inversion step. The recursive method does not accommodate convolutions as easily.

The Heckman–Meyers method has some technical difficulties when being applied to severity distributions that are of the discrete type or have some anomalies, such as "heaping" of losses at some round number (e.g., 1,000,000). At any "jump" in the severity distribution function, a very short interval containing the jump needs to be defined in setting up the points $(x_1, x_2, \ldots, x_n)$.

We save a discussion of simulation for last because it differs greatly from the other methods. The major advantage is a big one. If you can carefully articulate the model, you should be able to obtain the aggregate distribution by simulation. The programming effort may take a little time, but can be done in a straightforward manner. Today's computers will conduct the simulation in a reasonable amount of time. Most of the analytic methods were developed as a response to the excessive computing time that simulations used to require. That is less of a problem now. On the other hand, it is difficult to write a general purpose simulation program. Instead, it is possibly necessary to write a new routine as each problem occurs. Thus it is probably best to save the simulation approach for those problems that cannot be solved by the other methods. Then, of course, it is worth the effort because there is no alternative.

One other drawback of simulation occurs in extremely low frequency situations (which is where recursion excels). For example, consider an individual excess-of-loss reinsurance in which reinsurance benefits are paid on individual losses above 1,000,000, an event which occurs about 1 time in 100, but when it does, the tail is extremely long (for example, a Pareto distribution with small α). The simulation will have to discard 99% of the generated losses and then will need a large number of those that exceed the deductible (due to the large variation in losses). It may take a long time to obtain a reliable answer. One possible solution for simulation is to work with the conditional distribution of the loss variable, given that a payment has been made.

No method is clearly superior for all problems. Each method has both advantages and disadvantages when compared with the others. What we really have is an embarrassment of riches. Twenty-five years ago, actuaries wondered if there would ever be effective methods for determining aggregate distributions. Today we can choose from several.

Table 4.14 Hospitalizations, per family member, per year

No. of hospitalizations per family member	No. of family members
0	2,659
1	244
2	19
3	2
4 or more	0
Total	2,924

4.11 AN AGGREGATE LOSS EXAMPLE

The example to be covered in this section summarizes many of the techniques introduced to this point. The coverage is perhaps more complex than those found in practice, but that gives us a chance to work through a variety of tasks.

Example 4.23 *You are a consulting actuary and have been retained to assist in the pricing of a group hospitalization policy. Your task is to determine the expected payment to be made by the insurer. The terms of the policy (per covered employee) are as follows:*

1. *For each hospitalization of the employee or a member of the employee's family, the employee pays the first 500 plus any losses in excess of 50,500. On any one hospitalization, the insurance will pay at most 50,000.*
2. *In any calendar year, the employee will pay no more than 1,000 in deductibles, but there is no limit on how much the employee will pay in respect of losses exceeding 50,500.*
3. *Any particular hospitalization is assigned to the calendar year in which the individual entered the hospital. Even if hospitalization extends into subsequent years, all payments are made in respect to the policy year assigned.*
4. *The premium is the same, regardless of the number of family members.*

Experience studies have provided the data contained in Tables 4.14 and 4.16. The data in Table 4.15 represent the profile of the current set of employees.

The first step is to fit parametric models to each of the three data sets. For the data in Table 4.14, twelve distributions were fitted. The best one-parameter distribution is the geometric with a negative loglikelihood (NLL)

Table 4.15 Number of family members per employee

No. of family members per employee	No. of employees
1	84
2	140
3	139
4	131
5	73
6	42
7	27
8 or more	33
Total	669

Table 4.16 Losses per hospitalization

Loss per hospitalization	No. of hospitalizations
0–250	36
250–500	29
500–1,000	43
1,000–1,500	35
1,500–2,500	39
2,500–5,000	47
5,000–10,000	33
10,000–50,000	24
50,000–	2
Total	288

of 969.251 and a chi-square goodness-of-fit p-value of 0.5325. The best two-parameter model is the zero-modified geometric. The NLL improves to 969.058, but by the likelihood ratio test, this is not sufficient to justify the second parameter. The best three-parameter distribution is the zero-modified negative binomial which has a NLL of 969.056, again not enough to dislodge the geometric as our choice. For the two- and three-parameter models there were not enough degrees of freedom to conduct the chi-square test. We choose the geometric distribution with $\beta = 0.098495$.

For the data in Table 4.15, only zero-truncated distributions should be considered. The best one-parameter model is the zero-truncated Poisson with an NLL of 1,298.725 and a p-value near zero. The two-parameter zero-truncated negative binomial has an NLL of 1,292.532, a significant improvement. The

p-value is 0.2571, indicating that this is an acceptable choice. The parameters are $r = 13.207$ and $\beta = 0.25884$.

For the data in Table 4.16, fifteen continuous distributions were fitted. The four best models for a given number of parameters are listed below.

Name	No. of parameters	NLL	p-value
Inverse exponential	1	632.632	Near 0
Pareto	2	601.642	0.9818
Burr	3	601.612	0.9476
Transformed beta	4	601.553	0.8798

It should be clear that the best choice is the Pareto distribution. The parameters are $\alpha = 1.6693$ and $\theta = 3{,}053.0$.

The remaining calculations were done using the recursive method, but inversion or simulation would work equally well.

The first step is to determine the distribution of payments by the employee per family member with regard to the deductible. The frequency distribution is the geometric distribution while the individual loss distribution is the Pareto distribution, limited to the maximum deductible of 500. That is, any losses in excess of 500 are assigned to the value 500. With regard to discretization for recursion, the span should divide evenly into 500 and then all probability not accounted for by the time 500 is reached is placed there. For this example a span of 1 was used. The first few and last few values of the discretized distribution appear in Table 4.17. After applying the recursive formula, probability extended beyond 3,000. However, looking ahead, we know that, with regard to the employee aggregate deductible, payments beyond 1,000 have no impact. A few of these probabilities appear in Table 4.18.

We next must obtain the aggregate distribution of deductibles paid per employee per year. This is another compound distribution. The frequency

Table 4.17 Discretized Pareto distribution with 500 limit

Loss	Probability
0	0.000273
1	0.000546
2	0.000546
3	0.000545
⋮	⋮
498	0.000365
499	0.000365
500	0.776512

Table 4.18 Probabilities for aggregate deductibles per family member

Loss	Probability
0	0.910359
1	0.000045
2	0.000045
3	0.000045
⋮	⋮
499	0.000031
500	0.063386
501	0.000007
⋮	⋮
999	0.000004
1,000	0.004413
1,001	0.000001
⋮	⋮

distribution is the truncated negative binomial and the individual loss distribution is the one for losses per family member that was just obtained. Recursions can again be used to obtain this distribution. Because there is a 1,000 limit on deductibles, all probability not accounted for by 1,000 will be placed there. Selected values from this aggregate distribution are given in Table 4.19. Note that the chance that more than 1,000 in deductibles will be paid is very small. The cost to the insurer of limiting the insured's costs is also small. Using this discrete distribution it is easy to obtain the mean and standard deviation of aggregate deductibles. They are 150.02 and 274.42, respectively.

We next require the expected value of aggregate costs to the insurer for individual losses below the upper limit of 50,000. This can be found analytically. The expected payment per loss is $E(X \wedge 50{,}500) = 3{,}890.87$ for the Pareto distribution. The expected number of losses per family member is the mean of the geometric distribution which is the parameter, 0.098495. The expected number of family members per employee comes from the zero-truncated negative binomial distribution and is 3.59015. This implies that the expected number of losses per employee is $0.098495(3.59015) = 0.353612$. Then the expected aggregate dollars in payments up to the individual limit is $0.353612(3{,}890.87) = 1{,}375.86$.

Then the expected cost to the insurer is the difference, $1{,}375.86 - 150.02 = 1{,}225.84$. As a final note, it is not possible to use any method other than simulation if the goal is to obtain the probability distribution of the insurer's payments. This situation is similar to that of Example 4.13 where it is easy to

Table 4.19 Probabilities for aggregate deductibles per employee

Loss	Probability
0	0.725517
1	0.000116
2	0.000115
3	0.000115
⋮	⋮
499	0.000082
500	0.164284
501	0.000047
⋮	⋮
999	0.000031
1,000	0.042343

get the overall distribution as well as the distribution for the insured (in this case, if payments for losses over 50,500 are ignored), but not for the insurer. □

4.12 THE INDIVIDUAL RISK MODEL

4.12.1 Introduction and parametric approximation

Following Definition 1.15, the **individual risk model** represents the aggregate loss as a fixed sum of independent (but not necessarily identically distributed) random variables:

$$S = X_1 + X_2 + \cdots + X_n.$$

This is usually thought of as the sum of the losses from n insurance contracts, e.g., n persons covered under a group insurance policy.

The individual risk model was originally developed for life insurance in which the probability of death within a year is q_j and the fixed benefit paid for the death of the jth person is b_j. In this case, the distribution of the loss to the insurer for the jth policy is

$$f_{X_j}(x) = \begin{cases} 1 - q_j, & x = 0 \\ q_j, & x = b_j. \end{cases}$$

In this case the mean and variance of aggregate losses are

$$E(S) = \sum_{j=1}^{n} b_j q_j$$

and

$$Var(S) = \sum_{j=1}^{n} b_j^2 q_j(1 - q_j)$$

since the X_js are assumed to be independent. Then, the pgf of aggregate losses is

$$P_S(z) = \prod_{j=1}^{n}(1 - q_j + q_j z^{b_j}). \tag{4.31}$$

In the special case where all the risks are identical with $q_j = q$ and $b_j = 1$, the pgf reduces to

$$P_S(z) = (1 + q(z - 1))^n,$$

and in this case S has a binomial distribution.

The individual risk model can be generalized as follows. Let $X_j = I_j B_j$, where $I_1, \ldots, I_n$, $B_1, \ldots, B_n$ are independent. The random variable I_j is in indicator variable that takes on the value 1 with probability q_j and the value 0 with probability $1 - q_j$. This variable indicates whether or not the jth policy produced a payment. The random variable B_j can have any distribution and represents the amount of the payment in respect of the jth policy given that a payment was made. In the life insurance case, B_j is degenerate, with all probability on the value b_j. If we let $\mu_j = E(B_j)$ and $\sigma_j^2 = Var(B_j)$ then

$$E(S) = \sum_{j=1}^{n} q_j \mu_j \tag{4.32}$$

and

$$Var(S) = \sum_{j=1}^{n} [q_j \sigma_j^2 + q_j(1 - q_j)\mu_j^2]. \tag{4.33}$$

You are asked to verify these formulas in Exercise 4.59. The following example is a simple version of this situation.

Example 4.24 *Consider a group life insurance contract with an accidental death benefit. Assume that for all members the probability of death in the next year is .01 and that 30% of deaths are accidental. For 50 employees the benefit for an ordinary death is 50,000 and for an accidental death it is 100,000. For the remaining 25 employees the benefits are 75,000 and 150,000. Develop an individual risk model and determine its mean and variance.*

For all 75 employees $q_j = 0.01$. For 50 employees, B_j takes on the value 50,000 with probability 0.7 and 100,000 with probability 0.3. For them, $\mu_j =$ 65,000 and $\sigma_j^2 =$ 525,000,000. For the remaining 25 employees B_j takes on the value 75,000 with probability 0.7 and 150,000 with probability 0.3. For them, $\mu_j =$ 97,500 and $\sigma_j^2 =$ 1,181,250,000. Then

$$\begin{aligned} E(S) &= 50(0.01)(65{,}000) + 25(0.01)(97{,}500) \\ &= 56{,}875 \end{aligned}$$

Table 4.20 Employee data for Example 4.25

Employee, j	Age (years)	Sex	Benefit, b_j	Mortality rate, q_j
1	20	M	15,000	0.00149
2	23	M	16,000	0.00142
3	27	M	20,000	0.00128
4	30	M	28,000	0.00122
5	31	M	31,000	0.00123
6	46	M	18,000	0.00353
7	47	M	26,000	0.00394
8	49	M	24,000	0.00484
9	64	M	60,000	0.02182
10	17	F	14,000	0.00050
11	22	F	17,000	0.00050
12	26	F	19,000	0.00054
13	37	F	30,000	0.00103
14	55	F	55,000	0.00479
Total			373,000	

and

$$\begin{aligned} Var(S) &= 50(0.01)(525{,}000{,}000) + 50(0.01)(0.99)(65{,}000)^2 \\ &\quad +25(0.01)(1{,}181{,}250{,}000) + 25(0.01)(0.99)(97{,}500)^2 \\ &= 5{,}001{,}984{,}375. \end{aligned}$$

□

When the risks are different, the probabilities defined by pgf (4.31) can be computed exactly or approximately. A normal, gamma, lognormal, or any other distribution can be used to approximate the distribution. This is usually done by matching the first few moments. Since the normal, gamma, and lognormal distributions each have two parameters, the mean and variance are sufficient.

Example 4.25 (Group life insurance) *A small manufacturing business has a group life insurance contract on its 14 permanent employees. The actuary for the insurer has selected a mortality table to represent the mortality of the group. Each employee is insured for the amount of his or her salary rounded up to the next 1,000 dollars. The group's data are given in Table 4.20.*

If the insurer adds a 45% relative loading to the net (pure) premium, what are the chances that it will lose money in a given year? Use the normal and lognormal approximations.

The mean and variance of the aggregate losses for the group are

$$E(S) = \sum_{j=1}^{14} b_j q_j = 2{,}054.41$$

and

$$Var(S) = \sum_{j=1}^{14} b_j^2 q_j (1 - q_j) = 1.02534 \times 10^8.$$

The premium being charged is $1.45 \times 2,054.41 = 2{,}978.89$. For the normal approximation (in units of 1,000), $\mu = 2.05441$ and $\sigma^2 = 102.534$. Then the probability of a loss is

$$\begin{aligned} \Pr(S > 2.97889) &= \Pr\left[\frac{S-\mu}{\sigma} > \frac{2.97889 - 2.05441}{(102.534)^{1/2}}\right] \\ &\doteq \Pr(Z > 0.0913) \\ &= 0.46 \text{ or } 46\%. \end{aligned}$$

For the lognormal approximation (as in Example 4.3)

$$\mu + \sigma^2/2 = \log 2.05441 = 0.719989$$

and

$$2\mu + 2\sigma^2 = \log(102.534 + 4.221) = 4.670533.$$

From this $\mu = -0.895289$ and $\sigma^2 = 3.230555$. Then

$$\begin{aligned} \Pr(S > 2.97889) &= 1 - \Phi\left[\frac{\log 2.97889 + 0.895289}{(3.230555)^{1/2}}\right] \\ &= 1 - \Phi(1.105) \\ &= 0.13 \text{ or } 13\%. \end{aligned}$$

□

In the next subsection we present several ways of obtaining the exact distribution of S for the case where the benefit amounts are fixed.

4.12.2 Exact calculation of the aggregate distribution

4.12.2.1 Direct calculation The pf of aggregate losses is given by

$$f_S(x) = f_{X_1} * f_{X_2} * \cdots * f_{X_n}(x) \tag{4.34}$$

where

$$f_{X_j}(x) = \begin{cases} p_j = 1 - q_j, & x = 0 \\ q_j, & x = b_j. \end{cases}$$

The density (4.34) can be calculated recursively over the partial sums $S_j = S_{j-1} + X_j$ for $j = 2, 3, ..., n$ beginning with $S_1 = X_1$. Then

$$\begin{aligned} f_{S_j}(x) &= \begin{cases} f_{S_{j-1}}(x) f_{X_j}(0), & x < b_j \\ f_{S_{j-1}}(x) f_{X_j}(0) + f_{S_{j-1}}(x - b_j) f_{X_j}(b_j), & x \geq b_j \end{cases} \\ &= \begin{cases} p_j f_{S_{j-1}}(x), & x < b_j \\ p_j f_{S_{j-1}}(x) + q_j f_{S_{j-1}}(x - b_j), & x \geq b_j. \end{cases} \end{aligned}$$

If we wish to calculate the distribution of total claims up to some value r, the computer time involved, as measured by the number of multiplications, can be seen to be of order nr. If both r and n are large (e.g., $r = 10{,}000$ and $n = 10{,}000$), the number of computations can be prohibitive.

Example 4.26 (Example 4.25 continued) *Use the direct method to determine the pf of S as well as the probability required in Example 4.25.*

$$\begin{aligned} f_{S_1}(0) &= 0.99851 \\ f_{S_1}(15) &= 0.00149 \\ f_{S_2}(0) &= p_2 f_{S_1}(0) = 0.99709212 \\ f_{S_2}(15) &= p_2 f_{S_1}(15) = 0.00148788 \\ f_{S_2}(16) &= p_2 f_{S_1}(16) + q_2 f_{S_1}(0) = 0.00141788 \\ f_{S_2}(31) &= p_2 f_{S_1}(31) + q_2 f_{S_1}(15) = 0.0000021158. \end{aligned}$$

The final values of the cdf $F_S(x)$ are given below for $x = 0, ..., 79$ are given in Table 4.21. From Table 4.21, the probability of exceeding 2,978.89 is seen to be exactly 0.047 which shows that both approximations in Example 4.25 are poor. □

This approach is reasonable when n is not too large, but for larger groups an alternative method is needed.

4.12.2.2 Recursive calculation The following approach allows for the distribution to be calculated recursively based on De Pril [28]. We first divide the portfolio into subportfolios according to policy size and claim probability. Let n_{ij} be the number of policies with benefit i (where $i = 1, 2, ..., r$)[5] and claim probability q_j (where $j = 1, 2, ..., m$). Then the pgf of total claims may be written as

$$P_S(z) = \prod_{i=1}^{r} \prod_{j=1}^{m} (1 - q_j + q_j z^i)^{n_{ij}}.$$

[5] As in the discretization of severity distributions, it is necessary that the benefit amounts be in arithmetic progression. However, the monetary unit need not be 1. For example, $i = 1, 2, \ldots$ could represent benefit amounts of 5,000, 10,000, ...

Table 4.21 Cumulative probabilities for Example 4.26

x	$F_S(x)$	x	$F_S(x)$	x	$F_S(x)$	x	$F_S(x)$
0	0.95273905	20	0.96157969	40	0.97335098	60	0.99933062
1	0.95273905	21	0.96157969	41	0.97335892	61	0.99933187
2	0.95273905	22	0.96157969	42	0.97338128	62	0.99933191
3	0.95273905	23	0.96157969	43	0.97338740	63	0.99933193
4	0.95273905	24	0.96621337	44	0.97340884	64	0.99933198
5	0.95273905	25	0.96621337	45	0.97341351	65	0.99933202
6	0.95273905	26	0.96998201	46	0.97342561	66	0.99933206
7	0.95273905	27	0.96998201	47	0.97342840	67	0.99933209
8	0.95273905	28	0.97114577	48	0.97343397	68	0.99933217
9	0.95273905	29	0.97114648	49	0.97343866	69	0.99933450
10	0.95273905	30	0.97212950	50	0.97345889	70	0.99934141
11	0.95273905	31	0.97330507	51	0.97346040	71	0.99934796
12	0.95273905	32	0.97330747	52	0.97346606	72	0.99935031
13	0.95273905	33	0.97331344	53	0.97346608	73	0.99936659
14	0.95321566	34	0.97331962	54	0.97347547	74	0.99937973
15	0.95463736	35	0.97332386	55	0.97806678	75	0.99941735
16	0.95599217	36	0.97332585	56	0.97807068	76	0.99944759
17	0.95646878	37	0.97332829	57	0.97807536	77	0.99945823
18	0.95984386	38	0.97333493	58	0.97807660	78	0.99953355
19	0.96035862	39	0.97334251	59	0.97807808	79	0.99956734

The logarithm of the pgf is

$$\log P_S(z) = \sum_{i=1}^{r}\sum_{j=1}^{m} n_{ij} \log(1 - q_j + q_j z^i). \tag{4.35}$$

We now differentiate (4.35) to obtain

$$P_S'(z) = P_S(z)\left[\sum_{i=1}^{r}\sum_{j=1}^{m} i q_j n_{ij} z^{i-1}(1 - q_j + q_j z^i)^{-1}\right]. \tag{4.36}$$

Setting $z = 1$ in (4.36) yields the mean of the total claims distribution, namely

$$E(S) = P_S'(1) = \sum_{i=1}^{r}\sum_{j=1}^{m} i q_j n_{ij}.$$

Now, (4.36) may be rewritten as

$$\begin{aligned} zP_S'(z) &= P_S(z)\left[\sum_{i=1}^{r}\sum_{j=1}^{m} i n_{ij}\left(\frac{q_j}{1-q_j} z^i\right)\left(1 + \frac{q_j}{1-q_j} z^i\right)^{-1}\right] \\ &= P_S(z)\left[\sum_{i=1}^{r}\sum_{j=1}^{m} i n_{ij} \sum_{k=1}^{\infty} (-1)^{k-1}\left(\frac{q_j}{1-q_j}\right)^k z^{ik}\right] \end{aligned} \tag{4.37}$$

for $|z| < \min_{i,j}\{q_j^{-1}(1-q_j)\}^{1/i}$. The second term on the right-hand side of (4.37) may be rewritten as

$$\sum_{i=1}^{r}\sum_{k=1}^{\infty} h(i,k)z^{ik}$$

where

$$h(i,k) = i(-1)^{k-1}\sum_{j=1}^{m} n_{ij}\left(\frac{q_j}{1-q_j}\right)^k. \tag{4.38}$$

Thus, (4.37) may be written as

$$zP_S'(z) = P_S(z)\left[\sum_{i=1}^{r}\sum_{k=1}^{\infty} h(i,k)z^{ik}\right]. \tag{4.39}$$

The coefficient of z^x on the left-hand side of (4.39) is $xf_S(x)$, where $f_S(x)$ is the coefficient of z^x in $P_S(z)$. The right-hand side of (4.39) is a convolution, and the coefficient of z^x is thus given by

$$\sum_{ik\le x} h(i,k)f_S(x-ik). \tag{4.40}$$

A simpler way of writing (4.40) is

$$\sum_{i=1}^{x}\sum_{k=1}^{[\frac{x}{i}]} h(i,k)f_S(x-ik)$$

where $[\cdot]$ denotes the greatest integer function. Finally, since $h(i,k) = 0$ if $i > x$, one may equate coefficients of z^x on both sides of (4.39) and divide by x to obtain

$$f_S(x) = \frac{1}{x}\sum_{i=1}^{x\wedge r}\sum_{k=1}^{[\frac{x}{i}]} h(i,k)f_S(x-ik), \qquad n \ge 1. \tag{4.41}$$

Now,

$$f_S(0) = P_S(0) = \prod_{i=1}^{r}\prod_{j=1}^{m}(1-q_j)^{n_{ij}} \tag{4.42}$$

and from (4.41),

$$f_S(1) = h(1,1)f_S(0)$$

$$f_S(2) = \frac{1}{2}\{h(1,1)f_S(1) + [h(1,2)+h(2,1)]f_S(0)\}, \ldots.$$

The probabilities $\{f_S(x);\ x = 1, 2, ...,\}$ may be calculated recursively using (4.41), beginning with (4.42).

It can be seen from (4.38) that $h(i,k)$ is a weighted sum of the kth power of $q_j/(1-q_j)$, $j=1,2,...,m$. When q_j is close to zero, $[q_j/(1-q_j)]^k$ is small. Consequently, the magnitude $h(i,k)$ decreases rapidly as k increases. This suggests that the inner summation in (4.41) can be limited to a small number of terms without significant loss of accuracy in computations while speeding up the computations considerably.

If we limit k to a maximum of K terms, let

$$f_S^{(K)}(x) = \frac{1}{x}\sum_{i=1}^{x\wedge r}\sum_{k=1}^{K\wedge[\frac{x}{i}]} h(i,k) f_S^{(K)}(x-ik) \tag{4.43}$$

denote the approximation using at most K terms in (4.41). In a later paper, De Pril [29] shows that if $q_j < 1/2$, $j=1,2,...,m$, then

$$\sum_{x=0}^{M} |f_S(x) - f_S^{(K)}(x)| < e^{\delta(K)} - 1 \tag{4.44}$$

where

$$\delta(K) = \frac{1}{K+1}\sum_{i=1}^{r}\sum_{j=1}^{m} n_{ij}\frac{1-q_j}{1-2q_j}\left(\frac{q_j}{1-q_j}\right)^{K+1} \tag{4.45}$$

and $M = \sum_{i=1}^{r}\sum_{j=1}^{m} i n_{ij}$ is the maximum possible aggregate claim amount.

The value of $\delta(K)$ is easily calculated for any value of K. Equation (4.44) provides an upper bound on the sum of the absolute errors over the entire distribution of aggregate claims and can be used to guarantee accuracy of results when a limited number of terms is used in (4.43).

Example 4.27 (Example 4.25 continued) *Determine the exact distribution of aggregate losses using the recursive method.*

The values of q_j and the nonzero rows of the matrix of $n_{i,j}$ values are given in Table 4.22:

Using (4.45), we find that $\delta(1) = 5.947 \times 10^{-4}$, $\delta(2) = 3.900 \times 10^{-6}$, $\delta(3) = 6.369 \times 10^{-8}$, and $\delta(4) = 1.131 \times 10^{-9}$. Hence, equation (4.43) with $K=4$ will give us about 8 decimal place accuracy. The (nonzero) values of $h(i,k)$ computed using (4.38) are given in Table 4.23.

The values of $f_S(x)$ and the associated cdf $F_S(x)$ calculated using (4.43) with $K=4$ and (4.42) are given in Table 4.24.

It can be seen that the values in the last column of this table are identical to the corresponding values from Example 4.26 based on the direct method. This method is especially useful when there are a large number of lives in a group insurance contract. □

Example 4.28 (An expanded group) *For the purpose of illustrating the effect of the size of the portfolio, consider a portfolio consisting of 1,400 independent*

Table 4.22 Values of n_{ij} for Example 4.27

		i													
j	$1000q_j$	14	15	16	17	18	19	20	24	26	28	30	31	55	60
1	0.50	1	0	0	1	0	0	0	0	0	0	0	0	0	0
2	0.54	0	0	0	0	0	1	0	0	0	0	0	0	0	0
3	1.03	0	0	0	0	0	0	0	0	0	0	1	0	0	0
4	1.22	0	0	0	0	0	0	0	0	0	1	0	0	0	0
5	1.23	0	0	0	0	0	0	0	0	0	0	0	1	0	0
6	1.28	0	0	0	0	0	0	1	0	0	0	0	0	0	0
7	1.42	0	0	1	0	0	0	0	0	0	0	0	0	0	0
8	1.49	0	1	0	0	0	0	0	0	0	0	0	0	0	0
9	3.53	0	0	0	0	1	0	0	0	0	0	0	0	0	0
10	3.94	0	0	0	0	0	0	0	0	1	0	0	0	0	0
11	4.79	0	0	0	0	0	0	0	0	0	0	0	0	1	0
12	4.84	0	0	0	0	0	0	0	1	0	0	0	0	0	0
13	21.82	0	0	0	0	0	0	0	0	0	0	0	0	0	1

lives, with exactly 100 lives like each life in the group life portfolio of the previous examples. Determine the exact distribution of total losses.

The values of n_{ij} are now either 100 or 0. From (4.38), it can be seen that each $h(i,k)$ is now 100 times larger than that of the previous example. The distribution of total claims can be computed as in the previous example. In Table 4.25 we give some values of the distribution function of total claims (as before, x is measured in units of 1,000).

From Example 4.25, the mean and variance of total claims for this portfolio of 1400 lives are $\mu_1 = 205,441$ dollars and $\mu_2 = 1.025335632 \times 10^{10}$. Also, the coefficient of skewness is calculated as $\mu_3(\mu_2)^{-3/2} = 0.5267345$, which is exactly one-tenth of the value of 5.26734515 for the corresponding group of 14 lives. This indicates that the distribution is much more symmetric. □

There are numerous papers on the subject of the individual risk model. De Pril [30] develops a generalization of this method to the case where the loss can take on more than one value.

4.12.3 Compound Poisson approximation

Because of the computational complexity of calculating the distribution of total claims for a portfolio of n risks using the individual risk model, it has been popular to attempt to approximate the distribution by using the compound Poisson distribution. As was seen in Section 4.5, use of the compound Poisson allows calculation of the total claims distribution by using a very simple recursive procedure or by using the fast Fourier transform. The pgf (4.31) of

Table 4.23 Values of $h(i, k)$ for Example 4.27

	$h(i,k)$			
i	$k = 1$	$k = 2$	$k = 3$	$k = 4$
14	7.0035018E–03	–3.5035025E–06	1.7526276E–09	–8.7675218E–13
15	2.2383351E–02	–3.3400962E–05	4.9841694E–08	–7.4374944E–11
16	2.2752309E–02	–3.2354221E–05	4.6008325E–08	–6.5424723E–11
17	8.5042522E–03	–4.2542531E–06	2.1281907E–09	–1.0646277E–12
18	6.3765090E–02	–2.2588816E–04	8.0020991E–07	–2.8347477E–09
19	1.0265543E–02	–5.5463884E–06	2.9966680E–09	–1.6190750E–12
20	2.5632810E–02	–3.2852048E–05	4.2104515E–08	–5.3962851E–11
24	1.1672495E–01	–5.6769638E–04	2.7610139E–06	–1.3428300E–08
26	1.0284521E–01	–4.0681297E–04	1.6091833E–06	–6.3652612E–09
28	3.4201726E–02	–4.1777074E–05	5.1030287E–08	–6.2332996E–11
30	3.0931860E–02	–3.1892665E–05	3.2883315E–08	–3.3904736E–11
31	3.8176959E–02	–4.7015487E–05	5.7900266E–08	–7.1305033E–11
55	2.6471800E–01	–1.2741022E–03	6.1323232E–06	–2.9515207E–08
60	1.3384040E–00	–2.9855420E–02	6.6597689E–04	–1.4855768E–05

aggregate losses is

$$P_S(z) = \prod_{j=1}^{n}[1 + q_j(z^{b_j} - 1)].$$

By taking logarithms, we obtain

$$\log P_S(z) = \sum_{j=1}^{n}\sum_{k=1}^{\infty}\frac{(-1)^{k+1}}{k}[q_j(z^{b_j} - 1)]^k.$$

Retaining only the first term in the inner sum yields the approximation

$$\begin{aligned}\log P_S(z) &\doteq \sum_{j=1}^{n} q_j(z^{b_j} - 1) \\ &= \lambda\sum_{j=1}^{n}\frac{\lambda_j}{\lambda}(z^{b_j} - 1)\end{aligned} \tag{4.46}$$

where $\lambda_j = q_j$ and $\lambda = \sum_{j=1}^{n}\lambda_j$. This results in

$$P_S(z) = \exp\{\lambda[P_X(z) - 1]\}$$

which is the pgf of a compound Poisson distribution with individual loss distribution pgf

$$P_X(z) = \frac{1}{\lambda}\sum_{j=1}^{n}\lambda_j z^{b_j}. \tag{4.47}$$

Table 4.24 Aggregate probabilities for Example 4.27

x	$f_S(x)$	$F_S(x)$	x	$f_S(x)$	$F_S(x)$
0	9.5273905E–01	0.95273905	38	6.6436439E–06	0.97333493
14	4.7660783E–04	0.95321566	39	7.5742253E–06	0.97334251
15	1.4216995E–03	0.95463736	40	8.4744508E–06	0.97335098
16	1.3548133E–03	0.95599217	41	7.9416543E–06	0.97335892
17	4.7660783E–04	0.95646878	42	2.2356100E–05	0.97338128
18	3.3750829E–03	0.95984386	43	6.1253961E–06	0.97338740
19	5.1475706E–04	0.96035862	44	2.1435448E–05	0.97340884
20	1.2210690E–03	0.96157969	45	4.6721584E–06	0.97341351
24	4.6336840E–03	0.96621337	46	1.2100769E–05	0.97342561
26	3.7686403E–03	0.96998201	47	2.7915140E–06	0.97342840
28	1.1637614E–03	0.97114577	48	5.5621896E–05	0.97343397
29	7.1120536E–07	0.97114649	49	4.6964950E–06	0.97343866
30	9.8301077E–04	0.97212950	50	2.0226469E–05	0.97345889
31	1.1755723E–03	0.97330507	51	1.5103585E–06	0.97346040
32	2.3995910E–06	0.97330747	52	5.6661624E–06	0.97346607
33	5.9716305E–06	0.97331344	53	1.3705553E–08	0.97346608
34	6.1784053E–06	0.97331962	54	9.3923168E–06	0.97347547
35	4.2424878E–06	0.97332386	55	4.5913084E–03	0.97806678
36	1.9938909E–06	0.97332585	56	3.9003832E–06	0.97807068
37	2.4343694E–06	0.97332829	57	4.6823253E–06	0.97807536

This distribution has pf

$$\Pr(X = x) = \frac{1}{\lambda} \sum_{x=b_j} \lambda_j. \tag{4.48}$$

The numerator sums all probabilities associated with amount b_j.

Note that the means of the frequency distribution and the aggregate loss distribution match those of the exact distribution.

Example 4.29 (Example 4.25 continued) *Consider the group life case of Example 4.25. Derive a compound Poisson approximation.*

Using the compound Poisson approximation of this section with Poisson parameter $\lambda = \sum q_j = 0.04813$, the distribution function given in Table 4.26 is obtained.

When these values are compared to those of Example 4.26, it can be seen that the maximum error of 0.0002708 occurs at $x = 0$. □

Table 4.25 Aggregate distribution for Example 4.28

x	$F_S(x)$	x	$F_S(x)$	x	$F_S(x)$	x	$F_S(x)$
0	0.00789581	200	0.51793382	400	0.96031865	600	0.99927281
8	0.00789581	208	0.55208736	408	0.96528977	608	0.99939248
16	0.01059183	216	0.57594360	416	0.96999808	616	0.99950270
24	0.01906263	224	0.60583632	424	0.97360199	624	0.99958630
32	0.02561396	232	0.63412023	432	0.97694855	632	0.99965590
40	0.02979597	240	0.66687534	440	0.98031827	640	0.99971784
48	0.03774849	248	0.68632432	448	0.98297484	648	0.99976679
56	0.04770335	256	0.71292641	456	0.98515668	656	0.99980876
64	0.06976756	264	0.73984823	464	0.98728185	664	0.99984205
72	0.07610528	272	0.76061061	472	0.98914069	672	0.99987063
80	0.10013432	280	0.78007667	480	0.99068429	680	0.99989481
88	0.12399672	288	0.80016248	488	0.99194854	688	0.99991371
96	0.13839960	296	0.82073766	496	0.99321771	696	0.99992950
104	0.15945674	304	0.83672920	504	0.99421982	704	0.99994275
112	0.18004988	312	0.85121374	512	0.99504627	712	0.99995364
120	0.22162539	320	0.86849112	520	0.99580992	720	0.99996222
128	0.23569706	328	0.88253629	528	0.99644624	728	0.99996932
136	0.26335063	336	0.89343529	536	0.99701494	736	0.99997545
144	0.30172723	344	0.90530546	544	0.99745896	744	0.99998004
152	0.33041683	352	0.91610001	552	0.99785787	752	0.99998383
160	0.35497038	360	0.92599796	560	0.99821941	760	0.99998707
168	0.38635038	368	0.93340851	568	0.99850204	768	0.99998955
176	0.42177800	376	0.94180939	576	0.99873887	776	0.99999162
184	0.45476409	384	0.94903173	584	0.99895067	784	0.99999326
192	0.47881084	392	0.95482932	592	0.99912971	792	0.99999461

Some closely related approximations have been used. One popular one is to let λ_j in (4.46) be set to

$$\lambda_j = -\log(1 - q_j), \qquad j = 1, 2, ..., n. \tag{4.49}$$

This matches the "no loss" probability $1-q_j$ with the no loss probability of the Poisson distribution, $e^{-\lambda_j}$. This effectively replaces each life in the group by a Poisson distribution. This approximation is appropriate in the context of a group life insurance contract where a life is "replaced" upon death, leaving the Poisson intensity unchanged by the death. Naturally the expected number of losses is greater than $\sum_{j=1}^{n} q_j$. An alternative choice was proposed by Kornya [79]. It used $\lambda_j = q_j/(1 - q_j)$ in (4.46). It results in an expected number of losses that exceeds that using (4.49) (see Exercise 4.60).

4.12.3.1 More than one possible loss amount It was noted in the beginning of this section that there may be more than one possible loss amount. Again let B_j be the random variable that measures the amount of the loss,

Table 4.26 Aggregate distribution for Example 4.29

x	$F_S(x)$	x	$F_S(x)$	x	$F_S(x)$	x	$F_S(x)$
0	0.9530099	20	0.9618348	40	0.9735771	60	0.9990974
1	0.9530099	21	0.9618348	41	0.9735850	61	0.9990986
2	0.9530099	22	0.9618348	42	0.9736072	62	0.9990994
3	0.9530099	23	0.9618348	43	0.9736133	63	0.9990995
4	0.9530099	24	0.9664473	44	0.9736346	64	0.9990995
5	0.9530099	25	0.9664473	45	0.9736393	65	0.9990996
6	0.9530099	26	0.9702022	46	0.9736513	66	0.9990997
7	0.9530099	27	0.9702022	47	0.9736541	67	0.9990997
8	0.9530099	28	0.9713650	48	0.9736708	68	0.9990998
9	0.9530099	29	0.9713657	49	0.9736755	69	0.9991022
10	0.9530099	30	0.9723490	50	0.9736956	70	0.9991091
11	0.9530099	31	0.9735235	51	0.9736971	71	0.9991156
12	0.9530099	32	0.9735268	52	0.9737101	72	0.9991179
13	0.9530099	33	0.9735328	53	0.9737102	73	0.9991341
14	0.9534864	34	0.9735391	54	0.9737195	74	0.9991470
15	0.9549064	35	0.9735433	55	0.9782901	75	0.9991839
16	0.9562597	36	0.9735512	56	0.9782947	76	0.9992135
17	0.9567362	37	0.9735536	57	0.9782994	77	0.9992239
18	0.9601003	38	0.9735604	58	0.9783006	78	0.9992973
19	0.9606149	39	0.9735679	59	0.9783021	79	0.9993307

given that there was one and let $X_j = I_j B_j$. Then

$$P_{X_j}(z) = [1 - q_j + q_j P_{B_j}(z)].$$

The pgf corresponding to (4.31) is

$$P_S(z) = \prod_{j=1}^{n} [1 - q_j + q_j P_{B_j}(z)].$$

Although it is possible to extend the exact computational methods to this case, it is quite cumbersome. However, the compound Poisson approximation based on matching of moments (4.46) simply requires replacing z^{b_j} by $P_{B_j}(z)$. Then, the pgf of the severity distribution (4.47) becomes

$$P_X(z) = \frac{1}{\lambda} \sum_{j=1}^{n} \lambda_j P_{B_j}(z)$$

and so

$$f_X(x) = \frac{1}{\lambda} \sum_{j=1}^{n} \lambda_j f_{B_j}(x) \tag{4.50}$$

which is a weighted average of the n individual severity densities. Extensions to continuous severity distributions also satisfy (4.50) for all values of x.

Example 4.30 (Example 4.24 continued) *Develop compound Poisson approximations using all three methods suggested here. Compute the mean and variance for each approximation and compare it to the exact value.*

Using the method that matches the mean, we have $\lambda = 50(0.01)+25(0.01) = 0.75$. The severity distribution is

$$\begin{aligned} f_X(50{,}000) &= \frac{50(0.01)(0.7)}{0.75} = 0.4667 \\ f_X(75{,}000) &= \frac{25(0.01)(0.7)}{0.75} = 0.2333 \\ f_X(100{,}000) &= \frac{50(0.01)(0.3)}{0.75} = 0.2000 \\ f_X(150{,}000) &= \frac{25(0.01)(0.3)}{0.75} = 0.1000. \end{aligned}$$

The mean is $\lambda E(X) = 0.75(75{,}833.33) = 56{,}875$, which matches the exact value; and the variance is $\lambda E(X^2) = 0.75(6{,}729{,}166{,}667) = 5{,}046{,}875{,}000$, which exceeds the exact value.

For the method that preserves the probability of no losses, $\lambda = -75\log(0.99) = 0.753775$. For this method, the severity distribution turns out to be exactly the same as before (this is due to the fact that all individuals have the same value of q_j). Thus the mean is 57,161 and the variance is 5,072,278,876, both of which exceed the previous approximate values.

Using Kornya's method, $\lambda = 75(0.01)/0.99 = 0.757576$ and again the severity distribution is unchanged. The mean is 57,449 and the variance is 5,097,853,535, which are the largest values of all.

4.13 EXERCISES

Section 4.1

4.1 Show that the costs to the insured, the insurer, and the reinsurer in Example 4.1 sum to the aggregate loss.

Section 4.2

4.2 For pgfs satisfying (4.2), show that the mean is proportional to α.

4.3 Which of the distributions in Figures 3.2–3.5 satisfy (4.2) for any positive value of α?

Section 4.3

4.4 From (4.5), show that the relationships between the moments in (4.6) hold.

4.5 When an individual is admitted to the hospital, the hospital charges have the following characteristics:

1.

Charges	Mean	Standard deviation
Room	1,000	300
Other	500	300

2. The covariance between an individual's room charges and other charges is 100,000.

An insurer issues a policy that reimburses 100% for Room Charges and 80% for Other Charges. The number of hospital admissions has a Poisson distribution with parameter 4. Determine the mean and standard deviation of the insurer's payout for the policy.

4.6 Aggregate claims have been modeled by a compound negative binomial distribution with parameters $r = 15$ and $\beta = 5$. The claim amounts are uniformly distributed on the interval $(0, 10)$. Using the normal approximation, determine the premium such that the probability that claims will exceed premium is 0.05.

4.7 Automobile drivers can be divided into three homogeneous classes. The number of claims for each driver follows a Poisson distribution with parameter λ. Determine the variance of the number of claims for a randomly selected driver.

Class	Proportion of population	λ
1	0.25	5
2	0.25	3
3	0.50	2

4.8 X_1, X_2, and X_3 are mutually independent loss random variables with probability functions as given in Table 4.27. Determine the pf of $S = X_1 + X_2 + X_3$.

Table 4.27 Distributions for Exercise 4.8

x	$f_1(x)$	$f_2(x)$	$f_3(x)$
0	0.90	0.50	0.25
1	0.10	0.30	0.25
2	0.00	0.20	0.25
3	0.00	0.00	0.25

Table 4.28 Distributions for Exercise 4.9

x	$f_1(x)$	$f_2(x)$	$f_3(x)$
0	p	0.6	0.25
1	$1-p$	0.2	0.25
2	0	0.1	0.25
3	0	0.1	0.25

4.9 X_1, X_2, and X_3 are mutually independent random variables with probability functions as given in Table 4.28. If $S = X_1 + X_2 + X_3$ and $f_S(5) = 0.06$, determine p.

4.10 Consider the following information about AIDS patients.

1. The conditional distribution of an individual's medical care costs, given that the individual does not have AIDS, has mean 1,000 and variance 250,000.

2. The conditional distribution of an individual's medical care costs, given that the individual does have AIDS, has mean 70,000 and variance 1,600,000.

3. The number of individuals with AIDS in a group of m randomly selected adults has a binomial distribution with parameters m and $q = 0.01$.

An insurance company determines premiums for a group as the mean plus ten percent of the standard deviation of the group's aggregate claims distribution. P is the premium for a group of 10 independent lives for which all individuals have been proven *not* to have AIDS. Q is the premium for a group of 10 randomly selected adults. Determine P/Q.

4.11 You have been asked by a city planner to analyze office cigarette smoking patterns. The planner has provided the information in Table 4.29 about the distribution of the number of cigarettes smoked during a workday.

Table 4.29 Data for Exercise 4.11

	Male	Female
Mean	6	3
Variance	64	31

The number of male employees in a randomly selected office of N employees has a binomial distribution with parameters N and 0.4. Determine the mean plus the standard deviation of the number of cigarettes smoked during a workday in a randomly selected office of 8 employees.

4.12 For a certain group, aggregate claims are uniformly distributed over $(0, 10)$. Insurer A proposes stop-loss coverage with a deductible of 6 for a premium equal to the expected stop-loss claims. Insurer B proposes group coverage with a premium of 7 and a dividend (a premium refund) equal to the excess, if any, of $7k$ over claims. Calculate k such that the expected cost to the group is equal under both proposals.

4.13 For a group health contract, aggregate claims are assumed to have an exponential distribution where the mean of the distribution is estimated by the group underwriter. Aggregate stop-loss insurance for total claims in excess of 125% of the expected claims is provided by a gross premium that is twice the expected stop-loss claims. You have discovered an error in the underwriter's method of calculating expected claims. The underwriter's estimate is 90% of the correct estimate. Determine the actual percentage loading in the premium.

4.14 A random loss, X, has the following probability function. You are given that $E(X) = 4$ and $E[(X - d)_+] = 2$. Determine d.

x	$f(x)$
0	0.05
1	0.06
2	0.25
3	0.22
4	0.10
5	0.05
6	0.05
7	0.05
8	0.05
9	0.12

4.15 A reinsurer pays aggregate claim amounts in excess of d, and in return it receives a stop-loss premium $E[(S-d)_+]$. You are given $E[(S-100)_+] = 15$, $E[(S-120)_+] = 10$, and the probability that the aggregate claim amounts are greater than 80 and less than or equal to 120 is 0. Determine the probability that the aggregate claim amounts are less than or equal to 80.

4.16 A loss random variable X has pdf $f(x) = \frac{1}{100}$, $0 < x < 100$. Two policies can be purchased to alleviate the financial impact of the loss.

$$A = \begin{cases} 0, & x < 50k \\ \dfrac{x}{k} - 50, & x \geq 50k \end{cases}$$

and

$$B = kx, \qquad 0 < x < 100$$

where A and B are the amounts paid when the loss is x. Both policies have the same net premium, that is, $E(A) = E(B)$. Determine k.

4.17 For a nursing home insurance policy, you are given that the average length of stay is 440 days and 30% of the stays are terminated in the first 30 days. These terminations are distributed uniformly during that period. The policy pays 20 per day for the first 30 days and 100 per day thereafter. Determine the expected benefits payable for a single stay.

4.18 An insurance portfolio produces N claims, where

n	$\Pr(N = n)$
0	0.5
1	0.4
3	0.1

Individual claim amounts have the following distribution.

x	$f_X(x)$
0	0.9
10	0.1

Individual claim amounts and N are mutually independent. Calculate the probability that the ratio of aggregate claims to expected claims will exceed 3.0.

4.19 A company sells group travel-accident life insurance with m payable in the event of a covered individual's death in a travel accident. The gross premium for a group is set equal to the expected value plus the standard deviation of the group's aggregate claims. The standard premium is based on the following assumptions:

1. All individual claims within the group are mutually independent, and
2. $m^2q(1-q) = 2{,}500$, where q is the probability of death by travel accident for an individual.

In a certain group of 100 lives, the independence assumption fails because three specific individuals always travel together. If one dies in an accident, all three are assumed to die. Determine the difference between this group's premium and the standard premium.

4.20 A life insurance company covers 16,000 lives for 1-year term life insurance, as shown below.

Benefit amount	Number covered	Probability of claim
1	8,000	0.025
2	3,500	0.025
4	4,500	0.025

All claims are mutually independent. The insurance company's retention limit is 2 units per life. Reinsurance is purchased for 0.03 per unit. $\Pr\left[\dfrac{S-E(S)}{\sqrt{Var(S)}} > K\right]$ is equal to the probability that the insurance company's retained claims, S, plus cost of reinsurance will exceed 1,000. Determine K using a normal approximation.

4.21 The probability density function of individual losses Y is

$$f(y) = \begin{cases} 0.02\left(1 - \dfrac{y}{100}\right), & 0 < y < 100 \\ 0, & \text{elsewhere.} \end{cases}$$

The amount paid, Z, is 80% of that portion of the loss that exceeds a deductible of 10. Determine $E(Z)$.

4.22 An individual loss distribution is normal with $\mu = 100$ and $\sigma^2 = 9$. The distribution for the number of claims, N, is given in Table 4.30

Determine the probability that aggregate claims exceed 100.

4.23 An employer self-insures a life insurance program with the following characteristics.

Table 4.30 Distribution for Exercise 4.22

n	$\Pr(N = n)$
0	0.5
1	0.2
2	0.2
3	0.1

1. Given that a claim has occurred, the claim amount is 2,000 with probability 0.4, and 3,000 with probability 0.6; and

2. The number of claims has the distribution given in Table 4.31.

Table 4.31 Distribution for Exercise 4.23

n	$f(n)$
0	1/16
1	1/4
2	3/8
3	1/4
4	1/16

4.24 The employer purchases aggregate stop-loss coverage that limits the employer's annual claims cost to 5,000. The aggregate stop-loss coverage costs 1,472. Determine the employer's expected annual cost of the program, including the cost of stop-loss coverage.

4.25 The probability that an individual admitted to the hospital will stay k days or less is $1 - 0.8^k$ for $k = 0, 1, 2, \ldots$. A hospital indemnity policy provides a fixed amount per day for the fourth day through the tenth day (that is, for a maximum of 7 days). Determine the percentage increase in the expected cost per admission if the maximum number of days paid is increased from 7 to 14.

4.26 The probability density function of aggregate claims, S, is given by $f_S(x) = 3x^{-4}$, $x \geq 1$. The relative loading θ and the value λ are selected so that

$$\Pr[S \leq (1+\theta)E(S)] = \Pr\left[S \leq E(S) + \lambda\sqrt{Var(S)}\right] = 0.90.$$

Calculate λ and θ.

Section 4.4

4.27 The following questions concern closure under convolution.

(a) Show that the gamma and inverse Gaussian distributions are closed under convolution. Show that of the two distributions, only the gamma has the additional property mentioned in Example 4.9.

(b) Discrete distributions can also be used as severity distributions. Which of the distributions in Appendix B are closed under convolution? How can this information be used in simplifying calculation of compound probabilities of the form (3.28)?

4.28 A compound negative binomial distribution has parameters $\beta = 1$, $r = 2$, and severity distribution $\{f_X(x);\ x = 0, 1, 2, \ldots\}$. How do the parameters of the distribution change if the severity distribution is $\{g_X(x) = f_X(x)/[1 - f_X(0)];\ x = 1, 2, \ldots\}$ but the aggregate claims distribution remains unchanged?

4.29 Consider the compound logarithmic distribution with exponential severity distribution.

(a) Show that the density of aggregate losses may be expressed as

$$f_S(x) = \frac{1}{\log(1+\beta)} \sum_{n=1}^{\infty} \frac{1}{n!} \left[\frac{\beta}{\theta(1+\beta)}\right]^n x^{n-1} e^{-x/\theta}.$$

(b) Reduce this to

$$f_S(x) = \frac{\exp\left[-\dfrac{x}{\theta(1+\beta)}\right] - \exp\left(-\dfrac{x}{\theta}\right)}{x \log(1+\beta)}.$$

4.30 An insurance policy reimburses aggregate incurred expenses at the rate of 80% of the first 1,000 in excess of 100, 90% of the next 1,000, and 100% thereafter. Express the expected cost of this coverage in terms of $R_d = E[(S-d)_+]$ for different values of d.

4.31 The number of accidents incurred by an insured driver in a single year has a Poisson distribution with parameter $\lambda = 2$. If an accident occurs, the probability that the damage amount exceeds the deductible is 0.25. The number of claims and the damage amounts are independent. What is the probability that there will be no damages exceeding the deductible in a single year?

4.32 The aggregate loss distribution is modeled by an insurance company using an exponential distribution. However, the mean is uncertain. The company uses a uniform distribution (2,000,000, 4,000,000) to express its view of what the mean should be. Determine the expected aggregate losses.

4.33 A group hospital indemnity policy provides benefits at a continuous rate of 100 per day of hospital confinement for a maximum of 30 days. Benefits for partial days of confinement are pro-rated. The length of hospital confinement in days, T, has the following continuance (survival) function for $0 \leq t \leq 30$:

$$Pr(T \geq t) = \begin{cases} 1 - 0.04t, & 0 \leq t \leq 10 \\ 0.95 - 0.035t, & 10 < t \leq 20 \\ 0.65 - 0.02t, & 20 < t \leq 30. \end{cases}$$

For a policy period, each member's probability of a single hospital admission is .1 and of more than one admission is 0. Determine the pure premium per member, ignoring the time value of money.

4.34 Medical and dental claims are assumed to be independent with compound Poisson distributions as follows.

Claim type	Claim amount distribution	λ
Medical claims	Uniform $(0, 1{,}000)$	2
Dental claims	Uniform $(0, 200)$	3

Let X be the amount of a given claim under a policy which covers both medical and dental claims. Determine $E[(X - 100)_+]$, the expected cost (in excess of 100) of any given claim.

4.35 For a certain insured, the distribution of aggregate claims is binomial with parameters $m = 12$ and $q = 0.25$. The insurer will pay a dividend, D, equal to the excess of 80% of the premium over claims, if positive. The premium is 5. Determine $E[D]$.

4.36 Consider a severity distribution which is a finite mixture of gamma distributions with integer shape parameters, that is, which may be expressed as

$$f_X(x) = \sum_{k=1}^{r} q_k \frac{\theta^{-k} x^{k-1} e^{-x/\theta}}{(k-1)!}, \quad x > 0.$$

(a) Show that the moment-generating function may be written as

$$M_X(z) = Q\{(1 - \theta z)^{-1}\}$$

where

$$Q(z) = \sum_{k=1}^{r} q_k z^k$$

is the pgf of the distribution $\{q_1, q_2, ..., q_r\}$. Thus interpret $f_X(x)$ as the pf of a compound distribution.

(b) Show that the mgf of S is

$$M_S(z) = C\{(1-\theta z)^{-1}\}$$

where

$$C(z) = \sum_{k=0}^{\infty} c_k z^k = P_N\{Q(z)\}.$$

(c) Describe how the distribution $\{c_k;\ k = 0, 1, 2, ...\}$ may be calculated recursively if the number of claims distribution is a member of the $(a, b, 1)$ class (Section 3.6).

(d) Show that the distribution function of S is given by

$$\begin{aligned} F_S(x) &= 1 - \sum_{n=1}^{\infty} c_n \sum_{j=0}^{n-1} \frac{(x/\theta)^j e^{-x/\theta}}{j!} \\ &= 1 - e^{-x/\theta} \sum_{j=0}^{\infty} \bar{C}_j \frac{(x/\theta)^j}{j!}, \qquad x \geq 0 \end{aligned}$$

where $\bar{C}_j = \sum_{n=j+1}^{\infty} c_n$.

Section 4.6

4.37 Show that the method of local moment matching with $k = 1$ (matching total probability and the mean) using (4.21) and (4.22) results in

$$f_0 = m_0^0 f_0 = 1 - E[X \wedge h]/h$$

$$\begin{aligned} f_{i+1} &= m_1^i + m_0^{i+1} \\ &= \{2E[X \wedge ih] - E[X \wedge (i-1)h] - E[X \wedge (i+1)h]\}/h, \quad i = 0, 1, 2, ... \end{aligned}$$

and that $\{f_i;\ i = 0, 1, 2, ...\}$ forms a valid distribution with the same mean as the original severity distribution. Using the formula given here, verify the formula given in Example 4.10.

4.38 You are the agent for a baseball player who desires an incentive contract that will pay the amounts given in Table 4.32.

The number of times at bat has a Poisson distribution with $k = 200$. The parameter, x, is determined so that the probability of the player earning at least 4,000,000 is 0.95. Determine the player's expected compensation.

Table 4.32 Data for Exercise 4.38

Type of hit	Probability of hit per time at bat	Compensation per hit
Single	0.14	x
Double	0.05	$2x$
Triple	0.02	$3x$
Home run	0.03	$4x$

4.39 A weighted average of two Poisson distributions

$$p_k = w\frac{e^{-\lambda_1}\lambda_1^k}{k!} + (1-w)\frac{e^{-\lambda_2}\lambda_2^k}{k!}$$

has been used by some authors, e.g., Tröbliger [120], to treat drivers as either "good" or "bad" (see Example 3.26).

(a) Find the pgf $P_N(z)$ of the number of losses in terms of the two pgfs $P_1(z)$ and $P_2(z)$ of the number of losses of the two types of drivers.

(b) Let $f_X(x)$ denote a severity distribution defined on the non-negative integers. How can (4.16) be used to compute the distribution of aggregate claims for the entire group?

(c) Can this be extended to other frequency distributions?

4.40 A compound Poisson aggregate loss model has 5 expected claims per year. The severity distribution is defined on positive multiples of 1,000. Given that $f_S(1) = e^{-5}$ and $f_S(2) = \frac{5}{2}e^{-5}$, determine $f_X(2)$.

4.41 For a compound Poisson distribution, $\lambda = 6$ and individual losses have pf $f_X(1) = f_X(2) = f_X(4) = 1/3$. Some of the pf values for the aggregate distribution S are:

x	$f_S(x)$
3	0.0132
4	0.0215
5	0.0271
6	$f(6)$
7	0.0410

Determine $f_S(6)$.

4.42 Consider the $(a, b, 0)$ class of frequency distributions and any severity distribution defined on the positive integers $\{1, 2, ..., M < \infty\}$, where M is the maximum possible single loss.

(a) Show that for the compound distribution, the following "backwards recursion" holds:

$$f_X(x) = \frac{f_S(x+M) - \sum_{y=1}^{M-1}\left(a + b\frac{M-y}{x+M}\right) f_X(M-y) f_S(x+y)}{\left(a + b\frac{M}{x+M}\right) f_X(M)}$$

(b) For the binomial (m, q) frequency distribution, how can the above formula be used to obtain the distribution of aggregate losses? See Panjer and Wang [100].

4.43 Aggregate claims are compound Poisson with $\lambda = 2$, $f_X(1) = 1/4$, and $f_X(2) = 3/4$. For a premium of 6 an insurer covers aggregate claims and agrees to pay a dividend (a refund of premium) equal to the excess, if any, of 75% of the premium over 100% of the claims. Determine the excess of premium over expected claims and dividends.

4.44 On a given day, a physician provides medical care to N_A adults and N_C children. N_A and N_C have Poisson distributions with parameters 3 and 2, respectively. The distributions of length of care per patient are as follows.

	Adult	Child
1 hour	0.4	0.9
2 hour	0.6	0.1

N_A, N_C, and the lengths of care for all individuals are independent. The physician charges 200 per hour of patient care. Determine the probability that the office income on a given day is less than or equal to 800.

4.45 A group policyholder's aggregate claims, S, has a compound Poisson distribution with $\lambda = 1$ and all claim amounts equal to 2. The insurer pays the group the following dividend:

$$D = \begin{cases} 6 - S, & S < 6 \\ 0, & S \geq 6. \end{cases}$$

Determine $E[D]$.

4.46 You are given two independent compound Poisson random variables S_1 and S_2, where $f_j(x)$, $j = 1, 2$, are the two single-claim size distributions. You are given $\lambda_1 = \lambda_2 = 1$, $f_1(1) = 1$; and $f_2(1) = f_2(2) = 0.5$. Let $F_X(x)$ be the single-claim size distribution function associated with the compound distribution $S = S_1 + S_2$. Calculate $F_X^{*4}(6)$.

4.47 S has a compound Poisson claims distribution with the following:

1. Individual claim amounts equal to 1, 2, or 3.
2. $E(S) = 56$.
3. $Var(S) = 126$.
4. $\lambda = 29$.

Determine the expected number of claims of size 2.

4.48 For a compound Poisson distribution with positive integer claim amounts, the probability function follows:

$$f_S(x) = \frac{1}{x}[0.16 f_S(x-1) + k f_S(x-2) + 0.72 f_S(x-3)], \qquad x = 1, 2, 3, \ldots.$$

The expected value of aggregate claims is 1.68. Determine the expected number of claims.

4.49 For a portfolio of policies you are given the following:

1. The number of claims has a Poisson distribution.
2. Claim amounts can be 1, 2, or 3.
3. A stop-loss reinsurance contract has net premiums for various deductibles as follows:

Deductible	Net premium
4	0.20
5	0.10
6	0.04
7	0.02

Determine the probability that aggregate claims will be either 5 or 6.

4.50 For group disability income insurance, the expected number of disabilities per year is 1 per 100 lives covered. The continuance (survival) function for the length of a disability in days, Y, is

$$\Pr(Y > y) = 1 - \frac{y}{10}, \qquad y = 0, 1, \ldots, 10.$$

The benefit is 20 per day following a 5-day waiting period. Using a compound Poisson distribution, determine the variance of aggregate claims for a group of 1,500 independent lives.

4.51 A population has two classes of drivers. The number of accidents per individual driver has a geometric distribution. For a driver selected at random from Class I, the geometric distribution parameter has a uniform distribution over the interval $(0, 1)$. Twenty-five percent of the drivers are in Class I. All drivers in Class II have expected number of claims 0.25. For a driver selected at random from this population, determine the probability of exactly two accidents.

Note: The following two Exercises require the use of a computer.

4.52 A policy covers physical damage incurred by the trucks in a company's fleet. The number of losses in a year has a Poisson distribution with $\lambda = 5$. The amount of a single loss has a gamma distribution with $\alpha = 0.5$ and $\theta = 2{,}500$. The insurance contract pays a maximum annual benefit of 20,000. Determine the probability that the maximum benefit will be paid. Use a span of 100 and the method of rounding.

4.53 An individual has purchased health insurance for which he pays 10 for each physician visit and 5 for each prescription. The probability that a payment will be 10 is 0.25, and the probability that it will be 5 is 0.75. The total number of payments per year has the Poisson–Poisson (Neyman Type A) distribution with $\lambda_1 = 10$ and $\lambda_2 = 4$. Determine the probability that total payments in one year will exceed 400. Compare your answer to a normal approximation.

Section 4.7

4.54 Repeat Exercises 4.52 and 4.53 using the inversion method.

Section 4.8

4.55 Repeat Exercises 4.52 and 4.53 using simulation. Perform enough simulations so that there is a 90% probability that the relative error in your answer is less than 1%. Note that if a random variable, X, has the gamma distribution with $\alpha = 0.5$ and θ unspecified, then $X = \theta Z^2/2$ where Z has a standard normal distribution. Thus values of X can be simulated by first simulating a standard normal observation (see Appendix H) and then using the relationship to obtain X.

Section 4.9

4.56 Let the frequency (of losses) distribution be negative binomial with $r = 2$ and $\beta = 2$. Let the severity distribution (of losses) have the gamma distribution with $\alpha = 4$ and $\theta = 25$. Determine $F_S(200)$ and $E(S \wedge 200)$ for an ordinary per-loss deductible of 25 and then for a disappearing deductible with $d = 25$ and $d' = 100$. In both cases use the recursive formula to obtain the aggregate distribution and use a discretization interval of 5 with the method of rounding to discretize the severity distribution.

4.57 *(Exercise 4.52 continued)* Recall that the number of claims has a Poisson distribution with $\lambda = 5$ and the amount of a single claim has a gamma distribution with $\alpha = 0.5$ and $\theta = 2{,}500$. Determine the mean, standard deviation, and 90th percentile of payments by the insurance company under each of the following coverages. Any computational method may be used.

(a) A maximum aggregate payment of 20,000.

(b) A per claim ordinary deductible of 100 and a per claim maximum payment of 10,000. There is no aggregate maximum payment.

(c) A per claim ordinary deductible of 100 with no maximum payment. There is an aggregate ordinary deductible of 15,000, an aggregate coinsurance factor of .8, and a maximum insurance payment of 20,000. This corresponds to an aggregate reinsurance provision.

4.58 *(Exercise 4.53 continued)* Recall that the number of payments has the Poisson–Poisson distribution with $\lambda_1 = 10$ and $\lambda_2 = 4$ while the payment per claim by the insured is 5 with probability .75 and 10 with probability 0.25. Determine the mean, standard deviation, and 90th percentile of payments by the insured under each of the following situations. Any computational method may be used.

(a) A maximum payment of 400.

(b) A coinsurance arrangement where the insured pays 100% up to an aggregate total of 300 and then pays 20% of aggregate payments above 300.

Section 4.12

4.59 Derive (4.32) and (4.33).

4.60 Demonstrate that the compound Poisson model given by $\lambda_j = q_j$ and (4.48) produces a model with the same mean as the exact distribution but with a larger variance. Then show that the one using $\lambda_j = -\log(1 - q_j)$ must produce a larger mean and even larger variance, and finally show that the one using $\lambda_j = q_j/(1-q_j)$ must produce the largest mean and variance of all.

4.61 Individual members of an insured group have independent claims. The claim distribution has the statistics given in Table 4.33.

Table 4.33 Data for Exercise 4.61

	Mean	Variance
Males	2	4
Females	4	10

The premium for a group with future claims S is the mean of S plus 2 times the standard deviation of S. If the genders of the members of a group of m members are not known, the number of males is assumed to have a binomial distribution with parameters m and $q = 0.4$. A is the premium for a group of 100 for which the genders of the members are not known. B is the premium for a group of 40 males and 60 females. Determine A/B.

4.62 An insurance company assumes claim probabilities for persons covered by its group life insurance contracts as given in Table 4.34.

Table 4.34 Data for Exercise 4.62

Class	Probability of claim
Smoker	0.02
Nonsmoker	0.01

A group of mutually independent lives has coverage of 1,000 per life. The company assumes that 20% of the lives are smokers. Based on this assumption, the premium is set equal to 110% of expected claims. If 30% of the lives are smokers, the probability that claims will exceed the premium is less than 0.20. Using the normal approximation, determine the minimum number of lives which must be in the group.

4.63 Based on the individual risk model with independent claims, the cumulative distribution function of aggregate claims for a portfolio of life insurance policies is as in Table 4.35.

One policy with face amount 100 and probability of claim 0.20 is increased in face amount to 200. Determine the probability that aggregate claims for the revised portfolio will not exceed 500.

4.64 A group life insurance contract covering independent lives is rated in the three age groupings as given in Table 4.36.

Table 4.35 Distribution for Exercise 4.63

x	$F_S(x)$
0	0.40
100	0.58
200	0.64
300	0.69
400	0.70
500	0.78
600	0.96
700	1.00

Table 4.36 Data for Exercise 4.64

Age group	Number in age group	Probability of claim per life	Mean of the exponential distribution of claim amounts
18–35	400	0.03	5
36–50	300	0.07	3
51–65	200	0.10	2

The insurer prices the contract so that the probability that claims will exceed the premium is 0.05. Using the normal approximation, determine the premium that the insurer will charge.

4.65 The probability model for the distribution of annual claims per member in a health plan is shown in Table 4.37.

Table 4.37 Data for Exercise 4.65

Service	Probability of claim	Distribution of annual charges given that a claim occcurs	
		Mean	Variance
Office visits	0.7	160	4,900
Surgery	0.2	600	20,000
Other services	0.5	240	8,100

Independence of costs and occurrences among services and members is assumed. Using the normal approximation, determine the minimum number of members that a plan must have such that the probability that actual charges will exceed 115% of the expected charges is less than 0.10.

4.66 An insurer has a portfolio of independent risks as given in Table 4.38.

Table 4.38 Data for Exercise 4.66

Class	Probability of claim	Benefit	Number of risks
Standard	0.2	k	3,500
Substandard	0.6	αk	2,000

The insurer sets α and k such that aggregate claims have expected value 100,000 and minimum variance. Determine α.

4.67 An insurance company has a portfolio of independent one-year term life policies as given in Table 4.39.

Table 4.39 Data for Exercise 4.67

Class	Number in class	Benefit amount	Probability of a claim
1	500	x	0.01
2	500	$2x$	0.02

The actuary approximates the distribution of claims in the individual model using the compound Poisson model in which the expected number of claims is the same as in the individual model. Determine the maximum value of x such that the variance of the compound Poisson approximation is less than 4,500.

4.68 An insurance company sold one-year term life insurance on a group of 2,300 independent lives as given in Table 4.40.

Table 4.40 Data for Exercise 4.68

Class	Benefit amount	Probability of death	Number of policies
1	100,000	0.10	500
2	200,000	0.02	500
3	300,000	0.02	500
4	200,000	0.10	300
5	200,000	0.10	500

The insurance company reinsures amounts in excess of 100,000 on each life. The reinsurer wishes to charge a premium that is sufficient to guarantee that it will lose money 5% of the time on such groups. Obtain the appropriate premium each of the following ways.

(a) Using a normal approximation to the aggregate claims distribution.

(b) Using a lognormal approximation.

(c) Using a gamma approximation.

(d) Using the compound Poisson approximation which matches the means.

(e) Carrying out the calculations exactly (using the method developed by de Pril, or some other method). This requires a computer.

4.14 CASE STUDY

4.14.1 The case study continued

In Chapters 2 and 3 we constructed models for the distributions of the number of losses and the amount of loss for actuarial liability. In particular, in Chapter 2 we determined that losses for all three types of actuary have the lognormal distribution with $\mu = 10.5430$ and $\sigma = 2.31315$. We also determined that the number of losses has the Poisson distribution, regardless of the type of actuary. However, the Poisson parameter differs by type of actuary.

In this chapter we acquired the tools with which we can determine the distribution of aggregate payments under a number of reinsurance schemes. We begin by considering excess of loss reinsurance in which the reinsurance pays the excess over a deductible, d, up to a maximum payment $u - d$, where u is the limit established in the primary coverage. There are two approaches available to create the distribution of reinsurer payments. The first is to work with the distribution of payments per payment. On this basis, the severity distribution is mixed, with pdf

$$f_Y(x) = \frac{f_X(x+d)}{1 - F_X(d)}, \qquad 0 \leq x < u - d$$

and discrete probability

$$\Pr(Y = u - d) = \frac{1 - F_X(u)}{1 - F_X(d)}.$$

This distribution would then be discretized for use with the recursive formula or the FFT, or approximated by a histogram for use with the Heckman–Meyers method. Regardless, the frequency distribution must be adjusted to reflect the distribution of the number of payments as opposed to the number of losses. The new Poisson parameter will be $\lambda[1 - F_X(d)]$.

The other option is to work with the distribution of payments per loss. On this basis, the severity distribution is again mixed, with pdf

$$f_Y(x) = f_X(x+d), \qquad 0 < x < u - d$$

Table 4.41 Excess of loss reinsurance, one policy

Deductible (10^6)	Limit (10^6)	Mean	Standard Deviation	C.V.
0.5	1	778	18,858	24.24
0.5	5	2,910	94,574	32.50
0.5	10	3,809	144,731	38.00
0.5	25	4,825	229,284	47.52
0.5	50	5,415	306,359	56.58
1.0	5	2,132	80,354	37.69
1.0	10	3,031	132,516	43.72
1.0	25	4,046	219,475	54.24
1.0	50	4,636	298,101	64.30
5.0	10	899	62,556	69.58
5.0	25	1,914	162,478	84.89
5.0	50	2,504	249,752	99.74
10.0	25	1,015	111,054	109.41
10.0	50	1,605	205,939	128.71

and discrete probabilities

$$\begin{aligned}\Pr(Y=0) &= F_X(d) \\ \Pr(Y=u-d) &= 1-F_X(u).\end{aligned}$$

As before, this distribution is approximated for use with the selected algorithm. In this case the Poisson parameter is left unchanged.

4.14.1.1 *Distribution for a single policy* We consider the distribution of losses for a single covered actuary for various combinations of d and u. We use the Poisson parameter for the combined group and have employed the recursive algorithm with a discretization interval of 10,000 and the method of rounding. In all cases the 90th and 99th percentiles are zero, indicating that most of the time the excess of loss reinsurance will involve no payments. This is not surprising because the probability there will be no losses is $\exp(-0.0154578) = 0.985$ and with the deductible, this probability is even higher. The mean, standard deviation, and coefficient of variation for various combinations of d and u are given in Table 4.41.

It is not surprising that the risk (as measured by the coefficient of variation, c.v.) increases when either the deductible or the limit is increased. It is also clear that the risk of insuring a single actuary is extreme.

4.14.1.2 *One hundred policies—excess of loss* We next consider the possibility of reinsuring 100 actuaries. If we assume that the same deductible and limit apply to all of them, the aggregate distribution requires only that

Table 4.42 Excess of loss reinsurance, 100 policies

Deductible (10^6)	Limit (10^6)	Mean (10^3)	Standard deviation (10^3)	C.V.	Percentiles (10^3) 90	99
0.5	5	291	946	3.250	708	4,503
0.5	10	381	1,447	3.800	708	9,498
0.5	25	482	2,293	4.752	708	11,674
1.0	5	213	804	3.769	190	4,002
1.0	10	303	1,325	4.372	190	8,997
1.0	25	405	2,195	5.424	190	11,085
5.0	10	90	626	6.958	0	4,997
5.0	25	191	1,625	8.489	0	6,886
10.0	25	102	1,111	10.941	0	1,854

the frequency be changed. When 100 independent Poisson random variables are added the sum has a Poisson distribution with the original parameter multiplied by 100. The same process was repeated with the revised Poisson parameter. The results appear in Table 4.42.

As must be the case with independent policies, the mean is 100 times the mean for one policy and the standard deviation is 10 times the standard deviation for one policy. This implies that the coefficient of variation will be one-tenth of its previous value. In all cases the 99th percentile is now above zero. This may make it appear that there is more risk, but in reality it just indicates that it is now more likely that a claim will be paid.

4.14.1.3 One hundred policies—aggregate stop-loss We now turn to the aggregate reinsurance. Here policies have no individual deductible, but do have a policy limit. In the notation used in this section, $d = 0$ and u is variable. There are again one hundred policies and this time the reinsurer pays all aggregate losses in excess of an aggregate deductible. For a given limit, the severity distribution is modified as before, the Poisson parameter is multiplied by 100 and then some algorithm is used to obtain the aggregate distribution. Let this distribution have cdf $F_S(s)$ or, in the case of a discretized distribution (as will be the output from the recursive algorithm or the FFT), a pf $f_S(s_i)$ for $i = 1, \ldots, n$. If the deductible is a, then the corresponding functions for the reinsurance distribution S_r are

$$\begin{aligned} F_{S_r}(s) &= F_S(s+a), \quad s \geq 0 \\ f_{S_r}(0) &= F_S(a) = \sum_{s_i \leq a} f_S(s_i) \\ f_{S_r}(r_i) &= f_S(r_i + a), \quad r_i = s_i - a, \ i = 1, \ldots, n. \end{aligned}$$

Moments and percentiles may be determined in the usual manner.

Table 4.43 Aggregate stop-loss reinsurance, 100 policies

Deductible (10^6)	Limit (10^6)	Mean (10^3)	Standard deviation (10^3)	C.V.	Percentiles (10^3) 90	99
0.5	5	322	1,003	3.11	863	4,711
0.5	10	412	1,496	3.63	863	9,504
0.5	25	513	2,331	4.54	863	11,895
1.0	5	241	879	3.64	363	4,211
1.0	10	331	1,389	4.19	363	9,004
1.0	25	433	2,245	5.19	363	11,395
2.5	5	114	556	4.86	0	2,711
2.5	10	204	1,104	5.40	0	7,504
2.5	25	306	2,013	6.58	0	9,895
5.0	5	13	181	13.73	0	211
5.0	10	103	714	6.93	0	5,004
5.0	25	205	1,690	8.26	0	7,395

Using the recursive formula with an interval of 10,000, results for various stop-loss deductibles and individual limits are given in Table 4.43. The results are similar to those for the excess of loss coverage. For the most part, as either the individual limit or the aggregate deductible are increased, the risk, as measured by the coefficient of variation, increases. The exception is when both the limit and the deductible are 5,000,000. This is a risky setting because it is the only one in which two losses are required before the reinsurance will take effect.

In Chapter 3 it was determined that the three different types of actuary have different Poisson parameters. Suppose it was known that of the 100 covered actuaries, 30 are life/health, 50 are pension, and 20 are property/liability actuaries. Using the separate Poisson estimates and incorporating the sample size we have Poisson parameters $30(0.0162249) = 0.486747$ for life/health, $50(0.0174087) = 0.870435$ for pension, and $20(0.0096121) = 0.192242$ for property/liability. There are three methods for obtaining the distribution of the sum of the three separate aggregate distributions.

1. Because the sum of independent Poisson random variables is still Poisson, the total number of losses has the Poisson distribution with parameter 1.549424. The common severity distribution remains the lognormal. This reduces to a single compound distribution which can be evaluated by any method.

2. Obtain the three aggregate distributions separately. If the recursive or FFT algorithms are used, the result will be three discrete distributions. The distribution of their sum can be obtained by using convolutions.

3. If the FFT or Heckman–Meyers algorithms are used, the three transforms can be found and then multiplied. The inverse transform is then taken of the product.

Each of the methods has advantages and drawbacks. The first method is restricted to those frequency distributions for which the sum has a known form. If the severity distributions are not identical, it may not be possible to combine them to form a single model. The major advantage is that if it is available, this method requires only one aggregate calculation.

The advantage of method two is that there is no restriction on the frequency and severity components of the components. The drawback is the expansion of computer storage. For example, if the first distribution requires 3,000 points, the second one 5,000 points, and the third one 2,000 points (with the same discretization interval being used for the three distributions), the combined distribution will require 10,000 points. More will be said about this at the end of this section.

The third method also has no restriction on the separate models. It has the same drawback as the second model, but here the expansion must be done in advance. That is, in the example, all three components must work with 10,000 points. There is no way to avoid this.

4.14.1.4 Numerical convolutions The remaining problem is expansion of the number of points required when performing numerical convolutions. The problem arises when the individual distributions use a large number of discrete points, to the point where the storage capacity of the computer becomes an obstacle. The following example is a small-scale version of the problem and indicates a simple solution.

Example 4.31 *The probability functions for two discrete distributions are given below. Suppose the maximum vector allowed by the computer program being used is of length 6. Determine an approximation to the probability function for the sum of the two random variables.*

x	$f_1(x)$	$f_2(x)$
0	0.3	0.4
2	0.2	0.3
4	0.2	0.2
6	0.2	0.1
8	0.1	0.0

The maximum possible value for the sum of the two random variables is 14 and would require a vector of length 8 to store. Usual convolutions produces the answer as given below.

x	0	2	4	6	8	10	12	14
$f(x)$	0.12	0.17	0.20	0.21	0.16	0.09	0.04	0.01

Table 4.44 Allocation of probabilities for Example 4.31

x	$f(x)$	Lower point	Probabiity	Upper point	Probability
0	0.12	0	0.1200		
2	0.17	0	0.0486	2.8	0.1214
4	0.20	2.8	0.1143	5.6	0.0857
6	0.21	5.6	0.1800	8.4	0.0300
8	0.16	5.6	0.0229	8.4	0.1371
10	0.09	8.4	0.0386	11.2	0.0514
12	0.04	11.2	0.0286	14.0	0.0114
14	0.01	14.0	0.0100		

With 6 points available, the span must be increased to $14/5 = 2.8$. We then do a sort of reverse interpolation, taking the probability at each point that is not a multiple of 2.8 and allocating it to the two nearest multiples of 2.8. For example, the probability of 0.16 at $x = 8$ is allocated to the points 5.6 and 8.4. Because 8 is 2.4/2.8 of the way from 5.6 to 8.4, six-sevenths of the probability is placed at 8.4 and the remaining one-seventh is placed at 5.6. The complete allocation process appears in Table 4.44. The probabilities allocated to each multiple of 2.8 are then combined to produce the approximation to the true distribution of the sum. The approximating distribution is given below.

x	0	2.8	5.6	8.4	11.2	14.0
$f(x)$	0.1686	0.2357	0.2886	0.2057	0.0800	0.0214

This method preserves both the total probability of one and the mean (both the true distribution and the approximating distribution have a mean of 5.2). □

One refinement that can eliminate some of the need for storage is to note that when a distribution requires a large vector, the probabilities at the end are likely to be very small. When they are multiplied to create the convolution, the probabilities at the ends of the new, long vector may be so small that they can be ignored. Thus those cells need not be retained and do not add to the storage problem.

Many more refinements are possible. In the Appendix to the article by Bailey [8] a method which preserves the first three moments is presented.

Table 4.45 Exposures and deaths by benefit amount

Benefit	Exposure	Deaths
0–4,999	172,947	141
5,000	146,612	133
7,500	211,014	187
10,000	75,536	102
15,000	63,399	95
20,000	8,424	19
25,000	2,918	7
30,000	11,779	20
30,000+	3,884	12

He also provides guidance with regard to the elimination or combination of storage locations with exceptionally small probability.

4.14.2 Exercises

4.69 The following exposure and death data was collected for a 1935 paper by Crane [25]. For each death benefit range the number of life-years exposed and the number of deaths is given. Using the data in Table 4.45, develop an individual risk model for the cost per life-year of providing this benefit. Use an open group approach.

4.70 In the same paper which produced the data for Exercise 3.60, Weber [126] looked at the cost of a single accident. He presented values from the empirical distribution as given in Table 4.46. He recommended a two-point mixture of exponential distributions. Does that model fit the data? Can you determine a better one? Using your model for the 34.5-month period developed in Exercise 3.60, determine the expected total payments and their variance if the insurance is for a one-year period and has a deductible of 50.

4.71 In 1990, Stanard and John [115] investigated the incorporation of discounting into the collective risk model. They are interested in the distribution (and of course, moments and percentiles) of the discounted cash flow. The random variable may be written

$$V = \sum_{t=1}^{\infty} v^t (NCF)_t$$

where $v = (1+i)^{-1}$ is the one-year discount rate and $(NCF)_t = C_t - C_{t-1}$ is the net cash flow for year t with C_t being the cumulative cash flow at time t.

Table 4.46 Loss per accident

Loss	Empirical cdf
0	0.0000
50	0.1783
100	0.3238
250	0.6395
500	0.8328
1,000	0.9229
2,500	0.9702
5,000	0.9874
∞	1.0000

This can be written $C_t = R_t - P_t$, where R_t is the cumulative premium and P_t is the cumulative benefit payments as of time t. In the paper, the authors study a number of reinsurance schemes. Here are three of them.

1. Flat. The premium of r is paid once at the beginning, so $R_t = r$ for all t. Then $C_t = r - P_t$ and except for the first year, the net cash flow is the benefits paid in that year.

2. Paid loss retro. Here the primary insurer receives a dividend if the total claims are small. If, in the end, total payments exceed $r/(a+1)$, then the premium of r is paid in full. However, if payments are less than $r/(a+1)$ the primary insurer receives a dividend of $r-(a+1)P$, where P is the ultimate total payments. Rather than wait for the ultimate payments to be realized, the premium is paid as the losses are incurred. Thus no premium is paid until the first benefit payment is made. This results in the fairly simple cumulative cash flow formula

$$C_t = \min(aP_t, r - P_t).$$

3. Aggregate deductible. The premium is paid up front, but until payments exceed the deductible of d, none are received. The cash flow is

$$C_t = \min(r, r + d - P_t).$$

For an example, consider an individual layer in which the reinsurance pays individual losses in excess of 250,000 with a maximum of 250,000 per loss. Suppose losses have the single-parameter Pareto distribution with $\theta = 250{,}000$ and $\alpha = 1.5$. Further suppose the number of losses in excess of 250,000 has the negative binomial distribution with $\beta = 0.032$ and $r = 300$. Finally, suppose the time from when the policy is sold to when a claim is paid has the exponential distribution with mean 3 years. Conduct simulations to compare the three reinsurance policies as follows:

1. For the flat rate policy, set r so that $E(V) = 0.2r$.

2. For the paid loss retro with $a = 1/3$ and the aggregate deductible with $d = 500{,}000$, determine r so that $E(V)$ has the same value as for the flat rate policy.

3. Which of the three policies would you prefer to buy (as the primary insurer)?

4. Which of the three policies would you prefer to sell (as the reinsurer)?

5

Credibility Theory

5.1 INTRODUCTION

Credibility theory is a set of quantitative tools which allows an insurer to perform prospective experience rating (adjust future premiums based on past experience) on a risk or group of risks. If the experience of a policyholder is consistently better than that assumed in the underlying manual rate (sometimes called the **pure premium**), then the policyholder may demand a rate reduction.

The policyholder's argument is as follows: The manual rate is designed to reflect the expected experience of the entire rating class and implicitly assumes that the risks are homogeneous. However, no rating system is perfect, and there always remains some heterogeneity in the risk levels after all the underwriting criteria are accounted for. Consequently, some policyholders will be better risks than that assumed in the underlying manual rate. Of course, the same logic dictates that a rate increase should be applied to a poor risk, but the policyholder in this situation is certainly not going to ask for a rate increase! Nevertheless, an increase may be necessary, due to considerations of equity and the economics of the situation.

The insurer is then forced to answer the following question: How much of the difference in experience of a given policyholder is due to random variation in the underlying claims experience and how much is due to the fact that the policyholder really is a better or worse risk than average for the given rating class? In other words, how credible is the policyholder's own experience? Two facts must be considered in this regard:

1. The more past information the insurer has on a given policyholder, the more "credible" the policyholder's own experience, all else being equal. In the same vein, in group insurance the experience of larger groups is more credible than that of smaller groups.

2. Competitive considerations may force the insurer to give full (using the past experience of the policyholder only and not the manual rate) or nearly full credibility to a given policyholder in order to retain the business.

There are other considerations which motivate the use of credibility. A policyholder who takes steps to reduce losses should be rewarded. Such actions reduce total losses to society, clearly a desirable consequence. Conversely, a policyholder who fails to employ loss control techniques should be penalized. Credibility ideas promote an obvious mechanism for the implementation of risk management ideas.

Another use for credibility is in the setting of rates for classification systems. For example, in workers compensation insurance there may be hundreds of occupational classes, some of which may provide very little data. In order to accurately estimate the expected cost for insuring these classes, it may be appropriate to combine the limited actual experience with some other information, such as past rates, or the experience of occupations that are closely related.

From a statistical perspective, credibility theory leads to a result that would appear to be counterintuitive. If experience from an insured, or group of insureds, is available, our statistical training may convince us to use the sample mean or some other unbiased estimator. But credibility theory tells us that it is optimal to give only partial weight to this experience and give the remaining weight to an estimator produced from other information. We will discover that what we sacrifice in terms of bias we gain in terms of reducing the average (squared) error.

Credibility theory allows an insurer to quantitatively formulate the above problem, and this chapter provides an introduction to this theory. A few relevant statistical concepts are reviewed in the next section. Many of them were already discussed in Chapter 2, but are repeated here so that this chapter is relatively self-contained.

Section 5.3 deals with **limited fluctuation credibility theory**, a subject developed in the early part of the 20th century. This provides a mechanism for assigning full (Subsection 5.3.1) or partial (Subsection 5.3.2) credibility to a policyholder's experience. The difficulty with this approach is the lack of a sound underlying mathematical theory justifying the use of these methods. Nevertheless, this approach provided the original treatment of the subject.

A classic paper by Bühlmann in 1967 [18] provided a statistical framework within which credibility theory has developed and flourished. While this approach, termed **greatest accuracy credibility theory,**[1] was formalized by Bühlmann, the basic ideas were around for some time. This approach is introduced in Section 5.4. The simplest model, that of Bühlmann [18], is discussed in Subsection 5.4.3. Practical improvements were made by Bühlmann and Straub in 1970 [20]. Their model is discussed in Subsection 5.4.4. The concept of exact credibility is presented in Subsection 5.4.5.

Practical use of the theory requires that unknown model parameters be estimated from the data. Nonparametric estimation (where the problem is somewhat model-free and the parameters are generic, such as the mean and variance) is considered in Subsection 5.5.1, semiparametric estimation (where some of the parameters are based on assuming particular distributions) in Subsection 5.5.2, and finally the fully parametric situation (where all parameters come from assumed distributions) in Subsection 5.5.3.

We close with a quote from Arthur Bailey in 1950 [7, p. 8] that aptly summarizes much of the history of credibility. We, too, must tip our hats to the early actuaries, who, with unsophisticated mathematical tools at their disposal, were able to come up with formulas that not only worked, but were very similar to those we carefully develop in this chapter.

> "It is at this point in the discussion that the ordinary individual has to admit that, while there seems to be some hazy logic behind the actuaries' contentions, it is too obscure for him to understand. The trained statistician cries 'Absurd! Directly contrary to any of the accepted theories of statistical estimation.' The actuaries themselves have to admit that they have gone beyond anything that has been proven mathematically, that all of the values involved are still selected on the basis of judgment, and that the only demonstration they can make is that, in actual practice, it works. Let us not forget, however, that they have made this demonstration many times. It does work!"

5.2 STATISTICAL CONCEPTS

In this section various statistical concepts relevant to credibility theory are presented. Much of the material is of a review nature and hence may be quickly glossed over by a reader with a good background in statistics. Nevertheless, there may be some material which may not have been seen before, and so this section should not be completely ignored, and subsequent sections will refer back to this material.

[1]The terms "limited fluctuation" and "greatest accuracy" go back at least as far as a 1943 paper by Arthur Bailey [6].

5.2.1 Conditional and mixed distributions

Suppose that X and Y are two random variables with joint probability function (pf) or probability density function (pdf)[2] $f_{X,Y}(x,y)$ and marginal pfs $f_X(x)$ and $f_Y(y)$, respectively. The conditional pf of X given that $Y = y$ is

$$f_{X|Y}(x|y) = \frac{f_{X,Y}(x,y)}{f_Y(y)}. \tag{5.1}$$

If X and Y are discrete random variables, then (5.1) is the conditional probability of the event '$X = x$' under the hypothesis that '$Y = y$.' If X and Y are continuous, then (5.1) may be interpreted as a definition. When X and Y are independent random variables,

$$f_{X,Y}(x,y) = f_X(x)f_Y(y)$$

and in this case (5.1) yields

$$f_{X|Y}(x|y) = f_X(x)$$

and we observe that the conditional and marginal distributions of X are identical.

Example 5.1 *Suppose X and Z are independent Poisson random variables with means λ_1 and λ_2, respectively. Let $Y = X + Z$. Demonstrate that $X|Y = y$ is binomial with parameters $m = y$ and $q = \lambda_1/(\lambda_1 + \lambda_2)$ (see, for example, [65, p. 131]).*

The conditional distribution of X given that $Y = y$ is

$$\begin{aligned}
f_{X|Y}(x|y) &= \frac{f_{X,Y}(x,y)}{f_Y(y)} \\
&= \frac{\Pr(X = x, Y = y)}{\Pr(Y = y)} \\
&= \frac{\Pr(X = x, Z = y - x)}{\Pr(Y = y)} \\
&= \frac{\Pr(X = x)\Pr(Z = y - x)}{\Pr(Y = y)} \\
&= \frac{\dfrac{\lambda_1^x e^{-\lambda_1}}{x!}\,\dfrac{\lambda_2^{y-x} e^{-\lambda_2}}{(y-x)!}}{\dfrac{(\lambda_1 + \lambda_2)^y\, e^{-\lambda_1 - \lambda_2}}{y!}}
\end{aligned}$$

[2] When it is unclear, or when the random variable may be continuous, discrete, or a mixture of the two, the term **probability function** and abbreviation pf will be used. The term **probability density function** and the abbreviation pdf will be used only when the random variable is known to be continuous.

$$= \frac{y!}{x!(y-x)!}\left(\frac{\lambda_1}{\lambda_1+\lambda_2}\right)^x\left(\frac{\lambda_2}{\lambda_1+\lambda_2}\right)^{y-x}$$

for $x = 0, 1, 2, \cdots, y$. This is a binomial distribution with parameters $m = y$ and $q = \lambda_1/(\lambda_1 + \lambda_2)$. □

Note that (5.1) may be rewritten as

$$f_{X,Y}(x,y) = f_{X|Y}(x|y)f_Y(y) \tag{5.2}$$

demonstrating that joint distributions may be constructed from products of conditional and marginal distributions. Since the marginal distribution of X may be obtained by integrating (or summing) y out of the joint distribution,

$$f_X(x) = \int f_{X,Y}(x,y)dy$$

we find using (5.2) that

$$f_X(x) = \int f_{X|Y}(x|y)f_Y(y)dy. \tag{5.3}$$

Formula (5.3) has an interesting interpretation as a mixed distribution (see Subsection 2.7.3 or [66, Sec. 2.7]). To see this, assume that the conditional distribution $f_{X|Y}(x|y)$ is one of the usual parametric distributions where y is a parameter, itself with distribution $f_Y(y)$.

Example 5.2 *Suppose that, given $\Theta = \theta$, X is Poisson distributed with mean θ, that is,*

$$f_{X|\Theta}(x|\theta) = \frac{\theta^x e^{-\theta}}{x!}, \qquad x = 0, 1, 2, \ldots,$$

and Θ is gamma distributed with pdf

$$f_\Theta(\theta) = \frac{\theta^{\alpha-1}e^{-\theta/\beta}}{\Gamma(\alpha)\beta^\alpha}, \qquad \theta > 0$$

where $\beta > 0$, and $\alpha > 0$ are parameters. Determine the marginal pf of X.

From (5.3), the marginal pf of X is

$$\begin{aligned} f_X(x) &= \int_0^\infty \frac{\theta^x e^{-\theta}}{x!}\frac{\theta^{\alpha-1}e^{-\theta/\beta}}{\Gamma(\alpha)\beta^\alpha}d\theta \\ &= \frac{1}{x!\Gamma(\alpha)\beta^\alpha}\int_0^\infty \theta^{\alpha+x-1}e^{-\theta/[\beta/(1+\beta)]}d\theta \\ &= \frac{\Gamma(\alpha+x)[\beta/(1+\beta)]^{\alpha+x}}{x!\Gamma(\alpha)\beta^\alpha}\int_0^\infty \frac{\theta^{\alpha+x-1}e^{-\theta/[\beta/(1+\beta)]}}{\Gamma(\alpha+x)[\beta/(1+\beta)]^{\alpha+x}}d\theta. \end{aligned}$$

The integrand in the above expression is that of a gamma pdf with β replaced by $\beta/(1+\beta)$ and α replaced by $\alpha + x$. Hence the integral is 1 and so

$$\begin{aligned} f_X(x) &= \frac{\Gamma(\alpha+x)}{x!\Gamma(\alpha)}\left(\frac{1}{1+\beta}\right)^{\alpha}\left(\frac{\beta}{1+\beta}\right)^{x} \\ &= \binom{\alpha+x-1}{x}p^x(1-p)^{\alpha}, \qquad x = 0, 1, 2, \ldots \end{aligned}$$

where $p = \beta/(1+\beta)$. This is the pf of a negative binomial distribution. □

Example 5.3 *Suppose that, given* $\Theta = \theta$, X *is normally distributed with mean* θ *and variance* v, *so that*

$$f_{X|\Theta}(x|\theta) = \frac{1}{\sqrt{2\pi v}}\exp\left[-\frac{1}{2v}(x-\theta)^2\right], \qquad -\infty < x < \infty$$

and Θ *is itself normally distributed with mean* μ *and variance* a, *i.e.*

$$f_\Theta(\theta) = \frac{1}{\sqrt{2\pi a}}\exp\left[-\frac{1}{2a}(\theta-\mu)^2\right], \qquad -\infty < \theta < \infty.$$

Determine the marginal pdf of X.

Using (5.3), the marginal pdf of X is

$$\begin{aligned} f_X(x) &= \int_{-\infty}^{\infty}\frac{1}{\sqrt{2\pi v}}\exp\left[-\frac{1}{2v}(x-\theta)^2\right]\frac{1}{\sqrt{2\pi a}}\exp\left[-\frac{1}{2a}(\theta-\mu)^2\right]d\theta \\ &= \frac{1}{2\pi\sqrt{va}}\int_{-\infty}^{\infty}\exp\left[-\frac{1}{2v}(x-\theta)^2-\frac{1}{2a}(\theta-\mu)^2\right]d\theta. \end{aligned}$$

We leave as an exercise for the reader the verification of the algebraic identity

$$\frac{(x-\theta)^2}{v}+\frac{(\theta-\mu)^2}{a} = \frac{a+v}{va}\left(\theta-\frac{ax+v\mu}{a+v}\right)^2+\frac{(x-\mu)^2}{a+v}$$

obtained by completion of the square in θ. Thus,

$$f_X(x) = \frac{\exp\left[-\dfrac{(x-\mu)^2}{2(a+v)}\right]}{\sqrt{2\pi(a+v)}}\int_{-\infty}^{\infty}\sqrt{\frac{a+v}{2\pi va}}\exp\left[-\frac{a+v}{2va}\left(\theta-\frac{ax+v\mu}{a+v}\right)^2\right]d\theta.$$

We recognize the integrand as the pdf (as a function of θ) of a normal distribution with mean $(ax+v\mu)/(a+v)$ and variance $(va)/(a+v)$. Thus the integral is 1 and so

$$f_X(x) = \frac{\exp\left[-\dfrac{(x-\mu)^2}{2(a+v)}\right]}{\sqrt{2\pi(a+v)}}, \qquad -\infty < x < \infty$$

that is, X is normal with mean μ and variance $a+v$. □

Note that the roles of X and Y in (5.2) can be interchanged, yielding

$$f_{X|Y}(x|y)f_Y(y) = f_{Y|X}(y|x)f_X(x)$$

since both sides of this equation equal the joint distribution of X and Y. Division by $f_Y(y)$ yields Bayes' theorem, namely

$$f_{X|Y}(x|y) = \frac{f_{Y|X}(y|x)f_X(x)}{f_Y(y)}.$$

5.2.2 Conditional expectation

As in the previous subsection, assume that X and Y are two random variables and the conditional pf of X given that $Y=y$ is $f_{X|Y}(x|y)$. Clearly, this is a valid probability distribution, and its mean is denoted by

$$E(X|Y=y) = \int x\ f_{X|Y}(x|y)dx \tag{5.4}$$

with the integral replaced by a sum in the discrete case. Clearly, (5.4) is a function of y, and it is often of interest to view this conditional expectation as a random variable obtained by replacing y by Y in the right-hand side of (5.4). Thus we can write $E(X|Y)$ instead of the left-hand side of (5.4), and so $E(X|Y)$ is itself a random variable since it is a function of the random variable Y. The expectation of $E(X|Y)$ is given by

$$E[E(X|Y)] = E(X). \tag{5.5}$$

To see this, note that from (5.3) and (5.4)

$$\begin{aligned} E[E(X|Y)] &= \int E(X|Y=y)f_Y(y)dy \\ &= \int\int x f_{X|Y}(x|y)dx f_Y(y)dy \\ &= \int x \int f_{X|Y}(x|y)f_Y(y)dydx \\ &= \int x f_X(x)dx \\ &= E(X) \end{aligned}$$

with a similar proof in the discrete case.

Example 5.4 (Example 5.2 continued) *Derive the mean of the negative binomial distribution by conditional expectation.*

We have

$$E(X|\Theta) = \Theta$$

and so

$$E(X) = E[E(X|\Theta)] = E(\Theta).$$

From Appendix A the mean of the gamma distribution of Θ is $\alpha\beta$, and so $E(X) = \alpha\beta$. □

It is often convenient to replace X by an arbitrary function $h(X,Y)$ in (5.4), yielding the more general definition

$$E[h(X,Y)|Y = y] = \int h(x,y) f_{X|Y}(x|y)dx.$$

Similarly, $E[h(X,Y)|Y]$ is the conditional expectation viewed as a random variable which is a function of Y. Then, (5.5) generalizes to

$$E\{E[h(X,Y)|Y]\} = E[h(X,Y)]. \tag{5.6}$$

To see (5.6), note that

$$\begin{aligned} E\{E[h(X,Y)|Y]\} &= \int E[h(X,Y)|Y = y] f_Y(y)dy \\ &= \int\int h(x,y) f_{X|Y}(x|y)dx f_Y(y)dy \\ &= \int\int h(x,y)[f_{X|Y}(x|y) f_Y(y)]dxdy \\ &= \int\int h(x,y) f_{X,Y}(x,y)dxdy \\ &= E[h(X,Y)] \end{aligned}$$

from (5.2).

If we choose $h(X,Y) = [X - E(X|Y)]^2$, then its expected value, based on the conditional distribution of X given Y, is the variance of this conditional distribution,

$$Var(X|Y) = E\{[X - E(X|Y)]^2|Y\}. \tag{5.7}$$

Clearly, (5.7) is still a function of the random variable Y.

It is instructive now to analyze the variance of X where X and Y are two random variables. To begin, note that (5.7) may be written as

$$Var(X|Y) = E(X^2|Y) - [E(X|Y)]^2.$$

Thus,

$$\begin{aligned} E[Var(X|Y)] &= E\{E(X^2|Y) - [E(X|Y)]^2\} \\ &= E[E(X^2|Y)] - E\{[E(X|Y)]^2\} \\ &= E(X^2) - E\{[E(X|Y)]^2\}. \end{aligned}$$

Also, since $Var[h(Y)] = E\{[h(Y)]^2\} - \{E[h(Y)]\}^2$, we may use $h(Y) = E(X|Y)$ to obtain

$$\begin{aligned} Var[E(X|Y)] &= E\{[E(X|Y)]^2\} - \{E[E(X|Y)]\}^2 \\ &= E\{[E(X|Y)]^2\} - [E(X)]^2. \end{aligned}$$

Thus,

$$\begin{aligned} E[Var(X|Y)] + Var[E(X|Y)] &= E(X^2) - E\{[E(X|Y)]^2\} \\ &\quad + E\{[E(X|Y)]^2\} - [E(X)]^2 \\ &= E(X^2) - [E(X)]^2 \\ &= Var(X). \end{aligned}$$

Thus, we have established the important formula

$$Var(X) = E[Var(X|Y)] + Var[E(X|Y)]. \tag{5.8}$$

Formula (5.8) states that the variance of X is composed of the sum of two parts: the mean of the conditional variance plus the variance of the conditional mean.

Example 5.5 (Example 5.2 continued) *Derive the variance of the negative binomial distribution.*

The Poisson distribution has equal mean and variance, that is,

$$E(X|\Theta) = Var(X|\Theta) = \Theta,$$

and so from (5.8),

$$\begin{aligned} Var(X) &= E[Var(X|\Theta)] + Var[E(X|\Theta)] \\ &= E(\Theta) + Var(\Theta). \end{aligned}$$

Since Θ itself has a gamma distribution with parameters α and β, $E(\Theta) = \alpha\beta$ and $Var(\Theta) = \alpha\beta^2$. Thus the variance of the negative binomial distribution is

$$\begin{aligned} Var(X) &= E(\Theta) + Var(\Theta) \\ &= \alpha\beta + \alpha\beta^2 \\ &= \alpha\beta(1 + \beta). \end{aligned}$$

□

Example 5.6 (Example 5.3 continued) *It was shown that if $X|\Theta$ is normally distributed with mean Θ and variance v where Θ is itself normally distributed with mean μ and variance a, then X (unconditionally) is normally distributed* *with mean μ and variance $a+v$. Use (5.5) and (5.8) to obtain the mean and variance of X directly.*

For the mean we have

$$E(X) = E[E(X|\Theta)] = E(\Theta) = \mu$$

and for the variance we obtain

$$\begin{aligned} Var(X) &= E[Var(X|\Theta)] + Var[E(X|\Theta)] \\ &= E(v) + Var(\Theta) \\ &= v + a \end{aligned}$$

since v is a constant. □

Example 5.7 *Consider a compound Poisson distribution with Poisson mean λ, where $X = Y_1 + \cdots + Y_N$ with $E(Y_i) = \mu_Y$ and $Var(Y_i) = \sigma_Y^2$. Determine the mean and variance of X.*

Formula (5.8) was used in Chapter 4 to obtain the answers:

$$E(X) = \lambda\mu_Y \text{ and } Var(X) = \lambda(\mu_Y^2 + \sigma_Y^2).$$

□

5.2.3 Unbiased estimation

Suppose we have a model pf $f(x;\theta)$, which depends on a parameter θ. To begin, assume that θ is a scalar parameter, and it is of interest to use a random sample $X_1, X_2, \cdots, X_n$ (that is, $X_1, X_2, \ldots, X_n$ are assumed to be independent and identically distributed, each with pf $f(x;\theta)$) to estimate θ. Let $\mathbf{X} = (X_1, X_2, \cdots, X_n)'$. Recalling Definition 2.18, an estimator $\hat{\theta}$ is said to be unbiased for θ if

$$E(\hat{\theta}) = \theta \tag{5.9}$$

for all θ.

Example 5.8 *If X_j is exponentially distributed with pdf*

$$f(x;\beta) = \beta e^{-\beta x}, \qquad x > 0$$

show that the sample mean $\bar{X} = n^{-1}\sum_{j=1}^{n} X_j$ is unbiased for β^{-1}.

We have $E(\bar{X}) = n^{-1}\sum_{j=1}^{n} E(X_j) = n^{-1}\sum_{j=1}^{n} \beta^{-1} = \beta^{-1}$. □

While (5.9) seems intuitively to be a reasonable property for an estimator $\hat{\theta}$ to have (and it often is), there are some drawbacks.

1. As noted in Subsection 2.5.1 there is a limit to the accuracy of $\hat{\theta}$ as measured by $Var(\hat{\theta})$. In particular, if (5.9) is satisfied, then

$$Var(\hat{\theta}) \geq [I(\theta)]^{-1} \tag{5.10}$$

the CramérRao lower bound. However, a more useful measure of quality is the mean squared error (Definition 2.21) and from (2.9) we recall

$$\text{MSE}(\hat{\theta}) = Var(\hat{\theta}) + [b_\theta(\hat{\theta})]^2$$

where $b_\theta(\hat{\theta}) = E(\hat{\theta}) - \theta$ is the bias. Thus there is a trade-off between the bias and the variance of $\hat{\theta}$. For an unbiased estimator, $b_\theta(\hat{\theta}) = 0$, but the variance must satisfy (5.10). It may be possible to do better if we do not insist upon an unbiased estimator.

2. Unbiasedness is not preserved under parameter transformations. Thus, if (5.9) is satisfied and $h(\cdot)$ is a nonlinear function, then $E[h(\hat{\theta})] \neq h(\theta)$.

3. In some situations, (5.9) is satisfied but $\hat{\theta}$ is a silly estimator of θ.

The following examples illustrate points 2 and 3.

Example 5.9 (Example 5.8 continued) *It was shown that the single observation X (it is the sample mean for a sample of size one) is unbiased for $1/\beta$. Show that $1/X$ is biased for β.*

$$E(1/X) = \int_0^\infty x^{-1}\beta e^{-\beta x}dx = \infty \neq \beta.$$

□

Example 5.10 *Suppose X is a single observation from a Poisson distribution with mean θ,*

$$f_X(x;\theta) = \frac{\theta^x e^{-\theta}}{x!}, \qquad x = 0, 1, 2, \ldots.$$

Show that $(-1)^X$ is an unbiased estimator of $e^{-2\theta}$.

Unbiasedness is demonstrated by observing

$$\begin{aligned} E\left[(-1)^X\right] &= \sum_{x=0}^{\infty}(-1)^x \frac{\theta^x e^{-\theta}}{x!} \\ &= e^{-\theta}\sum_{x=0}^{\infty}\frac{(-\theta)^x}{x!} \\ &= e^{-\theta}(e^{-\theta}) \\ &= e^{-2\theta}. \end{aligned}$$

Clearly, $0 < e^{-2\theta} < 1$ yet $(-1)^X = -1$ if X is odd and 1 if X is even. In fact, e^{-2X} is a better estimate of $e^{-2\theta}$ even though it is biased. □

Despite the above shortcomings, unbiasedness may often be a reasonable property for an estimator to have, and we will often use this criterion in constructing estimators.

Suppose, in a slightly more general situation, that $X_1, \cdots, X_n$ have common mean

$$\mu = E(X_j),\ j = 1, \ldots, n$$

but are not necessarily independent or even identically distributed. We now derive a nonparametric (not dependent on the form of the distribution of the X_js) estimator of μ. Clearly,

$$\hat{\mu} = \bar{X} = n^{-1}(X_1 + \ldots + X_n) \tag{5.11}$$

is an unbiased estimator of μ because

$$\begin{aligned} E(\hat{\mu}) &= E[n^{-1}(X_1 + \cdots + X_n)] \\ &= n^{-1}\sum_{j=1}^{n} E(X_j) \\ &= n^{-1}\sum_{j=1}^{n} \mu \\ &= \mu \end{aligned} \tag{5.12}$$

regardless of the distribution of the X_js.

Example 5.11 *Let Y_j be the amount of claims per exposure in year j, $j = 1, \ldots, n$, for a particular policyholder. It has been observed that a trend factor of 1.07 is appropriate, that is, claim amounts are increasing at the rate of 7% per year. Then if μ is the underlying pure premium (expected claims per exposure) at the midpoint of the year before the first year, it may be reasonable to assume that*

$$E(Y_j) = (1.07)^j \mu.$$

Determine an unbiased estimator of μ.

Let $X_j = (1.07)^{-j} Y_j$, and so $E(X_j) = \mu$. Thus, $\bar{X} = n^{-1}\sum_{j=1}^{n}(1.07)^{-j}Y_j$ is an unbiased estimator of μ. Furthermore, the intuitively obvious quantity

$$(1.07)^{n+1}\bar{X} = n^{-1}\sum_{j=1}^{n}(1.07)^{n+1-j}Y_j$$

is an unbiased estimator of next year's pure premium, $E(Y_{n+1}) = (1.07)^{n+1}\mu$, in the present situation. □

If $X_1, \cdots, X_n$ are independent, but not necessarily identically distributed, with common mean $\mu = E(X_j)$ and common variance

$$v = Var(X_j)$$

then (5.11) is a (nonparametric) unbiased estimator of μ. In Example 2.12 the estimator

$$\hat{v} = \frac{1}{n-1}\sum_{j=1}^{n}(X_j - \bar{X})^2 \tag{5.13}$$

was shown to be unbiased. We provide a different derivation of that result here. The equalities produced along the way will be used again.

$$\begin{aligned}
\sum_{j=1}^{n}(X_j - \bar{X})^2 &= \sum_{j=1}^{n}(X_j - \mu + \mu - \bar{X})^2 \\
&= \sum_{j=1}^{n}(X_j - \mu)^2 + 2\sum_{j=1}^{n}(X_j - \mu)(\mu - \bar{X}) + \sum_{j=1}^{n}(\mu - \bar{X})^2 \\
&= \sum_{j=1}^{n}(X_j - \mu)^2 + 2(\mu - \bar{X})\sum_{j=1}^{n}(X_j - \mu) + n(\mu - \bar{X})^2 \\
&= \sum_{j=1}^{n}(X_j - \mu)^2 + 2(\mu - \bar{X})n(\bar{X} - \mu) + n(\mu - \bar{X})^2 \\
&= \sum_{j=1}^{n}(X_j - \mu)^2 - n(\bar{X} - \mu)^2. \qquad (5.14)
\end{aligned}$$

Consider the sample mean $\bar{X}$ given by (5.11). From (5.12), $E(\bar{X}) = \mu$, and by the independence of the X_js we have

$$\begin{aligned}
Var(\bar{X}) &= Var\left(\sum_{j=1}^{n} n^{-1}X_j\right) \\
&= \sum_{j=1}^{n} n^{-2}Var(X_j) \\
&= n^{-2}\sum_{j=1}^{n} v \\
&= v/n.
\end{aligned}$$

Take expectations in (5.14) to obtain

$$
\begin{aligned}
E\left[\sum_{j=1}^{n}(X_j-\bar{X})^2\right] &= E\left[\sum_{j=1}^{n}(X_j-\mu)^2\right]-nE[(\bar{X}-\mu)^2] \\
&= \sum_{j=1}^{n}E[(X_j-\mu)^2]-nVar(\bar{X}) \\
&= \sum_{j=1}^{n}Var(X_j)-n(v/n) \\
&= \sum_{j=1}^{n}v-v \\
&= (n-1)v.
\end{aligned}
$$

Dividing both sides by $n-1$ demonstrates that $\hat{v}$ is an unbiased estimate of v.

The following example generalizing these results may appear somewhat artificial at this point, but is important in connection with the Bühlmann–Straub model of Subsection 5.4.4.

Example 5.12 *Suppose $X_1,\cdots,X_n$ are independent with common mean $\mu=E(X_j)$ and variance $Var(X_j)=\beta+\alpha/m_j,\alpha,\beta>0$ and all $m_j\geq 1$. Let $m=\sum_{j=1}^{n}m_j$ and consider the three estimators*

$$\bar{X}=\frac{1}{m}\sum_{j=1}^{n}m_jX_j$$

$$\hat{\mu}_1=\frac{1}{n}\sum_{j=1}^{n}X_j$$

and

$$\hat{\mu}_2=\frac{\displaystyle\sum_{j=1}^{n}\frac{m_jX_j}{m_j\beta+\alpha}}{\displaystyle\sum_{j=1}^{n}\frac{m_j}{m_j\beta+\alpha}}.$$

Show that all three estimators are unbiased for μ and then rank them in order by mean squared error. Also obtain the expected value of a sum of squares that may be useful for estimating α and β.

First consider $\bar{X}$.

$$E(\bar{X})=m^{-1}\sum_{j=1}^{n}m_jE(X_j)=m^{-1}\sum_{j=1}^{n}m_j\mu=\mu.$$

$$\begin{aligned} Var(\bar{X}) &= m^{-2}\sum_{j=1}^{n} m_j^2 Var(X_j) \\ &= m^{-2}\sum_{j=1}^{n} m_j^2(\beta + \alpha/m_j) \\ &= \alpha m^{-1} + \beta m^{-2}\sum_{j=1}^{n} m_j^2. \end{aligned}$$

The estimator $\hat{\mu}_1$ is the one defined in (5.11) and has already been shown to be unbiased. We also have

$$\begin{aligned} Var(\hat{\mu}_1) &= n^{-2}\sum_{j=1}^{n} Var(X_j) \\ &= n^{-2}\sum_{j=1}^{n}(\beta + \alpha/m_j) \\ &= \beta n^{-1} + n^{-2}\alpha\sum_{j=1}^{n} m_j^{-1}. \end{aligned}$$

With regard to $\hat{\mu}_2$,

$$E(\hat{\mu}_2) = \frac{\sum_{j=1}^{n} \frac{m_j}{m_j\beta + \alpha} E(X_j)}{\sum_{j=1}^{n} \frac{m_j}{m_j\beta + \alpha}} = \frac{\sum_{j=1}^{n} \frac{m_j}{m_j\beta + \alpha}\mu}{\sum_{j=1}^{n} \frac{m_j}{m_j\beta + \alpha}} = \mu$$

and

$$\begin{aligned} Var(\hat{\mu}_2) &= \frac{\sum_{j=1}^{n} \left(\frac{m_j}{m_j\beta + \alpha}\right)^2 (\beta + \alpha/m_j)}{\left(\sum_{j=1}^{n} \frac{m_j}{m_j\beta + \alpha}\right)^2} \\ &= \frac{\sum_{j=1}^{n} \left(\frac{m_j}{m_j\beta + \alpha}\right)}{\left(\sum_{j=1}^{n} \frac{m_j}{m_j\beta + \alpha}\right)^2} \\ &= \left(\sum_{j=1}^{n} \frac{m_j}{m_j\beta + \alpha}\right)^{-1}. \end{aligned}$$

We now consider the relative ranking of these variances. To show that it is not possible to order $Var(\hat{\mu}_1)$ and $Var(\bar{X})$ examine their difference:

$$Var(\bar{X}) - Var(\hat{\mu}_1) = \alpha\left(m^{-1} - n^{-2}\sum_{j=1}^{n} m_j^{-1}\right) + \beta\left(m^{-2}\sum_{j=1}^{n} m_j^2 - n^{-1}\right).$$

The coefficient of β must be non-negative. To see this, note that

$$\frac{1}{n}\sum_{j=1}^{n} m_j^2 \geq \left(\frac{1}{n}\sum_{j=1}^{n} m_j\right)^2 = \frac{m^2}{n^2}$$

(the left-hand side is like a sample second moment and the right-hand side is like the square of the sample mean) and then multiply both sides by nm^{-2}. To show that the coefficient of α must be non-positive, note that

$$\frac{n}{\sum_{j=1}^{n} m_j^{-1}} \leq \frac{1}{n}\sum_{j=1}^{n} m_j = \frac{m}{n}$$

(the harmonic mean is always greater than or equal to the arithmetic mean) and then multiply both sides by n and then invert both sides. Therefore, by suitablechoice of α and β, the difference in the variances can be made positive or negative.

We can do more than just show that $\hat{\mu}_2$ has the smallest variance of the three. Consider an arbitrary estimator of the form $\hat{\mu} = \sum_{j=1}^{n} a_j X_j$ where $\sum_{j=1}^{n} a_j = 1$. All three estimators are of this type. Incorporating the constraint by using Lagrange multipliers, the smallest variance is found by minimizing

$$\sum_{j=1}^{n} a_j^2 Var(X_j) + \lambda\left(\sum_{j=1}^{n} a_j - 1\right).$$

The derivative with regard to a_i is

$$2a_i Var(X_i) + \lambda$$

and setting it equal to zero gives $a_i = -\lambda[2Var(X_i)]^{-1}$. In other words, the weights should be proportional to the reciprocal of the variance. These are precisely the weights used in $\hat{\mu}_2$ and therefore it must have the smallest variance of all linear estimators.

With regard to a sum of squares, consider

$$\begin{aligned}
\sum_{j=1}^{n} m_j (X_j - \bar{X})^2 &= \sum_{j=1}^{n} m_j (X_j - \mu + \mu - \bar{X})^2 \\
&= \sum_{j=1}^{n} m_j (X_j - \mu)^2 + 2 \sum_{j=1}^{n} m_j (X_j - \mu)(\mu - \bar{X}) \\
&\quad + \sum_{j=1}^{n} m_j (\mu - \bar{X})^2 \\
&= \sum_{j=1}^{n} m_j (X_j - \mu)^2 + 2(\mu - \bar{X}) \sum_{j=1}^{n} m_j (X_j - \mu) \\
&\quad + m(\mu - \bar{X})^2 \\
&= \sum_{j=1}^{n} m_j (X_j - \mu)^2 + 2(\mu - \bar{X}) m(\bar{X} - \mu) + m(\mu - \bar{X})^2 \\
&= \sum_{j=1}^{n} m_j (X_j - \mu)^2 - m(\bar{X} - \mu)^2. \qquad (5.15)
\end{aligned}$$

Taking expectations yields

$$\begin{aligned}
E\left[\sum_{j=1}^{n} m_j (X_j - \bar{X})^2\right] &= \sum_{j=1}^{n} m_j E[(X_j - \mu)^2] - mE[(\bar{X} - \mu)^2] \\
&= \sum_{j=1}^{n} m_j Var(X_j) - m Var(\bar{X}) \\
&= \sum_{j=1}^{n} m_j (\beta + \alpha/m_j) - \beta \left(m^{-1} \sum_{j=1}^{n} m_j^2 \right) - \alpha
\end{aligned}$$

and thus

$$E\left[\sum_{j=1}^{n} m_j (X_j - \bar{X})^2\right] = \beta \left(m - m^{-1} \sum_{j=1}^{n} m_j^2 \right) + \alpha(n-1). \qquad (5.16)$$

In addition to being of interest in its own right, (5.16) provides an unbiased estimator in situations more general than (5.13). The latter is recovered with the choice $\alpha = 0$ and $m_j = 1$ for $j = 1, 2, \cdots, n$, implying that $m = n$. Also, if $\beta = 0$, (5.16) allows us to derive an estimator of α when each X_j is the average of m_j independent observations, each with mean μ and variance α. In any event, it is usually the case that the m_js (and hence m) are known. □

5.2.4 Maximum likelihood estimation

While there are many ways to obtain an estimator, one method that is often successful is the method of moments (Definition 2.24). Also, recall from Theorem 2.2 that maximum likelihood estimators are asymptotically unbiased and are often unbiased. As well, they have a number of other superior statistical qualities. Here we present a brief review of maximum likelihood estimation and an example which will be used later in this chapter.

If $X_1, \ldots, X_n$ are independent and identically distributed with pdf or pf $f(x;\theta)$, then the mle is the value of θ which maximizes the likelihood function

$$L(\theta) = \prod_{j=1}^{n} f(x_j;\theta)$$

or, equivalently, the loglikelihood function

$$l(\theta) = \log L(\theta) = \sum_{j=1}^{n} \log f(x_j;\theta). \tag{5.17}$$

If (5.17) is differentiable with respect to θ, the mle, $\hat{\theta}$, satisfies

$$l'(\hat{\theta}) = 0.$$

Example 5.13 *Suppose X_j is normally distributed with mean μ and variance v. Determine the mle of μ and v.*

The likelihood function is

$$\begin{aligned} L(\mu, v) &= \prod_{j=1}^{n} (2\pi v)^{-1/2} \exp\left[-\frac{(x_j-\mu)^2}{2v}\right] \\ &= (2\pi v)^{-n/2} \exp\left[-\frac{1}{2v}\sum_{j=1}^{n}(x_j-\mu)^2\right] \end{aligned}$$

and so the loglikelihood function is

$$l(\mu, v) = -\frac{n}{2}\log(2\pi v) - \frac{1}{2v}\sum_{j=1}^{n}(x_j-\mu)^2.$$

Using (5.14), this may be rewritten as

$$l(\mu, v) = -\frac{n}{2}\log(2\pi v) - \frac{1}{2v}\left[\sum_{j=1}^{n}(x_j-\bar{x})^2 + n(\bar{x}-\mu)^2\right].$$

The partial derivative with respect to μ is

$$\frac{\partial}{\partial \mu} l(\mu, v) = \frac{n}{v}(\bar{x} - \mu).$$

Setting it equal to zero produces the mle $\hat{\mu} = \bar{x}$. The partial derivative with respect to v is (inserting the mle for μ)

$$\frac{\partial}{\partial v} l(\mu, v) = -\frac{n}{2v} + \frac{1}{2v^2} \sum_{j=1}^{n} (x_j - \bar{x})^2.$$

Setting it equal to zero produces the mle

$$\hat{v} = \frac{1}{n} \sum_{j=1}^{n} (x_j - \bar{x})^2.$$

This estimator is biased (recall that division by $n - 1$ is needed to yield an unbiased estimator). Also note that this is both the empirical and the method of moments estimator of the variance for any population. In addition, it is the mle for the normal distribution. □

In general, the likelihood function is the joint density function of the sample observations. Therefore, they need not be independent or be identically distributed. If dependent, the likelihood function is not the product of the individual densities, but is the probability of the observed data under the model.

5.2.5 Bayesian estimation

Bayesian estimation was discussed at length in Section 2.8. Because of its importance in credibility theory, a few of the details are repeated here. Assume that given a parameter (which may be a vector) θ, the observations $X_1, \ldots, X_n$ are independently and identically distributed with pf $f_{X|\Theta}(x|\theta)$. It is further assumed that the parameter θ is a realization of a random variable Θ with pf $\pi(\theta)$. In a typical Bayesian analysis the pf of Θ represents our subjective prior opinion about the unknown parameter. In the credibility setting this may still be the case, but Θ may also represent a real, though unobservable, random variable. For example, θ may indicate an automobile driver's propensity to have a claim and $\pi(\theta)$ may describe how that propensity is distributed throughout the population of insured drivers. With no additional information, $\pi(\theta)$ represents our prior opinion about a randomly selected driver's parameter.

The joint distribution of $X_1, \ldots, X_n, \Theta$ is obtained by first conditioning on Θ, and it is thus given by the likelihood multiplied by the prior density, that is,

$$f_{\mathbf{X},\Theta}(\mathbf{x},\theta) = \left[\prod_{j=1}^{n} f_{X_j|\Theta}(x_j|\theta)\right]\pi(\theta). \tag{5.18}$$

The marginal distribution of $X_1, \ldots, X_n$ is obtained by integrating θ (summing if $\pi(\theta)$ is discrete) out of the joint density of $X_1, \ldots, X_n, \Theta$, that is,

$$f_{\mathbf{X}}(\mathbf{x}) = \int \left[\prod_{j=1}^{n} f_{X_j|\Theta}(x_j|\theta)\right]\pi(\theta)d\theta. \tag{5.19}$$

The information about Θ "posterior" to observation of $\mathbf{X} = (X_1, \cdots, X_n)'$ is summarized in the posterior distribution of Θ. This is simply the conditional density of Θ given that $\mathbf{X}$ equals $\mathbf{x} = (x_1, \cdots, x_n)'$, which we express notationally as $\Theta|\mathbf{X} = \mathbf{x}$ or simply as $\Theta|\mathbf{X}$ or $\Theta|\mathbf{x}$, and by (5.1) this is the ratio of the joint density of $X_1, \cdots, X_n, \Theta$ to the marginal density of $X_1, \cdots, X_n$. In other words, the posterior density is

$$\pi_{\Theta|\mathbf{X}}(\theta|\mathbf{x}) = \frac{\left[\prod_{j=1}^{n} f_{X_j|\Theta}(x_j|\theta)\right]\pi(\theta)}{\int \left[\prod_{j=1}^{n} f_{X_j|\Theta}(x_j|\theta)\right]\pi(\theta)d\theta}. \tag{5.20}$$

From a practical viewpoint, the denominator in (5.20) does not depend on θ and simply serves as a normalizing constant. Thus, as a function of θ, $\pi_{\Theta|\mathbf{X}}(\theta|\mathbf{x})$ is proportional to

$$\left[\prod_{j=1}^{n} f_{X_j|\Theta}(x_j|\theta)\right]\pi(\theta) \tag{5.21}$$

and if the form of the terms involving θ in (5.21) is recognized as belonging to a particular distribution, then it is not necessary to evaluate the denominator in (5.21), but only to identify the appropriate normalizing constant.

A point estimate of θ derived from $\pi_{\Theta|\mathbf{X}}(\theta|\mathbf{x})$ requires the selection of a loss function, and the choice of squared error loss results in the posterior mean

$$E(\Theta|\mathbf{X} = \mathbf{x}) = \int \theta\pi_{\Theta|\mathbf{X}}(\theta|\mathbf{x})d\theta.$$

Example 5.14 *Suppose that $X_1, \cdots, X_n$ are (given $\Theta = \theta$) Poisson distributed with mean θ, and Θ itself is gamma distributed with parameters α and β. Determine the posterior mean of Θ.*

The model pf is

$$f_{X_j|\Theta}(x_j|\theta) = \frac{\theta^{x_j}e^{-\theta}}{x_j!}$$

and the prior pdf is

$$\pi(\theta) = \frac{\theta^{\alpha-1}e^{-\theta/\beta}}{\Gamma(\alpha)\beta^{\alpha}}.$$

Thus, from (5.21), $\pi_{\Theta|\mathbf{X}}(\theta|\mathbf{x})$ is proportional to

$$\left(\prod_{j=1}^{n} \frac{\theta^{x_j}e^{-\theta}}{x_j!}\right) \frac{\theta^{\alpha-1}e^{-\theta/\beta}}{\Gamma(\alpha)\beta^{\alpha}}$$

which is itself proportional (as a function of θ) to

$$\theta^{n\bar{x}+\alpha-1}e^{-\theta(n+1/\beta)}.$$

Writing $\alpha_* = \alpha + n\bar{x}$ and $\beta_* = \beta/(n\beta+1)$, this is $\theta^{\alpha_*-1}e^{-\theta/\beta_*}$, which can be seen to be the term corresponding to a gamma density with parameters α_* and β_*, that is,

$$\pi_{\Theta|\mathbf{X}}(\theta|\mathbf{x}) = \frac{\theta^{\alpha_*-1}e^{-\theta/\beta_*}}{\Gamma(\alpha_*)\,\beta_*^{\alpha_*}}.$$

The mean of this gamma distribution is

$$E(\Theta|\mathbf{X}=\mathbf{x}) = \alpha_*\beta_* = \frac{(\alpha+n\bar{x})\beta}{n\beta+1}.$$

It is worth noting that this may be expressed as

$$E(\Theta|\mathbf{X}=\mathbf{x}) = \frac{1}{n\beta+1}(\alpha\beta) + \frac{n\beta}{n\beta+1}\bar{x}$$

a weighted average of the prior mean $\alpha\beta$ and the sample mean $\bar{x}$. □

Example 5.15 *Suppose that $X_1, \cdots, X_n$ are (given $\Theta = \theta$) normally distributed with mean θ and variance v, that is,*

$$f_{X_j|\Theta}(x_j|\theta) = (2\pi v)^{-1/2}\exp\left[-\frac{1}{2v}(x_j-\theta)^2\right]$$

and Θ itself is normally distributed with mean μ and variance a, thus,

$$\pi(\theta) = (2\pi a)^{-1/2}\exp\left[-\frac{1}{2a}(\theta-\mu)^2\right].$$

Determine the posterior mean of Θ.

From (5.21), and ignoring all terms not involving θ, the posterior density of Θ given $\mathbf{X} = \mathbf{x}$ is proportional to

$$\exp\left[-\frac{1}{2v}\sum_{j=1}^{n}(x_j-\theta)^2 - \frac{1}{2a}(\theta-\mu)^2\right].$$

To simplify further, use (5.14) to rewrite this as

$$\exp\left\{-\frac{1}{2v}\left[\sum_{j=1}^{n}(x_j-\bar{x})^2+n(\bar{x}-\theta)^2\right]-\frac{1}{2a}(\theta-\mu)^2\right\}$$

which (as a function of θ) is proportional to

$$\exp\left[-\frac{n}{2v}(\bar{x}-\theta)^2-\frac{1}{2a}(\theta-\mu)^2\right].$$

Expanding the squared terms and collecting like terms in θ, we see that this is proportional to

$$\exp\left[-\frac{\theta^2}{2}\left(\frac{n}{v}+\frac{1}{a}\right)+\theta\left(\frac{n\bar{x}}{v}+\frac{\mu}{a}\right)\right]. \tag{5.22}$$

Now, if Θ has a normal distribution with mean μ_* and variance a_*, then the density is proportional to

$$\exp\left[-\frac{1}{2a_*}(\theta-\mu_*)^2\right]$$

which is itself proportional (as a function of θ) to

$$\exp\left[-\frac{\theta^2}{2a_*}+\frac{\mu_*}{a_*}\theta\right]. \tag{5.23}$$

But if

$$a_*=\left(\frac{n}{v}+\frac{1}{a}\right)^{-1} \tag{5.24}$$

and

$$\frac{\mu_*}{a_*}=\frac{n\bar{x}}{v}+\frac{\mu}{a}$$

that is,

$$\mu_*=\frac{\frac{n\bar{x}}{v}+\frac{\mu}{a}}{\frac{n}{v}+\frac{1}{a}} \tag{5.25}$$

it is clear that (5.23) and (5.22) are equal. This means that the posterior distribution of Θ must be normal with parameters μ_* and a_* as given by (5.25) and (5.24), respectively. The posterior mean $E(\Theta|\mathbf{X}=\mathbf{x})$ is μ_*, which may be expressed as

$$E(\Theta|\mathbf{X}=\mathbf{x}) = \left(\frac{\frac{v}{a}}{n+\frac{v}{a}}\right)\mu + \left(\frac{n}{n+\frac{v}{a}}\right)\overline{x}$$

again a weighted average of the prior mean μ and the sample mean $\overline{x}$. □

In each of the previous two examples, the prior distribution is said to be a conjugate prior distribution because the posterior distribution is from the same parametric family as the prior, but with different parameters. In Example 5.14, $\pi(\theta)$ and $\pi_{\Theta|\mathbf{X}}(\theta|\mathbf{x})$ are both gamma densities, the only difference being the parameters α and β. We say that the gamma distribution is a conjugate prior for the Poisson likelihood. Similarly, in Example 5.15, both $\pi(\theta)$ and $\pi_{\Theta|\mathbf{X}}(\theta|\mathbf{x})$ are normally distributed, but with different means and variances. Other examples of conjugate prior distributions are given in the exercises.

5.2.6 The linear exponential family

A large parametric family that includes many of the distributions we have encountered so far has a special use in credibility theory. The definition is as follows:

Definition 5.1 *A random variable X (discrete or continuous) has a distribution which is from the* ***linear exponential family*** *if its pf may be parameterized in terms of a parameter θ and expressed as*

$$f(x;\theta) = \frac{p(x)e^{-\theta x}}{q(\theta)}. \tag{5.26}$$

The function $p(x)$ depends only on x (not on θ), and the function $q(\theta)$ is a normalizing constant. Also, the support of the random variable must not depend on θ. The parameter θ is called the ***natural parameter*** *of the distribution.*

Example 5.16 *Show that the exponential distribution with pdf $f(x;\beta) = \beta^{-1}e^{-x/\beta}$ is of the form (5.26).*

The pdf is

$$f(x;\beta) = \beta^{-1}e^{-\beta^{-1}x}.$$

If we let $\theta = 1/\beta$, then the pdf is

$$f(x;\theta) = \frac{1e^{-\theta x}}{\theta^{-1}}$$

which is of the form (5.26) with $p(x) = 1$, and $q(\theta) = 1/\theta$. □

Example 5.17 *Show that the Poisson distribution (with mean λ) is a member of the linear exponential family.*

The pf is

$$f(x;\lambda) = \frac{\lambda^x e^{-\lambda}}{x!} = \frac{(1/x!)e^{-(-\log\lambda)x}}{e^{\lambda}}$$

If we let $\theta = -\log\lambda$, then the pf is

$$f(x;\theta) = \frac{(1/x!)e^{-\theta x}}{e^{e^{-\theta}}}$$

which is of the form (5.26) with $p(x) = 1/x!$, and $q(\theta) = e^{e^{-\theta}}$. Note that in this parameterization the Poisson mean is $e^{-\theta}$, whereas in examples elsewhere the mean is taken to be θ itself. □

Example 5.18 *Show that the normal distribution with mean μ and known variance v is a member of the linear exponential family.*

The pdf is

$$\begin{aligned} f(x;\mu,v) &= (2\pi v)^{-1/2}\exp\left[-\frac{1}{2v}(x-\mu)^2\right] \\ &= (2\pi v)^{-1/2}\exp\left(-\frac{x^2}{2v} + \frac{\mu}{v}x - \frac{\mu^2}{2v}\right) \\ &= \frac{(2\pi v)^{-1/2}\exp\left(-\frac{x^2}{2v}\right)\exp\left(\frac{\mu}{v}x\right)}{\exp\left(\frac{\mu^2}{2v}\right)}. \end{aligned}$$

If we let $\theta = -\mu/v$, the pdf is

$$f(x;\theta,v) = \frac{(2\pi v)^{-1/2}\exp\left(-\frac{x^2}{2v}\right)\exp(-\theta x)}{\exp\left(\frac{\theta^2 v}{2}\right)}$$

which is of the form (5.26) with $p(x) = (2\pi v)^{-1/2}\exp\left(-\frac{x^2}{2v}\right)$, and $q(\theta) = \exp\left(\frac{\theta^2 v}{2}\right)$. □

We now find the mean and variance of the distribution defined by (5.26). First, note that

$$\log f(x;\theta) = \log p(x) - \theta x - \log q(\theta).$$

Differentiate with respect to θ to obtain

$$\frac{\partial}{\partial\theta}f(x;\theta) = \left[-x - \frac{q'(\theta)}{q(\theta)}\right] f(x;\theta). \tag{5.27}$$

Integrate (or sum) over the range of x (known not to depend on θ) to obtain

$$\int \frac{\partial}{\partial\theta}f(x;\theta)dx = -\int xf(x;\theta)dx - \frac{q'(\theta)}{q(\theta)}\int f(x;\theta)dx.$$

On the left-hand side, interchange the order of differentiation and integration (or summation) to obtain

$$\frac{\partial}{\partial\theta}\left[\int f(x;\theta)dx\right] = -\int xf(x;\theta)dx - \frac{q'(\theta)}{q(\theta)}\int f(x;\theta)dx.$$

We know that $\int f(x;\theta)dx = 1$ and $\int xf(x;\theta)dx = E(X)$ and thus

$$\frac{\partial}{\partial\theta}(1) = -E(X) - \frac{q'(\theta)}{q(\theta)}.$$

In other words, the mean is

$$E(X) = \mu(\theta) = -\frac{q'(\theta)}{q(\theta)} = -\frac{d}{d\theta}\log q(\theta). \tag{5.28}$$

To obtain the variance, (5.27) may first be rewritten as

$$\frac{\partial}{\partial\theta}f(x;\theta) = -[x - \mu(\theta)]f(x;\theta).$$

Differentiate again with respect to θ to obtain

$$\begin{aligned}\frac{\partial^2}{\partial\theta^2}f(x;\theta) &= \mu'(\theta)f(x;\theta) - [x-\mu(\theta)]\frac{\partial}{\partial\theta}f(x;\theta)\\ &= \mu'(\theta)f(x;\theta) + [x-\mu(\theta)]^2 f(x;\theta).\end{aligned}$$

Again, integrate over the range of x to obtain

$$\int \frac{\partial^2}{\partial\theta^2}f(x,\theta)dx = \mu'(\theta)\int f(x;\theta)dx + \int [x-\mu(\theta)]^2 f(x;\theta)dx.$$

In other words

$$\int [x-\mu(\theta)]^2 f(x;\theta)dx = -\mu'(\theta) + \frac{\partial^2}{\partial\theta^2}\int f(x;\theta)dx.$$

Since $\mu(\theta)$ is the mean, the left-hand side is the variance (by definition), and then because the second term on the right hand side is zero we obtain

$$Var(X) = v(\theta) = -\mu'(\theta) = \frac{d^2}{d\theta^2}\log q(\theta). \tag{5.29}$$

One advantage of working with members of the linear exponential family is that conjugate prior distributions are easy to find. The result is expressed in the following theorem.

Theorem 5.1 *Suppose that given $\Theta = \theta$, the random variables $X_1, \ldots, X_n$ are independent and identically distributed with pf*

$$f_{X_j|\Theta}(X_j|\theta) = \frac{p(x_j)e^{-\theta x_j}}{q(\theta)}$$

where Θ has pdf

$$\pi(\theta) = \frac{[q(\theta)]^{-k}e^{-\theta\mu k}}{c(\mu, k)}$$

where k and μ are parameters of the distribution and $c(\mu, k)$ is the normalizing constant. Then the posterior pf $\pi_{\Theta|\mathbf{X}}(\theta|\mathbf{x})$ is of the same form as $\pi(\theta)$.

Proof: See Exercise 5.16.

5.3 LIMITED FLUCTUATION CREDIBILITY THEORY

This branch of credibility theory represents the first attempt to quantify the credibility problem. This approach was suggested in the early part of the century in connection with workers compensation insurance. The original paper on the subject was by Mowbray in 1914 [93], followed by Whitney [127] in 1917. Another classic paper on the subject was by Bailey [7] in 1950. The latter paper identified many of the ways in which classical statistical ideas are used in credibility theory, including conjugate Bayesian analysis, least squares approximation, and regression.

The problem may be formulated as follows. Suppose that a policyholder has experienced X_j claims or losses[3] in past experience period j, where $j \in \{1, 2, 3, \ldots, n\}$. Another view is that X_j is the experience from the jth policy in a group or from the jth member of a particular class in a rating scheme. Suppose that $E(X_j) = \xi$, that is, the mean is stable over time or across the members of a group or class.[4] This quantity would be the premium to charge (net of expenses, profits, and a provision for adverse experience) if only we knew its value. Also suppose $Var(X_j) = \sigma^2$, again, the same for all j. The past experience may be summarized by the average $\bar{X} = n^{-1}(X_1 + \cdots + X_n)$.

[3]As in Chapter 4, claims will refer to the number of claims and losses will refer to payment amounts. In many cases, such as in this introductory paragraph, the ideas apply equally whether we are counting claims or losses. In those situations, the experience may be used.
[4]The customary symbol for the mean, μ, is not used here because that symbol is used for a different, but related, mean in the next section. We have chosen this particular symbol ("Xi") because it is the most difficult Greek letter to write and pronounce. It is an unwritten rule of textbook writing that it appear at least once.

We know that $E(\bar{X}) = \xi$ and if the X_j are independent, $Var(\bar{X}) = \sigma^2/n$. The insurer's goal is to decide on the value of the ξ. One possibility is to ignore the past data (no credibility) and simply charge M a value obtained from experience on other, similar, but not identical, policyholders. This quantity is often called the **manual premium**, because it would come from a book (manual) of premiums. Another possibility is to ignore M and charge $\bar{X}$ (full credibility). A third possibility is to choose some combination of M and $\bar{X}$ (partial credibility).

From the insurer's standpoint, it seems sensible to "lean towards" the choice $\bar{X}$ if the experience is more "stable" (less variable, σ^2 small). This implies that $\bar{X}$ is of more use as a predictor of next year's results. Conversely, if the experience is more volatile (variable), then $\bar{X}$ is of less use as a predictor of next year's results and the choice M makes more sense.

Also, if we have an *a priori* reason to believe that the chances are great that this policyholder is unlike those who produced the manual premium M, then more weight should be given to $\bar{X}$. This is because as an unbiased estimator, $\bar{X}$ tells us something useful about ξ, while M is likely to be of little value. On the other hand, if all of our other policyholders have similar values of ξ, there is no point in relying on the (perhaps limited) experience of any one of them when M is likely to provide an excellent description of the propensity for claims or losses.

While reference is made to policyholders, the entity contributing to each X_j could arise from a single policyholder, a class of policyholders possessing similar underwriting characteristics, or a group of insureds assembled for some other reason. For example, for a given year X_j could be the number of claims filed in respect of a single automobile policy in one year, the average number of claims filed by all policyholders in a certain ratings class (e.g. single, male, under age 25, living in an urban area, driving over 7,500 miles per year), or the average amount of losses per vehicle for a fleet of delivery trucks owned by a food wholesaler.

We first present one approach to decide whether to assign full credibility (charge $\bar{X}$), and then we present an approach to assign partial credibility if it is felt that full credibility is inappropriate.

5.3.1 Full credibility

One method of quantifying the "stability" of $\bar{X}$ is to infer that $\bar{X}$ is stable if the difference between $\bar{X}$ and ξ is small relative to ξ with high probability. In statistical terms, this means that we should select two numbers $r > 0$ and $0 < p < 1$ (with r close to 0 and p close to 1, common choices being $r = 0.05$ and $p = 0.9$) and assign full credibility if

$$\Pr(-r\xi \le \bar{X} - \xi \le r\xi) \ge p. \tag{5.30}$$

It is convenient to restate (5.30) as

$$\Pr\left(\left|\frac{\bar{X}-\xi}{\sigma/\sqrt{n}}\right| \leq \frac{r\xi\sqrt{n}}{\sigma}\right) \geq p.$$

Now let y_p be defined by

$$y_p = \inf_y \left\{ \Pr\left(\left|\frac{\bar{X}-\xi}{\sigma/\sqrt{n}}\right| \leq y\right) \geq p \right\}. \tag{5.31}$$

That is, y_p is the smallest value of y which satisfies the probability statement in braces in (5.31). If $\bar{X}$ has a continuous distribution, the "$\geq$" sign in (5.31) may be replaced by an "$=$" sign and y_p satisfies

$$\Pr\left(\left|\frac{\bar{X}-\xi}{\sigma/\sqrt{n}}\right| \leq y_p\right) = p. \tag{5.32}$$

Then the condition for full credibility is $r\xi\sqrt{n}/\sigma \geq y_p$,

$$\frac{\sigma}{\xi} \leq \frac{r}{y_p}\sqrt{n} = \sqrt{\frac{n}{\lambda_0}} \tag{5.33}$$

where $\lambda_0 = (y_p/r)^2$. Condition (5.33) states that full credibility is assigned if the coefficient of variation σ/ξ is no larger than $\sqrt{n/\lambda_0}$, an intuitively reasonable result.

Also of interest is that (5.33) can be rewritten to show that full credibility occurs when

$$Var(\bar{X}) = \frac{\sigma^2}{n} \leq \frac{\xi^2}{\lambda_0}. \tag{5.34}$$

As indicated earlier, $\bar{X}$ should be used when the process is stable. This formula indicates the point at which the appropriate level of stability is deemed to have been achieved.

Alternatively, solving (5.33) for n gives the number of exposure units required for full credibility, namely

$$n \geq \lambda_0 \left(\frac{\sigma}{\xi}\right)^2. \tag{5.35}$$

In many situations it is reasonable to approximate the distribution of $\bar{X}$ by a normal distribution with mean ξ and variance σ^2/n. For example, central limit theorem arguments may be applicable if n is large. In that case $(\bar{X} - \xi)/(\sigma/\sqrt{n})$ has a standard normal distribution. Then (5.32) becomes (where Z has a standard normal distribution, and $\Phi(y)$ is its cdf)

$$\begin{aligned} p &= \Pr(|Z| \leq y_p) \\ &= \Pr(-y_p \leq Z \leq y_p) \\ &= \Phi(y_p) - \Phi(-y_p) \\ &= \Phi(y_p) - 1 + \Phi(y_p) \\ &= 2\Phi(y_p) - 1. \end{aligned}$$

Therefore $\Phi(y_p) = (1+p)/2$ and therefore y_p is the $(1+p)/2$ percentile of the standard normal distribution.

For example, if $p = 0.9$, then standard normal tables give $y_{0.9} = 1.645$. If, in addition, $r = 0.05$, then $\lambda_0 = (32.9)^2 = 1{,}082.41$ and (5.35) yields $n \geq 1{,}082.41\sigma^2/\xi^2$. Note that this answer assumes we know the coefficient of variation of X_j. It is possible we have some idea of its value, even though we do not know the value of ξ (remember, that is the quantity we want to estimate).

Example 5.19 *Suppose that past losses $X_1, \ldots, X_n$ are available for a particular policyholder, and it is reasonable to assume that the X_js are independent and compound Poisson distributed with Poisson parameter λ and claim size distribution with mean θ_Y and variance σ_Y^2. Determine the number of exposures, the number of expected claims, and the total claim dollars required for full credibility.*

We have $\xi = E(X_j) = \lambda\theta_Y$ and $\sigma^2 = Var(X_j) = \lambda\left(\theta_Y^2 + \sigma_Y^2\right)$, implying from (5.35) that

$$n \geq \frac{\lambda_0}{\lambda}\left[1 + \left(\frac{\sigma_Y}{\theta_Y}\right)^2\right].$$

The expected total number of claims for all past exposure units is $n\lambda$ and must satisfy

$$n\lambda \geq \lambda_0\left[1 + \left(\frac{\sigma_Y}{\theta_Y}\right)^2\right].$$

The expected total dollars of claims for all past exposure units is $n\lambda\theta_Y$ and must satisfy

$$n\lambda\theta_Y \geq \lambda_0\left[\theta_Y + \frac{\sigma_Y^2}{\theta_Y}\right].$$

In practice, when deciding whether to assign full credibility, the expected value $n\lambda$ is often estimated by the total observed number of past claims and $n\lambda\theta_Y$ is estimated by the total observed dollars of past losses. It can be shown that if λ is large, the compound Poisson distribution may be approximated by a normal distribution, and so y_p is readily obtainable. If $r = 0.05$ and $p = 0.9$, then $\lambda_0 = 1{,}082.41$ and we have

$$n \geq \frac{1{,}082.41}{\lambda}\left[1 + \left(\frac{\sigma_Y}{\theta_Y}\right)^2\right]$$

or

$$n\lambda \geq 1{,}082.41\left[1 + \left(\frac{\sigma_Y}{\theta_Y}\right)^2\right]$$

or

$$n\lambda\theta_Y \geq 1{,}082.41\left[\theta_Y + \frac{\sigma_Y^2}{\theta_Y}\right].$$

Note that only the middle formula depends exclusively on the coefficient of variation (of the individual losses), while the other two require more detailed knowledge. For that reason it has been the most popular formula. □

In all three cases the objective was to estimate the average dollars of claims per exposure. The standard for full credibility could be established in terms of any of three measurements: exposures, number of claims, or dollars of claims. In some situations the objective is to determine the number of claims (e.g., some types of automobile insurance) and not the amount of claims. The following example provides a convenient mechanism for handling such situations.

Example 5.20 *Suppose that one has past data $N_1, N_2, \ldots, N_n$ on the number of claims for a particular policyholder, and the N_js are independently Poisson distributed with mean λ. Obtain a standard for full credibility in terms of the number of expected claims.*

We have $\xi = \sigma^2 = \lambda$ and so (5.35) becomes

$$\lambda \geq \frac{\lambda_0}{n}.$$

Again, with $r = 0.05$ and $p = 0.9$, $\lambda_0 = 1{,}082.41$ and

$$n\lambda \geq 1{,}082.41.$$

It may be convenient to estimate $n\lambda$ by $N_1 + N_2 + \cdots + N_n$, in which case full credibility would be assigned if

$$N_1 + N_2 + \cdots + N_n \geq \lambda_0$$

that is, the number of observed claims exceeds λ_0, often taken to be 1,082.41.□

Standards for full credibility could also have been established in terms of number of exposures or dollars of claims. The latter seems unlikely to occur in practice (recall, the goal was to estimate the frequency of claims).

5.3.2 Partial credibility

If it is decided that full credibility is inappropriate, then for competitive reasons (or otherwise) it may be desirable to reflect the past experience $\bar{X}$ in the net premium, as well as the externally obtained mean, M. An intuitively appealing method for doing this is through a weighted average, that is, through the credibility premium

$$P_c = Z\bar{X} + (1 - Z)M \tag{5.36}$$

where the credibility factor $Z \in [0,1]$ needs to be chosen. There are many formulas for Z which have been suggested in the actuarial literature, usually justified on intuitive rather than theoretical grounds. One important choice is

$$Z = \frac{n}{n+k} \tag{5.37}$$

where k needs to be determined. This particular choice will be shown to be theoretically justified on the basis of a statistical model to be presented in the next section. Another choice, based on the same idea as full credibility (and including the full credibility case $Z = 1$), will now be discussed.

A variety of arguments have been used for developing the value of Z, many of which lead to the same answer. All of them are flawed in one way or another. The development we have chosen to present is also flawed, but is at least simple. Recall that the goal of the full credibility standard was to ensure that the difference between the net premium we are considering ($\bar{X}$) and what we should be using (ξ) is small with high probability. Because $\bar{X}$ is unbiased, this is essentially (and exactly if $\bar{X}$ has the normal distribution) equivalent to controlling the variance of the proposed net premium, $\bar{X}$, in this case. We see from (5.34) that there is no assurance that the variance of $\bar{X}$ will be small enough. However, it is possible to control the variance of the credibility premium, P_c, as follows:

$$\begin{aligned} \frac{\xi^2}{\lambda_0} &= Var(P_c) \\ &= Var[Z\bar{X} + (1-Z)M] \\ &= Z^2 Var(\bar{X}) \\ &= Z^2 \frac{\sigma^2}{n}. \end{aligned}$$

Thus $Z = \dfrac{\xi}{\sigma}\sqrt{\dfrac{n}{\lambda_0}}$, provided it is less than one. This can be written using the single formula

$$Z = \min\left\{\frac{\xi}{\sigma}\sqrt{\frac{n}{\lambda_0}}, 1\right\}. \tag{5.38}$$

One interpretation of (5.38) is that the credibility factor Z is the ratio of the coefficient of variation required for full credibility ($\sqrt{n/\lambda_0}$) to the actual coefficient of variation. For obvious reasons this is often called the square root rule for partial credibility.

So where is the flaw? Other than assuming that the variance captures the variability of $\bar{X}$ in the right way, all of the mathematics is correct. The flaw is in the goal. Unlike $\bar{X}$, P_c is not an unbiased estimator of ξ. In fact, one of the qualities that allows credibility to work is its use of biased estimators. But that means that the appropriate measure of the quality of P_c is not its variance, but its mean squared error. However, the mean squared error requires knowledge of the bias, and, in turn, that requires knowledge of the relationship of ξ and M. However, we know nothing about that relationship, and the data we have collected are of little help. As noted in the next subsection, this is not only a problem with our determination of Z, it is a problem that is characteristic of the limited fluctuation approach. A model for this relationship is introduced in the next section.

If, as before, the distribution of $\bar{X}$ can be approximated by a normal distribution with the same mean and variance, then y_p is easily obtained. If $r = 0.05$ and $p = 0.9$, then $\lambda_0 = 1{,}082.41$ as before.

Example 5.21 (Example 5.19 continued) *Suppose that X_j is compound Poisson with Poisson parameter λ and claim size mean θ_Y and variance σ_Y^2. Then $\xi = \lambda\theta_Y$ and $\sigma^2 = \lambda\left(\theta_Y^2 + \sigma_Y^2\right)$. Determine the credibility factor.*

From (5.38), the credibility factor is

$$Z = \min\left\{\sqrt{\frac{\lambda n/\lambda_0}{1+\left(\dfrac{\sigma_Y}{\theta_Y}\right)^2}}, 1\right\}$$

and if $r = 0.05$ and $p = 0.9$, this becomes

$$Z = \min\left\{\sqrt{\frac{\lambda n/1{,}082.41}{1+\left(\dfrac{\sigma_Y}{\theta_Y}\right)^2}}, 1\right\}.$$

Suppose the standard for full credibility was to be in terms of the expected number of claims (λn). Then the credibility factor is the square root of the ratio of the actual expected number of claims to the standard for full credibility. In practice, the actual expected number of claims would be replaced by the number of observed claims. In general, the credibility factor will be the square root of the ratio of "how much you have" to "how much you need." The latter is the standard for full credibility, and the former is how much of the standard was actually obtained. □

As was the case with full credibility, the situation involving partial credibility based solely on the number of claims is straightforward to implement, as the following example demonstrates.

Example 5.22 (Example 5.20 continued) *Suppose we have past data $N_1, \ldots, N_n$ on the number of claims where the N_js are independently Poisson distributed with mean λ. Determine the formula for partial credibility.*

In this case $\xi/\sigma = \sqrt{\lambda}$ and so

$$Z = \min\left\{\sqrt{\frac{\lambda n}{1{,}082.41}}, 1\right\}.$$

As with full credibility, we could estimate λn by $N_1 + N_2 + \cdots + N_n$, the total observed number of claims. □

Several examples using this theory, where we assume that $r = 0.05$ and $p = 0.9$, are now presented.

Example 5.23 *For group dental insurance, historical experience on many groups has revealed that annual losses per life insured have a mean of 175 and a standard deviation of 140. A particular group has been covered for 2 years with 100 lives insured in year 1 and 110 in year 2 and has experienced average claims of 150 over that period. Determine if full or partial credibility is appropriate, and determine the credibility premium for next year's losses if there will be 125 lives insured.*

We will apply the credibility on a per life insured basis. We have observed $100+110 = 210$ exposure units (assume experience is independent for different lives and years), and $\bar{X} = 150$. Now $\xi = 175$ and $\sigma = 140$. Thus, with $n = 210$ and $\lambda_0 = 1{,}082.41$ we use (5.38) to obtain

$$Z = (175/140)\sqrt{210/1{,}082.41} = 0.55$$

(note that $\bar{X}$ is the average of 210 claims, so approximate normality is assumed by the central limit theorem). Thus, the net premium per life insured is

$$P_c = (0.55)(150) + (0.45)(175) = 161.25.$$

The net premium for the whole group is $125(161.25) = 20{,}156.25$. □

Example 5.24 *An insurance coverage involves credibility based on number of claims only. For a particular group, 715 claims have been observed. Determine an appropriate credibility factor, assuming that the number of claims is Poisson distributed.*

From Example 5.22, estimate λn by the observed number of claims. Then

$$Z = (715/1{,}082.41)^{1/2} = 0.813.$$ □

Example 5.25 *Past data on a particular group are $\mathbf{X} = (X_1, X_2, \ldots, X_n)'$, where the X_j are independent and identically distributed compound Poisson random variables with exponentially distributed claim sizes. If the credibility factor based on claim numbers is 0.8, determine the appropriate credibility factor based on total claims.*

From Example 5.22, $Z = 0.8$ implies that $\lambda n/\lambda_0 = (0.8)^2 = 0.64$. For exponentially distributed claim sizes $\sigma_Y^2 = \theta_Y^2$. Hence from Example 5.21, $Z = (0.64/2)^{1/2} = 0.566$. Note that the normal approximation is not needed in this case. □

5.3.3 Problems with the approach

While the limited fluctuation approach yields simple solutions to the problem, there are theoretical difficulties. First, there is no underlying theoretical model for the distribution of the X_js, which suggests that a premium of the form (5.36) is appropriate and preferable to M. Why not just estimate ξ from a collection of homogeneous policyholders and charge all policyholders the same rate? While there is a practical reason for using (5.36), no model has been presented to suggest that this may be appropriate. Consequently, the choice of Z (and hence P_c) is completely arbitrary.

Second, even if (5.36) were "appropriate" for a particular model, there is no guidance for the selection of r and p.

Finally, the limited fluctuation approach does not examine the difference between ξ and M. When (5.36) is employed we are essentially stating that the value of M is accurate as a representation of the expected value given no information about this particular policyholder. However, it is usually the case that M is also an estimate and therefore unreliable in itself. The correct credibility question should be "how much more reliable is $\bar{X}$ compared to M?" and not "how reliable is $\bar{X}$?"

In the remainder of this chapter, a systematic modeling approach is presented for the claims experience of a particular policyholder which suggests that the past experience of the policyholder is relevant for prospective rate making. Furthermore, the intuitively appealing formula (5.36) is a consequence of this approach, and Z is often obtained from relations of the form (5.37).

5.3.4 Notes and References

The limited fluctuation approach is discussed by Herzog [59] and Longley-Cook [85]. See also Norberg [96].

5.4 GREATEST ACCURACY CREDIBILITY THEORY

In this and the following section, we consider a model-based approach to the solution of the problem addressed in the previous section. This approach, referred to as greatest accuracy credibility theory, is the outgrowth of a classic 1967 paper by Bühlmann [18]. Many of the ideas are also found in Bailey [7].

We return to the basic problem. For a particular policyholder, we have observed n exposure units of past claims $\mathbf{X} = (X_1, \ldots, X_n)'$. We have a manual rate μ (We no longer use M for the manual rate) which is applicable to this policyholder, but the past experience indicates that this may not be appropriate ($\bar{X} = n^{-1}(X_1 + \cdots + X_n)$, as well as $E(X)$, could be quite different from μ). This raises the question of whether next year's (per exposure unit) net premium should be based on μ, on $\bar{X}$, or on a combination of the two.

The insurer needs to consider the following question: Is the policyholder really different from what has been assumed in the calculation of μ or has it just been random chance which has been responsible for the differences between μ and $\bar{X}$?

While it is difficult to definitively answer the above question, it is clear that no underwriting system is perfect. The manual rate μ has presumably been obtained by (a) evaluation of the underwriting characteristics of the policyholder and (b) assignment of the rate on the basis of inclusion of the policyholder in a rating class. Such a class should include risks with similar underwriting characteristics. In other words, the rating class is viewed as homogeneous with respect to the underwriting characteristics used. Surely, not all risks in the class are truly homogeneous, however. No matter how detailed the underwriting procedure, there still remains some heterogeneity with respect to risk characteristics within the rating class (good and bad risks, relatively speaking).

Thus, it is possible that the given policyholder may be "different" from what has been assumed. If this is the case, how should one proceed in order to choose an appropriate rate for the policyholder?

To proceed, let us assume that the risk level of each policyholder in the rating class may be characterized by a risk parameter θ (possibly vector-valued), but the value of θ varies by policyholder. This allows us to quantify the differences between policyholders with respect to the risk characteristics. Since all observable underwriting characteristics have already been used, θ may be viewed as representative of the residual, unobserved, factors which affect the risk level. Consequently, we shall assume the existence of θ, but we shall further assume that it is not observable and that we can never know its true value.

Since θ varies by policyholder, there is a probability distribution with pf $\pi(\theta)$ of these values across the rating class. Thus if θ is a scalar parameter, the cumulative distribution function $\Pi(\theta)$ may be interpreted as the proportion of policyholders in the rating class with risk parameter Θ less than or equal to θ. (In statistical terms, Θ is a random variable with distribution function $\Pi(\theta) = \Pr(\Theta \leq \theta)$.) Stated another way, $\Pi(\theta)$ represents the probability that a policyholder picked at random from the rating class has a risk parameter less than or equal to θ (to accommodate the possibility of new insureds, we slightly generalize the "rating class" interpretation to include the population of all potential risks, either insured or not).

While the θ value associated with an individual policyholder is not (and cannot be) known, we assume (for this section) that $\pi(\theta)$ is known. That is, the structure of the risk characteristics within the population is known. This assumption can be relaxed, and we shall decide later how to estimate the relevant characteristics of $\pi(\theta)$, because this is needed in order to implement the theory.

Since risk levels vary within the population, it is clear that the experience of the policyholder varies in a systematic way with θ. Imagine that the experience of a policyholder picked (at random) from the population arises from a two-stage process. First, the risk parameter θ is selected from the distribution $\pi(\theta)$. Then the claims or losses X arise from the conditional distribution of X given θ, $f_{X|\Theta}(x|\theta)$. Thus the experience varies with θ via the distribution given the risk parameter θ. The distribution of claims thus differs from policyholder to policyholder to reflect the differences in the risk parameters.

Example 5.26 *Consider a rating class for automobile insurance, where θ represents the expected number of claims for a policyholder with risk parameter θ. To accommodate the variability in claims incidence, we assume that the values of θ vary across the rating class. Relatively speaking, the good drivers are those with small values of θ, whereas the poor drivers are those with larger values of θ. It is convenient mathematically in this case to assume that the number of claims for a policyholder with risk parameter θ is Poisson distributed with mean θ. The random variable Θ may also be assumed to be gamma distributed with parameters α and β. Suppose it is known that the average number of expected claims for this rating class is 0.15 ($E(\Theta) = 0.15$), and 95% of the policyholders have expected claims between 0.10 and 0.20. Determine α and β.*

Assuming the normal approximation to the gamma, where it is known that 95% of the probability lies within about 2 standard deviations of the mean, it follows that Θ has standard deviation 0.025. Thus $E(\Theta) = \alpha\beta = 0.15$ and $Var(\Theta) = \alpha\beta^2 = (0.025)^2$. Solving for α and β yields $\beta = 1/240$ and $\alpha = 36$. □

Example 5.27 *A simple model to describe both frequency and severity is as follows. For the frequency component, two dice are used. The first die D_1 has 2 "marked" faces and 4 "unmarked" faces. The second die D_2 has 3 "marked" faces and 3 "unmarked" faces. On each die, the probability of any 1 face showing is 1/6. An individual's propensity for claims is governed by either die D_1 or D_2 (which never changes once assigned to an individual) and the rolling of a "marked" face indicates that a claim has occurred. For the severity component, two spinners are used. Each spinner has 5 equally spaced sectors. The first spinner S_1 has 3 sectors marked 5 and 2 sectors marked 10. The second spinner S_2 has 1 sector marked 5 and 4 sectors marked 10. Again, each individual is assigned either S_1 or S_2 and if there is a claim, a* *spin determines the amount. Describe the risk process and how it involves an unknown risk parameter.*

The "risk characteristics" of a policyholder consist of one die and one spinner. That is, Θ takes on four values, $\theta_{11} = (D_1, S_1)$, $\theta_{12} = (D_1, S_2)$, $\theta_{21} = (D_2, S_1)$, and $\theta_{22} = (D_2, S_2)$. If each spinner and each die are assumed to be independently and equally likely selected, then $\Pr(\Theta = \theta_{ij}) = \frac{1}{2} \cdot \frac{1}{2} = \frac{1}{4}$ for $i = 1, 2$ and $j = 1, 2$. The underlying population consists of 4 equally likely risk types: θ_{11}, θ_{12}, θ_{21}, and θ_{22}.

To simulate the claims experience for a particular policyholder, the die is tossed first. If a "marked" face shows up, a claim occurs. If an "unmarked" face shows up, no claim occurs. If a claim occurs, the spinner is spun to determine the amount or severity of the loss, corresponding to the number in the sector. Let X be the total losses of a policyholder according to this model. The loss distributions for each of the four policyholder types are as follows:

For θ_{11}:

$$\begin{aligned}
\Pr(X = 0|\Theta = \theta_{11}) &= \frac{4}{6} = \frac{10}{15} \\
\Pr(X = 5|\Theta = \theta_{11}) &= \frac{2}{6} \cdot \frac{3}{5} = \frac{3}{15} \\
\Pr(X = 10|\Theta = \theta_{11}) &= \frac{2}{6} \cdot \frac{2}{5} = \frac{2}{15}.
\end{aligned}$$

For θ_{12}:

$$\begin{aligned}
\Pr(X = 0|\Theta = \theta_{12}) &= \frac{4}{6} = \frac{10}{15} \\
\Pr(X = 5|\Theta = \theta_{12}) &= \frac{2}{6} \cdot \frac{1}{5} = \frac{1}{15} \\
\Pr(X = 10|\Theta = \theta_{12}) &= \frac{2}{6} \cdot \frac{4}{5} = \frac{4}{15}.
\end{aligned}$$

For θ_{21}:

$$\begin{aligned}
\Pr(X = 0|\Theta = \theta_{21}) &= \frac{3}{6} = \frac{5}{10} \\
\Pr(X = 5|\Theta = \theta_{21}) &= \frac{3}{6} \cdot \frac{3}{5} = \frac{3}{10} \\
\Pr(X = 10|\Theta = \theta_{21}) &= \frac{3}{6} \cdot \frac{2}{5} = \frac{2}{10}.
\end{aligned}$$

For θ_{22}:

$$\begin{aligned}
\Pr(X = 0|\Theta = \theta_{22}) &= \frac{3}{6} = \frac{5}{10} \\
\Pr(X = 5|\Theta = \theta_{22}) &= \frac{3}{6} \cdot \frac{1}{5} = \frac{1}{10} \\
\Pr(X = 10|\Theta = \theta_{22}) &= \frac{3}{6} \cdot \frac{4}{5} = \frac{4}{10}.
\end{aligned}$$

□

5.4.1 The Bayesian methodology

As in the previous section, assume that the distribution of the risk characteristics in the population may be represented by $\pi(\theta)$, and the experience of a particular policyholder with risk parameter θ arises from the conditional distribution $f_{X|\Theta}(x|\theta)$ of claims or losses given θ.

We now return to the problem introduced in Section 5.3. That is, for a particular policyholder, we have observed $\mathbf{X} = \mathbf{x}$ where $\mathbf{X} = (X_1, \ldots, X_n)'$ and $\mathbf{x} = (x_1, \ldots, x_n)'$ and are interested in setting a rate to cover X_{n+1}. We assume that the risk parameter associated with the policyholder is θ (which is unknown). Furthermore, the experience of the policyholder corresponding to different exposure periods are assumed to be independent. In statistical terms, conditional on θ, the claims or losses $X_1, \ldots, X_n, X_{n+1}$ are independent (although not necessarily identically distributed).

Let X_j have conditional pf

$$f_{X_j|\Theta}(x_j|\theta); \qquad j = 1, \ldots, n, n+1.$$

Note that if the X_j are identically distributed (conditional on $\Theta = \theta$), then $f_{X_j|\Theta}(x_j|\theta)$ does not depend on j. Ideally, we are interested in the conditional distribution of X_{n+1} given $\Theta = \theta$ in order to predict the claims experience X_{n+1} of the same policyholder (whose value of θ has been assumed not to have changed). If we knew θ, we could use $f_{X_{n+1}|\Theta}(x_{n+1}|\theta)$. Unfortunately, we do not know θ, but we do know $\mathbf{x}$ for the same policyholder. The obvious next step is to condition on $\mathbf{x}$ rather than θ. Consequently, we will calculate the conditional distribution of X_{n+1} given $\mathbf{X} = \mathbf{x}$, termed the **predictive distribution**.

The predictive distribution of X_{n+1} given $\mathbf{X} = \mathbf{x}$ is the relevant distribution for risk analysis, management, and decision making. It combines the uncertainty about the claims losses with that of the parameters associated with the risk process.

The joint distribution of $X_1, \ldots, X_n, \Theta$ is obtained by first conditioning on Θ. Because the X_js are independent conditional on $\Theta = \theta$, we have (as in (5.18))

$$f_{\mathbf{X},\Theta}(\mathbf{x}, \theta) = f(x_1, \ldots, x_n|\theta)\pi(\theta) = \left[\prod_{j=1}^{n} f_{X_j|\Theta}(x_j|\theta)\right]\pi(\theta).$$

The joint distribution of $\mathbf{X}$ is thus the marginal distribution obtained by integrating θ out, as in (5.19):

$$f_{\mathbf{X}}(\mathbf{x}) = \int \left[\prod_{j=1}^{n} f_{X_j|\Theta}(x_j|\theta)\right]\pi(\theta)d\theta \qquad (5.39)$$

The integral in (5.39) is replaced by a sum if $\pi(\theta)$ is a discrete pf. Similarly, the joint distribution of $X_1, \ldots, X_{n+1}$ is simply the right-hand side of (5.39) with n replaced by $n+1$. Finally, the conditional density of X_{n+1} given $\mathbf{X} = \mathbf{x}$ is the joint density of $(X_1, \ldots, X_{n+1})$ divided by that of $\mathbf{X}$, namely

$$f_{X_{n+1}|\mathbf{X}}(x_{n+1}|\mathbf{x}) = \frac{1}{f_{\mathbf{X}}(\mathbf{x})} \int \left[\prod_{j=1}^{n+1} f_{X_j|\Theta}(x_j|\theta) \right] \pi(\theta) d\theta. \tag{5.40}$$

There is a hidden mathematical structure underlying (5.40) which may often be exploited. The posterior density of Θ given $\mathbf{X}$ is

$$\pi_{\Theta|\mathbf{X}}(\theta|\mathbf{x}) = \frac{1}{f_{\mathbf{X}}(\mathbf{x})} \left[\prod_{j=1}^{n} f_{X_j|\Theta}(x_j|\theta) \right] \pi(\theta).$$

In other words, $\left[\prod_{j=1}^{n} f_{X_j|\Theta}(x_j|\theta) \right] \pi(\theta) = \pi_{\Theta|\mathbf{X}}(\theta|\mathbf{x}) f_{\mathbf{X}}(\mathbf{x})$, and substitution in the numerator of (5.40) yields

$$f_{X_{n+1}|\mathbf{X}}(x_{n+1}|\mathbf{x}) = \int f_{X_{n+1}|\Theta}(x_{n+1}|\theta) \pi_{\Theta|\mathbf{X}}(\theta|\mathbf{x}) d\theta. \tag{5.41}$$

Equation (5.41) provides the additional insight that the conditional distribution of X_{n+1} given $\mathbf{X}$ may be viewed as a mixed distribution, with mixing distribution the posterior distribution $\pi_{\Theta|\mathbf{X}}(\theta|\mathbf{x})$.

The posterior distribution combines and summarizes the information about θ contained in the prior distribution and the likelihood, and consequently (5.41) reflects this information. As noted in Theorem 5.1, the posterior distribution admits a convenient form when the likelihood is derived from the linear exponential family and $\pi(\theta)$ is the natural conjugate prior. This provides an easy method to evaluate the conditional distribution of X_{n+1} given $\mathbf{X}$ in these cases.

Example 5.28 (Example 5.27 continued) *For a particular policyholder suppose we have observed $X_1 = x_1$ (with $x_1 = 0$, 5, or 10). Determine the conditional distribution of $X_2|X_1$.*

First, we obtain the marginal distribution of X_1. The distribution of Θ is $\pi(\theta)$ and in this case $\pi(\theta_{ij}) = 1/4$ for $i = 1, 2$ and $j = 1, 2$. Then from (5.39) with the integral replaced by a sum,

$$\begin{aligned} f_{X_1}(x_1) &= \sum_{i=1}^{2} \sum_{j=1}^{2} f_{X_1|\Theta}(x_1|\theta_{ij}) \pi(\theta_{ij}) \\ &= \frac{1}{4} \sum_{i=1}^{2} \sum_{j=1}^{2} \Pr(X_1 = x_1 | \Theta = \theta_{ij}). \end{aligned}$$

Thus,

$$\begin{aligned}
f_{X_1}(0) &= \frac{1}{4}\left(\frac{10}{15}+\frac{10}{15}+\frac{5}{10}+\frac{5}{10}\right)=\frac{7}{12}\\
f_{X_1}(5) &= \frac{1}{4}\left(\frac{3}{15}+\frac{1}{15}+\frac{3}{10}+\frac{1}{10}\right)=\frac{2}{12}\\
f_{X_1}(10) &= \frac{1}{4}\left(\frac{2}{15}+\frac{4}{15}+\frac{2}{10}+\frac{4}{10}\right)=\frac{3}{12}.
\end{aligned}$$

We now construct the joint distribution of X_1 and X_2 by first conditioning on Θ. This yields

$$\begin{aligned}
f_{X_1,X_2}(x_1,x_2) &= \sum_{i=1}^{2}\sum_{j=1}^{2} f_{X_1|\Theta}(x_1|\theta_{ij})f_{X_2|\Theta}(x_2|\theta_{ij})\pi(\theta_{ij})\\
&= \frac{1}{4}\sum_{i=1}^{2}\sum_{j=1}^{2}\Pr(X_1=x_1|\Theta=\theta_{ij})\Pr(X_2=x_2|\Theta=\theta_{ij}).
\end{aligned}$$

Thus,

$$\begin{aligned}
f_{X_1,X_2}(0,0) &= \frac{1}{4}\left[\left(\frac{10}{15}\right)^2+\left(\frac{10}{15}\right)^2+\left(\frac{5}{10}\right)^2+\left(\frac{5}{10}\right)^2\right]=\frac{125}{360}\\
f_{X_1,X_2}(0,5) &= \frac{1}{4}\left[\left(\frac{10}{15}\right)\left(\frac{3}{15}\right)+\left(\frac{10}{15}\right)\left(\frac{1}{15}\right)+\left(\frac{5}{10}\right)\left(\frac{3}{10}\right)\right.\\
&\qquad\left.+\left(\frac{5}{10}\right)\left(\frac{1}{10}\right)\right]=\frac{34}{360}\\
f_{X_1,X_2}(0,10) &= \frac{1}{4}\left[\left(\frac{10}{15}\right)\left(\frac{2}{15}\right)+\left(\frac{10}{15}\right)\left(\frac{4}{15}\right)+\left(\frac{5}{10}\right)\left(\frac{2}{10}\right)\right.\\
&\qquad\left.+\left(\frac{5}{10}\right)\left(\frac{4}{10}\right)\right]=\frac{51}{360}\\
f_{X_1,X_2}(5,5) &= \frac{1}{4}\left[\left(\frac{3}{15}\right)^2+\left(\frac{1}{15}\right)^2+\left(\frac{3}{10}\right)^2+\left(\frac{1}{10}\right)^2\right]=\frac{13}{360}\\
f_{X_1,X_2}(5,10) &= \frac{1}{4}\left[\left(\frac{3}{15}\right)\left(\frac{2}{15}\right)+\left(\frac{1}{15}\right)\left(\frac{4}{15}\right)+\left(\frac{3}{10}\right)\left(\frac{2}{10}\right)\right.\\
&\qquad\left.+\left(\frac{1}{10}\right)\left(\frac{4}{10}\right)\right]=\frac{13}{360}\\
f_{X_1,X_2}(10,10) &= \frac{1}{4}\left[\left(\frac{2}{15}\right)^2+\left(\frac{4}{15}\right)^2+\left(\frac{2}{10}\right)^2+\left(\frac{4}{10}\right)^2\right]=\frac{26}{360}
\end{aligned}$$

and by symmetry,

$$f_{X_1,X_2}(5,0) = f_{X_1,X_2}(0,5)=\frac{34}{360}$$

$$\begin{aligned} f_{X_1,X_2}(10,0) &= f_{X_1,X_2}(0,10) = \frac{51}{360} \\ f_{X_1,X_2}(10,5) &= f_{X_1,X_2}(5,10) = \frac{13}{360}. \end{aligned}$$

The conditional distributions are thus given by

For $X_1 = 0$:

$$\begin{aligned} f_{X_2|X_1}(0|0) &= \frac{f_{X_1,X_2}(0,0)}{f_{X_1}(0)} = \frac{\frac{125}{360}}{\frac{7}{12}} = \frac{125}{210} \\ f_{X_2|X_1}(5|0) &= \frac{f_{X_1,X_2}(0,5)}{f_{X_1}(0)} = \frac{\frac{34}{360}}{\frac{7}{12}} = \frac{34}{210} \\ f_{X_2|X_1}(10|0) &= \frac{f_{X_1,X_2}(0,10)}{f_{X_1}(0)} = \frac{\frac{51}{360}}{\frac{7}{12}} = \frac{51}{210}. \end{aligned}$$

For $X_1 = 5$:

$$\begin{aligned} f_{X_2|X_1}(0|5) &= \frac{f_{X_1,X_2}(5,0)}{f_{X_1}(5)} = \frac{\frac{34}{360}}{\frac{2}{12}} = \frac{34}{60} \\ f_{X_2|X_1}(5|5) &= \frac{f_{X_1,X_2}(5,5)}{f_{X_1}(5)} = \frac{\frac{13}{360}}{\frac{2}{12}} = \frac{13}{60} \\ f_{X_2|X_1}(10|5) &= \frac{f_{X_1,X_2}(5,10)}{f_{X_1}(5)} = \frac{\frac{13}{360}}{\frac{2}{12}} = \frac{13}{60}. \end{aligned}$$

For $X_1 = 10$:

$$\begin{aligned} f_{X_2|X_1}(0|10) &= \frac{f_{X_1,X_2}(10,0)}{f_{X_1}(10)} = \frac{\frac{51}{360}}{\frac{3}{12}} = \frac{51}{90} \\ f_{X_2|X_1}(5|10) &= \frac{f_{X_1,X_2}(10,5)}{f_{X_1}(10)} = \frac{\frac{13}{360}}{\frac{3}{12}} = \frac{13}{90} \\ f_{X_2|X_1}(10|10) &= \frac{f_{X_1,X_2}(10,10)}{f_{X_1}(10)} = \frac{\frac{26}{360}}{\frac{3}{12}} = \frac{26}{90}. \end{aligned}$$

Although the approach given is direct and straightforward, it is often computationally simpler to use (5.41). We will demonstrate this by evaluation of $f_{X_2|X_1}(x_2|5)$. Now, by Bayes' theorem,

$$\pi_{\Theta|X_1}(\theta|5) = \frac{f_{X_1|\Theta}(5|\theta)\pi(\theta)}{f_{X_1}(5)}.$$

We know that $\pi(\theta_{ij}) = 1/4$ for $i, j = 1, 2$. Also, we have calculated $f_{X_1}(5) = 2/12$. Then from Example 5.27,

$$\pi_{\Theta|X_1}(\theta_{11}|5) = \left(\frac{3}{15}\right)\left(\frac{1}{4}\right) / \left(\frac{2}{12}\right) = \frac{6}{20}$$

$$\begin{aligned}
\pi_{\Theta|X_1}(\theta_{12}|5) &= \left(\frac{1}{15}\right)\left(\frac{1}{4}\right)/\left(\frac{2}{12}\right) = \frac{2}{20} \\
\pi_{\Theta|X_1}(\theta_{21}|5) &= \left(\frac{3}{10}\right)\left(\frac{1}{4}\right)/\left(\frac{2}{12}\right) = \frac{9}{20} \\
\pi_{\Theta|X_1}(\theta_{22}|5) &= \left(\frac{1}{10}\right)\left(\frac{1}{4}\right)/\left(\frac{2}{12}\right) = \frac{3}{20}.
\end{aligned}$$

In this example, (5.41) becomes

$$f_{X_2|X_1}(x_2|5) = \sum_{i=1}^{2}\sum_{j=1}^{2} f_{X_2|\Theta}(x_2|\theta_{ij})\pi_{\Theta|X_1}(\theta_{ij}|5).$$

Hence, using Example 5.27 and the posterior distribution given above, we find

$$\begin{aligned}
f_{X_2|X_1}(0|5) &= \left(\frac{10}{15}\right)\left(\frac{6}{20}\right) + \left(\frac{10}{15}\right)\left(\frac{2}{20}\right) + \left(\frac{5}{10}\right)\left(\frac{9}{20}\right) \\
&\quad + \left(\frac{5}{10}\right)\left(\frac{3}{20}\right) = \frac{34}{60} \\
f_{X_2|X_1}(5|5) &= \left(\frac{3}{15}\right)\left(\frac{6}{20}\right) + \left(\frac{1}{15}\right)\left(\frac{2}{20}\right) + \left(\frac{3}{10}\right)\left(\frac{9}{20}\right) \\
&\quad + \left(\frac{1}{10}\right)\left(\frac{3}{20}\right) = \frac{13}{60} \\
f_{X_2|X_1}(10|5) &= \left(\frac{2}{15}\right)\left(\frac{6}{20}\right) + \left(\frac{4}{15}\right)\left(\frac{2}{20}\right) + \left(\frac{2}{10}\right)\left(\frac{9}{20}\right) \\
&\quad + \left(\frac{4}{10}\right)\left(\frac{3}{20}\right) = \frac{13}{60}.
\end{aligned}$$

This agrees with the distribution given above, as it must. □

Example 5.29 *Suppose $X_j|\Theta$ is normally distributed with mean Θ and variance v for $j = 1, 2, \cdots, n+1$. Further suppose Θ is normally distributed with mean μ and variance a. Thus,*

$$f_{X_j|\Theta}(x_j|\theta) = (2\pi v)^{-1/2} \exp\left[-\frac{1}{2v}(x_j - \theta)^2\right], \qquad -\infty < x_j < \infty$$

and

$$\pi(\theta) = (2\pi a)^{-1/2} \exp\left[-\frac{1}{2a}(\theta - \mu)^2\right], \qquad -\infty < \theta < \infty.$$

Determine the predictive distribution of $X_{n+1}|\mathbf{X}$.

From Example 5.15, $\Theta|\mathbf{x}$ has posterior density $\pi_{\Theta|\mathbf{X}}(\theta|\mathbf{x})$ which is normal with mean

$$\mu_* = \frac{\frac{n\overline{x}}{v} + \frac{\mu}{a}}{\frac{n}{v} + \frac{1}{a}}$$

and variance

$$a_* = \left(\frac{n}{v} + \frac{1}{a}\right)^{-1}.$$

Then (5.41) implies that $X_{n+1}|\mathbf{x}$ is a mixture with $X_{n+1}|\Theta$ having a normal distribution with mean Θ, and $\Theta|\mathbf{x}$ is normal with mean μ_* and variance a_*. From Example 5.3, this means that $X_{n+1}|\mathbf{x}$ is normally distributed with mean μ_* and variance $a_* + v$, that is

$$f_{X_{n+1}|\mathbf{X}}(x_{n+1}|\mathbf{x}) = [2\pi(a_* + v)]^{-1/2} \exp\left[-\frac{(x_{n+1} - \mu_*)^2}{2(a_* + v)}\right], \quad -\infty < x_{n+1} < \infty.$$

□

As illustrated in Example 5.29, conjugate prior distributions play an important role in evaluation of the (predictive) distribution of $X_{n+1}|\mathbf{x}$. With a conjugate prior, when using (5.41), $\pi(\theta|\mathbf{x})$ and $\pi(\theta)$ are from the same parametric family. This also implies that $X_{n+1}|\mathbf{x}$ is a mixture distribution with a simple mixing distribution, facilitating evaluation of the density of $X_{n+1}|\mathbf{x}$. Further examples of this idea are found in the exercises.

To return to the original problem, we have observed $\mathbf{X} = \mathbf{x}$ for a particular policyholder and we wish to predict X_{n+1} (or its mean). An obvious choice would be the hypothetical mean (or individual premium)

$$\mu_{n+1}(\theta) = E(X_{n+1}|\Theta = \theta) = \int x_{n+1} f_{X_{n+1}|\Theta}(x_{n+1}|\theta)dx_{n+1} \tag{5.42}$$

if we knew θ. Note that replacement of θ by Θ in (5.42) yields, upon taking the expectation,

$$\mu_{n+1} = E(X_{n+1}) = E[E(X_{n+1}|\Theta)] = E[\mu_{n+1}(\Theta)]$$

so that the pure, or collective, premium is the mean of the hypothetical mean. This is the premium we would use if we knew nothing about the individual. It does not depend on the individual's risk parameter, θ, nor does it use $\mathbf{x}$, the data collected from the individual. Since θ is unknown, the best we can do is try to use the data. This leads to the Bayesian premium (the mean of the predictive distribution)

$$E(X_{n+1}|\mathbf{X} = \mathbf{x}) = \int x_{n+1} f_{X_{n+1}|\mathbf{X}}(x_{n+1}|\mathbf{x})dx_{n+1}. \tag{5.43}$$

A computationally more convenient form is

$$E\left(X_{n+1}|\mathbf{X} = \mathbf{x}\right) = \int \mu_{n+1}(\theta)\pi_{\Theta|\mathbf{X}}(\theta|\mathbf{x})d\theta. \tag{5.44}$$

In other words, the Bayesian premium is the expected value of the hypothetical mean, with expectation taken over the posterior distribution $\pi_{\Theta|\mathbf{X}}(\theta|\mathbf{x})$. We remind the reader that the integrals are replaced by sums in the discrete case. To prove (5.44), we see from (5.41) that

$$\begin{aligned}
E(X_{n+1}|\mathbf{X}=\mathbf{x}) &= \int x_{n+1} f_{X_{n+1}|\mathbf{X}}(x_{n+1}|\mathbf{x})dx_{n+1} \\
&= \int x_{n+1} \left[\int f_{X_{n+1}|\Theta}(x_{n+1}|\theta)\pi_{\Theta|\mathbf{X}}(\theta|\mathbf{x})d\theta \right] dx_{n+1} \\
&= \int \left[\int x_{n+1} f_{X_{n+1}|\Theta}(x_{n+1}|\theta)dx_{n+1} \right] \pi_{\Theta|\mathbf{X}}(\theta|\mathbf{x})d\theta \\
&= \int \mu_{n+1}(\theta)\pi_{\Theta|\mathbf{X}}(\theta|\mathbf{x})d\theta.
\end{aligned}$$

Example 5.30 (Example 5.27 continued) *Determine the Bayesian premium using both (5.43) and (5.44).*

The (unobservable) hypothetical means are

$$\begin{aligned}
\mu_{n+1}(\theta_{11}) &= (0)\frac{10}{15} + (5)\frac{3}{15} + (10)\frac{2}{15} = \frac{7}{3} \\
\mu_{n+1}(\theta_{12}) &= (0)\frac{10}{15} + (5)\frac{1}{15} + (10)\frac{4}{15} = 3 \\
\mu_{n+1}(\theta_{21}) &= (0)\frac{5}{10} + (5)\frac{3}{10} + (10)\frac{2}{10} = \frac{7}{2} \\
\mu_{n+1}(\theta_{22}) &= (0)\frac{5}{10} + (5)\frac{1}{10} + (10)\frac{4}{10} = \frac{9}{2}.
\end{aligned}$$

If, as in Example 5.28, we have observed $X_1 = x_1$, we have the Bayesian premiums obtained directly from (5.43):

$$\begin{aligned}
E(X_2|X_1=0) &= (0)\frac{125}{210} + (5)\frac{34}{210} + (10)\frac{51}{210} = \frac{68}{21} \\
E(X_2|X_1=5) &= (0)\frac{34}{60} + (5)\frac{13}{60} + (10)\frac{13}{60} = \frac{13}{4} \\
E(X_2|X_1=10) &= (0)\frac{51}{90} + (5)\frac{13}{90} + (10)\frac{26}{90} = \frac{65}{18}.
\end{aligned}$$

The (unconditional) pure premium is

$$\mu_2 = E(X_2) = \sum_{i=1}^{2}\sum_{j=1}^{2} \mu_2(\theta_{ij})\pi(\theta_{ij}) = \frac{1}{4}\left(\frac{7}{3} + 3 + \frac{7}{2} + \frac{9}{2}\right) = \frac{10}{3}.$$

To verify (5.44) with $x_1 = 5$, for example, we have the posterior distribution $\pi_{\Theta|X_1}(\theta|5)$ from Example 5.28.

Thus, (5.44) yields

$$\begin{aligned}
E(X_2|X_1=5) &= \left(\frac{7}{3}\right)\left(\frac{6}{20}\right) + (3)\left(\frac{2}{20}\right) + \left(\frac{7}{2}\right)\left(\frac{9}{20}\right) \\
&\quad + \left(\frac{9}{2}\right)\left(\frac{3}{20}\right) = \frac{13}{4}
\end{aligned}$$

in agreement with that obtained above. In general, the latter approach utilizing (5.44) is simpler than the direct approach using the conditional distribution of $X_{n+1}|\mathbf{X}=\mathbf{x}$. □

Example 5.31 (Example 5.29 continued) *Determine the Bayesian premium.*

From Example 5.29, we have $\mu_{n+1}(\theta) = \theta$, and $\Theta|\mathbf{X}=\mathbf{x}$ is normal with mean μ_* and variance a_* (i.e., $N(\mu_*, a_*)$). Then, (5.44) yields

$$E(X_{n+1}|\mathbf{X}=\mathbf{x}) = \int_{-\infty}^{\infty} \theta \pi_{\Theta|\mathbf{X}}(\theta|\mathbf{x})d\theta = \mu_*$$

(as it must since $X_{n+1}|\mathbf{X} = \mathbf{x} \sim N(\mu_*, a_* + v)$) because $\pi_{\Theta|\mathbf{X}}(\theta|\mathbf{x})$ is the $N(\mu_*, a_*)$ density.

Note that

$$E(X_{n+1}|\mathbf{X}=\mathbf{x}) = \mu_* = Z\overline{x} + (1-Z)\mu_{n+1}$$

with $Z = n(n+v/a)^{-1}$ and $\mu_{n+1} = E(X_{n+1}) = E[\mu_{n+1}(\Theta)] = E(\Theta) = \mu$. This formula is of the credibility weighted type (5.36). □

In the following examples, the conditional distributions of X_j given θ are independent but not identically distributed.

Example 5.32 *Two urns contain balls marked with 0s and 1s in the following proportions.*

	Proportion of 0s	Proportion of 1s
Urn I	0.6	0.4
Urn II	0.8	0.2

An urn is selected at random and then three balls are selected at random with replacement from the urn. The total of the marked numbers is 2. Calculate the expected value of the total of the marked numbers on the next two balls selected with replacement from that same urn using the Bayesian premium.

The two "policyholder" types are the two urns. Thus, we have $\pi(\theta_I) = \pi(\theta_{II}) = 1/2$. It is convenient to (a) think of the initial selection of the three balls as generating the first exposure unit and hence the observation X_1 and (b) think of the selection of two more balls as generating the second exposure X_2. Let Y_j be the marked number on the jth ball drawn. Then

$$X_1 = Y_1 + Y_2 + Y_3 \quad \text{and} \quad X_2 = Y_4 + Y_5.$$

In this case the Y_js and thus the X_js are independent, given Θ. Furthermore, the conditional distributions are binomial. Thus

$$\begin{aligned} f_{X_1|\Theta}(2|\theta_I) &= \Pr(X_1 = 2|\Theta = \theta_I) = \binom{3}{2}(0.4)^2(0.6) = 0.288 \\ f_{X_1|\Theta}(2|\theta_{II}) &= \Pr(X_1 = 2|\Theta = \theta_{II}) = \binom{3}{2}(0.2)^2(0.8) = 0.096 \end{aligned}$$

and so

$$\begin{aligned} f_{X_1}(2) &= \Pr(X_1 = 2|\Theta = \theta_I)\pi(\theta_I) + \Pr(X_1 = 2|\Theta = \theta_{II})\pi(\theta_{II}) \\ &= (0.288)\left(\frac{1}{2}\right) + (0.096)\left(\frac{1}{2}\right) = 0.192. \end{aligned}$$

The posterior distribution of Θ given $X_1 = 2$ is given by

$$\begin{aligned} \pi_{\Theta|X_1}(\theta_I|2) &= \frac{f_{X_1|\Theta}(2|\theta_I)\pi(\theta_I)}{f_{X_1}(2)} = \frac{(0.288)(1/2)}{0.192} = 0.75 \\ \pi_{\Theta|X_1}(\theta_{II}|2) &= \frac{f_{X_1|\Theta}(2|\theta_{II})\pi(\theta_{II})}{f_{X_1}(2)} = \frac{(0.096)(1/2)}{0.192} = 0.25. \end{aligned}$$

Now,

$$\begin{aligned} E(Y_4|\Theta = \theta_I) &= 0(0.6) + 1(0.4) = 0.4 \\ E(Y_4|\Theta = \theta_{II}) &= 0(0.8) + 1(0.2) = 0.2 \end{aligned}$$

and similarly for Y_5, implying that

$$\begin{aligned} \mu_2(\theta_I) &= E(X_2|\Theta = \theta_I) = E(Y_4|\Theta = \theta_I) + E(Y_5|\Theta = \theta_I) \\ &= 0.4 + 0.4 = 0.8 \\ \mu_2(\theta_{II}) &= E(X_2|\Theta = \theta_{II}) = E(Y_4|\Theta = \theta_{II}) + E(Y_5|\Theta = \theta_{II}) \\ &= 0.2 + 0.2 = 0.4. \end{aligned}$$

Then, from (5.44),

$$\begin{aligned} E(X_2|X_1 = 2) &= \mu_2(\theta_I)\pi_{\Theta|X_1}(\theta_I|2) + \mu_2(\theta_{II})\pi_{\Theta|X_1}(\theta_{II}|2) \\ &= (0.8)(0.75) + (0.4)(0.25) = 0.7. \end{aligned}$$

A more tedious calculation gives the predictive density of $X_2|X_1 = 2$, namely

$$f_{X_2|X_1}(0|2) = 0.43,\ f_{X_2|X_1}(1|2) = 0.44,\ f_{X_2|X_1}(2|2) = 0.13$$

with mean $0(0.43) + 1(0.44) + 2(0.13) = 0.7$ as before. □

Example 5.33 *Suppose, as in Example 5.26, that the number of claims N_j in year j for a group policyholder with (unknown) risk parameter θ and m_j*

individuals in the group is Poisson distributed with mean $m_j\theta$, that is, for $j = 1, \ldots, n$ we obtain

$$\Pr(N_j = x|\Theta = \theta) = \frac{(m_j\theta)^x e^{-m_j\theta}}{x!}, \qquad x = 0, 1, 2, \ldots.$$

This would be the case if, per individual, the number of claims were independently Poisson distributed with mean θ. If claims inflation is at annual rate $100r\%$, then a loss of c at time 0 costs $c(1+r)^t$ at time t. Thus, let us assume that a loss in year j is for fixed amount $(1+r)^j c$. Determine the Bayesian premium for the m_{n+1} individuals to be insured in year $n+1$.

With these assumptions, the average losses per individual in year j are

$$X_j = c(1+r)^j N_j/m_j, \qquad j = 1, \ldots, n.$$

Therefore,

$$f_{X_j|\Theta}(x_j|\theta) = \Pr[N_j = c^{-1}(1+r)^{-j}m_j x_j|\Theta = \theta].$$

If Θ is gamma distributed with parameters α and β,

$$\pi(\theta) = \frac{\theta^{\alpha-1}e^{-\theta/\beta}}{\Gamma(\alpha)\beta^\alpha}, \qquad \theta > 0$$

then the posterior distribution $\pi_{\Theta|\mathbf{X}}(\theta|\mathbf{x})$ is proportional (as a function of θ) to

$$\left[\prod_{j=1}^n f_{X_j|\Theta}(x_j|\theta)\right] \pi(\theta)$$

which is itself proportional to

$$\left[\prod_{j=1}^n \theta^{c^{-1}(1+r)^{-j}m_j x_j} e^{-m_j\theta}\right] \theta^{\alpha-1}e^{-\theta/\beta}$$

$$= \theta^{\alpha + c^{-1}\sum_{j=1}^n (1+r)^{-j}m_j x_j - 1} e^{-\theta\left(\frac{1}{\beta} + \sum_{j=1}^n m_j\right)}.$$

But this is proportional to a gamma density with parameters $\alpha_* = \alpha + c^{-1}\sum\limits_{j=1}^n (1+r)^{-j}m_j x_j$ and $\beta_* = (1/\beta + \sum_{j=1}^n m_j)^{-1}$, and so $\Theta|\mathbf{X}$ is also gamma, but with α and β replaced by α_* and β_*, respectively.

Now,

$$E(X_j|\Theta = \theta) = E\left[\frac{c(1+r)^j}{m_j}N_j|\Theta = \theta\right] = \frac{c(1+r)^j}{m_j}E(N_j|\Theta = \theta) = c(1+r)^j\theta.$$

Thus $\mu_{n+1}(\theta) = E(X_{n+1}|\Theta = \theta) = c(1+r)^{n+1}\theta$ and $\mu_{n+1} = E(X_{n+1}) = E[\mu_{n+1}(\Theta)] = c(1+r)^{n+1}\alpha\beta$ since Θ is gamma distributed with parameters α

and β. From (5.44) and since $\Theta|\mathbf{X}$ is also gamma distributed with parameters α_* and β_*,

$$\begin{aligned} E(X_{n+1}|\mathbf{X}=\mathbf{x}) &= \int_0^\infty \mu_{n+1}(\theta)\pi_{\Theta|\mathbf{X}}(\theta|\mathbf{x})d\theta \\ &= E[\mu_{n+1}(\Theta)|\mathbf{X}=\mathbf{x}] \\ &= E[c(1+r)^{n+1}\Theta|\mathbf{X}=\mathbf{x}] \\ &= c(1+r)^{n+1}\alpha_*\beta_*. \end{aligned}$$

Define the total number of lives observed to be $m = \sum_{j=1}^n m_j$.

Then,

$$E(X_{n+1}|\mathbf{X}=\mathbf{x}) = Z\bar{x} + (1-Z)\mu_{n+1}$$

(where $Z = m/(m+\beta^{-1})$ and $\overline{x} = m^{-1}\sum_{j=1}^{n}(1+r)^{n+1-j}m_jx_j)$ is the past average claims per individual, expressed in year $n+1$ monetary units, again an expression of the form (5.36).

The total Bayesian premium for m_{n+1} individuals in the group for the next year would be $m_{n+1}E(X_{n+1}|\mathbf{X}=\mathbf{x})$. The analysis based on independent and identically distributed Poisson claim counts are obtained with $c=1, r=0$, and $m_j=1$. Then $X_j \equiv N_j$ for $j=1,2,\ldots,n$ are independent (given θ) Poisson random variables with mean θ. In this case

$$E(X_{n+1}|\mathbf{X}=\mathbf{x}) = Z\overline{x} + (1-Z)\mu$$

where $Z = n/(n+\beta^{-1})$, $\overline{x} = n^{-1}\sum_{j=1}^n x_j$, and $\mu = \alpha\beta$. □

In each of Examples 5.31 and 5.33 the Bayesian premium was a weighted average of the sample mean $\bar{x}$ and the pure premium μ_{n+1}. This is appealing from a credibility standpoint. Furthermore, the credibility factor Z in each case is an increasing function of the number of exposure units. The greater the amount of past data observed, the closer Z is to 1, consistent with our intuition.

5.4.2 The credibility premium

In the previous section a systematic approach was suggested for treatment of the past data of a particular policyholder. Ideally, rather than the pure premium $\mu_{n+1} = E(X_{n+1})$, one would like to charge the individual premium (or hypothetical mean) $\mu_{n+1}(\theta)$, where θ is the (hypothetical) parameter associated with the policyholder. Since θ is unknown this is impossible, but we could instead condition on $\mathbf{x}$, the past data from the policyholder. This leads to the Bayesian premium $E(X_{n+1}|\mathbf{x})$.

One difficulty with this approach is that it may be difficult to evaluate the Bayesian premium. Of course, in simple numerical examples such as in the previous subsection, the Bayesian premium is not difficult to evaluate numerically. But these examples can hardly be expected to capture the essential features of a realistic insurance scenario. More realistic models may well introduce analytic difficulties with respect to evaluation of $E\left(X_{n+1}|\mathbf{x}\right)$, whether one uses (5.43) or (5.44). Often, numerical integration may be required. There are exceptions such as Examples 5.31 and 5.33.

We now present an alternative suggested by Bühlmann in 1967 [18]. Recall the basic problem: We wish to use the conditional distribution $f_{X_{n+1}|\Theta}(x_{n+1}|\theta)$ or the hypothetical mean $\mu_{n+1}(\theta)$ for estimation of next year's claims. Since we have observed $\mathbf{x}$, one suggestion is to approximate $\mu_{n+1}(\theta)$ by a linear function of the past data. (After all, the formula $Z\overline{X}+(1-Z)\mu$ is of this form.) Thus, let us restrict ourselves to estimators of the form $\alpha_0+\sum_{j=1}^{n}\alpha_jX_j$, where $\alpha_0,\alpha_1,\ldots,\alpha_n$ need to be chosen. To this end, we will choose the αs to minimize squared error loss, that is,

$$Q=E\left\{\left[\mu_{n+1}(\Theta)-\alpha_0-\sum_{j=1}^{n}\alpha_jX_j\right]^2\right\} \tag{5.45}$$

and the expectation is over the joint distribution of $X_1,\ldots,X_n$, and Θ. To minimize Q we take derivatives. Thus,

$$\frac{\partial Q}{\partial\alpha_0}=E\left\{2\left[\mu_{n+1}(\Theta)-\alpha_0-\sum_{j=1}^{n}\alpha_jX_j\right](-1)\right\}.$$

We shall denote by $\tilde{\alpha}_0,\tilde{\alpha}_1,\ldots,\tilde{\alpha}_n$ the values of $\alpha_0,\alpha_1,\ldots,\alpha_n$ which minimize (5.45). Then equating $\partial Q/\partial\alpha_0$ to 0 yields

$$E[\mu_{n+1}(\Theta)]=\tilde{\alpha}_0+\sum_{j=1}^{n}\tilde{\alpha}_jE(X_j).$$

But $E(X_{n+1})=E[E(X_{n+1}|\Theta)]=E[\mu_{n+1}(\Theta)]$, and so $\partial Q/\partial\alpha_0=0$ implies that

$$E(X_{n+1})=\tilde{\alpha}_0+\sum_{j=1}^{n}\tilde{\alpha}_jE(X_j). \tag{5.46}$$

Equation (5.46) may be termed the **unbiasedness equation** since it requires that the estimate $\tilde{\alpha}_0+\sum_{j=1}^{n}\tilde{\alpha}_jX_j$ be unbiased for $E(X_{n+1})$. For $i=1,\ldots,n$, we have

$$\frac{\partial Q}{\partial\alpha_i}=E\left\{2\left[\mu_{n+1}(\Theta)-\alpha_0-\sum_{j=1}^{n}\alpha_jX_j\right](-X_i)\right\}$$

and setting this equal to 0 yields

$$E[\mu_{n+1}(\Theta)X_i]=\tilde{\alpha}_0E\left(X_i\right)+\sum_{j=1}^{n}\tilde{\alpha}_jE\left(X_iX_j\right).$$

The left-hand side of this equation may be reexpressed as

$$\begin{aligned} E[\mu_{n+1}(\Theta)X_i] &= E\{E[X_i\mu_{n+1}(\Theta)|\Theta]\} \\ &= E\{\mu_{n+1}(\Theta)E[X_i|\Theta]\} \\ &= E[E(X_{n+1}|\Theta)E(X_i|\Theta)] \\ &= E[E(X_{n+1}X_i|\Theta)] \\ &= E(X_iX_{n+1}) \end{aligned}$$

where the second from last step follows by independence of X_i and X_{n+1} conditional on Θ. Thus $\partial Q/\partial\alpha_i = 0$ implies

$$E(X_iX_{n+1}) = \tilde{\alpha}_0E(X_i) + \sum_{j=1}^{n}\tilde{\alpha}_jE(X_iX_j). \tag{5.47}$$

Next multiply (5.46) by $E(X_i)$ and subtract from (5.47) to obtain

$$Cov(X_i, X_{n+1}) = \sum_{j=1}^{n}\tilde{\alpha}_j\ Cov(X_i, X_j), \quad i = 1, \ldots, n. \tag{5.48}$$

Equation (5.46) and the n equations (5.48) together are called the **normal equations**. These equations may be solved for $\tilde{\alpha}_0, \tilde{\alpha}_1, \ldots, \tilde{\alpha}_n$ to yield the credibility premium

$$\tilde{\alpha}_0 + \sum_{j=1}^{n}\tilde{\alpha}_jX_j. \tag{5.49}$$

While it is straightforward to express the solution $\tilde{\alpha}_0, \tilde{\alpha}_1, \ldots, \tilde{\alpha}_n$ to the normal equations in matrix notation (if the covariance matrix of the X_js is nonsingular), we shall be content with solutions for some special cases.

Note that exactly one of the terms on the right hand side of (5.48) is a variance term, that is, $Cov(X_i, X_i) = Var(X_i)$. The other $n-1$ terms are true covariance terms.

As an added bonus, the values $\tilde{\alpha}_0, \tilde{\alpha}_1, \ldots, \tilde{\alpha}_n$ also minimize

$$Q_1 = E\left\{\left[E(X_{n+1}|\mathbf{X}) - \alpha_0 - \sum_{j=1}^{n}\alpha_jX_j\right]^2\right\} \tag{5.50}$$

and

$$Q_2 = E\left\{\left[X_{n+1} - \alpha_0 - \sum_{j=1}^{n}\alpha_jX_j\right]^2\right\}. \tag{5.51}$$

To see, this, differentiate (5.50) or (5.51) with respect to $\alpha_0, \alpha_1, \ldots, \alpha_n$ and observe that the solutions still satisfy the normal equations (5.46) and (5.48).

Thus the credibility premium (5.49) is the best linear estimator of each of the hypothetical mean $E(X_{n+1}|\Theta)$, the Bayesian premium $E(X_{n+1}|\mathbf{X})$, and X_{n+1}.

Example 5.34 *If $E(X_j) = \mu$, $Var(X_j) = \sigma^2$, and for $i \neq j$ $Cov(X_i, X_j) = \rho\sigma^2$ where the correlation coefficient ρ satisfies $-1 < \rho < 1$, determine the credibility premium $\tilde{\alpha}_0 + \sum_{j=1}^{n} \tilde{\alpha}_j X_j$.*

The unbiasedness equation (5.46) yields

$$\mu = \tilde{\alpha}_0 + \mu \sum_{j=1}^{n} \tilde{\alpha}_j$$

or

$$\sum_{j=1}^{n} \tilde{\alpha}_j = 1 - \tilde{\alpha}_0/\mu.$$

The n equations (5.48) become, for $i = 1, \ldots, n$,

$$\rho = \sum_{\substack{j=1 \\ j\neq i}}^{n} \tilde{\alpha}_j \rho + \tilde{\alpha}_i$$

or stated another way,

$$\rho = \sum_{j=1}^{n} \tilde{\alpha}_j \rho + \tilde{\alpha}_i (1-\rho), \qquad i = 1, \ldots, n.$$

Thus

$$\tilde{\alpha}_i = \frac{\rho\left(1 - \sum_{j=1}^{n} \tilde{\alpha}_j\right)}{1-\rho} = \frac{\rho\tilde{\alpha}_0}{\mu(1-\rho)}$$

using the unbiasedness equation. Summation over i from 1 to n yields

$$\sum_{i=1}^{n} \tilde{\alpha}_i = \sum_{j=1}^{n} \tilde{\alpha}_j = \frac{n\rho\tilde{\alpha}_0}{\mu(1-\rho)}$$

which, combined with the unbiasedness equation gives an equation for $\tilde{\alpha}_0$, namely

$$1 - \tilde{\alpha}_0/\mu = \frac{n\rho\tilde{\alpha}_0}{\mu(1-\rho)}.$$

Solving for $\tilde{\alpha}_0$ yields

$$\tilde{\alpha}_0 = \frac{(1-\rho)\mu}{1-\rho+n\rho}.$$

Thus,

$$\tilde{\alpha}_j = \frac{\rho\tilde{\alpha}_0}{\mu(1-\rho)} = \frac{\rho}{1-\rho+n\rho}.$$

The credibility premium is then

$$\begin{aligned} \tilde{\alpha}_0 + \sum_{j=1}^{n} \tilde{\alpha}_j X_j &= \frac{(1-\rho)\mu}{1-\rho+n\rho} + \sum_{j=1}^{n} \frac{\rho X_j}{1-\rho+n\rho} \\ &= (1-Z)\mu + Z\bar{X} \end{aligned}$$

where $Z = n\rho/(1-\rho+n\rho)$ and $\bar{X} = n^{-1}\sum_{j=1}^{n} X_j$. Thus if $0 < \rho < 1$, then $0 < Z < 1$ and the credibility premium is a weighted average of $\mu = E(X_{n+1})$ and $\bar{X}$, that is, is of the form (5.36). □

We now turn to some models which specify the conditional means and variances of $X_j|\Theta$ and hence the means $E(X_j)$, variances $Var(X_j)$, and covariances $Cov(X_i, X_j)$.

5.4.3 The Bühlmann model

This, the first and simplest credibility model, specifies that for each policyholder (conditional on Θ), past losses $X_1, \ldots, X_n$ have the same mean and variance and are independent and identically distributed, conditional on Θ.

Thus, define

$$\mu(\theta) = E(X_j|\Theta = \theta)$$

and

$$v(\theta) = Var(X_j|\Theta = \theta).$$

As discussed previously, $\mu(\theta)$ is referred to as the **hypothetical mean** whereas $v(\theta)$ is called the **process variance**. Define

$$\mu = E[\mu(\Theta)] \tag{5.52}$$

$$v = E[v(\Theta)] \tag{5.53}$$

and

$$a = Var[\mu(\Theta)]. \tag{5.54}$$

The quantity μ in (5.52) is the **expected value of the hypothetical means**, v in (5.53) is the **expected value of the process variance**, and a in (5.54) is the **variance of the hypothetical means**. Note that μ is the estimate to use if we have no information about θ (and thus no information about $\mu(\theta)$). It will also be referred to as the **collective premium**.

The mean, variance, and covariance of the X_js may now be obtained. First,

$$E(X_j) = E[E(X_j|\Theta)] = E[\mu(\Theta)] = \mu. \tag{5.55}$$

Second,

$$\begin{aligned} Var(X_j) &= E[Var(X_j|\Theta)] + Var[E(X_j|\Theta)] \\ &= E[v(\Theta)] + Var[\mu(\Theta)] \\ &= v + a. \end{aligned} \tag{5.56}$$

Finally, for $i \neq j$,

$$\begin{aligned} Cov(X_i, X_j) &= E(X_iX_j) - E(X_i)E(X_j) \\ &= E[E(X_iX_j|\Theta)] - \mu^2 \\ &= E[E(X_i|\Theta)E(X_j|\Theta)] - \{E[\mu(\Theta)]\}^2 \\ &= E\left\{[\mu(\Theta)]^2\right\} - \{E[\mu(\Theta)]\}^2 \\ &= Var[\mu(\Theta)] \\ &= a. \end{aligned} \tag{5.57}$$

This is exactly of the form of Example 5.34 with parameters $\mu, \sigma^2 = v + a$, and $\rho = a/(v + a)$. Thus the credibility premium is

$$\tilde{\alpha}_0 + \sum_{j=1}^{n} \tilde{\alpha}_j X_j = Z\bar{X} + (1 - Z)\mu \tag{5.58}$$

where

$$Z = \frac{n}{n + k} \tag{5.59}$$

and

$$k = \frac{v}{a} = \frac{E[Var(X_j|\Theta)]}{Var[E(X_j|\Theta)]}. \tag{5.60}$$

The credibility factor Z in (5.59) with k given by (5.60) is referred to as the **Bühlmann credibility factor**. Note that (5.58) is of the form (5.36), and (5.59) is exactly (5.37). Now, however, we know how to obtain k, namely from (5.60).

Formula (5.58) has many appealing features. First, the credibility premium (5.58) is a weighted average of the sample mean $\bar{X}$ and the collective premium μ, a formula which we found desirable. Furthermore, Z approaches 1 as n increases, giving more credit to $\bar{X}$ rather than μ as more past data accumulates, a feature which agrees with intuition. Also, if the population is fairly homogeneous with respect to the risk parameter Θ, then (relatively speaking) the hypothetical means $\mu(\Theta) = E(X_j|\Theta)$ do not vary greatly with Θ (i.e. they are close in value) and hence have small variability. Thus a is small relative to v, that is, k is large and Z is closer to 0. But this agrees with intuition, since for a homogeneous population the overall mean μ is of more value in helping to predict next year's claims for a particular policyholder. Conversely, for a heterogeneous population, the hypothetical means $E(X_j|\Theta)$ are more variable, that is, a is large and k is small, and so Z is closer to 1.

Again this makes sense, since in a heterogeneous population the experience of other policyholders is of less value in predicting the future experience of a particular policyholder than is the past experience of that policyholder.

We now present some examples.

Example 5.35 (Examples 5.27, 5.28, and 5.30 continued) *Determine the three Bühlmann premiums.*

In this case $\pi(\theta_{ij}) = 1/4$ for $i = 1, 2$, and $j = 1, 2$. Then $\mu = 10/3$, and since $\mu_{n+1}(\theta) = \mu(\theta)$,

$$\begin{aligned} a &= Var[\mu(\Theta)] = E\left\{[\mu(\Theta)]^2\right\} - \mu^2 \\ &= \sum_{i=1}^{2}\sum_{j=1}^{2} \mu(\theta_{ij})^2 \pi(\theta_{ij}) - (10/3)^2 \\ &= \frac{1}{4}\left[\left(\frac{7}{3}\right)^2 + (3)^2 + \left(\frac{7}{2}\right)^2 + \left(\frac{9}{2}\right)^2\right] - \left(\frac{10}{3}\right)^2 \\ &= \frac{5}{8}. \end{aligned}$$

Also,

$$\begin{aligned} v(\theta_{11}) &= Var(X_j|\Theta = \theta_{11}) = E(X_j^2|\Theta = \theta_{11}) - [\mu(\theta_{11})]^2 \\ &= 0^2\frac{10}{15} + 5^2\frac{3}{15} + 10^2\frac{2}{15} - \left(\frac{7}{3}\right)^2 = \frac{116}{9} \\ v(\theta_{12}) &= 0^2\frac{10}{15} + 5^2\frac{1}{15} + 10^2\frac{4}{15} - (3)^2 = \frac{58}{3} \\ v(\theta_{21}) &= 0^2\frac{5}{10} + 5^2\frac{3}{10} + 10^2\frac{2}{10} - \left(\frac{7}{2}\right)^2 = \frac{61}{4} \\ v(\theta_{22}) &= 0^2\frac{5}{10} + 5^2\frac{1}{10} + 10^2\frac{4}{10} - \left(\frac{9}{2}\right)^2 = \frac{89}{4}. \end{aligned}$$

Thus,

$$\begin{aligned} v &= E[v(\Theta)] = \sum_{i=1}^{2}\sum_{j=1}^{2} v(\theta_{ij})\pi(\theta_{ij}) \\ &= \frac{1}{4}\left(\frac{116}{9} + \frac{58}{3} + \frac{61}{4} + \frac{89}{4}\right) = \frac{1255}{72}. \end{aligned}$$

Then (5.60) gives

$$k = \frac{v}{a} = \frac{251}{9}$$

and (5.59) gives

$$Z = \frac{n}{n + \frac{251}{9}}.$$

With $n = 1$ we obtain the credibility premium for a single observation, namely

$$ZX_1 + (1 - Z)\,\mu = \frac{9}{260}X_1 + \frac{251}{260}\frac{10}{3}$$

which yields the values $\frac{251}{78} = 3.22$, $\frac{529}{156} = 3.39$, and $\frac{139}{39} = 3.56$ for $X_1 = 0, 5$ and 10 respectively. These are the best linear approximations to the Bayesian premiums (given in Example 5.30) $E(X_2|X_1 = 0) = \frac{68}{21} = 3.24$, $E(X_2|X_1 = 5) = \frac{13}{4} = 3.25$, and $E(X_2|X_1 = 10) = \frac{65}{18} = 3.61$, respectively. □

Example 5.36 *Suppose that* $\{X_j|\Theta;\ j = 1, \ldots, n\}$ *are independently and identically Poisson distributed with (given) mean* Θ*, and* Θ *is gamma distributed with parameters* α *and* β*. Determine the Bühlmann premium.*

We have

$$\begin{aligned} \mu(\theta) &= E(X_j|\Theta = \theta) = \theta \\ v(\theta) &= Var(X_j|\Theta = \theta) = \theta \end{aligned}$$

and so

$$\begin{aligned} \mu &= E[\mu(\Theta)] = E(\Theta) = \alpha\beta \\ v &= E[v(\Theta)] = E(\Theta) = \alpha\beta \end{aligned}$$

and

$$a = Var[\mu(\Theta)] = Var(\Theta) = \alpha\beta^2.$$

Then

$$\begin{aligned} k &= \frac{v}{a} = \frac{\alpha\beta}{\alpha\beta^2} = \frac{1}{\beta} \\ Z &= \frac{n}{n+k} = \frac{n}{n+1/\beta} = \frac{n\beta}{n\beta+1} \end{aligned}$$

and the credibility premium is

$$Z\bar{X} + (1 - Z)\,\mu = \frac{n\beta}{n\beta+1}\bar{X} + \frac{1}{n\beta+1}\alpha\beta.$$

But as shown at the end of Example 5.33, this is also the Bayesian premium $E(X_{n+1}|\mathbf{X})$. Thus, the credibility premium equals the Bayesian premium in this case. More on this idea is presented in Subsection 5.4.5. □

Example 5.37 (Example 5.32 continued) *As before, an urn is selected at random and three balls are selected at random from that urn and the total observed is 2. Calculate the Bühlmann credibility premium for the number on the next ball selected from the same urn.*

As before, let Y_i be the number on the ith ball drawn. Then,

$$\begin{aligned}
&\Pr(Y_i = 0|\Theta = \theta_I) = 0.6 \\
&\Pr(Y_i = 1|\Theta = \theta_I) = 0.4 \\
&\mu(\theta_I) = E(Y_i|\Theta = \theta_I) = 0(0.6) + 1(0.4) = 0.4 \\
&v(\theta_I) = E(Y_i^2|\Theta = \theta_I) - [\mu(\theta_I)]^2 \\
&\qquad = 0^2(0.6) + 1^2(0.4) - (0.4)^2 = (0.4)(0.6) = 0.24.
\end{aligned}$$

Similarly,

$$\begin{aligned}
&\Pr(Y_i = 0|\Theta = \theta_{II}) = 0.8 \\
&\Pr(Y_i = 1|\Theta = \theta_{II}) = 0.8 \\
&\mu(\theta_{II}) = 0(0.8) + 1(0.2) = 0.2 \\
&v(\theta_{II}) = 0^2(0.8) + 1^2(0.2) - (0.2)^2 = (0.2)(0.8) = 0.16.
\end{aligned}$$

Then

$$\begin{aligned}
\mu &= E[\mu(\Theta)] = \mu(\theta_I)\pi(\theta_I) + \mu(\theta_{II})\pi(\theta_{II}) \\
&= (0.4)(0.5) + (0.2)(0.5) = 0.3 \\
v &= E[v(\Theta)] = v(\theta_I)\pi(\theta_I) + v(\theta_{II})\pi(\theta_{II}) \\
&= (0.24)(0.5) + (0.16)(0.5) = 0.2 \\
a &= Var[\mu(\Theta)] = E\left\{[\mu(\Theta)]^2\right\} - \mu^2 \\
&= (0.4)^2(0.5) + (0.2)^2(0.5) - (0.3)^2 = 0.01.
\end{aligned}$$

Thus

$$\begin{aligned}
k &= v/a = 0.2/0.01 = 20 \\
Z &= 3/(3+k) = 3/23
\end{aligned}$$

and the credibility premium is

$$Z\bar{Y} + (1-Z)\mu = \frac{3}{23}\left(\frac{2}{3}\right) + \frac{20}{23}(0.3) = \frac{8}{23} = 0.348.$$

This is the best linear approximation to the Bayesian premium which is, using (5.44) and Example 5.32,

$$\mu(\theta_I)\pi_{\Theta|X_1}(\theta_I|2) + \mu(\theta_{II})\pi_{\Theta|X_1}(\theta_{II}|2) = (0.4)(0.75) + (0.2)(0.25) = 0.35.$$

Suppose that we wish to compute the credibility premium for the total on the next three balls selected from the same urn, rather than the next ball. While it is correct to simply multiply the credibility premium by three since $E(Y_4 + Y_5 + Y_6|\Theta) = 3E(Y_4|\Theta)$ and the credibility premium approximates the conditional mean, let us calculate the credibility factor Z assuming that each exposure unit consists of three draws. Then, in the notation of Example 5.32, we have observed $X_1 = 2$, and we are interested in predicting X_2^*, the total on the next three draws. In this case note that since $X_2^* = Y_4 + Y_5 + Y_6$ and the Y_is are independent and identically distributed, conditional on Θ, it follows that

$$E(X_2^*|\Theta) = 3E(Y_i|\Theta) = 3\mu(\Theta)$$

and

$$Var(X_2^*|\Theta) = 3Var(Y_i|\Theta) = 3v(\Theta).$$

The credibility factor Z is thus

$$\frac{1}{1+\dfrac{E[Var(X_2^*|\Theta)]}{Var[E(X_2^*|\Theta)]}} = \frac{1}{1+\dfrac{E[3v(\Theta)]}{Var[3\mu(\Theta)]}} = \frac{1}{1+\dfrac{3v}{(3)^2a}} = \frac{3}{3+\dfrac{v}{a}}. \tag{5.61}$$

This means that the credibility factor Z is the same whether one assumes that either (a)three exposure units of 1 draw each have been observed or (b) one exposure unit of 3 draws has been observed. □

5.4.4 The Bühlmann–Straub model

The Bühlmann model of the previous section is the simplest of the credibility models because it effectively requires that the past claims experience of a policyholder comprise independent and identically distributed components with respect to each past year. One difficulty with this assumption is that it does not allow for variations in exposure or size.

For example, what if the first year's claims experience of a policyholder reflected only a portion of a year due to an unusual policyholder anniversary? What if a benefit change occurred part way through a policy year? For group insurance, what if the size of the group changed over time?

To handle these variations, we consider the following generalization of the Bühlmann model. Assume that $X_1, \ldots, X_n$ are independent, conditional on Θ, with common mean (as before)

$$\mu(\theta) = E(X_j|\Theta = \theta)$$

but with conditional variances

$$Var(X_j|\Theta = \theta) = v(\theta)/m_j$$

where m_j is a known constant measuring exposure. Note that m_j need only be proportional to the size of the risk. This model would be appropriate if each X_j were the average of m_j independent (conditional on Θ) random variables, each with mean $\mu(\theta)$ and variance $v(\theta)$. In the above situations, m_j could be the number of months the policy was in force in past year j, or the number of individuals in the group in past year j, or the amount of premium income for the policy in past year j.

As in the Bühlmann model, let

$$\begin{aligned}\mu &= E[\mu(\Theta)]\\ v &= E[v(\Theta)]\end{aligned}$$

and

$$a = Var[\mu(\Theta)].$$

Then, for the unconditional moments, $E(X_j) = \mu$ from (5.55) and $Cov(X_i, X_j) = a$ from (5.57), but

$$\begin{aligned}Var(X_j) &= E[Var(X_j|\Theta)] + Var[E(X_j|\Theta)]\\ &= E[v(\Theta)/m_j] + Var[\mu(\Theta)]\\ &= v/m_j + a.\end{aligned}$$

To obtain the credibility premium (5.49), we will solve the normal equations (5.46) and (5.48) to obtain $\tilde{\alpha}_0, \tilde{\alpha}_1, \ldots, \tilde{\alpha}_n$. For notational convenience, define

$$m = m_1 + m_2 + \cdots + m_n$$

to be the total exposure. Then using (5.55), the unbiasedness equation (5.46) becomes

$$\mu = \tilde{\alpha}_0 + \sum_{j=1}^{n} \tilde{\alpha}_j \mu$$

which implies

$$\sum_{j=1}^{n} \tilde{\alpha}_j = 1 - \tilde{\alpha}_0/\mu. \tag{5.62}$$

For $i = 1, \ldots, n$, (5.48) becomes

$$a = \sum_{\substack{j=1\\ j\neq i}}^{n} \tilde{\alpha}_j a + \tilde{\alpha}_i\,(a + v/m_i) = \sum_{j=1}^{n} \tilde{\alpha}_j a + v\tilde{\alpha}_i/m_i$$

which may be rewritten as

$$\tilde{\alpha}_i = \frac{a}{v} m_i \left(1 - \sum_{j=1}^{n} \tilde{\alpha}_j\right) = \frac{a}{v}\frac{\tilde{\alpha}_0}{\mu} m_i, \qquad i = 1, \ldots, n. \tag{5.63}$$

Then, using (5.62) and (5.63),

$$1 - \frac{\tilde{\alpha}_0}{\mu} = \sum_{j=1}^{n} \tilde{\alpha}_j = \sum_{i=1}^{n} \tilde{\alpha}_i = \frac{a}{v}\frac{\tilde{\alpha}_0}{\mu} \sum_{i=1}^{n} m_i = \frac{a\tilde{\alpha}_0 m}{\mu v},$$

and so

$$\tilde{\alpha}_0 = \frac{\mu}{1 + \frac{am}{v}} = \frac{v/a}{m + v/a}\mu.$$

But this means that

$$\tilde{\alpha}_j = \frac{a\tilde{\alpha}_0}{\mu v} \cdot m_j = \frac{m_j}{m + v/a}.$$

The credibility premium (5.49) becomes

$$\tilde{\alpha}_0 + \sum_{j=1}^{n} \tilde{\alpha}_j X_j = Z\bar{X} + (1 - Z)\mu \tag{5.64}$$

where with $k = v/a$ from (5.60)

$$Z = \frac{m}{m + k}$$

and

$$\bar{X} = \sum_{j=1}^{n} \frac{m_j}{m} X_j. \tag{5.65}$$

Clearly, the credibility premium (5.64) is still of the form (5.36). In this case, m is the total exposure associated with the policyholder, and the Bühlmann–Straub credibility factor Z depends on m. Furthermore, $\bar{X}$ is a weighted average of the X_j, with weights proportional to m_j. Following the group interpretation, X_j is the average loss of the m_j group members in year j and so $m_j X_j$ is the total loss of the group in year j. Then $\bar{X}$ is the overall average loss per group member over the n years. The credibility premium to be charged to the group in year $n + 1$ would thus be $m_{n+1}[Z\bar{X} + (1 - Z)\,\mu]$ for m_{n+1} members in the next year.

Had we known that (5.65) would be the correct weighting of the X_j to receive the credibility weight Z, the rest would have been easy. For the single observation $\bar{X}$ the process variance is

$$Var(\bar{X}|\theta) = \sum_{j=1}^{n} \frac{m_j^2}{m^2} \frac{v(\theta)}{m_j} = \frac{v(\theta)}{m}$$

and so the expected process variance is v/m. The variance of the hypothetical means is still a and therefore $k = v/(am)$. There is only one observation of $\bar{X}$ and so the credibility factor is

$$Z = \frac{1}{1 + v/(am)} = \frac{m}{m + v/a} \tag{5.66}$$

as before. To further note the similarity of the Bühlmann and Bühlmann–Straub developments, compare (5.61) and (5.66). Equation (5.65) should not have been surprising because the weights are simply inversely proportional to the (conditional) variance of each X_j.

Example 5.38 *Suppose that, given Θ, the X_js are independent and normally distributed with mean Θ and variance v/m_j for $j = 1, \ldots, n$. Furthermore,*

Θ is normally distributed with mean μ and variance a. (By the central limit theorem, this may be a reasonable assumption in the group insurance context with m_j representing the number of lives in year j.) Determine the credibility and Bayesian premiums.

This model is a special case of Bühlmann–Straub where the process variance is known. The credibility premium is thus (5.64).

Now consider the Bayesian premium. The posterior distribution $\pi_{\Theta|\mathbf{X}}(\theta|\mathbf{x})$ is proportional to

$$\left[\prod_{j=1}^{n}(2\pi v/m_j)^{-1/2}\exp\left[-\frac{m_j}{2v}(x_j-\theta)^2\right]\right](2\pi a)^{-1/2}\exp\left[-\frac{1}{2a}(\theta-\mu)^2\right]$$

which is proportional to

$$\begin{aligned}
&\exp\left[-\frac{1}{2v}\sum_{j=1}^{n}m_j(x_j-\theta)^2-\frac{1}{2a}(\theta-\mu)^2\right]\\
=\;&\exp\left\{-\frac{1}{2v}\left[\sum_{j=1}^{n}m_j(x_j-\overline{x})^2+m(\overline{x}-\theta)^2\right]-\frac{1}{2a}(\theta-\mu)^2\right\}
\end{aligned}$$

using the identity (5.15). But this is proportional to

$$\exp\left[-\frac{m}{2v}(\overline{x}-\theta)^2-\frac{1}{2a}(\theta-\mu)^2\right]$$

which is proportional to

$$\exp\left[-\frac{\theta^2}{2}\left(\frac{m}{v}+\frac{1}{a}\right)+\theta\left(\frac{m\overline{x}}{v}+\frac{\mu}{a}\right)\right]=\exp\left(-\frac{\theta^2}{2a_*}+\frac{\mu_*}{a_*}\theta\right) \tag{5.67}$$

with

$$a_*=\left(\frac{m}{v}+\frac{1}{a}\right)^{-1}$$

and

$$\mu_*=\frac{\dfrac{m\overline{x}}{v}+\dfrac{\mu}{a}}{\dfrac{m}{v}+\dfrac{1}{a}}=\frac{m\bar{x}+\left(\dfrac{v}{a}\right)\mu}{m+\dfrac{v}{a}}.$$

From Example 5.15, (5.67) is proportional to the density of a $N(\mu_*,a_*)$ random variable and thus the posterior density of $\Theta|\mathbf{x}$ is

$$\pi_{\Theta|\mathbf{X}}(\theta|\mathbf{x})=(2\pi a_*)^{-1/2}\exp\left[-\frac{1}{2a_*}(\theta-\mu_*)^2\right].$$

Then (5.44) yields the Bayesian premium, namely with $\mu(\theta) = \theta$,

$$E(X_{n+1}|\mathbf{x}) = \int_{-\infty}^{\infty} \theta \pi_{\Theta|\mathbf{X}}(\theta|\mathbf{x}) d\theta = \mu_*$$

and $\mu_* = Z\bar{x} + (1 - Z)\mu$ with $Z = m/(m + v/a)$. Again the credibility premium is equal to the Bayesian premium. □

Example 5.39 (Example 5.32 continued) *An urn has been chosen and 3 balls totaling 2 have been drawn. Predict the total on the next 2 balls drawn using a credibility premium.*

Let

$$\begin{aligned} W_1 &= X_1/3 = (Y_1 + Y_2 + Y_3)/3 \\ W_2 &= X_2/2 = (Y_4 + Y_5)/2. \end{aligned}$$

Then W_1 is the average on the first three balls ($W_1 = 2/3$ was observed) and W_2 is the average on the next two. Let $\mu(\theta) = E(Y_j|\Theta = \theta)$ and $v(\theta) = Var(Y_j|\Theta = \theta)$. Then $E(W_1|\Theta = \theta) = E(W_2|\Theta = \theta) = \mu(\theta)$, $Var(W_1|\Theta = \theta) = v(\theta)/3$, and $Var(W_2|\Theta = \theta) = v(\theta)/2$. This means that the Bühlmann–Straub formulation applies to the prediction of W_2 from W_1. In this case $n = 1, m_1 = 3, m_2 = 2$, and $m = m_1 = 3$. Also, from Example 5.37, it was found that $\mu = 0.3$, $v = 0.2$, and $a = 0.01$. The credibility premium is

$$\begin{aligned} ZW_1 + (1 - Z)\mu &= \frac{m}{m+k}W_1 + \frac{k}{m+k}\mu \\ &= \frac{3}{3 + 0.2/0.01}\left(\frac{2}{3}\right) + \frac{0.2/0.01}{3 + 0.2/0.01}(0.3) = \frac{8}{23} = 0.348 \end{aligned}$$

which predicts the average W_2 of the next 2 balls. Thus the total $X_2 = 2W_2$ is predicted by $2(8/23) = 16/23 = 0.696$. Note that this is just twice the answer to Example 5.37. From Example 5.32, the Bayesian premium was 0.7. □

The assumptions underlying the Bühlmann–Straub model may be too restrictive to represent reality. In a 1967 paper, Hewitt [62] observed that large risks do not behave the same as an independent aggregation of small risks and, in fact, are more variable than would be indicated by independence. The data underlying this conclusion and Hewitt's arguments are presented in Exercise 2.142. A model that reflects this observation is created in the following example.

Example 5.40 *Let the conditional mean be $E(X_j|\Theta) = \mu(\Theta)$ and the conditional variance be $Var(X_j|\Theta) = w(\Theta) + v(\Theta)/m_j$. Further assume that $X_1, \ldots, X_n$ are conditionally independent, given Θ. Show that this model supports Hewitt's observation and determine the credibility premium.*

Consider independent risks i and j with exposures m_i and m_j and with a common value of Θ. When aggregated, the variance of the average loss is

$$
\begin{aligned}
Var\left(\frac{m_iX_i + m_jX_j}{m_i + m_j}\middle|\Theta\right) &= \left(\frac{m_i}{m_i+m_j}\right)^2 Var(X_i|\Theta) \\
&\quad + \left(\frac{m_j}{m_i+m_j}\right)^2 Var(X_j|\Theta) \\
&= \frac{m_i^2 + m_j^2}{(m_i+m_j)^2} w(\Theta) + \frac{1}{m_i+m_j} v(\Theta)
\end{aligned}
$$

while a single risk with exposure $m_i + m_j$ has variance $w(\Theta) + v(\Theta)/(m_i + m_j)$, which is larger.

With regard to the credibility premium, we have

$$
\begin{aligned}
E(X_j) &= E[E(X_j|\Theta)] = E[\mu(\Theta)] = \mu \\
Var(X_j) &= E[Var(X_j|\Theta)] + Var[E(X_j|\Theta)] \\
&= E[w(\Theta) + v(\Theta)/m_j] + Var[\mu(\Theta)] \\
&= w + v/m_j + a
\end{aligned}
$$

and for $i \neq j$, $Cov(X_i, X_j) = a$ as in (5.57). The unbiasedness equation is still

$$\mu = \tilde{\alpha}_0 + \sum_{j=1}^{n} \tilde{\alpha}_j \mu$$

and so

$$\sum_{j=1}^{n} \tilde{\alpha}_j = 1 - \tilde{\alpha}_0/\mu.$$

Equation (5.48) becomes

$$
\begin{aligned}
a &= \sum_{j=1}^{n} \tilde{\alpha}_j a + \tilde{\alpha}_i(w + v/m_i) \\
&= a(1 - \tilde{\alpha}_0/\mu) + \tilde{\alpha}_i(w + v/m_i), \qquad i = 1, \ldots, n.
\end{aligned}
$$

Therefore,

$$\tilde{\alpha}_i = \frac{a\tilde{\alpha}_0/\mu}{w + v/m_i}.$$

Summing both sides yields

$$\frac{a\tilde{\alpha}_0}{\mu} \sum_{j=1}^{n} \frac{m_j}{v + wm_j} = \sum_{j=1}^{n} \tilde{\alpha}_j = 1 - \tilde{\alpha}_0/\mu$$

and so

$$\tilde{\alpha}_0 = \frac{1}{\dfrac{a}{\mu}\displaystyle\sum_{j=1}^{n} \frac{m_j}{v + wm_j} + \dfrac{1}{\mu}} = \frac{\mu}{1 + am^*}$$

where $m^* = \sum_{j=1}^{n} \frac{m_j}{v+wm_j}$. Then

$$\tilde{\alpha}_j = \frac{am_j}{v+wm_j}\frac{1}{1+am^*}.$$

The credibility premium is

$$\frac{\mu}{1+am^*} + \frac{a}{1+am^*}\sum_{j=1}^{n}\frac{m_jX_j}{v+wm_j}.$$

The sum can be made to define a weighted average of the observations by letting

$$\bar{X} = \frac{\sum_{j=1}^{n}\frac{m_j}{v+wm_j}X_j}{\sum_{j=1}^{n}\frac{m_j}{v+wm_j}} = \frac{1}{m^*}\sum_{j=1}^{n}\frac{m_j}{v+wm_j}X_j.$$

If we now set

$$Z = \frac{am^*}{1+am^*}$$

the credibility premium is

$$Z\bar{X} + (1-Z)\mu.$$

Observe what happens as the exposures m_j go to infinity. The credibility factor becomes

$$Z \to \frac{an/w}{1+an/w} < 1.$$

Contrast this to the Bühlmann–Straub model where the limit is 1. Thus, no matter how large the risk, there is a limit to its credibility. A further generalization of this result is provided in Exercise 5.51. □

Another generalization is provided by letting the variance of $\mu(\Theta)$ depend on the exposure. This may be reasonable if we believe that the extent to which a given risk's propensity to produce claims differs from the mean is related to its size. For example, larger risks may be underwritten more carefully. In this case, extreme variations from the mean are less likely because we ensure that the risk not only meets the underwriting requirements, but appears to be exactly what it claims to be.

Example 5.41 (Example 5.40 continued) *In addition to the specification presented in Example 5.40, let* $Var[\mu(\Theta)] = a + b/m$, *where* $m = \sum_{j=1}^{n} m_j$ *is the total exposure for the group. Develop the credibility formula.*

We now have

$$\begin{aligned} E(X_j) &= E[E(X_j|\Theta)] = E[\mu(\Theta)] = \mu \\ Var(X_j) &= E[Var(X_j|\Theta)] + Var[E(X_j|\Theta)] \\ &= E[w(\Theta) + v(\Theta)/m_j] + Var[\mu(\Theta)] \\ &= w + v/m_j + a + b/m \end{aligned}$$

and for $i \neq j$,

$$\begin{aligned} Cov(X_i, X_j) &= E[E(X_i X_j|\Theta)] - \mu^2 \\ &= E[\mu(\Theta)^2] - \mu^2 \\ &= a + b/m. \end{aligned}$$

It can be seen that all the calculations used in Example 5.40 apply here with a replaced by $a + b/m$. The credibility factor is

$$Z = \frac{(a + b/m)m^*}{1 + (a + b/m)m^*}$$

and the credibility premium is

$$Z\bar{X} + (1 - Z)\mu$$

with $\bar{X}$ and m^* defined as in Example 5.40. This particular credibility formula has been used in workers compensation experience rating. One example of this is presented in detail in [51]. □

5.4.5 Exact credibility

In Examples 5.36 and 5.38 we found that the credibility premium and the Bayesian premium were equal. From (5.50), one may view the credibility premium as the best linear approximation to the Bayesian premium in the sense of squared error loss. In these examples the approximation is exact since the two premiums are equal. The term **exact credibility** is used to describe the situation when the credibility premium equals the Bayesian premium.

In fact, it is not hard to see that one can ascertain whether credibility is exact without even calculating the credibility premium. If the Bayesian premium is a linear function of $X_1, \ldots, X_n$,

$$E(X_{n+1}|\mathbf{X}) = a_0 + \sum_{j=1}^{n} a_j X_j$$

then it is clear that in (5.50) the quantity Q_1 attains its minimum value of zero with $\tilde{\alpha}_j = a_j$ for $j = 0, 1, \ldots, n$. Thus the credibility premium is $\tilde{\alpha}_0 + \sum_{j=1}^{n} \tilde{\alpha}_j X_j = a_0 + \sum_{j=1}^{n} a_j X_j = E(X_{n+1}|\mathbf{X})$ and credibility is exact.

This phenomenon occurs fairly generally in connection with linear exponential family members (Subsection 5.2.6) and their conjugate priors. Suppose, for example, that $X_j|\Theta = \theta$ is independently (conditional on $\Theta = \theta$) distributed with pf for $j = 1, \ldots, n+1$

$$f_{X_j|\Theta}(x_j|\theta) = \frac{p(x_j)e^{-\theta x_j}}{q(\theta)}$$

and Θ has pdf

$$\pi(\theta) = \frac{[q(\theta)]^{-k}e^{-\mu k\theta}}{c(\mu, k)}, \qquad \theta_0 < \theta < \theta_1 \tag{5.68}$$

where $-\infty \leq \theta_0 < \theta_1 \leq \infty$. It is also assumed that $\pi(\theta_0) = \pi(\theta_1)$. For the moment, μ and k are simply parameters of $\pi(\theta)$. We will now demonstrate that the choice of symbols was no coincidence.

In Subsection 5.2.6 it was shown that

$$\mu(\theta) = E(X_j|\Theta = \theta) = -\frac{q'(\theta)}{q(\theta)}.$$

We wish to find $E[\mu(\Theta)]$. From (5.68),

$$\log \pi(\theta) = -k \log q(\theta) - \mu k\theta - \log c(\mu, k)$$

and differentiating with respect to θ gives

$$\frac{\pi'(\theta)}{\pi(\theta)} = -\frac{kq'(\theta)}{q(\theta)} - \mu k.$$

In other words,

$$\pi'(\theta) = k[\mu(\theta) - \mu]\pi(\theta) \tag{5.69}$$

and integrating from θ_0 to θ_1 gives

$$\pi(\theta_1) - \pi(\theta_0) = k\int_{\theta_0}^{\theta_1} \mu(\theta)\pi(\theta)d\theta - k\mu\int_{\theta_0}^{\theta_1} \pi(\theta)d\theta.$$

This says that $0 = kE[\mu(\Theta)] - k\mu$, that is, that

$$E[\mu(\Theta)] = \mu. \tag{5.70}$$

Now consider the posterior distribution $\pi_{\Theta|\mathbf{X}}(\theta|\mathbf{x})$. It is proportional to

$$\left[\prod_{j=1}^{n} f_{X_j|\Theta}(x_j|\theta)\right]\pi(\theta)$$

itself proportional to

$$\begin{aligned}
&\left[\prod_{j=1}^{n}\frac{e^{-\theta x_j}}{q(\theta)}\right][q(\theta)]^{-k}e^{-\mu k\theta} \\
&= [q(\theta)]^{-(n+k)}e^{-\theta(\mu k+n\bar{x})} \\
&= [q(\theta)]^{-k_*}e^{-\mu_* k_* \theta} \qquad (5.71)
\end{aligned}$$

where

$$k_* = n + k$$

and

$$\mu_* = \frac{\mu k + n\bar{x}}{k+n} = \frac{n}{n+k}\bar{x} + \frac{k}{n+k}\mu.$$

Observe that (5.71) is proportional to a density of the form (5.68) with μ and k replaced by μ_* and k_* respectively. Hence

$$\pi_{\Theta|\mathbf{X}}(\theta|\mathbf{x}) = \frac{[q(\theta)]^{-k_*}e^{-\mu_* k_* \theta}}{c(\mu_*, k_*)}, \quad \theta_0 < \theta < \theta_1.$$

From (5.44) and using the same development that led to (5.70) the Bayesian premium is

$$\begin{aligned}
E(X_{n+1}|\mathbf{x}) &= \int_{\theta_0}^{\theta_1} \mu(\theta)\pi_{\Theta|\mathbf{X}}(\theta|\mathbf{x})d\theta \\
&= \mu_* \\
&= Z\bar{x} + (1-Z)\mu
\end{aligned}$$

where $Z = n/(n+k)$. This is of the form (5.36), and since it is a linear function of the x_js, credibility must be exact, that is, the credibility premium is

$$\tilde{\alpha}_0 + \sum_{j=1}^{n}\tilde{\alpha}_j X_j = Z\bar{X} + (1-Z)\mu = E(X_{n+1}|\mathbf{x}).$$

Since the $X_j|\Theta$ are also identically distributed for $j = 1, \ldots, n$, the Bühlmann model applies and so (5.58) also applies; that is, k must also satisfy (5.60). To see this directly, recall from Subsection 5.2.6 that

$$v(\theta) = Var(X_j|\Theta = \theta) = -\mu'(\theta).$$

Differentiation of (5.69) yields

$$\begin{aligned}
\pi''(\theta) &= k\mu'(\theta)\pi(\theta) + k^2[\mu(\theta) - \mu]^2\pi(\theta) \\
&= -kv(\theta)\pi(\theta) + k^2[\mu(\theta) - \mu]^2\pi(\theta).
\end{aligned}$$

Integration with respect to θ from θ_0 to θ_1 yields

$$\begin{aligned}
\pi'(\theta_1) - \pi'(\theta_0) &= -kE[v(\Theta)] + k^2E\{[\mu(\Theta) - \mu]^2\} \\
&= -kv + k^2 a
\end{aligned}$$

since $\mu(\Theta)$ has mean μ and $E\{[\mu(\Theta) - \mu]^2\} = Var[\mu(\Theta)] = a$. If $\pi'(\theta_1) = \pi'(\theta_0)$ (again, $\pi'(\theta_0)$ and $\pi'(\theta_1)$ are usually both 0), this implies that $k = v/a$, and so (5.60) is satisfied.

5.4.6 Linear versus Bayesian credibility

In Subsection 5.4.2 it was demonstrated that the credibility premium is the best linear estimator in the sense of minimizing expected squared error with respect to the next observation, X_{n+1}. In Exercise 5.83 you are asked to demonstrate that the Bayesian premium is the best estimator with no restrictions, in the same least squares sense. It was also demonstrated in Subsection 5.4.2 that the credibility premium is the linear estimator that is closest to the Bayesian estimator, again in the mean squared error sense. Finally, we have seen that in a number of cases the credibility and Bayesian premiums are the same. This leaves two questions. Is the additional error caused by using the credibility premium in place of the Bayesian premium worth worrying about? Is it worthwhile to go through the bother of using credibility in the first place? While the exact answer to these questions depends on the underlying distributions, we can obtain some feel for the answers by considering a few examples.

We begin with the second question and use a common situation that has already been discussed. What makes credibility work is that we expect to perform numerous estimations. As a result, we are willing to be biased in any one estimation, provided that the biases cancel out over the numerous estimations. This allows us to reduce variability and, therefore, squared error. The following example shows the power of credibility in this setting.

Example 5.42 *Suppose there are 50 occasions on which we obtain a random sample of size 10 from a Poisson distribution with unknown mean. The samples are from different Poisson populations and therefore may involve different means. Let the true means be $\theta_1, \ldots, \theta_{50}$. Compare overall the mean squared error using linear credibility to that using the sample mean.*

Using the sample mean the total squared error is

$$S_1 = \sum_{j=1}^{50} (\bar{X}_j - \theta_j)^2$$

and the mean squared error is

$$E(S_1) = \sum_{j=1}^{50} Var(\bar{X}_j) = \sum_{j=1}^{50} \theta_j/10.$$

Using a linear credibility estimator, the squared error is

$$S_2 = \sum_{j=1}^{50} [Z\bar{X}_j + (1-Z)\mu - \theta_j]^2$$

where Z and μ are chosen by some method such as the ones presented in the next section. The mean squared error is

$$\begin{aligned} E(S_2) &= \sum_{j=1}^{50} E[Z^2\bar{X}_j^2 + (1-Z)^2\mu^2 + \theta_j^2 + 2Z(1-Z)\bar{X}_j\mu \\ &\quad -2Z\bar{X}_j\theta_j - 2(1-Z)\mu\theta_j] \\ &= \sum_{j=1}^{50}[Z^2(\theta_j/10+\theta_j^2) + (1-Z)^2\mu^2 + \theta_j^2 + 2Z(1-Z)\theta_j\mu \\ &\quad -2Z\theta_j^2 - 2(1-Z)\mu\theta_j] \\ &= Z^2\sum_{j=1}^{50}\theta_j/10 + (1-Z)^2\sum_{j=1}^{50}(\theta_j-\mu)^2. \end{aligned}$$

Linear credibility will produce a smaller mean squared error provided $E(S_2) < E(S_1)$. That is,

$$Z^2\sum_{j=1}^{50}\theta_j/10 + (1-Z)^2\sum_{j=1}^{50}(\theta_j-\mu)^2 < \sum_{j=1}^{50}\theta_j/10.$$

The inequality is satisfied for

$$\frac{\sum_{j=1}^{50}(\theta_j-\mu)^2 - \sum_{j=1}^{50}\theta_j/10}{\sum_{j=1}^{50}(\theta_j-\mu)^2 + \sum_{j=1}^{50}\theta_j/10} < Z < 1$$

and $E(S_2)$ is minimized when $\mu = \sum_{j=1}^{50}\theta_j/50$ and

$$Z = \frac{\sum_{j=1}^{50}(\theta_j-\mu)^2}{\sum_{j=1}^{50}(\theta_j-\mu)^2 + \sum_{j=1}^{50}\theta_j/10}.$$

Of course, this does not provide a way to obtain Z and μ because the values of θ_j are not known.

To get a feel for the possibility of a reduction in mean squared error, suppose $\theta_1, \ldots, \theta_{50}$ were known to be the result of a random sample from a gamma distribution with $\alpha = 50$ and $\theta = 0.1$. This distribution has a mean of 5 and a second moment of 25.5. Then, we can take a second expectation over all such possible samples. When the sample mean is used, the unconditional expectation of S_1 is

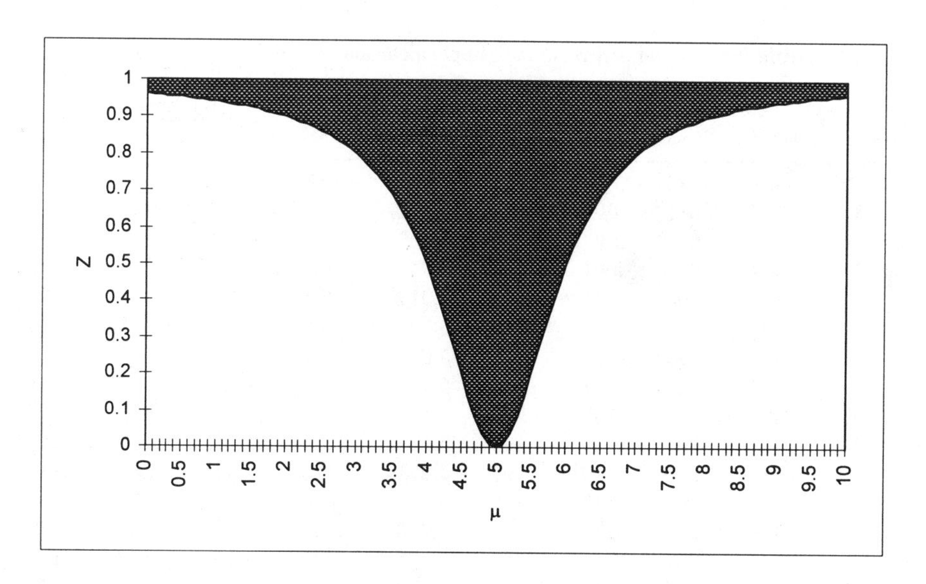

Fig. 5.1 Values of μ and Z for which the credibility estimator has a smaller mean squared error than the sample mean

$$E\left(\sum_{j=1}^{50}\theta_j/10\right) = 25$$

while for the credibility estimator the unconditional expectation of S_2 is

$$\begin{aligned} E&\left[Z^2\sum_{j=1}^{50}\theta_j/10 + (1-Z)^2\sum_{j=1}^{50}(\theta_j-\mu)^2\right] \\ &= 25Z^2 + (1-Z)^2(1{,}275 - 500\mu + 50\mu^2). \end{aligned}$$

The graph in Figure 5.1 illustrates the set of Z and μ for which the credibility estimator produces a smaller mean squared error. This is a fairly broad range of values, so even if we chose suboptimal values (the expected squared error of S_2 is minimized at $\mu = 5$ and $Z = 0.5$), the credibility estimator is still superior to the sample mean.

Finally, consider a specific set of 50 values of θ_j. The ones presented in Table 5.1 are a random sample from a gamma distribution with $\alpha = 50$ and $\theta = 0.1$, sorted in increasing order. The next column provides the mean squared error of the sample mean ($\theta_j/10$). The final three columns provide the bias, variance, and mean squared error for the credibility estimator based on $Z = 0.5$ and $\mu = 5$. The sample mean is always unbiased and therefore

Table 5.1 A comparison of the sample mean and the credibility estimator

	$\bar{X}$	$0.5\bar{X}+2.5$				$\bar{X}$	$0.5\bar{X}+2.5$		
θ	MSE	Bias	Var.	MSE	θ	MSE	Bias	Var.	MSE
3.510	.351	.745	.088	.643	4.875	.488	.062	.122	.126
3.637	.364	.681	.091	.555	4.894	.489	.053	.122	.125
3.742	.374	.629	.094	.489	4.900	.490	.050	.123	.125
3.764	.376	.618	.094	.476	4.943	.494	.028	.124	.124
3.793	.379	.604	.095	.459	4.977	.498	.012	.124	.125
4.000	.400	.500	.100	.350	5.002	.500	−.001	.125	.125
4.151	.415	.424	.104	.284	5.013	.501	−.006	.125	.125
4.153	.415	.424	.104	.283	5.108	.511	−.054	.128	.131
4.291	.429	.354	.107	.233	5.172	.517	−.086	.129	.137
4.405	.440	.298	.110	.199	5.198	.520	−.099	.130	.140
4.410	.441	.295	.110	.197	5.231	.523	−.116	.131	.144
4.413	.441	.293	.110	.196	5.239	.524	−.120	.131	.145
4.430	.443	.285	.111	.192	5.263	.526	−.132	.132	.149
4.438	.444	.281	.111	.190	5.300	.530	−.150	.132	.155
4.471	.447	.264	.112	.182	5.338	.534	−.169	.133	.162
4.491	.449	.254	.112	.177	5.400	.540	−.200	.135	.175
4.495	.449	.253	.112	.176	5.407	.541	−.203	.135	.176
4.505	.451	.247	.113	.174	5.431	.543	−.215	.136	.182
4.547	.455	.227	.114	.165	5.459	.546	−.229	.136	.189
4.606	.461	.197	.115	.154	5.510	.551	−.255	.138	.203
4.654	.465	.173	.116	.146	5.538	.554	−.269	.138	.211
4.758	.476	.121	.119	.134	5.646	.565	−.323	.141	.246
4.763	.476	.118	.119	.133	5.837	.584	−.419	.146	.321
4.766	.477	.117	.119	.133	5.937	.594	−.468	.148	.368
4.796	.480	.102	.120	.130	6.263	.626	−.631	.157	.555
					Mean	.482	.091	.120	.222

the variance matches the mean squared error and so these two quantities are not presented. For the credibility estimator,

$$\begin{aligned} \text{bias} &= E(0.5\bar{X}+2.5-\theta_j) = 2.5 - 0.5\theta_j \\ \text{variance} &= Var(0.5\bar{X}+2.5) = 0.25\theta_j/10 = 0.025\theta_j \\ \text{mean squared error} &= \text{bias}^2 + \text{variance} = 0.25\theta_j^2 - 2.475\theta_j + 6.25. \end{aligned}$$

We see that, as expected, the average mean squared error is much lower for the credibility estimator, and this is achieved by allowing for some bias in the individual estimators. □

We have seen that there is real value in using credibility. Also, even if Z is selected by a relatively poor method (such as the limited fluctuation approach), the results will still be superior to using a noncredibility approach (such as maximum likelihood for each of the separate samples). Our next task is to compare the linear credibility estimator to the Bayesian estimator. In most examples, this is difficult because the Bayesian estimates must be obtained by approximate integration. An alternative would be to explore the mean squared errors by simulation. This approach is taken in an illustration presented in *Foundations of Casualty Actuarial Science* [26, p. 467]. In the following example we use the same illustration but employ an approximation that avoids approximate integration. It should also be noted that the linear credibility approach requires only assumptions or estimation of the first two moments while the Bayesian approach requires the distributions to be completely specified. This nonparametric feature makes the linear approach more robust, which may compensate for any loss of accuracy.

Example 5.43 *Individual observations are samples of size 25 from an inverse gamma distribution with* $\alpha = 4$ *and unknown scale parameter* Θ. *The prior distribution for* Θ *is gamma with* $\alpha = 0.5$ *and* $\theta = 100$. *Compare the linear credibility and Bayesian estimators.*

For the Bühlmann linear credibility estimator we have

$$\begin{aligned}\mu &= E[\mu(\Theta)] = E(\Theta/3) = 50/3\\ a &= Var[\mu(\Theta)] = Var(\Theta/3) = 0.5(100)^2/9 = 5{,}000/9\\ v &= E[v(\Theta)] = E(\Theta^2/18) = 0.5(1.5)(100^2)/18 = 7{,}500/18\end{aligned}$$

and so

$$Z = \frac{25}{25 + \dfrac{7,500}{18}\dfrac{9}{5,000}} = \frac{100}{103}$$

and the credibility estimator is $\hat{\mu}_{cred} = (100\bar{X} + 50)/103$.

For the Bayesian estimator, the posterior density is

$$\begin{aligned}\pi_{\Theta|\mathbf{x}}(\theta|\mathbf{x}) &\propto e^{-\theta\sum_{j=1}^{25} x_j^{-1}}\theta^{100}\theta^{-0.5}e^{-\theta/100}\\ &\propto \theta^{99.5}e^{-\theta(0.01+\sum_{j=1}^{25} x_j^{-1})}\end{aligned}$$

which is a gamma density with parameters 100.5 and $\left(0.01 + \sum_{j=1}^{25} x_j^{-1}\right)^{-1}$. The posterior mean is

$$\hat{\theta}_{\text{Bayes}} = \frac{100.5}{0.01 + \sum_{j=1}^{25} x_j^{-1}} \quad \text{and so} \quad \hat{\mu}_{\text{Bayes}} = \frac{33.5}{0.01 + \sum_{j=1}^{25} x_j^{-1}}$$

which is clearly a nonlinear estimator.

With regard to accuracy, we can also consider the sample mean. Given the value of θ, the sample mean is unbiased with variance and mean squared error

$\theta^2/(18 \times 25) = \theta^2/450$. For the credibility estimator the bias is

$$\begin{aligned}\text{bias}_\theta(\hat{\mu}_{\text{cred}}) &= E[100\bar{X}/103 + 50/103 - \theta/3] \\ &= 100\theta/309 + 50/103 - \theta/3 \\ &= 50/103 - \theta/103\end{aligned}$$

the variance is

$$Var_\theta(\hat{\mu}_{\text{cred}}) = (100/103)^2\theta^2/450$$

and the mean squared error is

$$\text{MSE}_\theta(\hat{\mu}_{\text{cred}}) = \frac{1}{103^2}\left(2{,}500 - 100\theta + 10{,}450\theta^2/450\right).$$

For the Bayes estimate we observe that given θ, $1/X$ has a gamma distribution with parameters 4 and $1/\theta$. Therefore, $\sum_{j=1}^{25} X_j^{-1}$ has a gamma distribution with parameters 100 and $1/\theta$. We note that in the denominator of $\hat{\mu}_{\text{Bayes}}$ the term 0.01 will most always be small relative to the sum. An approximation can be created by ignoring this term, in which case $\hat{\mu}_{\text{Bayes}}$ has approximately an inverse gamma distribution with parameters 100 and 33.5θ. Then

$$\begin{aligned}\text{bias}_\theta(\hat{\mu}_{\text{Bayes}}) &= \frac{33.5\theta}{99} - \frac{\theta}{3} = \frac{0.5\theta}{99} \\ Var_\theta(\hat{\mu}_{\text{Bayes}}) &= \frac{33.5^2\theta^2}{99^2(98)} \\ \text{MSE}_\theta(\hat{\mu}_{\text{Bayes}}) &= \frac{33.5^2 + 49/2}{99^2(98)}\theta^2 = 0.00119391\theta^2.\end{aligned}$$

If we compare the coefficients of θ^2 in the MSE for the three estimators we see that they are 0.00222 for the sample mean, 0.00219 for the credibility estimator, and 0.00122 for the Bayesian estimator. Thus for large θ, the credibility estimator is not much of an improvement over the sample mean, but the Bayesian estimator cuts the mean squared error about in half. Calculated values of these quantities for various percentiles from the gamma prior distribution appear in Table 5.2. □

The inferior behavior of the credibility estimator when compared with the Bayes estimator is due to the heavy tails of the two distributions. One way to lighten the tail is to work with the logarithm of the data. This idea was proposed in *Foundations of Casualty Actuarial Science* [26] and evaluated for the above example. The idea is to work with the logarithms of the data and use linear credibility to estimate the mean of the distribution of logarithms. The result is then exponentiated. Because this procedure is sure to introduce bias, a multiplicative adjustment is made. The results are presented in the following example with many of the details left for Exercise 5.81.

Table 5.2 A comparison of the sample mean, credibility, and Bayes estimators

		$\bar{X}$	$\hat{\mu}_{\text{cred}}$		$\hat{\mu}_{\text{Bayes}}$	
Percentile	θ	MSE	Bias	MSE	Bias	MSE
1	0.008	0.000	0.485	0.236	0.000	0.000
5	0.197	0.000	0.484	0.234	0.001	0.000
10	0.790	0.001	0.478	0.230	0.004	0.001
25	5.077	0.057	0.436	0.244	0.026	0.031
50	22.747	1.150	0.265	1.154	0.115	0.618
75	66.165	9.729	−0.157	9.195	0.334	5.227
90	135.277	40.667	−0.828	39.018	0.683	21.849
95	192.072	81.982	−1.379	79.178	0.970	44.046
99	331.746	244.568	−2.735	238.011	1.675	131.397

Example 5.44 (Example 5.43 continued) *Obtain the log-credibility estimator and evaluate its bias and mean squared error.*

Let $W_j = \log X_j$. Then for the credibility on the logarithms,

$$\begin{aligned}
\mu(\Theta) &= E(W|\Theta) \\
&= \int_0^\infty (\log x)\Theta^4 x^{-5} e^{-\Theta/x} \frac{1}{6} dx \\
&= \int_0^\infty (\log \Theta - \log y) y^3 e^{-y} \frac{1}{6} dy \\
&= \log \Theta - \Psi(4)
\end{aligned}$$

where the second integral was obtained using the substitution $y = \Theta/x$. The last line follows from observing that the term $y^3 e^{-y}/6$ is a gamma density and thus integrates to one while the second term is the **digamma** function (see Exercise 5.81) and using tables in [1] we have $\Psi(4) = 1.25612$. The next required quantity is

$$\begin{aligned}
v(\Theta) &= E(W^2|\theta) - \mu(\Theta)^2 \\
&= \int_0^\infty (\log x)^2 \Theta^4 x^{-5} e^{-\Theta/x} \frac{1}{6} dx - [\log \Theta - \Psi(4)]^2 \\
&= \int_0^\infty (\log \Theta - \log y)^2 y^3 e^{-y} \frac{1}{6} dy - [\log \Theta - \Psi(4)]^2 \\
&= \Psi'(4)
\end{aligned}$$

where $\Psi'(4) = 0.283823$ is the **trigamma** function (see Exercise 5.81). Then

$$\begin{aligned}
\mu &= E[\log\Theta - \Psi(4)] \\
&= \int_0^\infty (\log\theta)\theta^{-.5}e^{-\theta/100}100^{-.5}\frac{1}{\Gamma(0.5)}d\theta - \Psi(4) \\
&= \int_0^\infty (\log 100 + \log\lambda)\lambda^{-.5}e^{-\lambda}\frac{1}{\Gamma(.5)}d\lambda - \Psi(4) \\
&= \log 100 + \Psi(0.5) - \Psi(4) = 1.38554.
\end{aligned}$$

Also,

$$\begin{aligned}
v &= E[\Psi'(4)] = \Psi'(4) = 0.283823 \\
a &= Var[\log\Theta - \Psi(4)] \\
&= \Psi'(0.5) = 4.934802 \\
Z &= \frac{25}{25 + \frac{0.283823}{4.934802}} = 0.997705.
\end{aligned}$$

The log-credibility estimate is

$$\hat{\mu}_{\text{log-cred}} = c\exp(0.997705\bar{W} + 0.00318024).$$

The value of c is obtained by setting

$$\begin{aligned}
E(X) = \frac{50}{3} &= cE[\exp(0.997705\bar{W} + 0.00318024)] \\
&= ce^{0.00318024}E\left[\exp\left(\frac{0.997705}{25}\sum_{j=1}^{25}\log X_j\right)\right] \\
&= ce^{0.00318024}E\left[E\left(\prod_{j=1}^{25} X_j^{0.997705/25}|\Theta\right)\right].
\end{aligned}$$

Given Θ, the X_js are independent and so the expected product is the product of the expected values. From Appendix A, the kth moment of the inverse gamma distribution produces

$$\begin{aligned}
\frac{50}{3} &= ce^{0.00318024}E\left\{\left[\frac{\Theta^{0.997705/25}\Gamma\left(4 - \frac{0.997705}{25}\right)}{6}\right]^{25}\right\} \\
&= ce^{0.00318024}\left[\frac{\Gamma\left(4 - \frac{0.997705}{25}\right)}{6}\right]^{25}\frac{100^{.997705}\Gamma(0.5 + 0.997705)}{\Gamma(0.5)}
\end{aligned}$$

which produces $c = 1.169318$ and

$$\hat{\mu}_{\text{log-cred}} = 1.173043(2.712051)^{\bar{W}}.$$

In order to evaluate the bias and mean squared error for a given value of Θ, we must obtain

$$\begin{aligned}
E(\hat{\mu}_{\text{log-cred}}|\Theta = \theta) &= 1.173043E\left(e^{\bar{W}\log 2.712051}|\Theta = \theta\right) \\
&= 1.173043E\left[\prod_{j=1}^{25} X_j^{(\log 2.712051)/25}|\Theta = \theta\right] \\
&= 1.173043\left[\frac{\theta^{(\log 2.712051)/25}\Gamma\left(4 - \frac{\log 2.712051}{25}\right)}{6}\right]^{25}
\end{aligned}$$

and

$$\begin{aligned}
E(\hat{\mu}^2_{\text{log-cred}}|\Theta = \theta) &= 1.173043^2 E\left(e^{2\bar{W}\log 2.712051}|\Theta = \theta\right) \\
&= 1.173043^2\left[\frac{\theta^{(2\log 2.712051)/25}\Gamma\left(4 - \frac{2\log 2.712051}{25}\right)}{6}\right]^{25}
\end{aligned}$$

The measures of quality are then

$$\begin{aligned}
\text{bias}_\theta(\hat{\mu}_{\text{log-cred}}) &= E(\hat{\mu}_{\text{log-cred}}|\Theta = \theta) - \theta/3 \\
\text{MSE}_\theta(\hat{\mu}_{\text{log-cred}}) &= E(\hat{\mu}^2_{\text{log-cred}}|\Theta = \theta) - [E(\hat{\mu}_{\text{log-cred}}|\Theta = \theta)]^2 \\
&\quad +[\text{bias}_\theta(\hat{\mu}_{\text{log-cred}})]^2.
\end{aligned}$$

Values of these quantities are calculated for various values of θ in Table 5.3. A comparison with Table 5.2 indicates that the log-credibility estimator is almost as good as the Bayes estimator. □

In practice, log-credibility is as easy to use as ordinary credibility. In either case, one of the computational methods of the next section would be used. For log-credibility, the logarithms of the observations are substituted for the observed values and then the final estimate is exponentiated. The bias is corrected by multiplying all the estimates by a constant such that the sample mean of the estimates matches the sample mean of the original data.

5.4.7 Notes and References

In this section, one of the two major criticisms of limited fluctuation credibility has been addressed. Through the use of the variance of the hypothetical

Table 5.3 Bias and mean squared error for the log-credibility estimator

Percentile	θ	Bias	MSE
1	0.008	0.000	0.000
5	0.197	0.001	0.000
10	0.790	0.003	0.001
25	5.077	0.012	0.034
50	22.747	0.026	0.666
75	66.165	0.023	5.604
90	135.277	−0.028	23.346
95	192.072	−0.091	46.995
99	331.746	−0.295	139.908

means, we now have a means of relating the mean of the group of interest, $\mu(\theta)$, to the manual, or collective, premium, μ. The development was also mathematically sound in that the results followed directly from a specific model and objective. We have also seen that the additional restriction of a linear solution was not as bad as it might be in that often we still obtain the exact Bayesian solution. There has subsequently been a great deal of effort expended to generalize the model. With a sound basis for obtaining a credibility premium, we have but one remaining obstacle: how to apply numbers to the quantities a and v in the Bühlmann formulation, or how to specify the prior distribution in the Bayesian formulation. Those matters are addressed in the final section of this chapter.

A historical review of credibility theory including a description of the limited fluctuation and greatest accuracy approaches is provided by Norberg [96]. Since the classic paper of Bühlmann [18], there has developed a vast literature on credibility theory in the actuarial literature. Other elementary introductions are given by Herzog [59] and Waters [125]. Other more advanced treatments are Goovaerts and Hoogstad [52] and Sundt [118]. An important generalization of the Bühlmann–Straub model is the Hachemeister [53] regression model, which was not discussed here. See also Klugman [75]. The material on exact credibility is taken from Jewell [71]. See also Ericson [37]. A special issue of *Insurance: Abstracts and Reviews* (Sundt [117]) contains an extensive list of papers on credibility.

5.5 EMPIRICAL BAYES PARAMETER ESTIMATION

In the previous section a modeling methodology was proposed which suggested the use of either the Bayesian or credibility premium as a way to incorporate past data into the prospective rate. There is a practical problem associated with the use of these models which has not yet been addressed.

In the die-spinner and urn model examples, we were able to obtain numerical values for the quantities of interest since the input distributions $f_{X_j|\Theta}(x_j|\theta)$ and $\pi(\theta)$ were assumed to be known. These examples, while useful for illustration of the methodology, can hardly be expected to accurately represent the business of an insurance portfolio. More practical models of necessity involve the use of parameters which must be chosen to ensure a close agreement between the model and reality. Examples of this include: the Poisson–gamma model (Example 5.26), where the gamma parameters α and β need to be selected, the normal–normal model (Example 5.29), where μ, v, and a need to be chosen; or the Bühlmann or Bühlmann–Straub parameters μ, v, and a. Assignment of numerical values to the Bayesian or credibility premium requires that these parameters be replaced by numerical values.

In general, the unknown parameters are those associated with the structure density $\pi(\theta)$, and hence we refer to these as **structural parameters**. The terminology we use follows the Bayesian framework of the previous section. Strictly speaking, in the Bayesian context all structural parameters are assumed known and there is no need for estimation. An example of this is the Poisson–gamma where our prior information about the structural density was quantified by the choice of $\alpha = 36$ and $\beta = 1/240$. For our purposes, this fully Bayesian approach is often unsatisfactory (e.g., when there is little or no prior information available, such as with a new line of insurance) and we may need to use the data at hand to estimate the structural (prior) parameters. This approach is called **empirical Bayes estimation**.

We refer to the situation where $\pi(\theta)$ and $f_{X_j|\Theta}(x_j|\theta)$ are left largely unspecified (for example, in the Bühlmann or Bühlmann–Straub models where only the first two moments need by known) as the nonparametric case. This situation is dealt with in Subsection 5.5.1. If $f_{X_j|\Theta}(x_j|\theta)$ is assumed to be of parametric form (e.g., Poisson, normal, etc.) but not $\pi(\theta)$, then we refer to the problem as being of a semiparametric nature, and this is considered in Subsection 5.5.2. Finally, the (technically more difficult) fully parametric case where both $f_{X_j|\Theta}(x_j|\theta)$ and $\pi(\theta)$ are assumed to be of parametric form is briefly discussed in Subsection 5.5.3.

This decision as to whether to select a parametric model or not depends partially on the situation at hand and partially on the judgment and knowledge of the person doing the analysis. For example, an analysis based on claim counts might involve the assumption that $f_{X_j|\Theta}(x_j|\theta)$ is of Poisson form, whereas the choice of a parametric model for $\pi(\theta)$ may not be reasonable. A strong case for the use of parametric models was made in Chapter 2.

Any parametric assumptions should be reflected (as far as possible) in parametric estimation. For example, in the Poisson case, since the mean and variance are equal, the same estimate would normally be used for both. Nonparametric estimators would normally be no more efficient than estimators appropriate for the parametric model selected, assuming that the model selected is appropriate. This notion is relevant for the decision as to whether to select a parametric model.

Finally, nonparametric models have the advantage of being appropriate for a wide variety of situations, a fact which may well eliminate the extra burden of a parametric assumption (often a stronger assumption than is reasonable).

In this section the data are assumed to be of the following form. For each of $r \geq 1$ policyholders we have the observed losses per unit of exposure $\mathbf{X}_i = (X_{i1}, \ldots, X_{in_i})'$ for $i = 1, \ldots, r$. The random vectors $\{\mathbf{X}_i, i = 1, \ldots, r\}$ are assumed to be statistically independent (experience of different policyholders is assumed to be independent). The (unknown) risk parameter for the ith policyholder is θ_i, $i = 1, \ldots, r$, and it is assumed further that $\theta_1, \ldots, \theta_r$ are realizations of the independent and identically distributed random variables Θ_i with structural density $\pi(\theta_i)$. For fixed i, the (conditional) random variables $X_{ij}|\Theta_i$ are assumed to be independent with pf $f_{X_{ij}|\Theta}(x_{ij}|\theta_i)$, $j = 1, \ldots, n_i$.

Two particularly common cases produce this data format. The first is classification rate making or experience rating. In either, i indexes the classes or groups and j indexes the individual members. The second case is like the first where i continues to index the class or group, but now j is the year and the observation is the average loss for that year. An example of the second setting is Meyers [88], where $i = 1, \ldots, 319$ employment classifications are studied over $j = 1, 2, 3$ years. Regardless of the potential settings, we will refer to the r entities as policyholders.

There may also be a known exposure vector $\mathbf{m}_i = (m_{i1}, m_{i2}, \cdots, m_{in_i})'$ for policyholder i, where $i = 1, \ldots, r$. If not (and if it is appropriate) one may set $m_{ij} = 1$ in what follows for all i and j. For notational convenience let

$$m_i = \sum_{j=1}^{n_i} m_{ij}, \qquad i = 1, \ldots, r$$

be the total past exposure for policyholder i, and let

$$\bar{X}_i = \frac{1}{m_i} \sum_{j=1}^{n_i} m_{ij} X_{ij}, \qquad i = 1, \ldots, r$$

be the past average loss experience. Furthermore, the total exposure is

$$m = \sum_{i=1}^{r} m_i = \sum_{i=1}^{r} \sum_{j=1}^{n_i} m_{ij}$$

and the overall average losses are

$$\bar{X} = \frac{1}{m} \sum_{i=1}^{r} m_i \bar{X}_i = \frac{1}{m} \sum_{i=1}^{r} \sum_{j=1}^{n_i} m_{ij} X_{ij}. \tag{5.72}$$

The parameters which need to be estimated depend on what is assumed about the distributions $f_{X_{ij}|\Theta}(x_{ij}|\theta_i)$ and $\pi(\theta)$.

For the Bühlmann–Straub formulation there are additional quantities of interest. The hypothetical mean (assumed not to depend on j) is

$$E(X_{ij}|\Theta_i = \theta_i) = \mu(\theta_i)$$

and the process variance is

$$Var(X_{ij}|\Theta_i = \theta_i) = v(\theta_i)/m_{ij}.$$

The structural parameters are

$$\begin{aligned} \mu &= E[\mu(\Theta_i)], \\ v &= E[v(\Theta_i)], \end{aligned}$$

and

$$a = Var[\mu(\Theta_i)].$$

The approach to be followed in this section is to estimate μ, v, and a (when unknown) from the data. The credibility premium for next year's losses (per exposure unit) for policyholder i is

$$Z_i\bar{X}_i + (1 - Z_i)\mu, \qquad i = 1, \ldots, r \tag{5.73}$$

where

$$Z_i = m_i/(m_i + k)$$

and

$$k = v/a.$$

If estimators of μ, v, and a are denoted by $\hat{\mu}, \hat{v}$, and $\hat{a}$ respectively, then one would replace the credibility premium (5.73) by its estimator

$$\hat{Z}_i\bar{X}_i + (1 - \hat{Z}_i)\hat{\mu} \tag{5.74}$$

where

$$\hat{Z}_i = m_i/(m_i + \hat{k})$$

and

$$\hat{k} = \hat{v}/\hat{a}.$$

Note that even if $\hat{v}$ and $\hat{a}$ are unbiased estimators of v and a, the same cannot be said of $\hat{k}$ and $\hat{Z}_i$. Finally, the credibility premium to cover all m_{i,n_i+1} exposure units for policyholder i in the next year would be (5.74) multiplied by m_{i,n_i+1}.

5.5.1 Nonparametric estimation

In this section we consider unbiased estimation of μ, v, and a. To illustrate ideas, let us begin with the following simple Bühlmann-type example.

Example 5.45 *Suppose that $n_i = n > 1$ independently of i and $m_{ij} = 1$ for all i and j. That is, for policyholder i, we have the loss vector*

$$\mathbf{X}_i = (X_{i1}, \ldots, X_{in})', \qquad i = 1, \ldots, r.$$

Furthermore, conditional on $\Theta_i = \theta_i$, X_{ij} has mean $\mu(\theta_i) = E(X_{ij}|\Theta_i = \theta_i)$ and variance $v(\theta_i) = Var(X_{ij}|\Theta_i = \theta_i)$, and $X_{i1}, \ldots, X_{in}$ are independent (conditionally). Also, different policyholders' past data are independent, so that if $i \neq l$, then X_{ij} and X_{lt} are independent. In this case $\bar{X}_i = n^{-1}\sum_{j=1}^{n} X_{ij}$ and $\bar{X} = r^{-1}\sum_{i=1}^{r} \bar{X}_i = (rn)^{-1}\sum_{i=1}^{r}\sum_{j=1}^{n} X_{ij}$. Determine unbiased estimators of the Bühlmann quantities.

An unbiased estimator of μ is

$$\hat{\mu} = \bar{X}$$

because

$$\begin{aligned} E(\hat{\mu}) &= (rn)^{-1}\sum_{i=1}^{r}\sum_{j=1}^{n} E(X_{ij}) = (rn)^{-1}\sum_{i=1}^{r}\sum_{j=1}^{n} E[E(X_{ij}|\Theta_i)] \\ &= (rn)^{-1}\sum_{i=1}^{r}\sum_{j=1}^{n} E[\mu(\Theta_i)] = (rn)^{-1}\sum_{i=1}^{r}\sum_{j=1}^{n} \mu = \mu. \end{aligned}$$

To estimate v, consider

$$\hat{v}_i = \frac{1}{n-1}\sum_{j=1}^{n}(X_{ij} - \bar{X}_i)^2.$$

Recall that for fixed i, the random variables $X_{i1}, \ldots, X_{in}$ are independent, conditional on $\Theta_i = \theta_i$. Thus, $\hat{v}_i$ is an unbiased estimate of $Var(X_{ij}|\Theta_i = \theta_i) = v(\theta_i)$. Unconditionally,

$$E(\hat{v}_i) = E[E(\hat{v}_i|\Theta_i)] = E[v(\Theta_i)] = v$$

and $\hat{v}_i$ is unbiased for v. Hence an unbiased estimator of v is

$$\hat{v} = \frac{1}{r}\sum_{i=1}^{r}\hat{v}_i = \frac{1}{r(n-1)}\sum_{i=1}^{r}\sum_{j=1}^{n}(X_{ij} - \bar{X}_i)^2. \tag{5.75}$$

We now turn to estimation of the parameter a. Begin with

$$E(\bar{X}_i|\Theta_i = \theta_i) = n^{-1}\sum_{j=1}^{n} E(X_{ij}|\Theta_i = \theta_i) = n^{-1}\sum_{j=1}^{n}\mu(\theta_i) = \mu(\theta_i).$$

Thus,

$$E(\bar{X}_i) = E[E(\bar{X}_i|\Theta_i)] = E[\mu(\Theta_i)] = \mu$$

and

$$\begin{aligned} Var(\bar{X}_i) &= Var[E(\bar{X}_i|\Theta_i)] + E[Var(\bar{X}_i|\Theta_i)] \\ &= Var[\mu(\Theta_i)] + E[v(\Theta_i)/n] = a + v/n. \end{aligned}$$

Therefore, $\bar{X}_1, \ldots, \bar{X}_r$ are independent with common mean μ and common variance $a + v/n$. Their sample average is $\bar{X} = r^{-1}\sum_{i=1}^r \bar{X}_i$. Consequently, an unbiased estimator of $a + v/n$ is $(r-1)^{-1}\sum_{i=1}^r \left(\bar{X}_i - \bar{X}\right)^2$. Since we already have an unbiased estimator of v given above, an unbiased estimator of a is given by

$$\begin{aligned} \hat{a} &= \frac{1}{r-1}\sum_{i=1}^r \left(\bar{X}_i - \bar{X}\right)^2 - \frac{\hat{v}}{n} \\ &= \frac{1}{r-1}\sum_{i=1}^r \left(\bar{X}_i - \bar{X}\right)^2 - \frac{1}{rn(n-1)}\sum_{i=1}^r\sum_{j=1}^n (X_{ij} - \bar{X}_i)^2. \end{aligned} \quad (5.76)$$

□

These estimators should look familiar. Consider a one-factor analysis of variance in which each policyholder represents a treatment. The estimator for v (5.75) is the within (also called the error) mean square. The first term in the estimator for a (5.76) is the between (also called the treatment) mean square divided by n. The hypothesis that all treatments have the same mean is accepted when the between mean square is small relative to the within mean square—that is, when $\hat{a}$ is small relative to $\hat{v}$. But that implies $\hat{Z}$ will be near zero and little credibility will be given to each $\bar{X}_i$. This is as it should be when the policyholders are essentially identical. Due to the subtraction in (5.76) it is possible that $\hat{a}$ could be negative. When that happens it is customary to set $\hat{a} = \hat{Z} = 0$. This case is equivalent to the F test statistic in the analysis of variance being less than one, a case that always leads to an acceptance of the hypothesis of equal means.

Example 5.46 (Example 5.45 continued) *As a numerical illustration, suppose we have $r = 2$ policyholders with $n = 3$ years experience for each. Let the losses be $\mathbf{x}_1 = (3, 5, 7)'$ and $\mathbf{x}_2 = (6, 12, 9)'$. Estimate the Bühlmann credibility premiums for each policyholder.*

We have

$$\bar{X}_1 = (3 + 5 + 7)/3 = 5, \ \bar{X}_2 = (6 + 12 + 9)/3 = 9$$

and so $\bar{X} = (5 + 9)/2 = 7$. Then $\hat{\mu} = 7$. We next have

$$\begin{aligned} \hat{v}_1 &= [(3-5)^2 + (5-5)^2 + (7-5)^2]/2 = 4 \\ \hat{v}_2 &= [(6-9)^2 + (12-9)^2 + (9-9)^2]/2 = 9 \end{aligned}$$

and so $\hat{v} = (4+9)/2 = 13/2$. Then $\hat{a} = [(5-7)^2 + (9-7)^2]/1 - \hat{v}/3 = 35/6$. Next, $\hat{k} = \hat{v}/\hat{a} = 39/35$ and the estimated credibility factor is $\hat{Z} = 3/(3+\hat{k}) = 35/48$. The estimated credibility premium for policyholder 1 is $\hat{Z}\bar{X}_1 + (1-\hat{Z})\hat{\mu} = \left(\frac{35}{48}\right)(5) + \left(\frac{13}{48}\right)(7) = 133/24$ whereas for policyholder 2 the estimated credibility premium is $\hat{Z}\bar{X}_2 + (1-\hat{Z})\hat{\mu} = \left(\frac{35}{48}\right)(9) + \left(\frac{13}{48}\right)(7) = 203/24$. □

We now turn to the more general Bühlmann–Straub setup described earlier in this section. We have $E(X_{ij}) = E[E(X_{ij}|\Theta_i)] = E[\mu(\Theta_i)] = \mu$. Thus,

$$E(\bar{X}_i|\Theta_i) = \sum_{j=1}^{n_i} \frac{m_{ij}}{m_i} E(X_{ij}|\Theta_i) = \sum_{j=1}^{n_i} \frac{m_{ij}}{m_i} \mu(\Theta_i) = \mu(\Theta_i)$$

implying that

$$E(\bar{X}_i) = E[E(\bar{X}_i|\Theta_i)] = E[\mu(\Theta_i)] = \mu.$$

Finally,

$$E(\bar{X}) = \frac{1}{m}\sum_{i=1}^{r} m_i E(\bar{X}_i) = \frac{1}{m}\sum_{i=1}^{r} m_i \mu = \mu$$

and so an obvious unbiased estimator of μ is

$$\hat{\mu} = \bar{X}. \tag{5.77}$$

Now, $E(X_{ij}|\Theta_i) = \mu(\Theta_i)$ and $Var(X_{ij}|\Theta_i) = v(\Theta_i)/m_{ij}$ for $j = 1, \ldots, n_i$. Consider

$$\hat{v}_i = \frac{\sum_{j=1}^{n_i} m_{ij}(X_{ij} - \bar{X}_i)^2}{n_i - 1}, \qquad i = 1, \ldots, r. \tag{5.78}$$

Condition on Θ_i and use (5.16) with $\beta = 0$ and $\alpha = v(\Theta_i)$. Then $E(\hat{v}_i|\Theta_i) = v(\Theta_i)$. But this means that, unconditionally,

$$E(\hat{v}_i) = E[E(\hat{v}_i|\Theta_i)] = E[v(\Theta_i)] = v$$

and so $\hat{v}_i$ is unbiased for v for $i = 1, \ldots, r$. Another unbiased estimator for v is then the weighted average $\hat{v} = \sum_{i=1}^{r} w_i \hat{v}_i$, where $\sum_{i=1}^{r} w_i = 1$. If we choose weights proportional to $n_i - 1$, we weight the original X_{ij}s by m_{ij}. That is, with $w_i = (n_i - 1)/\sum_{i=1}^{r}(n_i - 1)$, we obtain an unbiased estimator of v, namely,

$$\hat{v} = \frac{\sum_{i=1}^{r}\sum_{j=1}^{n_i} m_{ij}(X_{ij} - \bar{X}_i)^2}{\sum_{i=1}^{r}(n_i - 1)}. \tag{5.79}$$

We now turn to estimation of a. Recall that for fixed i, the random variables $X_{i1}, \ldots, X_{in_i}$ are independent conditional on Θ_i. Thus,

$$\begin{aligned} Var(\bar{X}_i|\Theta_i) &= \sum_{j=1}^{n_i} \left(\frac{m_{ij}}{m_i}\right)^2 Var(X_{ij}|\Theta_i) = \sum_{j=1}^{n_i} \left(\frac{m_{ij}}{m_i}\right)^2 \frac{v(\Theta_i)}{m_{ij}} \\ &= \frac{v(\Theta_i)}{m_i^2} \sum_{j=1}^{n_i} m_{ij} = \frac{v(\Theta_i)}{m_i}. \end{aligned}$$

But this means that, unconditionally,

$$\begin{aligned} Var(\bar{X}_i) &= Var[E(\bar{X}_i|\Theta_i)] + E[Var(\bar{X}_i|\Theta_i)] \\ &= Var[\mu(\Theta_i)] + E[v(\Theta_i)/m_i] = a + v/m_i. \end{aligned} \tag{5.80}$$

To summarize, $\bar{X}_1, \ldots, \bar{X}_r$ are independent with common mean μ and variances $Var(\bar{X}_i) = a + v/m_i$. Furthermore, $\bar{X} = \frac{1}{m}\sum_{i=1}^r m_i \bar{X}_i$. Now, (5.16) may again be used with $\beta = a$ and $\alpha = v$, to yield

$$E\left[\sum_{i=1}^r m_i \left(\bar{X}_i - \bar{X}\right)^2\right] = a\left(m - m^{-1}\sum_{i=1}^r m_i^2\right) + v(r-1).$$

An unbiased estimator for a may be obtained by replacing v by an unbiased estimator $\hat{v}$ and "solving" for a. That is, an unbiased estimator of a is

$$\hat{a} = \left(m - m^{-1}\sum_{i=1}^r m_i^2\right)^{-1} \left[\sum_{i=1}^r m_i(\bar{X}_i - \bar{X})^2 - \hat{v}(r-1)\right] \tag{5.81}$$

with $\hat{v}$ given by (5.79). An alternative form of (5.81) is given in Exercise 5.91.

Some remarks are in order at this point. Equations (5.77), (5.79), and (5.81) provide unbiased estimators for μ, v, and a, respectively. They are nonparametric, requiring no distributional assumptions. They are certainly not the only (unbiased) estimators which could be used, and it is possible that $\hat{a} < 0$. In this case, a is likely to be close to 0, and it makes sense to set $\hat{Z} = 0$. Furthermore, the ordinary Bühlmann estimators of Example 5.45 are recovered with $m_{ij} = 1$ and $n_i = n$. Also, it is assumed that $n_i > 1$ and $r > 1$. If $n_i = 1$, so that there is only one exposure unit's experience for policyholder i, it is difficult to obtain information on the process variance $v(\Theta_i)$ and thus v. Similarly, if $r = 1$ there is only one policyholder and it is difficult to obtain information on the variance of the hypothetical means a. In these situations stronger assumptions are needed such as knowledge of one or more of the parameters (e.g., the pure premium or manual rate μ, discussed below) or parametric assumptions which imply functional relationships between the parameters (discussed in Subsection 5.5.2).

There is one problem using the formulas developed above. In the past, the data from the ith policyholder was collected on an exposure of m_i. Total losses on all policyholders was $TL = \sum_{i=1}^{r} m_i \bar{X}_i$. If we had charged the credibility premium as given above, the total premium would have been

$$\begin{aligned} TP &= \sum_{i=1}^{r} m_i[\hat{Z}_i \bar{X}_i + (1 - \hat{Z}_i)\hat{\mu}] \\ &= \sum_{i=1}^{r} m_i(1 - \hat{Z}_i)(\hat{\mu} - \bar{X}_i) + \sum_{i=1}^{r} m_i \bar{X}_i \\ &= \sum_{i=1}^{r} m_i \frac{\hat{k}}{m_i + \hat{k}}(\hat{\mu} - \bar{X}_i) + \sum_{i=1}^{r} m_i \bar{X}_i. \end{aligned}$$

It is often desirable for TL to equal TP. The reason is that any premium increases that will meet the approval of regulators will be based on the total claim level from past experience. While credibility adjustments make both practical and theoretical sense, it is usually a good idea to keep the total unchanged. For this to happen we need

$$0 = \sum_{i=1}^{r} m_i \frac{\hat{k}}{m_i + \hat{k}}(\hat{\mu} - \bar{X}_i)$$

or

$$\hat{\mu} \sum_{i=1}^{r} \hat{Z}_i = \sum_{i=1}^{r} \hat{Z}_i \bar{X}_i$$

or

$$\hat{\mu} = \frac{\sum_{i=1}^{r} \hat{Z}_i \bar{X}_i}{\sum_{i=1}^{r} \hat{Z}_i}. \tag{5.82}$$

That is, rather than using (5.77) to compute $\hat{\mu}$, use a *credibility-weighted average* of the individual sample means. Either method provides an unbiased estimator, but this latter one has the advantage of preserving total claims. It can also be derived by least-squares arguments. It should be noted that when using (5.81) the value of $\bar{X}$ from (5.72) should still be used. Finally, from Example 5.12 and noting the form of $Var(\bar{X}_j)$ in (5.80), the weights in (5.82) provide the smallest unconditional variance for $\hat{\mu}$.

Example 5.47 *Past data on two group policyholders are available and are given in Table 5.4. Determine the estimated credibility premium to be charged to each group in year 4.*

We first need to determine the average claims per person for each group in each past year. We have $n_1 = 2$ years experience for group 1 and $n_2 = 3$ for group 2. It is immaterial which past years' data we have for policyholder 1, so for notational purposes we will choose $m_{11} = 50$ and $X_{11} = 10{,}000/50 = 200$. Similarly, $m_{12} = 60$ and $X_{12} = 13{,}000/60 = 216.67$. Then $m_1 = m_{11} + m_{12} = 50 + 60 = 110$ and $\bar{X}_1 = (10{,}000 + 13{,}000)/110 = 209.09$. For

Table 5.4 Data for Example 5.47

	Policyholder	Year 1	Year 2	Year 3	Year 4
Total claims	1	–	10,000	13,000	–
No. in group		–	50	60	75
Total claims	2	18,000	21,000	17,000	–
No. in group		100	110	105	90

policyholder 2, we obtain $m_{21} = 100$, $X_{21} = 18{,}000/100 = 180$, $m_{22} = 110$, $X_{22} = 21{,}000/110 = 190.91$, $m_{23} = 105$, $X_{23} = 17{,}000/105 = 161.90$. Then $m_2 = m_{21} + m_{22} + m_{23} = 100 + 110 + 105 = 315$ and $\bar{X}_2 = (18{,}000 + 21{,}000 + 17{,}000)/315 = 177.78$. Now, $m = m_1 + m_2 = 110 + 315 = 425$. The overall mean is $\hat{\mu} = \bar{X} = (10{,}000 + 13{,}000 + 18{,}000 + 21{,}000 + 17{,}000)/425 = 185.88$. The alternative estimate of μ (5.82) cannot be computed until later.

Now,

$$\begin{aligned}\hat{v} &= \frac{\begin{array}{c}50(200-209.09)^2 + 60(216.67-209.09)^2 + 100(180-177.78)^2 \\ +110(190.91-177.78)^2 + 105(161.90-177.78)^2\end{array}}{(2-1)+(3-1)} \\ &= 17{,}837.87\end{aligned}$$

and so

$$\begin{aligned}\hat{a} &= \frac{110(209.09-185.88)^2 + 315(177.78-185.88)^2 - (17{,}837.87)(1)}{425 - (110^2 + 315^2)/425} \\ &= 380.76.\end{aligned}$$

Then $\hat{k} = \hat{v}/\hat{a} = 46.85$. The estimated credibility factor for policyholder 1 is $\hat{Z}_1 = 110/(110 + 46.85) = 0.70$, and that for policyholder 2 is $\hat{Z}_2 = 315/(315 + 46.85) = 0.87$. Per individual the estimated credibility premium for policyholder 1 is

$$\hat{Z}_1\bar{X}_1 + (1 - \hat{Z}_1)\hat{\mu} = (0.70)(209.09) + (0.30)(185.88) = 202.13$$

and so the total estimated credibility premium for the whole group is

$$75(202.13) = 15{,}159.75.$$

For policyholder 2,

$$\hat{Z}_2\bar{X}_2 + (1 - \hat{Z}_2)\hat{\mu} = (0.87)(177.78) + (0.13)(185.88) = 178.83$$

and the total estimated credibility premium is

$$90(178.83) = 16{,}094.70.$$

For the alternative estimator we would use

$$\hat{\mu} = \frac{0.70(209.09) + 0.87(177.78)}{0.70 + 0.87} = 191.74.$$

The credibility premiums are

$$0.70(209.09)+0.30(191.74) = 203.89, \qquad 0.87(177.78)+0.13(191.74) = 179.59.$$

The total past credibility premium is $110(203.89) + 315(179.59) = 78{,}998.75$. Except for rounding error, this matches the actual total losses of 79,000. □

The above analysis assumes that the parameters μ, v and a are all unknown and need to be estimated. This may not always be the case. For example, the manual rate μ may be already known, but estimates of a and v may be needed. In that case, (5.79) can still be used to estimate v as it is unbiased whether μ is known or not. (Why is $\left[\sum_{j=1}^{n_i} m_{ij}(X_{ij} - \mu)^2\right]/n_i$ not unbiased for v in this case?) Similarly, (5.81) is still an unbiased estimator for a. However, if μ is known, an alternative unbiased estimator for a is

$$\tilde{a} = \sum_{i=1}^{r} \frac{m_i}{m}(\bar{X}_i - \mu)^2 - \frac{r}{m}\hat{v}$$

where $\hat{v}$ is given by (5.79). To see this, note that

$$\begin{aligned} E(\tilde{a}) &= \sum_{i=1}^{r} \frac{m_i}{m} E[(\bar{X}_i - \mu)^2] - \frac{r}{m} E(\hat{v}) \\ &= \sum_{i=1}^{r} \frac{m_i}{m} Var(\bar{X}_i) - \frac{r}{m} v \\ &= \sum_{i=1}^{r} \frac{m_i}{m} \left(a + \frac{v}{m_i}\right) - \frac{r}{m} v = a. \end{aligned}$$

If there are data on only one policyholder, an approach like this is necessary. Clearly, (5.78) provides an estimator for v based on data from policyholder i alone, and an unbiased estimator for a based on data from policyholder i alone is

$$\tilde{a}_i = (\bar{X}_i - \mu)^2 - \hat{v}_i/m_i = (\bar{X}_i - \mu)^2 - \frac{\sum_{j=1}^{n_i} m_{ij}(X_{ij} - \bar{X}_i)^2}{m_i\,(n_i - 1)}$$

which is unbiased since $E[(\bar{X}_i - \mu)^2] = Var(\bar{X}_i) = a + v/m_i$ and $E(\hat{v}_i) = v$.

Example 5.48 *For a group policyholder, we have the following data available:*

	Year 1	Year 2	Year 3
Total claims	60,000	70,000	–
No. in group	125	150	200

If the manual rate per person is 500 per year, estimate the total credibility premium for year 3.

In the above notation, we have (assuming for notational purposes that this group is policyholder i) $m_{i1} = 125$, $X_{i1} = 60{,}000/125 = 480$, $m_{i2} = 150$, $X_{i2} = 70{,}000/150 = 466.67$, $m_i = m_{i1} + m_{i2} = 275$, and $\bar{X}_i = (60{,}000 + 70{,}000)/275 = 472.73$. Then

$$\hat{v}_i = \frac{125(480 - 472.73)^2 + 150(466.67 - 472.73)^2}{2 - 1} = 12{,}115.15$$

and with $\mu = 500$, $\tilde{a}_i = (472.73 - 500)^2 - (12{,}115.15/275) = 699.60$. We then estimate k by $\hat{v}_i/\tilde{a}_i = 17.32$. The estimated credibility factor is $m_i/(m_i + \hat{v}_i/\tilde{a}_i) = 275/(275 + 17.32) = 0.94$. The estimated credibility premium per person is then $0.94(472.73) + 0.06(500) = 474.37$ and the estimated total credibility premium for year 3 is $200(474.37) = 94{,}874$. □

It is instructive to note that estimation of the parameters a and v based on data from a single policyholder as in the example above is not advised unless there is no alternative, since the estimators $\hat{v}_i$ and $\tilde{a}_i$ have high variability. In particular, we are effectively estimating a from one observation ($\bar{X}_i$). It is strongly suggested that an attempt be made to obtain more data.

5.5.2 Semiparametric estimation

In some situations it may be reasonable to assume a parametric form for the conditional distribution $f_{X_{ij}|\Theta}(x_{ij}|\theta_i)$. The situation at hand may suggest that such an assumption is reasonable, or prior information may imply its appropriateness.

For example, in dealing with numbers of claims, it may be reasonable to assume that the number of claims $m_{ij}X_{ij}$ for policyholder i in year j is Poisson distributed with mean $m_{ij}\theta_i$, given $\Theta_i = \theta_i$. Thus $E(m_{ij}X_{ij}|\Theta_i) = Var(m_{ij}X_{ij}|\Theta_i) = m_{ij}\Theta_i$, implying that $\mu(\Theta_i) = v(\Theta_i) = \Theta_i$ and so $\mu = v$ in this case. Rather than use (5.79) to estimate v, we could use $\hat{\mu} = \bar{X}$ to estimate v.

Example 5.49 *In the past year, the distribution of automobile insurance policyholders by number of claims is given below.*

No. of claims	No. of insureds
0	1,563
1	271
2	32
3	7
4	2
Total	1,875

For each policyholder, obtain a credibility estimate for the number of claims next year based on the past year's experience, assuming a (conditional) Poisson distribution of number of claims for each policyholder.

Assume that we have r = 1,875 policyholders, $n_i = 1$ year experience on each, and exposures $m_{ij} = 1$. For policyholder i (where $i = 1, \ldots, 1{,}875$) assume that $X_{i1}|\Theta_i = \theta_i$ is Poisson distributed with mean θ_i so that $\mu(\theta_i) = v(\theta_i) = \theta_i$ and $\mu = v$. As in Example 5.45,

$$\begin{aligned}\bar{X} &= \frac{1}{1{,}875}\left(\sum_{i=1}^{1{,}875} X_{i1}\right) \\ &= \frac{0(1{,}563) + 1(271) + 2(32) + 3(7) + 4(2)}{1{,}875} = 0.194.\end{aligned}$$

Now,

$$\begin{aligned}Var(X_{i1}) &= Var[E(X_{i1}|\Theta_i)] + E[Var(X_{i1}|\Theta_i)] \\ &= Var[\mu(\Theta_i)] + E[v(\Theta_i)] = a + v = a + \mu.\end{aligned}$$

Thus an unbiased estimator of $a + v$ is the sample variance

$$\begin{aligned}\frac{\sum_{i=1}^{1{,}875}\left(X_{i1} - \bar{X}\right)^2}{1{,}874} &= \frac{\begin{array}{c}1{,}563(0 - 0.194)^2 + 271(1 - 0.194)^2 \\ +32(2 - 0.194)^2 + 7(3 - 0.194)^2 + 2(4 - 0.194)^2\end{array}}{1{,}874} \\ &= 0.226.\end{aligned}$$

Thus $\hat{a} = 0.226 - 0.194 = 0.032$, and $\hat{k} = 0.194/0.032 = 6.06$ and the credibility factor Z is $1/(1 + 6.06) = 0.14$. The estimated credibility premium for the number of claims for each policyholder is $(0.14)X_{i1} + (0.86)(0.194)$, where X_{i1} is 0, 1, 2, 3, or 4, depending on the policyholder. □

Note that in this case, $v = \mu$ identically, so that only one year's experience per policyholder is needed.

Example 5.50 *Suppose we are interested in the probability that an individual in a group makes a claim (e.g. group life insurance), and the probability is believed to vary by policyholder. Then $m_{ij}X_{ij}$ could represent the number of the m_{ij} individuals in year j for policyholder i who made a claim. Develop a credibility model for this situation.*

If the claim probability is θ_i for policyholder i, then a reasonable model to describe this effect is that $m_{ij}X_{ij}$ is binomially distributed with parameters m_{ij} and θ_i, given $\Theta_i = \theta_i$. Then

$$E(m_{ij}X_{ij}|\Theta_i) = m_{ij}\Theta_i \quad \text{and} \quad Var(m_{ij}X_{ij}|\Theta_i) = m_{ij}\Theta_i(1-\Theta_i)$$

and so $\mu(\Theta_i) = \Theta_i$ with $v(\Theta_i) = \Theta_i(1-\Theta_i)$. Thus

$$\begin{aligned} \mu &= E(\Theta_i), \quad v = \mu - E[(\Theta_i)^2], \quad \text{and} \\ a &= Var(\Theta_i) = E[(\Theta_i)^2] - \mu^2 = \mu - v - \mu^2. \end{aligned}$$

□

In these and other examples, there is a functional relationship between the parameters μ, v, and a which follows from the parametric assumptions made, and this often facilitates estimation of parameters.

5.5.3 Parametric estimation

If fully parametric assumptions are made with respect to $f_{X_{ij}|\Theta}(x_{ij}|\theta_i)$ and $\pi(\theta_i)$ for $i = 1, \ldots, r$ and $j = 1, \ldots, n_i$, then the full battery of parametric estimation techniques are available in addition to the nonparametric methods discussed earlier. In particular, maximum likelihood estimation is straightforward (at least in principle) and is now discussed. For policyholder i, the joint density of $\mathbf{X}_i = (X_{i1}, \ldots, X_{in_i})'$ is, by conditioning on Θ_i, given for $i = 1, \ldots, r$ by

$$f_{\mathbf{X}_i}(\mathbf{x}_i) = \int \left\{ \prod_{j=1}^{n_i} f_{X_{ij}|\Theta}(x_{ij}|\theta_i) \right\} \pi(\theta_i) d\theta_i. \tag{5.83}$$

The likelihood function is given by

$$L = \prod_{i=1}^{r} f_{\mathbf{X}_i}(\mathbf{x}_i). \tag{5.84}$$

Maximum likelihood estimators of the parameters are then chosen to maximize L or equivalently $\log L$.

Example 5.51 *As a simple example, suppose that $n_i = n$ for $i = 1, \ldots, r$ and $m_{ij} = 1$. Let $X_{ij}|\Theta_i$ be Poisson distributed with mean Θ_i, that is*

$$f_{X_{ij}|\Theta}(x_{ij}|\theta_i) = \frac{\theta_i^{x_{ij}} e^{-\theta_i}}{x_{ij}!}, \qquad x_{ij} = 0, 1, \ldots$$

and let Θ_i be exponentially distributed with mean μ,

$$\pi(\theta_i) = \frac{1}{\mu} e^{-\theta_i/\mu}, \qquad \theta_i > 0.$$

Determine the mle of μ.

Equation (5.83) becomes

$$\begin{aligned}
f_{\mathbf{X}_i}(\mathbf{x}_i) &= \int_0^\infty \left(\prod_{j=1}^n \frac{\theta_i^{x_{ij}} e^{-\theta_i}}{x_{ij}!} \right) \frac{1}{\mu} e^{-\theta_i/\mu} d\theta_i \\
&= \left(\prod_{j=1}^n x_{ij}! \right)^{-1} \frac{1}{\mu} \int_0^\infty \theta_i^{\sum_{j=1}^n x_{ij}} e^{-\theta_i(n+1/\mu)} d\theta_i \\
&= C(\mathbf{x}_i)\mu^{-1} \left(n + \frac{1}{\mu} \right)^{-\sum_{j=1}^n x_{ij} - 1} \int_0^\infty \frac{\beta \left(\beta\theta_i\right)^{\alpha-1} e^{-\beta\theta_i}}{\Gamma(\alpha)} d\theta_i
\end{aligned}$$

where

$$C(\mathbf{x}_i) = \binom{\sum_{j=1}^n x_{ij}}{x_{i1}\ x_{i2}\ \cdots\ x_{in}}$$

in combinatorial notation,

$$\beta = n + \frac{1}{\mu}$$

and

$$\alpha = \sum_{j=1}^n x_{ij} + 1.$$

The integral is that of a gamma density with parameters α and $1/\beta$ and therefore equals 1, and so

$$f(\mathbf{x}_i) = C(\mathbf{x}_i)\mu^{-1} \left(n + \frac{1}{\mu} \right)^{-\sum_{j=1}^n x_{ij} - 1}.$$

Substitution into (5.84) yields

$$L(\mu) \propto \mu^{-r} \left(n + \frac{1}{\mu} \right)^{-\sum_{i=1}^r \sum_{j=1}^n x_{ij} - r}.$$

Thus

$$l(\mu) = \log L(\mu) = -r \log \mu - \left(r + \sum_{i=1}^r \sum_{j=1}^n x_{ij} \right) \log \left(n + \frac{1}{\mu} \right) + c$$

where c is a constant which does not depend on μ. Differentiating yields

$$l'(\mu) = -\frac{r}{\mu} - \frac{r + \sum_{i=1}^{r}\sum_{j=1}^{n} x_{ij}}{n + \frac{1}{\mu}}\left(-\frac{1}{\mu^2}\right).$$

The mle $\hat{\mu}$ of μ is found by setting $l'(\hat{\mu}) = 0$, which yields

$$\frac{r}{\hat{\mu}} = \frac{r + \sum_{i=1}^{r}\sum_{j=1}^{n} x_{ij}}{\hat{\mu}(\hat{\mu}n + 1)}$$

and so

$$\hat{\mu}n + 1 = 1 + \frac{1}{r}\sum_{i=1}^{r}\sum_{j=1}^{n} x_{ij}$$

or

$$\hat{\mu} = \frac{1}{nr}\sum_{i=1}^{r}\sum_{j=1}^{n} x_{ij}.$$

But this is the same as the nonparametric estimate obtained in Example 5.45. An explanation is in order. We have $\mu(\theta_i) = \theta_i$ by the Poisson assumption and so $E[\mu(\Theta_i)] = E(\Theta_i)$ which is the same μ as was used in the exponential distribution $\pi(\theta_i)$.

Furthermore, $v(\theta_i) = \theta_i$ as well (by the Poisson assumption), and so $v = E[v(\Theta_i)] = \mu$. Also, $a = Var[\mu(\Theta_i)] = Var(\Theta_i) = \mu^2$ by the exponential assumption for $\pi(\theta_i)$. Thus the mles of v and a are $\hat{\mu}$ and $\hat{\mu}^2$ by the invariance of maximum likelihood estimation under a parameter transformation. Similarly, the mles of $k = v/a$, the credibility factor Z, and the credibility premium $Z\bar{X}_i + (1-Z)\mu$ are $\hat{k} = \hat{\mu}^{-1} = \bar{X}^{-1}$, $\hat{Z} = n/(n+\hat{\mu}^{-1})$, and $\hat{Z}\bar{X}_i + (1-\hat{Z})\hat{\mu}$, respectively. We mention also that credibility is exact in this model so that the Bayesian premium is equal to the credibility premium. □

Example 5.52 *Suppose that $n_i = n$ for all i and $m_{ij} = 1$. Assume that $X_{ij}|\Theta_i \sim N(\Theta_i, v)$,*

$$f_{X_{ij}|\Theta}(x_{ij}|\theta_i) = (2\pi v)^{-1/2}\exp\left[-\frac{1}{2v}(x_{ij} - \theta_i)^2\right], \qquad -\infty < x_{ij} < \infty$$

and $\Theta_i \sim N(\mu, a)$, so that

$$\pi(\theta_i) = (2\pi a)^{-1/2}\exp\left[-\frac{1}{2a}(\theta_i - \mu)^2\right], \qquad -\infty < \theta_i < \infty.$$

Determine the maximum likelihood estimators of the parameters.

We have $\mu(\theta_i) = \theta_i$ and $v(\theta_i) = v$. Thus $\mu = E[\mu(\Theta_i)]$, $v = E[v(\Theta_i)]$, and $a = Var[\mu(\Theta_i)]$, consistent with previous use of μ, v, and a. We shall now derive maximum likelihood estimators of μ, v, and a. To begin with, consider $\bar{X}_i = n^{-1}\sum_{j=1}^{n} X_{ij}$. Conditional on Θ_i, the X_{ij} are independent $N(\Theta_i, v)$ random variables, implying that $\bar{X}_i|\Theta_i \sim N(\Theta_i, v/n)$. Since $\Theta_i \sim N(\mu, a)$, it follows from Example 5.3 that unconditionally $\bar{X}_i \sim N(\mu, a + v/n)$. Hence the density of $\bar{X}_i$ is, with $w = a + v/n$,

$$f(\bar{x}_i) = (2\pi w)^{-1/2} \exp\left[-\frac{1}{2w}(\bar{x}_i - \mu)^2\right], \qquad -\infty < \bar{x}_i < \infty.$$

On the other hand, by conditioning on Θ_i, we have

$$\begin{aligned} f(\bar{x}_i) &= \int_{-\infty}^{\infty} (2\pi v/n)^{-1/2} \exp\left[-\frac{n}{2v}(\bar{x}_i - \theta_i)^2\right] \\ &\quad \times (2\pi a)^{-1/2} \exp\left[-\frac{1}{2a}(\theta_i - \mu)^2\right] d\theta_i. \end{aligned}$$

Ignoring terms not involving μ, v, or a, this means that $f(\bar{x}_i)$ is proportional to

$$v^{-1/2}a^{-1/2} \int_{-\infty}^{\infty} \exp\left[-\frac{n}{2v}(\bar{x}_i - \theta_i)^2 - \frac{1}{2a}(\theta_i - \mu)^2\right] d\theta_i.$$

Now (5.83) yields

$$\begin{aligned} f(\mathbf{x}_i) &= \int_{-\infty}^{\infty} \left\{\prod_{j=1}^{n} (2\pi v)^{-1/2} \exp\left[-\frac{1}{2v}(x_{ij} - \theta_i)^2\right]\right\} (2\pi a)^{-1/2} \\ &\quad \times \exp\left[-\frac{1}{2a}(\theta_i - \mu)^2\right] d\theta_i \end{aligned}$$

which is proportional to

$$v^{-n/2}a^{-1/2} \int_{-\infty}^{\infty} \exp\left[-\frac{1}{2v}\sum_{j=1}^{n}(x_{ij} - \theta_i)^2 - \frac{1}{2a}(\theta_i - \mu)^2\right] d\theta_i.$$

Now use the identity (5.14) restated as

$$\sum_{j=1}^{n}(x_{ij} - \theta_i)^2 = \sum_{j=1}^{n}(x_{ij} - \bar{x}_i)^2 + n(\bar{x}_i - \theta_i)^2$$

which means that $f(\mathbf{x}_i)$ is proportional to

$$v^{-n/2}a^{-1/2} \int_{-\infty}^{\infty} \exp\left\{-\frac{1}{2v}\left[\sum_{j=1}^{n}(x_{ij} - \bar{x}_i)^2 + n(\bar{x}_i - \theta_i)^2\right] - \frac{1}{2a}(\theta_i - \mu)^2\right\} d\theta_i$$

itself proportional to

$$v^{-(n-1)/2} \exp\left[-\frac{1}{2v}\sum_{j=1}^{n}(x_{ij}-\bar{x}_i)^2\right] f(\bar{x}_i)$$

using the second expression for the density $f(\bar{x}_i)$ of $\bar{X}_i$ given above. Then (5.84) yields

$$L \propto v^{-r(n-1)/2} \exp\left[-\frac{1}{2v}\sum_{i=1}^{r}\sum_{j=1}^{n}(x_{ij}-\bar{x}_i)^2\right] \prod_{i=1}^{r} f(\bar{x}_i).$$

Let us now invoke the invariance of mles under a parameter transformation and use μ, v, and $w = a + v/n$ rather than μ, v, and a. This means that

$$L \propto L_1(v)L_2(\mu, w)$$

where

$$L_1(v) = v^{-r(n-1)/2} \exp\left[-\frac{1}{2v}\sum_{i=1}^{r}\sum_{j=1}^{n}(x_{ij}-\bar{x}_i)^2\right]$$

and

$$L_2(\mu, w) = \prod_{i=1}^{r} f(\bar{x}_i) = \prod_{i=1}^{r}\left\{(2\pi w)^{-1/2} \exp\left[-\frac{1}{2w}(\bar{x}_i-\mu)^2\right]\right\}.$$

The mle $\hat{v}$ of v can be found by maximizing $L_1(v)$ alone and the mle $(\hat{\mu}, \hat{w})$ of (μ, w) can be found by maximizing $L_2(\mu, w)$. Taking logarithms, we obtain

$$\begin{aligned} l_1(v) &= -\frac{r(n-1)}{2}\log v - \frac{1}{2v}\sum_{i=1}^{r}\sum_{j=1}^{n}(x_{ij}-\bar{x}_i)^2 \\ l_1'(v) &= -\frac{r(n-1)}{2v} + \frac{1}{2v^2}\sum_{i=1}^{r}\sum_{j=1}^{n}(x_{ij}-\bar{x}_i)^2 \end{aligned}$$

and with $l'(\hat{v}) = 0$ we have

$$\hat{v} = \frac{\sum_{i=1}^{r}\sum_{j=1}^{n}(X_{ij}-\bar{X}_i)^2}{r(n-1)}.$$

Since $L_2(\mu, w)$ is the usual normal likelihood, the mles are simply the empirical mean and variance (as shown in Subsection 5.2.4). That is,

$$\hat{\mu} = \frac{1}{r}\sum_{i=1}^{r}\bar{X}_i = \frac{1}{nr}\sum_{i=1}^{r}\sum_{j=1}^{n}X_{ij} = \bar{X}$$

and

$$\hat{w} = \frac{1}{r}\sum_{i=1}^{r}(\bar{X}_i - \bar{X})^2.$$

But $a = w - v/n$ and so the mle of a is

$$\hat{a} = \frac{1}{r}\sum_{i=1}^{r}(\bar{X}_i - \bar{X})^2 - \frac{1}{rn(n-1)}\sum_{i=1}^{r}\sum_{j=1}^{n}(X_{ij} - \bar{X}_i)^2.$$

It is instructive to note that the mles $\hat{\mu}$ and $\hat{v}$ are exactly the nonparametric unbiased estimators in the Bühlmann model of Example 5.45. The mle $\hat{a}$ is almost the same as the nonparametric unbiased estimator, the only difference being the divisor r rather than $r-1$ in the first term. □

5.5.4 Notes and References

In this section a simple approach was employed to find parameter estimates. No attempt was made to find optimum estimators in the sense of minimum variance. A good deal of research has been done on this problem. See Goovaerts and Hoogstad [52] for more details and further references.

5.6 EXERCISES

Section 5.2

5.1 Suppose X is binomially distributed with parameters n_1 and p, that is

$$f_X(x) = \binom{n_1}{x}p^x(1-p)^{n_1-x}, \qquad x = 0, 1, 2, \ldots, n_1.$$

Suppose also that Z is binomially distributed with parameters n_2 and p independently of X. Then $Y = X + Z$ is binomially distributed with parameters $n_1 + n_2$ and p. Find the conditional distribution of X given that $Y = y$.

5.2 Let X and Y have joint probability distribution as follows

	y		
x	0	1	2
0	0.20	0	0.10
1	0	0.15	0.25
2	0.05	0.15	0.10

(a) Compute the marginal distributions of X and Y.

(b) Compute the conditional distribution of X given $Y = y$ for $y = 0, 1, 2$.

(c) Compute $E(X|y), E(X^2|y)$, and $Var(X|y)$ for $y = 0, 1, 2$.

(d) Compute $E(X)$ and $Var(X)$ using (5.5), (5.8), and (c).

5.3 Suppose that X and Y are two random variables with bivariate normal joint density function

$$f_{X,Y}(x,y) = \frac{1}{2\pi\sigma_1\sigma_2\sqrt{1-\rho^2}} \times \exp\left\{-\frac{1}{2(1-\rho^2)}\left[\left(\frac{x-\mu_1}{\sigma_1}\right)^2 - 2\rho\left(\frac{x-\mu_1}{\sigma_1}\right)\left(\frac{y-\mu_2}{\sigma_2}\right) + \left(\frac{y-\mu_2}{\sigma_2}\right)^2\right]\right\}.$$

Show that

(a) The conditional density function is

$$f_{X|Y}(x|y) = \frac{1}{\sqrt{2\pi}\sigma_1\sqrt{1-\rho^2}} \exp\left\{-\frac{1}{2}\left[\frac{x-\mu_1-\rho\frac{\sigma_1}{\sigma_2}(y-\mu_2)}{\sigma_1\sqrt{1-\rho^2}}\right]^2\right\}.$$

Hence $E(X|Y=y) = \mu_1 + \rho\frac{\sigma_1}{\sigma_2}(y-\mu_2)$.

(b) The marginal pdf is

$$f_X(x) = \frac{1}{\sqrt{2\pi}\sigma_1} \exp\left[-\frac{1}{2}\left(\frac{x-\mu_1}{\sigma_1}\right)^2\right].$$

(c) X and Y are independent if and only if $\rho = 0$.

5.4 Suppose that the random variables $Y_1, \cdots, Y_n$ are independent with

$$E(Y_j) = \gamma \quad \text{and} \quad Var(Y_j) = a_j + \sigma^2/b_j, \qquad j = 1, 2, \ldots, n.$$

Define $b = b_1 + b_2 + \cdots + b_n$ and $\bar{Y} = \sum_{j=1}^{n} \frac{b_j}{b} Y_j$. Prove that

$$E\left[\sum_{j=1}^{n} b_j (Y_j - \overline{Y})^2\right] = (n-1)\sigma^2 + \sum_{j=1}^{n} a_j\left(b_j - \frac{b_j^2}{b}\right).$$

5.5 Consider the inverse Gaussian distribution with density given by

$$f_X(x) = \left(\frac{\theta}{2\pi x^3}\right)^{1/2} \exp\left[-\frac{\theta}{2x}\left(\frac{x-\mu}{\mu}\right)^2\right], \qquad x > 0.$$

(a) Show that

$$\sum_{j=1}^{n} \frac{(x_j-\mu)^2}{x_j} = \mu^2 \sum_{j=1}^{n}\left(\frac{1}{x_j} - \frac{1}{\overline{x}}\right) + \frac{n}{\overline{x}}(\overline{x}-\mu)^2$$

where $\overline{x} = \frac{1}{n}\sum_{j=1}^{n} x_j$.

(b) For a sample $(x_1, \cdots, x_n)$, show that the maximum likelihood estimates of θ and μ are

$$\hat{\mu} = \bar{x}$$

and

$$\hat{\theta} = \frac{n}{\sum_{j=1}^{n}\left(\frac{1}{x_j} - \frac{1}{\overline{x}}\right)}.$$

5.6 Suppose that $X_1, \ldots, X_n$ are independent and normally distributed with mean $E(X_j) = \mu$ and $Var(X_j) = (\theta m_j)^{-1}$, where $m_j > 0$ is a known constant. Prove that the maximum likelihood estimates of μ and θ are

$$\hat{\mu} = \bar{X}$$

and

$$\hat{\theta} = n\left[\sum_{j=1}^{n} m_j (X_j - \bar{X})^2\right]^{-1}$$

where $\bar{X} = \frac{1}{m}\sum_{j=1}^{n} m_j X_j$ and $m = \sum_{j=1}^{n} m_j$.

5.7 Suppose that given $\Theta = \theta$, the random variables $X_1, \ldots, X_n$ are independent and binomially distributed with pf

$$f_{X_j|\Theta}(x_j|\theta) = \binom{K_j}{x_j}\theta^{x_j}(1-\theta)^{K_j - x_j}, \qquad x_j = 0, 1, \ldots, K_j$$

and Θ itself is beta distributed with parameters a and b and pdf

$$\pi(\theta) = \frac{\Gamma(a+b)}{\Gamma(a)\Gamma(b)}\theta^{a-1}(1-\theta)^{b-1}, \qquad 0 < \theta < 1.$$

(a) Verify that the marginal pf of X_j is

$$f_{X_j}(x_j) = \frac{\binom{-a}{x_j}\binom{-b}{K_j - x_j}}{\binom{-a-b}{K_j}}, \qquad x_j = 0, 1, \ldots, K_j$$

and $E(X_j) = aK_j/(a+b)$. This distribution is termed the binomial–beta or negative hypergeometric distribution.

(b) Determine the posterior pdf $\pi_{\Theta|\mathbf{X}}(\theta|\mathbf{x})$ and the posterior mean $E(\Theta|\mathbf{x})$.

5.8 Suppose that given $\Theta = \theta$, the random variables $X_1, \ldots, X_n$ are independent and identically exponentially distributed with pdf

$$f_{X_j|\Theta}(x_j|\theta) = \theta e^{-\theta x_j}, \qquad x_j > 0$$

and Θ is itself gamma distributed with parameters $\alpha > 1$ and $\beta > 0$,

$$\pi(\theta) = \frac{\theta^{\alpha-1} e^{-\theta/\beta}}{\Gamma(\alpha)\beta^{\alpha}}, \qquad \theta > 0.$$

(a) Verify that the marginal pdf of X_j is

$$f_{X_j}(x_j) = \alpha\beta^{-\alpha}(\beta^{-1} + x_j)^{-\alpha-1}, \qquad x_j > 0$$

and

$$E(X_j) = \frac{1}{\beta(\alpha - 1)}.$$

This distribution is one form of the Pareto distribution.

(b) Determine the posterior pdf $\pi_{\Theta|\mathbf{X}}(\theta|\mathbf{x})$ and the posterior mean $E(\Theta|\mathbf{x})$.

5.9 Suppose that given $\Theta = \theta$, the random variables $X_1, \ldots, X_n$ are independent and identically negative binomially distributed with parameters r and θ with pf

$$f_{X_j|\Theta}(x_j|\theta) = \binom{r + x_j - 1}{x_j}\theta^r(1-\theta)^{x_j}, \qquad x_j = 0, 1, 2, \ldots$$

and Θ itself is beta distributed with parameters a and b and pdf

$$\pi(\theta) = \frac{\Gamma(a+b)}{\Gamma(a)\Gamma(b)}\theta^{a-1}(1-\theta)^{b-1}, \quad 0 < \theta < 1.$$

(a) Verify that the marginal pf of X_j is

$$f_{X_j}(x_j) = \frac{\Gamma(r + x_j)}{\Gamma(r)x_j!}\,\frac{\Gamma(a+b)}{\Gamma(a)\Gamma(b)}\,\frac{\Gamma(a+r)\Gamma(b+x_j)}{\Gamma(a+r+b+x_j)}, \quad x_j = 0, 1, 2, \ldots$$

and

$$E(X_j) = \frac{rb}{a-1}.$$

This distribution is termed the **generalized Waring distribution.** The special case where $b = 1$ is the **Waring distribution** and the **Yule distribution** if $r = 1$ and $b = 1$.

(b) Determine the posterior pdf $f_{\Theta|\mathbf{X}}(\theta|\mathbf{x})$ and the posterior mean $E(\Theta|\mathbf{x})$.

5.10 Suppose that given $\Theta = \theta$, the random variables $X_1, \ldots, X_n$ are independent and identically normally distributed with mean μ and variance θ^{-1} and Θ is gamma distributed with parameters α and (θ replaced by) $1/\beta$.

(a) Verify that the marginal pdf of X_j is

$$f_{X_j}(x_j) = \frac{\Gamma(\alpha + \frac{1}{2})}{\sqrt{2\pi/\beta}\,\Gamma(\alpha)} \left[1 + \frac{\beta}{2}(x_j - \mu)^2\right]^{-\alpha-1/2}, \qquad -\infty < x_j < \infty$$

which is a form of the t-distribution.

(b) Determine the posterior pdf $f_{\Theta|\mathbf{X}}(\theta|\mathbf{x})$ and the posterior mean $E(\theta|\mathbf{x})$.

5.11 Suppose that given $\Theta_1 = \theta_1$ and $\Theta_2 = \theta_2$, the random variables $X_1, \ldots,$ X_n are independent and identically normally distributed with mean θ_1 and variance θ_2^{-1}. Suppose also that the conditional distribution of Θ_1 given $\Theta_2 = \theta_2$ is a normal distribution with mean μ and variance σ^2/θ_2, and Θ_2 is gamma distributed with parameters α and $\theta = 1/\beta$.

(a) Show that the posterior conditional distribution of Θ_1 given $\Theta_2 = \theta_2$ is normally distributed with mean

$$\mu_* = \frac{1}{1 + n\sigma^2}\mu + \frac{n\sigma^2}{1 + n\sigma^2}\overline{x}$$

and variance

$$\sigma_*^2 = \frac{\sigma^2}{\theta_2(1 + n\sigma^2)}$$

and the posterior marginal distribution of Θ_2 is gamma distributed with parameters

$$\alpha_* = \alpha + \frac{n}{2}$$

and

$$\beta_* = \beta + \frac{1}{2}\sum_{i=1}^{n}(x_i - \overline{x})^2 + \frac{n\,(\overline{x} - \mu)^2}{2\,(1 + n\sigma^2)}.$$

(b) Find the posterior marginal means $E(\Theta_1|\mathbf{x})$ and $E(\Theta_2|\mathbf{x})$.

5.12 Prove that the binomial distribution with pf

$$f(x;p) = \binom{n}{x} p^x (1-p)^{n-x}$$

is of the form (5.26), and identify θ, $p(x)$, and $q(\theta)$.

5.13 Consider the negative binomial distribution with pf

$$f(x;\alpha,\beta) = \frac{\Gamma(\alpha+x)}{\Gamma(\alpha)x!}\left(\frac{\beta}{1+\beta}\right)^{\alpha}\left(\frac{1}{1+\beta}\right)^{x}$$

If α is fixed, show that $f(x;\alpha,\beta)$ is of the form (5.26) and identify θ, $p(x)$, and $q(\theta)$.

5.14 Suppose $X_1, \ldots, X_n$ are independent and identically distributed with distribution (5.26). Prove that the maximum likelihood estimate of the mean is the sample mean. In other words, if $\hat{\theta}$ is the mle of θ, prove that

$$\widehat{\mu(\theta)} = \mu(\hat{\theta}) = \bar{X}.$$

5.15 Consider the generalization of (5.26) given by

$$f(x;\theta) = \frac{p(m,x)e^{-m\theta x}}{[q(\theta)]^m}$$

where m is also a parameter. Prove that the mean is still given by (5.28), but the variance is given by $v(\theta)/m$ where $v(\theta)$ is given by (5.29).

5.16 Prove Theorem 5.1 by completing the following two steps.

(a) Show that the marginal pf of X_j is

$$f_{X_j}(x_j) = p(x_j)\frac{c\left(\dfrac{x_j+\mu k}{k+1}, k+1\right)}{c(\mu,k)}.$$

(b) Show that the posterior distribution $\pi_{\Theta|\mathbf{X}}(\theta|\mathbf{x})$ is of the same form as $\pi(\theta)$ and identify the parameters.

5.17 Suppose that given $\mathbf{\Theta} = (\Theta_1, \Theta_2)$, the random variable X is normally distributed with mean Θ_1 and variance Θ_2.

(a) Show that $E(X) = E(\Theta_1)$ and $Var(X) = E(\Theta_2) + Var(\Theta_1)$.

(b) If Θ_1 and Θ_2 are independent, show that X has the same distribution as $\Theta_1 + Y$, where Θ_1 and Y are independent and Y conditional on Θ_2 is normally distributed with mean 0 and variance Θ_2.

5.18 Suppose that Θ has pdf $\pi(\theta)$, $\theta > 0$ and Θ_1 has pdf $\pi_1(\theta) = \pi(\theta - \alpha)$, $\theta > \alpha > 0$. If, given Θ_1, X is Poisson distributed with mean Θ_1, show that X has the same distribution as $Y + Z$, where Y and Z are independent, Y is Poisson distributed with mean α, and $Z|\Theta$ is Poisson distributed with mean Θ.

5.19 Let $X_1, \ldots, X_n$ be independent and identically distributed conditional on Θ, with pf

$$f_{X_j|\Theta}(x_j|\theta) = \frac{p(x_j)e^{-\theta x_j}}{q(\theta)}.$$

Let $S = X_1 + \cdots + X_n$.

(a) Show that, conditional on Θ, S has pf of the form

$$f_{S|\Theta}(s|\theta) = \frac{p_n(s)e^{-\theta s}}{[q(\theta)]^n}$$

where $p_n(s)$ does not depend on θ.

(b) Prove that the posterior distribution $\pi_{\Theta|\mathbf{X}}(\theta|\mathbf{x})$ is the same as the (conditional) distribution of $\Theta|S$,

$$\pi_{\Theta|\mathbf{X}}(\theta|\mathbf{x}) = \frac{f_{S|\Theta}(s|\theta)\pi(\theta)}{f_S(s)}$$

where $\pi(\theta)$ is the pf of Θ and $f_S(s)$ is the marginal pf of S.

5.20 Suppose that given N, the random variable X is binomially distributed with parameters N and p.

(a) Show that if N is Poisson distributed, so is X (unconditionally) and identify the parameters.

(b) Show that if N is binomially distributed, so is X (unconditionally) and identify the parameters.

(c) Show that if N is negative binomially distributed, so is X (unconditionally) and identify the parameters.

5.21 A die is selected at random from an urn that contains two six-sided dice. Die number 1 has three faces with the number 2 while one face each has the numbers 1, 3, and 4. Die number 2 has three faces with the number 4 while one face each has the numbers 1, 2, and 3. The first five rolls of the die yielded the numbers 2, 3, 4, 1, and 4 in that order. Determine the probability that the selected die was die number 2. (89-28)

5.22 The number of claims in a year, Y, has a distribution which depends on a parameter θ. As a random variable, Θ has the uniform distribution on the interval $(0, 1)$. The unconditional probability that Y is 0 is greater than

0.35. For each conditional pf given below, determine if it is possible that it is the true conditional pf of Y. (89-31)

(a) $\Pr(Y = y|\theta) = e^{-\theta}\theta^y/y!$.
(b) $\Pr(Y = y|\theta) = (y+1)\theta^2(1-\theta)^y$.
(c) $\Pr(Y = y|\theta) = \binom{2}{y}\theta^y(1-\theta)^{2-y}$.

5.23 Your prior distribution concerning the unknown value of H is $\Pr(H = 1/4) = 4/5$ and $\Pr(H = 1/2) = 1/5$. The observation from a single experiment has distribution $\Pr(D = d|H = h) = h^d(1-h)^{1-d}$ for $d = 0, 1$. The result of a single experiment is $d = 1$. Determine the posterior distribution of H. (89-38)

5.24 The number of claims in one year, Y, has the Poisson distribution with parameter θ. The parameter θ has the exponential distribution with pdf, $\pi(\theta) = e^{-\theta}$. A particular insured had no claims in one year. Determine the posterior distribution of θ for this insured. (89-40)

5.25 The number of claims in one year, Y, has the Poisson distribution with parameter θ. The prior distribution has the gamma distribution with pdf $\pi(\theta) = \theta e^{-\theta}$. There was one claim in one year. Determine the posterior pdf of θ. (90-46)

5.26 Each individual car's claim count has a Poisson distribution with parameter λ. All individual cars have the same parameter. The prior distribution is gamma with parameters $\alpha = 50$ and $\theta = 1/500$. In a two year period, the insurer covers 750 and 1,100 cars for years one and two, respectively. There were 65 and 112 claims in the two years, respectively. Determine the coefficient of variation of the posterior gamma distribution. (90-48)

5.27 The number of claims, r, made by an individual in one year has the binomial distribution with pf $f(r) = \binom{3}{r}\theta^r(1-\theta)^{3-r}$. The prior distribution for θ has pdf $\pi(\theta) = 6(\theta - \theta^2)$. There was one claim in a one-year period. Determine the posterior pdf of θ. (91-32)

5.28 The number of claims for an individual in one year has a Poisson distribution with parameter λ. The prior distribution for λ has the gamma distribution with mean .14 and variance 0.0004. During the past two years a total of 110 claims has been observed. In each year there were 310 policies in force. Determine the expected value and variance of the posterior distribution of λ. (M92-11)

5.29 The number of claims for an individual in one year has a Poisson distribution with parameter λ. The prior distribution for λ is exponential with an expected value of 2. There were three claims in the first year. Determine the posterior distribution of λ. (M92-28)

5.30 The number of claims in one year has the binomial distribution with $n = 3$ and θ unknown. The prior distribution for θ is beta with pdf $\pi(\theta) = 280\theta^3(1-\theta)^4, \quad 0 < \theta < 1$. Two claims were observed. Determine each of the following. (N92-27)

(a) The posterior distribution of θ.

(b) The expected value of θ from the posterior distribution.

5.31 An individual risk has exactly one claim each year. The amount of the single claim has an exponential distribution with pdf $f(x) = te^{-tx}, \ x > 0$. The parameter t has a prior distribution with pdf $\pi(t) = te^{-t}$. A claim of 5 has been observed. Determine the posterior pdf of t. (N93-6)

Section 5.3

5.32 An insurance company has decided to establish its full credibility requirements for an individual state rate filing. The full credibility standard is to be set so that the observed total amount of claims underlying the rate filing would be within 5% of the true value with probability 0.95. The claim frequency follows a Poisson distribution and the severity distribution has pdf

$$f(x) = \frac{100 - x}{5{,}000}, \qquad 0 \le x \le 100.$$

Determine the expected number of claims necessary to obtain full credibility using the normal approximation.

5.33 For a particular policyholder, the past total claims experience is given by $X_1, \ldots, X_n$, where the X_js are independent and identically distributed compound random variables with Poisson parameter λ and gamma claim size distribution with pdf

$$f_Y(y) = \frac{y^{\alpha-1}e^{-y/\beta}}{\Gamma(\alpha)\beta^\alpha}, \qquad y > 0.$$

You also know the following:

- The credibility factor based on numbers of claims is 0.9.
- The expected claim size $\alpha\beta = 100$.
- The credibility factor based on total claims is 0.8.

Determine α and β.

5.34 For a particular policyholder, the manual premium is 600 per year. The past claims experience is as follows:

Year	1	2	3
Claims	475	550	400

Assess whether full or partial credibility is appropriate and determine the net premium for next year's claims assuming the normal approximation. Use $r = 0.05$ and $p = 0.9$.

5.35 Redo Example 5.21 assuming that X_j is a compound negative binomial distribution rather than compound Poisson.

5.36 The total number of claims for a group of insureds is Poisson with mean λ. Determine the value of λ such that the observed number of claims will be within 3% of λ with a probability of 0.975 using the normal approximation. (89-29)

5.37 An insurance company is revising rates based on old data. The expected number of claims for full credibility is selected so that observed total claims will be within 5% of the true value 90% of the time. Individual claim amounts have pdf $f(x) = 1/200,000$, $0 < x < 200,000$ and the number of claims has the Poisson distribution. The recent experience consists of 1,082 claims. Determine the credibility, Z, to be assigned to the recent experience. Use the normal approximation. (89-30)

5.38 The average claim size for a group of insureds is 1,500 with a standard deviation of 7,500. Assume that claim counts have the Poisson distribution. Determine the expected number of claims so that the total loss will be within 6% of the expected total loss with probability 0.90. (91-22)

5.39 A group of insureds had 6,000 claims and a total loss of 15,600,000. The prior estimate of the total loss was 16,500,000. Determine the limited fluctuation credibility estimate of the total loss for the group. Use the standard for full credibility determined in Exercise 5.38. (91-23)

5.40 The full credibility standard is set so that the total number of claims is within 5% of the true value with probability p. This standard is 800 claims. The standard is then altered so that the total cost of claims is to be within 10% of the true value with probability p. The claim frequency has a Poisson distribution and the claim severity distribution has pdf $f(x) = 0.0002(100 - x)$, $0 < x < 100$. Determine the expected number of claims necessary to obtain full credibility under the new standard. (91-39)

5.41 A standard for full credibility of 1,000 claims has been selected so that the actual pure premium will be within 10% of the expected pure premium 95% of the time. The number of claims has the Poisson distribution. Determine the coefficient of variation of the severity distribution. (M92-1)

5.42 For a group of insureds you are given the following information:

- The prior estimate of expected total losses is 20,000,000.
- The observed total losses are 25,000,000.
- The observed number of claims is 10,000.
- The number of claims required for full credibility is 17,500.

Determine the credibility estimate of the group's expected total losses based upon all the above information. Use the credibility factor that is appropriate if the goal is to estimate the expected number of losses. (M92-6)

5.43 A full credibility standard is determined so that the total number of claims is within 5% of the expected number with probability 98%. If the same expected number of claims for full credibility is applied to the total cost of claims, the actual total cost would be within $100K$% of the expected cost with 95% probability. Individual claims have severity pdf $f(x) = 2.5x^{-3.5}$, $x > 1$. Determine K. (M92-16)

5.44 The number of claims has the Poisson distribution. The number of claims and the claim severity are independent. Individual claim amounts can be for 1, 2, or 10 with probabilities 0.5, 0.3, and 0.2, respectively. Determine the number of claims needed so that the total cost of claims is within 10% of the expected cost with 90% probability. (N92-1)

5.45 The number of claims has the Poisson distribution. The coefficient of variation of the severity distribution is 2. The standard for full credibility in estimating total claims is 3,415. With this standard the observed pure premium will be within k% of the expected pure premium 95% of the time. Determine k. (M93-10)

5.46 You are given the following:

- $P =$ Prior estimate of pure premium for a particular class of business.
- $O =$ Observed pure premium during the latest experience period for the same class of business.
- $R =$ Revised estimate of pure premium for the same class following the observations.
- $F =$ Number of claims required for full credibility of the pure premium.

Express the observed number of claims as a function of these four items. (N93-20)

Section 5.4

5.47 Consider a die-spinner model as in Example 5.27. The first die has 1 "marked" face and 5 "unmarked" faces whereas the second die has 4 "marked" faces and 2 "unmarked" faces. There are 3 spinners, each with 5 equally spaced sectors marked 3 or 8. The first spinner has 1 sector marked 3 and 4 marked 8, the second has 2 marked 3 and 3 marked 8, and the third has 4 marked 3 and 1 marked 8. One die and one spinner are selected at random. Claims are generated as in Example 5.27.

(a) Determine $\pi(\theta)$.

(b) Determine the conditional distributions $f_{X|\Theta}(x|\theta)$ for the claim sizes for each die-spinner combination.

(c) Determine the hypothetical means $\mu(\theta)$ and the process variances $v(\theta)$.

(d) Determine the marginal probability that the claim X_1 on the first iteration equals 3.

(e) Determine the posterior distribution $\pi_{\Theta|X_1}(\theta|3)$ of Θ using Bayes theorem.

(f) Use (5.41) to determine the conditional distribution $f_{X_2|X_1}(x_2|3)$ of the claims X_2 on the second iteration given that $X_1 = 3$ was observed on the first iteration.

(g) Use (5.44) to determine the Bayesian premium $E(X_2|X_1 = 3)$.

(h) Determine the joint probability that $X_2 = x_2$ and $X_1 = 3$ for $x_2 = 0$, 3, and 8.

(i) Determine the conditional distribution $f_{X_2|X_1}(x_2|3)$ directly using (5.40) and compare your answer to that of (f).

(j) Determine the Bayesian premium directly using (5.43) and compare your answer to that of (g).

(k) Determine the structural parameters μ, v, and a.

(l) Compute the Bühlmann credibility factor and the Bühlmann credibility premium to approximate the Bayesian premium $E(X_2|X_1 = 3)$.

5.48 Three urns have balls marked 0, 1, and 2 in the proportions given in Table 5.5.

An urn is selected at random, and 2 balls are drawn from that urn with replacement. A total of 2 on the 2 balls is observed. Two more balls are then

Table 5.5 Data for Exercise 5.48

Urn	0s	1s	2s
1	0.40	0.35	0.25
2	0.25	0.10	0.65
3	0.50	0.15	0.35

drawn with replacement from the same urn, and it is of interest to predict the total on these next 2 balls.

(a) Determine $\pi(\theta)$.

(b) Determine the conditional distributions $f_{X|\Theta}(x|\theta)$ for the totals on the 2 balls for each urn.

(c) Determine the hypothetical means $\mu(\theta)$ and the process variances $v(\theta)$.

(d) Determine the marginal probability that the total X_1 on the first 2 balls equals 2.

(e) Determine the posterior distribution $\pi_{\Theta|X_1}(\theta|2)$ using Bayes' theorem.

(f) Use (5.41) to determine the conditional distribution $f_{X_2|X_1}(x_2|2)$ of the total X_2 on the next 2 balls drawn given that $X_1 = 2$ was observed on the first 2 draws.

(g) Use (5.44) to determine the Bayesian premium $E(X_2|X_1 = 2)$.

(h) Determine the joint probability that the total X_2 on the next two balls equals x_2 and the total X_1 on the first two balls equals 2 for $x_2 = 0, 1, 2, 3, 4$.

(i) Determine the conditional distribution $f_{X_2|X_1}(x_2|2)$ directly using (5.40) and compare your answer to that of (f).

(j) Determine the Bayesian premium directly using (5.43) and compare your answer to that of (g).

(k) Determine the structural parameters μ, v, and a.

(l) Determine the Bühlmann credibility factor and the Bühlmann credibility premium.

(m) Show that the Bühlmann credibility factor is the same if each "exposure unit" consists of one draw from the urn rather than two draws.

5.49 Suppose that there are two types of policyholder: type A and type B. Two-thirds of the total number of the policyholders are of type A and one-third are of type B. For each type, the information on annual claim numbers and severity are given as follows:

Type	Number of claims		Severity	
	Mean	Variance	Mean	Variance
A	0.2	0.2	200	4,000
B	0.7	0.3	100	1,500

A policyholder has a total claim amount of 500 in the last four years. Determine the credibility factor Z and the credibility premium for next year for this policyholder.

5.50 Let Θ_1 represent the risk factor for claim numbers and let Θ_2 represent the risk factor for the claim severity for a line of insurance. Suppose that Θ_1 and Θ_2 are independent. Suppose also that given $\Theta_1 = \theta_1$, the claim number N is Poisson distributed and given $\Theta_2 = \theta_2$, the severity Y is exponentially distributed. The expectations of the hypothetical mean and process variance for the claim number and severity as well as the variance of the hypothetical means for frequency are respectively:

$$\begin{array}{lll} \mu_N = 0.1 & v_N = 0.1 & a_N = 0.05 \\ \mu_Y = 100 & v_Y = 25{,}000 & \end{array}$$

Three observations are made on a particular policyholder and we observe total claims of 200. Determine the Bühlmann credibility factor and the Bühlmann premium for this policyholder.

5.51 Suppose that $X_1, \ldots, X_n$ are independent (conditional on Θ) and that

$$E(X_j|\Theta) = \beta_j \mu(\Theta) \text{ and } Var(X_j|\Theta) = \tau_j(\Theta) + \psi_j v(\Theta), \qquad j = 1, \ldots, n.$$

Let

$$\mu = E[\mu(\Theta)], \quad v = E[v(\Theta)], \quad \tau_j = E[\tau_j(\Theta)], \quad \text{and} \quad a = Var[\mu(\Theta)].$$

(a) Show that

$$E(X_j) = \beta_j \mu, \; Var(X_j) = \tau_j + \psi_j v + \beta_j^2 a$$

and

$$Cov(X_i, X_j) = \beta_i \beta_j a, \qquad i \neq j.$$

(b) Solve the normal equations for $\tilde{\alpha}_0, \tilde{\alpha}_1, \ldots, \tilde{\alpha}_n$ to show that the credibility premium satisfies

$$\tilde{\alpha}_0 + \sum_{j=1}^{n} \tilde{\alpha}_j X_j = (1 - Z)\, E(X_{n+1}) + Z \beta_{n+1} \bar{X}$$

where

$$m_j = \beta_j^2(\tau_j + \psi_j v)^{-1}, \qquad j = 1, \ldots, n$$

$$m = m_1 + \cdots + m_n$$

$$Z = am(1 + am)^{-1}$$

$$\text{and } \bar{X} = \sum_{j=1}^{n} \frac{m_j}{m} \frac{X_j}{\beta_j}.$$

5.52 For the situation described in Exercise 5.7, determine $\mu(\theta)$ and the Bayesian premium $E(X_{n+1}|\mathbf{x})$. Why is the Bayesian premium equal to the credibility premium?

5.53 For the situation described in Exercise 5.8, determine $\mu(\theta)$ and the Bayesian premium $E(X_{n+1}|\mathbf{x})$ and verify directly that the credibility premium equals the Bayesian premium.

5.54 For the situation described in Exercise 5.9, determine $\mu(\theta)$ and the Bayesian premium $E(X_{n+1}|\mathbf{x})$ and verify directly that the credibility premium equals the Bayesian premium.

5.55 Consider the generalization of the linear exponential family given by

$$f(x; \theta, m) = \frac{p(m, x)e^{-m\theta x}}{[q(\theta)]^m}.$$

If m is a parameter, this is called the **exponential dispersion family**. In Exercise 5.15 it was shown that the mean of this random variable is $-q'(\theta)/q(\theta)$. For this exercise, assume that m is known.

(a) Consider the prior distribution

$$\pi(\theta) = \frac{[q(\theta)]^{-k} \exp(-\theta\mu k)}{c(\mu, k)}, \qquad \theta_0 < \theta < \theta_1 \text{ with } \pi(\theta_0) = \pi(\theta_1).$$

Determine the Bayesian premium.

(b) Using the same prior, determine the Bühlmann premium.

(c) Show that the inverse Gaussian distribution is a member of the exponential dispersion family.

5.56 Suppose that $X_1 \ldots, X_n$ are independent (conditional on Θ) and

$$E(X_j|\Theta) = \tau^j \mu(\Theta) \quad \text{and} \quad Var(X_j|\Theta) = \tau^{2j} v(\Theta)/m_j, \qquad j = 1, \ldots, n.$$

Let $\mu = E[\mu(\Theta)]$, $v = E[v(\Theta)]$, $a = Var[\mu(\Theta)]$, $k = v/a$, and $m = m_1 + \cdots + m_n$.

(a) Discuss when these assumptions may be appropriate.

(b) Show that

$$E(X_j) = \tau^j \mu, \qquad Var(X_j) = \tau^{2j}(a + v/m_j),$$

and

$$Cov(X_i, X_j) = \tau^{i+j} a, \qquad i \neq j.$$

(c) Solve the normal equations for $\tilde{\alpha}_0, \tilde{\alpha}_1, \ldots, \tilde{\alpha}_n$ to show that the credibility premium satisfies

$$\tilde{\alpha}_0 + \sum_{j=1}^{n} \tilde{\alpha}_j X_j = \frac{k}{k+m} \tau^{n+1} \mu + \frac{m}{k+m} \sum_{j=1}^{n} \frac{m_j}{m} \tau^{n+1-j} X_j.$$

(d) Give a verbal interpretation of the formula in (c).

(e) Suppose that

$$f_{X_j|\Theta}(x_j|\theta) = \frac{p(x_j, m_j, \tau) e^{-m_j \tau^{-j} x_j \theta}}{[q(\theta)]^{m_j}}.$$

Show that $E(X_j|\Theta) = \tau^j \mu(\Theta)$ and that $Var(X_j|\Theta) = \tau^{2j} v(\Theta)/m_j$ where $\mu(\theta) = -\frac{d}{d\theta} \log q(\theta)$ and $v(\theta) = -\mu'(\theta)$.

(f) Prove that credibility is exact if Θ has pdf

$$\pi(\theta) = \frac{[q(\theta)]^{-k} e^{-\theta \mu k}}{c(\mu, k)}, \qquad \theta_0 < \theta < \theta_1$$

which satisfies $\pi(\theta_0) = \pi(\theta_1)$.

5.57 Suppose that given $\Theta = \theta$, the random variables $X_1, \cdots, X_n$ are independent with Poisson pf

$$f_{X_j|\Theta}(x_j|\theta) = \frac{\theta^{x_j} e^{-\theta}}{x_j!}, \qquad x_j = 0, 1, 2, \ldots.$$

(a) Let $S = X_1 + \cdots + X_n$. Show that S has pf

$$f_S(s) = \int_0^\infty \frac{(n\theta)^s e^{-n\theta}}{s!} \pi(\theta) d\theta, \qquad s = 0, 1, 2, \ldots,$$

where Θ has pdf $\pi(\theta)$.

(b) Show that the Bayesian premium is

$$E(X_{n+1}|\mathbf{x}) = \frac{s+1}{n} \frac{f_S(s+1)}{f_S(s)}$$

where $s = \sum_{j=1}^{n} x_j$.

(c) Evaluate the distribution of S in (a) when $\pi(\theta)$ is a gamma distribution. What type of distribution is this?

5.58 Your friend selected at random one of two urns and then she pulled a ball with number 4 on it from the urn. Then she replaced the ball in the urn. One of the urns contains four balls, numbered 1 through 4. The other urn contains six balls, numbered 1 through 6. Your friend will make another random selection from the same urn.

(a) Estimate the expected value of the number on the next ball using the Bayesian method. (89-36)

(b) Estimate the expected number on the next ball using Bühlmann credibility. (89-37)

5.59 The number of claims for a randomly selected insured has the Poisson distribution with parameter θ. The parameter θ is distributed across the population with pdf $\pi(\theta) = 3\theta^{-4}$, $\theta > 1$. For an individual, the parameter does not change over time. A particular insured experienced a total of 20 claims in the previous two years.

(a) Determine the Bühlmann credibility estimate for the future expected claim frequency for this particular insured. (89-41).

(b) Determine the Bayesian credibility estimate for the future expected claim frequency for this particular insured.

5.60 The distribution of payments to an insured is constant over time. If the Bühlmann credibility assigned for one-half year of observation is .5, determine the Bühlmann credibility to be assigned for three years. (90-35)

5.61 Three urns contain balls marked either zero or one. In urn A, 10% are marked zero; in urn B, 60% are marked zero; and in urn C, 80% are marked zero. An urn is selected at random and three balls selected with replacement. The total of the values is 1. Three more balls are selected with replacement from the same urn.

(a) Determine the expected total of the three balls using Bayes' theorem. (90-39)

(b) Determine the expected total of the three balls using Bühlmann credibility. (90-40)

5.62 The number of claims follows the Poisson distribution with parameter λ. A particular insured had three claims in the past three years.

(a) The value of λ has pdf $f(\lambda) = 4\lambda^{-5}$, $\lambda > 1$. Determine the value of K used in Bühlmann's credibility formula. (90-41) Then use Bühlmann credibility to estimate the claim frequency for this insured.

(b) The value of λ has pdf $f(\lambda) = 1, \quad 0 < \lambda < 1$. Determine the value of K used in Bühlmann's credibility formula. Then use Bühlmann credibility to estimate the claim frequency for this insured. (90-52)

5.63 The number of claims follows the Poisson distribution with parameter h. The value of h has the gamma distribution with pdf $f(h) = he^{-h}, \quad h > 0$. Determine the Bühlmann credibility to be assigned to a single observation. (The Bayes solution was obtained in Exercise 5.25.) (90-47)

5.64 Consider the situation of Exercise 5.27.

(a) Determine the expected number of claims in the second year using Bayesian credibility.

(b) Determine the expected number of claims in the second year using Bühlmann credibility. (91-33)

5.65 One spinner is selected at random from a group of three spinners. Each spinner is divided into six equally likely sectors. The number of sectors marked 0, 12, and 48, respectively on each spinner is as follows: Spinner A: 2,2,2, Spinner B: 3,2,1, Spinner C: 4,1,1. A spinner is selected at random and a zero is obtained on the first spin.

(a) Determine the Bühlmann credibility estimate of the expected value of the second spin using the same spinner. (91-37)

(b) Determine the Bayesian credibility estimate of the expected value of the second spin using the same spinner. (91-38)

5.66 The number of claims in a year has the Poisson distribution with mean λ. The parameter λ has the uniform distribution over the interval $(1, 3)$.

(a) Determine the probability that a randomly selected individual will have no claims. (91-42)

(b) If an insured had one claim during the first year, estimate the expected number of claims for the second year using Bühlmann credibility. (91-43).

(c) If an insured had one claim during the first year, estimate the expected number of claims for the second year using Bayesian credibility.

5.67 Each of two classes, A and B, have the same number of risks. In class A the number of claims per risk per year has mean 0.1667 and variance 0.1389 while the amount of a single claim has mean 4 and variance 20. In class B the number of claims per risk per year has mean 0.8333 and variance 0.1389 while the amount of a single claim has mean 2 and variance 5. A risk is selected at random from one of the two classes and is observed for four years.

(a) Determine the value of Z for Bühlmann credibility for the observed pure premium. (M92-18)

(b) Suppose the pure premium calculated from the four observations is .25. Determine the Bühlmann credibility estimate for the risk's pure premium. (M92-19)

5.68 Let X_1 be the outcome of a single trial and let $E(X_2|X_1)$ be the expected value of the outcome of a second trial. You are given the following information:

Outcome, T	$\Pr(X_1 = T)$	Bühlmann estimate of $E(X_2\|X_1 = T)$	Bayesian estimate of $E(X_2\|X_1 = T)$
1	1/3	3.4	2.6
8	1/3	7.6	7.8
12	1/3	10.0	–

Determine the Bayesian estimate for $E(X_2|X_1 = 12)$. (M92-24)

5.69 Consider the situation of Exercise 5.29.

(a) Determine the expected number of claims in the second year using Bayesian credibility.

(b) Determine the expected number of claims in the second year using Bühlmann credibility. (M92-29)

5.70 Consider the situation of Exercise 5.30.

(a) Determine the expected number of claims in the second year using Bayesian credibility.

(b) Determine the expected number of claims in the second year using Bühlmann credibility.

5.71 Two spinners, A_1 and A_2, are used to determine the number of claims. For spinner A_1 there is a 0.15 probability of one claim and 0.85 of no claim. For spinner A_2 there is a 0.05 probability of one claim and 0.95 of no claim. If there is a claim, one of two spinners, B_1 and B_2, is used to determine the amount. Spinner B_1 produces a claim of 20 with probability 0.8 and 40 with probability 0.2. Spinner B_2 produced a claim of 20 with probability 0.3 and 40 with probability 0.7. A spinner is selected at random from each of A_1, A_2 and from B_1, B_2. Three observations from the selected pair yields claims amounts of 0, 20, and 0.

(a) Use Bühlmann credibility to separately estimate the expected number of claims and the expected severity. Use these estimates to estimate the expected value of the next observation from the same pair of spinners. (N92-6)

(b) Use Bühlmann credibility once on the three observations to estimate the expected value of the next observation from the same pair of spinners.

(c) Repeat parts (a) and (b) using Bayesian estimation. (N92-7)

(d) For the same selected pair of spinners, determine

$$\lim_{n\to\infty} E(X_n|X_1 = X_2 = \cdots = X_{n-1} = 0).$$

(M93-14)

5.72 A portfolio of risks is such that all risks are normally distributed. Those of type A have a mean of 0.1 and a standard deviation of 0.03. Those of type B have a mean of 0.5 and a standard deviation of 0.05. Those of type C have a mean of 0.9 and a standard deviation of 0.01. There are an equal number of each type of risk. The observed value for a single risk is 0.12. Determine the Bayesian estimate of the same risk's expected value. (N92-24)

5.73 You are given the following:

- The conditional distribution $f_{X|\Theta}(x|\theta)$ is a member of the linear exponential family.
- The prior distribution $\pi(\theta)$ is a conjugate prior for $f_{X|\Theta|}(x|\theta)$.
- $E(X) = 1$.
- $E(X|X_1 = 4) = 2$, where X_1 is the value of a single observation.
- The expected value of the process variance, $E[Var(X|\Theta)] = 3$.

Determine the variance of the hypothetical means, $Var[E(X|\Theta)]$. (N92-28)

5.74 You are given the following:

- X is a random variable with mean μ and variance v.
- μ is a random variable with mean 2 and variance 4.
- v is a random variable with mean 8 and variance 32.

Determine the value of the Bühlmann credibility factor Z, after three observations of X. (M93-3)

5.75 The amount of an individual claim has the exponential distribution with pdf $f_{Y|\Lambda}(y|\lambda) = \lambda^{-1}e^{-y/\lambda}$, $y, \lambda > 0$. The parameter λ has the inverse gamma distribution with pdf $\pi(\lambda) = 400\lambda^{-3}e^{-20/\lambda}$.

(a) Determine the unconditional expected value, $E(X)$. (M93-19)

(b) Suppose two claims were observed, with values 15 and 25. Determine the Bühlmann credibility estimate of the expected value of the next claim from the same insured.

(c) Repeat part (b), but determine the Bayesian credibility estimate.

5.76 The distribution of the number of claims is binomial with $n = 1$ and θ unknown. The parameter θ is distributed with mean 0.25 and variance 0.07. Determine the value of Z for a single observation using Bühlmann's credibility formula.

5.77 Consider four marksmen. Each is firing at a target that is 100 feet away. The four targets are two feet apart (that is, they lie on a straight line at positions 0, 2, 4, and 6 in feet). The marksmen miss to the left or right, never high or low. Each marksman's shot follows a normal distribution with mean at his target and a standard deviation that is a constant times the distance to the target. At 100 feet the standard deviation is 3 feet. By observing where an unknown marksman's shot hits the straight line, you are to estimate the location of the next shot by the same marksman.

(a) Determine the Bühlmann credibility assigned to a single shot of a randomly selected marksman. (N93-3)

(b) Which of the following will increase Bühlmann credibility the most? (N93-4)

i. Revise the targets to 0, 4, 8, and 12.

ii. Move the marksmen to 60 feet from the targets.

iii. Revise targets to 2, 2, 10, 10.

iv. Increase the number of observations from the same marksman to three.

v. Move two of the marksmen to 50 feet from the targets and increase the number of observations from the same marksman to two.

5.78 Risk one produces claims of amounts 100, 1,000, and 20,000 with probabilities 0.5, 0.3, and 0.2 respectively. For risk two the probabilities are 0.7, 0.2, and 0.1. Risk one is twice as likely as risk two of being observed. A claim of 100 is observed, but the observed risk is unknown.

(a) Determine the Bayesian credibility estimate of the expected value of the second claim amount from the same risk. (N93-17)

(b) Determine the Bühlmann credibility estimate of the expected value of the second claim amount from the same risk. (N93-18)

5.79 You are given the following:

- The number of claims for a single insured follows a Poisson distribution with mean μ.
- The amount of a single claim has an exponential distribution with pdf $f_{X|\Lambda}(x|\lambda) = \lambda^{-1}e^{-x/\lambda},\ x, \lambda > 0$.
- μ and λ are independent.
- $E(\mu) = 0.10$ and $Var(\mu) = 0.0025$.
- $E(\lambda) = 1{,}000$ and $Var(\lambda) = 640{,}000$.
- The number of claims and the claim amounts are independent.

(a) Determine the expected value of the pure premium's process variance for a single risk. (M93-21)

(b) Determine the variance of the hypothetical means for the pure premium. (M93-22)

5.80 In Example 5.34, if $\rho = 0$, then $Z = 0$ and the estimator is μ. That is, the data should be ignored. However, as ρ increases toward 1, Z increases to 1, and the sample mean becomes the preferred predictor of X_{n+1}. Explain why this is a reasonable result.

5.81 In this exercise you are asked to derive a number of the items from Example 5.44.

(a) The **digamma function** is formally defined as $\Psi(\alpha) = \Gamma'(\alpha)/\Gamma(\alpha)$. From this definition, show that

$$\Psi(\alpha) = \frac{1}{\Gamma(\alpha)} \int_0^\infty (\log x) x^{\alpha-1} e^{-x} dx.$$

(b) The **trigamma function** is formally defined as $\Psi'(\alpha)$. Derive an expression for

$$\int_0^\infty (\log x)^2 x^{\alpha-1} e^{-x} dx$$

in terms of trigamma, digamma, and gamma functions.

5.82 Consider the following situation, which is similar to Examples 5.43 and 5.44. Individual observations are samples of size 25 from a lognormal distribution with μ unknown and $\sigma = 2$. The prior distribution for Θ (using Θ to represent the unknown value of μ) is normal with mean 5 and standard deviation 1. Determine the Bayes, credibility, and log-credibility estimators and compare their mean squared errors, evaluating them at the same percentiles as used in Examples 5.43 and 5.44.

5.83 In the following, let the random vector $\mathbf{X}$ represent all the past data and let X_{n+1} represent the next observation. Let $g(\mathbf{X})$ be any function of the past data.

(a) Prove that the following is true.

$$E\left\{[X_{n+1} - g(\mathbf{X})]^2\right\} = E\{[X_{n+1} - E(X_{n+1}|\mathbf{X})]^2\} + E\{[E(X_{n+1}|\mathbf{X}) - g(\mathbf{X})]^2\}$$

where the expectation is taken over $(X_{n+1}, \mathbf{X})$.

(b) Show that setting $g(\mathbf{X})$ equal to the Bayesian premium (the mean of the predictive distribution) minimizes the expected squared error, $E\left\{[X_{n+1} - g(\mathbf{X})]^2\right\}$.

(c) Show that if $g(\mathbf{X})$ is restricted to be a linear function of the past data, then the expected squared error is minimized by the credibility premium.

Section 5.5

5.84 Past claims data on a portfolio of policyholders are given in Table 5.6.

Table 5.6 Data for Exercise 5.84

	Year		
Policyholder	1	2	3
1	750	800	650
2	625	600	675
3	900	950	850

Estimate the Bühlmann credibility premium for each of the three policyholders for year 4.

5.85 Past data on a portfolio of group policyholders are given in Table 5.7. Estimate the Bühlmann–Straub credibility premiums to be charged to each group in year 4.

5.86 For the situation in Exercise 5.34, estimate the Bühlmann credibility premium for the next year for the policyholder.

Table 5.7 Data for Exercise 5.85

	Policyholder	Year 1	2	3	4
Claims	1	–	20,000	25,000	–
No. in group		–	100	120	110
Claims	2	19,000	18,000	17,000	–
No. in group		90	75	70	60
Claims	3	26,000	30,000	35,000	–
No. in group		150	175	180	200

5.87 Consider the Bühlmann model in Example 5.45.

(a) Prove that $Var(X_{ij}) = a + v$.

(b) If $\{X_{ij}: \; i = 1, \ldots, r \text{ and } j = 1, \ldots, n\}$ are unconditionally independent for all i and j, argue that an unbiased estimator of $a + v$ is

$$\frac{1}{nr-1}\sum_{i=1}^{r}\sum_{j=1}^{n}(X_{ij}-\bar{X})^2.$$

(c) Prove the algebraic identity

$$\sum_{i=1}^{r}\sum_{j=1}^{n}(X_{ij}-\bar{X})^2 = \sum_{i=1}^{r}\sum_{j=1}^{n}(X_{ij}-\bar{X}_i)^2 + n\sum_{i=1}^{r}(\bar{X}_i-\bar{X})^2.$$

(d) Show that, conditionally,

$$E\left[\frac{1}{nr-1}\sum_{i=1}^{r}\sum_{j=1}^{n}(X_{ij}-\bar{X})^2\right] = (v+a) - \frac{n-1}{nr-1}a.$$

(e) Comment on the implications of (b) and (d).

5.88 The distribution of automobile insurance policyholders by number of claims is given in Table 5.8

Assuming a (conditional) Poisson distribution for the number of claims per policyholder, estimate the Bühlmann credibility premiums for the number of claims next year.

5.89 Suppose that given Θ, $X_1, \ldots, X_n$ are independently geometrically distributed with pf

$$f_{X_j|\Theta}(x_j|\theta) = \frac{1}{1+\theta}\left(\frac{\theta}{1+\theta}\right)^{x_j}, \qquad x_j = 0, 1, \ldots .$$

Table 5.8 Data for Exercise 5.88

No. of claims	No. of insureds
0	2,500
1	250
2	30
3	5
4	2
Total	2,787

(a) Show that $\mu(\theta) = \theta$ and $v(\theta) = \theta(1+\theta)$.

(b) Prove that $a = v - \mu - \mu^2$.

(c) Rework Exercise 5.88 assuming a (conditional) geometric distribution.

5.90 Suppose that

$$\Pr(m_{ij}X_{ij} = t_{ij}|\Theta_i = \theta_i) = \frac{(m_{ij}\theta_i)^{t_{ij}} e^{-m_{ij}\theta_i}}{t_{ij}!}$$

and

$$\pi(\theta_i) = \frac{1}{\mu}e^{-\theta_i/\mu}, \qquad \theta_i > 0.$$

Write down the equation satisfied by the mle $\hat{\mu}$ of μ for Bühlmann–Straub-type data.

5.91 (a) Prove the algebraic identity

$$\sum_{i=1}^{r}\sum_{j=1}^{n_i} m_{ij}(X_{ij}-\bar{X})^2 = \sum_{i=1}^{r}\sum_{j=1}^{n_i} m_{ij}(X_{ij}-\bar{X}_i)^2 + \sum_{i=1}^{r} m_i(\bar{X}_i-\bar{X})^2.$$

(b) Use part (a) and (5.79) to show that (5.81) may be expressed as

$$\hat{a} = m_*^{-1}\left\{\frac{\sum_{i=1}^{r}\sum_{j=1}^{n_i} m_{ij}(X_{ij}-\bar{X})^2}{\sum_{i=1}^{r} n_i - 1} - \hat{v}\right\}$$

where

$$m_* = \frac{\sum_{i=1}^{r} m_i\left(1-\frac{m_i}{m}\right)}{\sum_{i=1}^{r} n_i - 1}.$$

Table 5.9 Data for Exercise 5.92

Number of claims	Number of insureds
0	200
1	80
2	50
3	10

5.92 A group of 340 insureds in a high crime area submit the 210 theft claims in a one-year period as given in Table 5.9

Each insured is assumed to have a Poisson distribution for the number of thefts, but the mean of such distribution may vary from one insured to another. If a particular insured experienced two claims in the observation period, determine the Bühlmann credibility estimate for the number of claims for this insured in the next period. (89-39)

5.7 CASE STUDY

5.7.1 The case study continued

Credibility theory may be used in two ways for this problem. The first is to reconcile the experience of the three types of actuary. The second is to create an experience rating plan.

5.7.1.1 Combining experience In Section 2.13.1 it was determined that the severity distributions for the three types of actuary are the same. However, in Section 3.12.1 it was learned that while the Poisson distribution applies to all three types, a different parameter is needed for each one. Credibility theory tells us that the three point estimates can be improved by combining the results in an appropriate way. This should be especially helpful for the estimate for casualty actuaries as there were only 2,913 exposures. The frequency data in Table 1.5 are repeated below as Table 5.10. There are four different types of credibility that may be used on these data. They are explored in turn in the next three subsections.

Limited fluctuation credibility We have the option of basing the credibility factor on either the number of claims, the number of exposures, or the number of dollars. As well, we may apply the credibility factor to the expected claim frequency or the expected number of dollars. We choose to apply it to the expected claim frequency for two reasons. First, we have already determined that the three populations have the same expected claim size, so any fluctuation in the observed claim amounts should not impact our final results.

Table 5.10 Exposures and loss counts

Year	Life/health		Pension		Property/liability	
	Exposure	Claims	Exposure	Claims	Exposure	Claims
1990	853	20	1,446	27	639	5
1991	1,105	14	1,780	35	725	8
1992	1,148	16	1,717	36	685	4
1993	1,270	21	2,065	24	864	11
Total	4,376	71	7,008	122	2,913	28

Table 5.11 Limited fluctuation credibility estimates

Group	m_i	c_i	c_i/m_i	Based on losses		Based on exposures	
				Z_i	$\hat{\lambda}_i$	Z_i	$\hat{\lambda}_i$
1	4,376	71	0.01622	0.2561	0.01565	0.2500	0.01565
2	7,008	122	0.01741	0.3357	0.01611	0.3164	0.01608
3	2,913	28	0.00961	0.1608	0.01452	0.2040	0.01427

Second, we plan to investigate the effects of various reinsurance arrangements and so it is useful to keep the frequency and severity components separate. For the same reason, it makes sense to not base the standard for full credibility on the number of dollars paid in benefits.

For the standard based on the number of claim, we have from Example 5.20 that the standard is $\lambda_0 = (y_p/r)^2$ and with $r = 0.05$ and $p = 0.9$ the standard is 1,082.41 claims. From that same example we can infer that the required number of exposures is $1{,}082.41/\lambda$. Because we do not know λ, we can estimate it by the combined mean of $221/14,297 = 0.015458$ for a required exposure of 70,023.60. For partial credibility, the required result is in Example 5.22. Provided it is less than one, the credibility factor is $Z = (\lambda n/1{,}082.41)^{1/2}$. If credibility is based on the number of claims, λn is replaced by the number of observed claims. If credibility is based on exposure, λ is replaced by 0.015458 and n is replaced by the exposure. Once the partial credibility factor is determined, the final estimate of the Poisson parameter is $Z(c_i/m_i) + (1 - Z)(0.015458)$ where c_i is the number of losses for group i and m_i is the exposure for group i. The calculations appear in Table 5.11.

Both sets of estimates are unbalanced in the sense that the number of expected losses based on the exposures does not match the 221 actual losses. For the estimates based on the number of claims, the number of expected losses is $\sum_{i=1}^{3} m_i\hat{\lambda}_i = 223.71$ and so it is reasonable to multiply each estimate by $221/223.71 = 0.98789$. This results in estimates of 0.01546, 0.01592, and 0.01434. Similarly, for the estimates based on exposure, the expected total is 222.69 for a factor of 0.99240. The revised estimates are 0.01553, 0.01596, and 0.01416.

Empirical Bayes credibility For this analysis no distributional assumptions are made. In the notation of Section 5.5 we have m_{ij} is the exposure for group i in year j. The observations are $X_{ij} = c_{ij}/m_{ij}$, the relative frequency for group i in year j. This is the number of losses divided by the exposure. Similarly, $\bar{X}_i = c_i/m_i$ is the relative frequency for group i. We begin with the estimation of v using (5.79). With $n_i = 4$ and $r = 3$ (in the notation of that section), (5.79) becomes

$$
\begin{aligned}
\hat{v} &= \frac{1}{9}\sum_{i=1}^{3}\sum_{j=1}^{4} m_{ij}(c_{ij}/m_{ij} - c_i/m_i)^2 \\
&= \frac{1}{9}\sum_{i=1}^{3}\sum_{j=1}^{4} c_{ij}^2/m_{ij} - 2c_{ij}c_i/m_i + m_{ij}c_i^2/m_i^2 \\
&= \frac{1}{9}\left(\sum_{i=1}^{3}\sum_{j=1}^{4} c_{ij}^2/m_{ij} - \sum_{i=1}^{3} c_i^2/m_i\right) \\
&= \frac{1}{9}(3.733444 - 3.544962) \\
&= 0.0209424.
\end{aligned}
$$

We next estimate a using (5.81). This formula is (where $c = 221$ is the total number of losses and $m = 14{,}297$ is the total number of exposures)

$$
\begin{aligned}
\hat{a} &= \left(m - m^{-1}\sum_{i=1}^{3} m_i^2\right)^{-1}\left[\sum_{i=1}^{3} m_i(c_i/m_i - c/m)^2 - \hat{v}(3-1)\right] \\
&= (8{,}928.95)^{-1}\left(\sum_{i=1}^{3} c_i^2/m_i - c^2/m - 0.0418848\right) \\
&= (8{,}928.95)^{-1}(0.086906) \\
&= 0.00000973306 \qquad (5.85)
\end{aligned}
$$

and so $\hat{k} = \hat{v}/\hat{a} = 2{,}151.68$ and the credibility factor is $Z_i = m_i/(2{,}151.68 + m_i)$. The overall mean, μ, is estimated using (5.82). This calculation (producing $\hat{\mu} = 0.02972/2.0107 = 0.01478$), as well as the credibility estimates, is given in Table 5.12. As expected, $\sum m_i\hat{\lambda}_i = 221$, the observed number of claims.

Semiparametric credibility We discovered in Subsection 3.12.1 that the Poisson distribution is appropriate for each of the three groups. What is not known is the manner in which the Poisson parameter varies from group to group. We can modify the example in Section 5.5.2 to handle this problem.

Table 5.12 Empirical Bayes credibility estimates

Group	m_i	Z_i	$\bar{X}_i$	$Z_i\bar{X}_i$	$\hat{\lambda}_i$
1	4,376	0.6704	0.01622	0.01087	0.01575
2	7,008	0.7651	0.01741	0.01332	0.01679
3	2,913	0.5752	0.00961	0.00553	0.01181
Total		2.0107		0.02972	

We have assumed that each group is stable over time and that all members have the same Poisson parameter. Then the number of losses, c_i, has the Poisson distribution with parameter $m_i\lambda_i$. The three key credibility quantities are

$$
\begin{aligned}
\mu &= E\left(\frac{c_i}{m_i}\right) = E\left[E\left(\frac{c_i}{m_i}|\lambda_i\right)\right] = E(\lambda_i) \\
v_i &= E\left[Var\left(\frac{c_i}{m_i}|\lambda_i\right)\right] = E\left(\frac{\lambda_i}{m_i}\right) = \mu/m_i \\
a &= Var\left[E\left(\frac{c_i}{m_i}|\lambda_i\right)\right] = Var(\lambda_i) = \sigma^2 \\
Z_i &= \frac{a}{a+v_i}
\end{aligned}
$$

We must now determine the expected value of the two quantities most likely to prove useful in the estimation process. The mean for the entire sample is $\bar{X} = (\sum_{i=1}^{3} c_i / \sum_{i-1}^{3} m_i)$ and its expected value is

$$
\begin{aligned}
E(\bar{X}) &= E\left(\frac{\sum_{i=1}^{3} c_i}{\sum_{i=1}^{3} m_i}\right) = E\left[E\left(\frac{\sum_{i=1}^{3} c_i}{\sum_{i=1}^{3} m_i}|\lambda_1,\lambda_2,\lambda_3\right)\right] \\
&= E\left(\frac{\sum_{i=1}^{3} m_i\lambda_i}{\sum_{i=1}^{3} m_i}\right) = E(\lambda_i) = \mu
\end{aligned}
$$

where we are assuming the same prior mean for all three groups. With regard to the sum of squares used in the sample variance, consider the quantity used in (5.85),

$$
SS = \sum_{i=1}^{3} m_i\left(\frac{c_i}{m_i} - \bar{X}\right)^2 = \sum_{i=1}^{3}\frac{c_i^2}{m_i} - \frac{\left(\sum_{i=1}^{3} c_i\right)^2}{\sum_{i=1}^{3} m_i}.
$$

Its expected value is

$$
E(SS) = E\left[E\left(\sum_{i=1}^{3}\frac{c_i^2}{m_i} - \frac{\left(\sum_{i=1}^{3} c_i\right)^2}{\sum_{i=1}^{3} m_i}|\lambda_1,\lambda_2,\lambda_3\right)\right]
$$

Table 5.13 Empirical Bayes credibility estimates

Group	m_i	Z_i	$\bar{X}_i$	$Z_i\bar{X}_i$	$\hat{\lambda}_i$
1	4,376	0.7563	0.01622	0.01227	0.01585
2	7,008	0.8325	0.01741	0.01449	0.01695
3	2,913	0.6738	0.00961	0.00648	0.01127
Total		2.2626		0.03324	

$$
\begin{aligned}
&= E\left(\sum_{i=1}^{3} \frac{m_i\lambda_i + m_i^2\lambda_i^2}{m_i} - \frac{\sum_{i=1}^{3}(m_i\lambda_i + m_i^2\lambda_i^2) + \sum_{i\neq j} m_i m_j \lambda_i \lambda_j}{\sum_{i=1}^{3} m_i}\right) \\
&= 3\mu + (\sum_{i=1}^{3} m_i)(\sigma^2+\mu^2) - \mu \\
&\quad - \frac{1}{\sum_{i=1}^{3} m_i}\left[(\sum_{i=1}^{3} m_i^2)(\sigma^2+\mu^2) + \sum_{i\neq j} m_i m_j \mu^2\right] \\
&= 2\mu + (\sum_{i=1}^{3} m_i)(\sigma^2+\mu^2) - \sigma^2\frac{\sum_{i=1}^{3} m_i^2}{\sum_{i=1}^{3} m_i} - \mu^2\sum_{i=1}^{3} m_i \\
&= 2\mu + \sigma^2\left(\sum m_i - \frac{\sum m_i^2}{\sum m_i}\right).
\end{aligned}
$$

For the problem at hand, $\hat{\mu} = 221/14{,}297 = 0.015458$ and $SS = 0.128791$. The above equation yields $0.128791 - 2(0.015458) - \sigma^2(8{,}928.95)$ for $\sigma^2 = 0.0000109616$. The credibility factor is

$$
Z_i = \frac{0.0000109616}{0.0000109616 + 0.015458/m_i} = \frac{m_i}{m_i + 1{,}410.18}.
$$

The calculation then proceeds as in the empirical Bayes analysis. The key quantities are in Table 5.13. The credibility weighted mean is $0.03324/2.2626 = 0.01469$ and again the expected claims match the 221 observed claims.

Parametric estimation We must now make a parametric assumption concerning the manner in which λ is distributed among the population of groups. To keep this simple, assume the distribution is exponential with mean μ. The work for this case was done in Exercise 5.90. The maximum likelihood estimate of μ was determined to satisfy the following equation.

$$
3\mu = \sum_{i=1}^{3} \frac{c_i + 1}{m_i + \mu^{-1}}.
$$

Trial and error produces the estimate, $\hat{\mu} = 0.014434$. Now that the prior distribution is established, Bayes' theorem is used to supply the revised estimate of each λ_i. The posterior mean is

$$\hat{\lambda}_i = \frac{c_i + 1}{m_i + \mu^{-1}}$$

and the three estimates are 0.01620, 0.01738, and 0.00972 for the three groups, respectively.

5.7.1.2 *Experience rating* The final request from the KPWV assignment is to develop an experience rating formula. That is, many of the actuaries who have purchased liability insurance (in fact, nearly all of them) are convinced that they are extremely unlikely to have a claim and are certain that they are being overcharged. However, they have been convinced to wait a year and prove their superiority. An appropriate method was developed in Section 5.5.2, but cannot be used here. The reason is that we were not informed which, if any, of the 221 claims represented a second or third claim in a year by the same actuary.

The only chance to obtain an answer is to use the fully parametric approach. As above, but with only one group, the estimate of the prior mean μ for the life/health group is the solution to

$$\mu = \frac{71 + 1}{4,376 + \mu^{-1}}$$

which is the sample mean, $71/4{,}376 = 0.01622$. If a particular actuary has c claims in a one-year period, the experience rated premium for the next year should be based on an expected number of claims of

$$\hat{\lambda} = \frac{c + 1}{1 + .01622^{-1}} = \frac{c + 1}{62.6338}.$$

For each of the other two groups, the experience rating formula would be slightly different, with the value of μ being based on that group's experience over the three prior years.

5.7.2 Exercises

5.93 In 1958, Bondy [15] reported on automobile bodily injury liability claims in New York from 1952 to 1954. The data were presented by rating territory and consisted of the number of claims and the number of exposures in each territory. The results appear in Table 5.14. Use greatest accuracy credibility to estimate the expected number of claims per exposure for each territory.

5.94 In 1979, Woll [128] obtained some North Carolina data in which the same drivers were followed for four years. He recorded the number of accidents

Table 5.14 Automobile bodily injury liability claims[a]

Terr.	Exp. (10^3)	No. of claims	Terr.	Exp. (10^3)	No. of claims
1	15	875	19	314	10,411
2	58	3,385	20	233	7,319
3	18	784	21	64	1,711
4	562	30,047	22	149	4,264
5	149	6,377	23	77	2,261
6	23	856	24	61	2,019
7	92	3,271	25	33	856
8	21	765	26	33	937
9	550	22,663	27	216	5,753
10	71	2,804	28	44	1,043
11	209	6,863	29	77	1,986
12	341	12,935	30	47	1,338
13	16	585	31	107	2,724
14	20	670	32	51	1,410
15	18	573	33	447	10,754
16	20	798	34	22	567
17	134	5,082	35	220	4,786
18	33	1,022			

[a]Terr., Territory; Exp., Exposures.

each driver had in the first three years and then the number that same driver had in the fourth year. The results are in Table 5.15. Use only the data from the first three years (the column totals) to determine empirical Bayes credibility estimates for the expected number of accidents in the fourth year. Next determine the mean squared prediction error for year four based on these estimates. Because we have data for year four, it is possible to obtain estimates of quantities such as $Cov(X_4, Y)$ where Y is the average number of accidents in the first three years. Determine the credibility formula using linear algebra and formulas (5.56) and (5.57).

Table 5.15 North Carolina accident data

No. of accidents	Losses in years 1–3							
	0	1	2	3	4	5	6	7
0	2,002,577	295,414	45,203	7,666	1 ,441	300	82	25
1	104,048	26,776	6,255	1,577	375	83	20	4
2	5,931	2,362	811	247	80	30	13	7
3	438	231	102	34	11	10	0	3
4	30	16	12	2	3	1	0	1
5	5	9	3	2	0	0	0	0
Total	2,113,029	324,808	52,386	9,528	1,910	424	115	40

6

Long-Term Models

6.1 INTRODUCTION

The risk assumed with a portfolio of insurance contracts is difficult to assess, but it is nevertheless important to attempt to do so in order to ensure the viability of an insurance operation. The distribution of total claims over a fixed period of time is an obvious input parameter to such a process, and this quantity has been the subject of the first five chapters.

In this chapter we take a multiperiod approach in which the fortunes of the policy, portfolio, or company are followed over time. The most common use of this approach is **ruin theory**, in which the quantity of interest is the amount of surplus, with ruin occurring when the surplus becomes negative. In order to track surplus we must model more than the claim payments. We must include premiums, investment income, and expenses, along with any other item that impacts the cash flow.

The models described in this chapter are quite simple and idealized in order to maintain mathematical simplicity. Consequently the output from the analysis should not be viewed as a representation of absolute reality, but rather as important additional information on the risk associated with the portfolio of business. Such information is extremely useful for long-run financial planning and maintenance of the insurer's solvency.

This chapter is organized into three parts. The first part (Section 6.2) introduces process models. The appropriate definitions are made and the terms of ruin theory defined. The second part (Section 6.3) analyzes discrete-time models. This can be done with the tools presented in the first five chapters. The final sections (6.4–6.10) present an analysis of continuous time models. This requires an introduction to stochastic processes. Two processes are analyzed: the compound Poisson process and Brownian motion. The compound Poisson process has been the standard model for ruin analysis in actuarial science, while the Brownian motion has found considerable use in modern financial theory and also can be used as an approximation to the compound Poisson process.

6.2 PROCESS MODELS FOR INSURANCE

6.2.1 Processes

The major difference between this chapter and the earlier ones is that we now want to view the evolution of the portfolio over time. With that in mind, we define two kinds of process. We note that while processes that involve random events are usually called **stochastic processes**, we will not employ the modifier "stochastic" and instead trust that the context will make it clear which processes are random and which are not.

Definition 6.1 *A* ***continuous time process*** *is denoted $\{X_t; t \geq 0\}$. If there are random elements, it is sufficient to specify the joint distribution of $(X_{t_1}, \ldots, X_{t_n})$ for all $t_1, \ldots, t_n$ and all n.*

In general, it is insufficient to describe the process by specifying the distribution of X_t for arbitrary t. Many processes have correlations between the values observed at different times.

Example 6.1 *Let $\{S_t; t \geq 0\}$ be the total losses paid from time 0 to time t. Indicate how the collective risk model of Chapter 4 may be used to describe this process.*

For the joint distribution of $(S_{t_1}, \ldots, S_{t_n})$, suppose $t_1 < \cdots < t_n$. Let $W_j = S_{t_j} - S_{t_{j-1}}$ with $S_{t_0} = S_0 = 0$. Let the W_j have independent distributions given by the collective risk model. The individual loss distributions could be identical while the frequency distribution would have a mean that is proportional to the length of the time period, $t_j - t_{j-1}$. An example of a realization of this process (called a **sample path**) is given in Figure 6.1. □

It is usually easier to describe a process if it does not change much over time. Two specific ways in which this can happen are defined below.

Definition 6.2 *A process has* ***independent increments*** *if the random variables $X_t - X_s$ and $X_u - X_v$ are independent whenever $s < t \leq v < u$.*

This property indicates that the movement in the process in any one period is independent of the movement in a different, nonoverlapping, period.

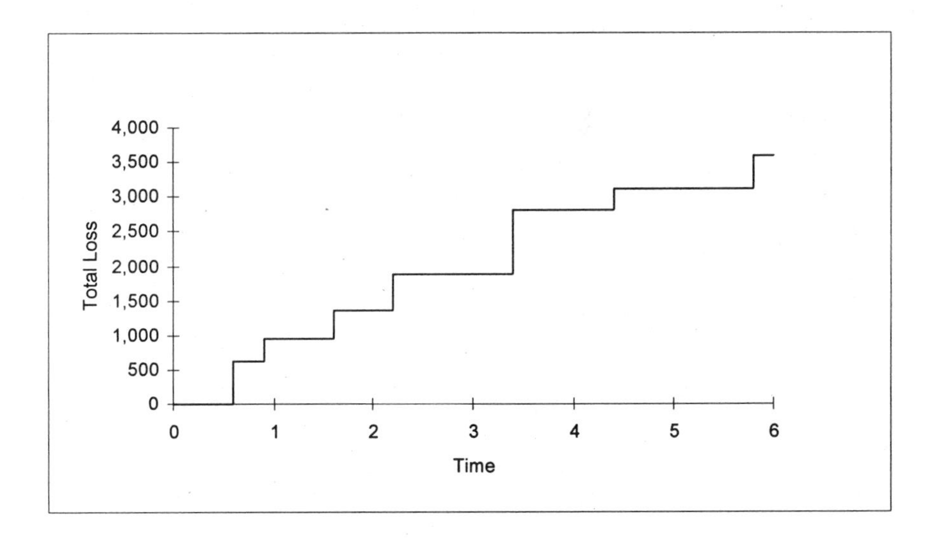

Fig. 6.1 Continuous total loss process, S_t

Definition 6.3 *A process has* ***stationary increments*** *if the distribution of $X_t - X_s$ depends only on t and s through the difference $t - s$.*

This property implies that the movement does not depend on the date. In other words, you cannot tell what time it is by looking at increments in the process.

Most business organizations do not continuously monitor their status. Instead, it is checked at regular intervals. This leads to the other kind of process.

Definition 6.4 *A* ***discrete time process*** *is denoted by $\{X_t; t = 0, 1, 2, \ldots\}$. If there are random elements, it is sufficient to specify the joint distribution of $(X_{t_1}, \ldots, X_{t_n})$ for integer t_i and any n.*

A discrete time process can be derived from a continuous time process by just writing down the values of X_t at integral times. In this chapter, all discrete time processes will take measurements at the end of each observation period, such as a month, quarter, or year.

Example 6.2 (Example 6.1 continued) *Convert the process to a discrete time process with stationary, independent increments.*

Let $X_1, X_2, \ldots$ be the amount of the total losses in each period where the X_js are iid and each X_j has a compound distribution. Then let the total loss process be $S_t = X_1 + \cdots + X_t$. The process has stationary increments because $S_t - S_s = X_{s+1} + \cdots + X_t$ and its distribution depends only on the number of X_js, which is $t - s$. The property of independent increments follows directly from the independence of the X_js. Figure 6.2 is the discrete time version of Figure 6.1. □

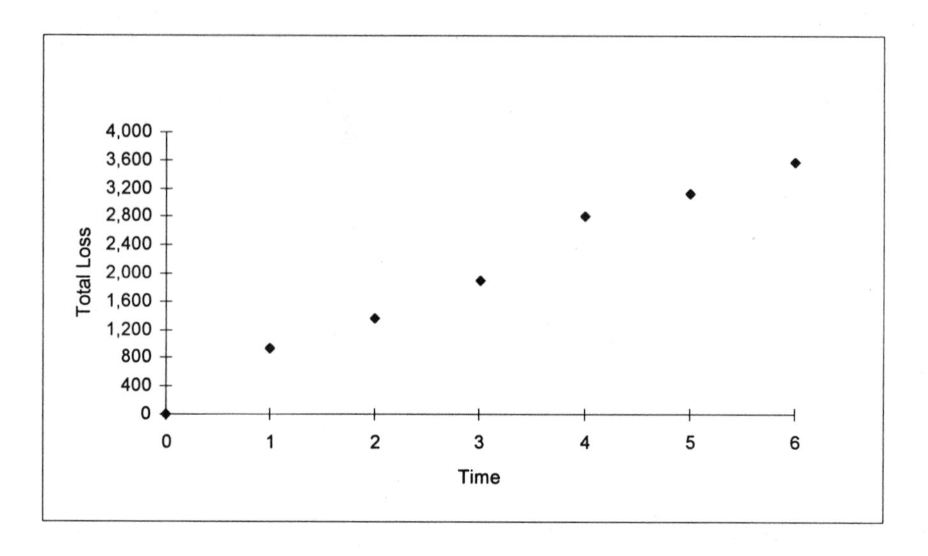

Fig. 6.2 Discrete total loss process, $\tilde{S}_t$

6.2.2 An insurance model

The earlier examples in this section have already illustrated most of the model we will use for the insurance process. We are interested in the **surplus process** $\{U_t; t \geq 0\}$ (or perhaps its discrete time version, $\{U_t; t = 0, 1, \ldots\}$) which measures the surplus of the portfolio at time t. We begin at time zero with $u = U_0$, the **initial surplus**. We think of the surplus in the accounting sense in that it represents excess funds that would not be needed if the portfolio terminated today. For an ongoing concern, a positive value provides protection against adversity. For the rest of this subsection, all continuous time processes could just as easily be discrete time processes. The surplus at time t is

$$U_t = U_0 + P_t - S_t$$

where $\{P_t; t \geq 0\}$ is the premium process which measures all premiums (net of expenses) collected up to time t and $\{S_t; t \geq 0\}$ is the loss process, which measures all losses paid up to time t. We make the following observations.

1. P_t may be written or earned premiums, as appropriate.

2. S_t may be paid or incurred losses, again, as appropriate.

3. P_t may depend on S_u for $u < t$. For example, dividends based on favorable past loss experience may reduce the current premium.

It is possible, though not necessary, to separate the frequency and severity components of S_t. Let $\{N_t; t \geq 0\}$ be the claims process which records the number of claims as of time t. Then let $S_t = X_1 + \cdots + X_{N_t}$. While the sequence $\{X_1, X_2, \ldots\}$ need not consist of iid variables, if it does, and they are independent of all N_t, then S_t will have a compound distribution.

We now look at two special cases of the surplus process. These are the only ones that will be studied in this chapter.

6.2.2.1 *A discrete time model* Let the increment in the surplus process in year t be defined as

$$W_t = P_t - P_{t-1} - S_t + S_{t-1},\ t = 1, 2, \ldots.$$

Then the progression of surplus is

$$U_t = U_{t-1} + W_t,\ t = 1, 2, \ldots.$$

It will be relatively easy to learn about the distribution of $\{U_t; t = 1, 2, \ldots\}$ provided that the random variable W_t either is independent of the other W_ts or only depends on the value of U_{t-1}. The dependency of W_t on U_{t-1} allows us to pay a dividend based on the surplus at the end of the previous year (since W_t depends on P_t).

In Section 6.3, two methods of determining the distribution of the U_ts will be presented. These will be computationally intensive, but given enough time and resources, the answers are easy to obtain.

6.2.2.2 *A continuous time model* In most cases it is extremely difficult to analyze continuous time models. This is because the joint distribution must be developed at every time point, not just at a countable set of time points. One model that has been extensively analyzed is the compound Poisson claim process where premiums are collected at a constant continuous nonrandom rate,

$$P_t = (1 + \theta)E(S_1)t$$

and the total loss process is $S_t = X_1 + \cdots + X_{N_t}$, where $\{N_t; t \geq 0\}$ is the Poisson process. This process is discussed in detail in Section 6.4. It suffices for now to note that for any period, the number of losses has the Poisson distribution with a mean that is proportional to the length of the time period.

Because these models are more difficult to work with, several sections in this chapter will be devoted to their development and analysis. We are now ready to define the quantity of interest, the one that measures the portfolio's chance for success.

6.2.3 Ruin

The main purpose for building a process model is to determine if the portfolio will survive over time. The probability of survival can be defined in four different ways.

Definition 6.5 *The* ***continuous time, infinite horizon survival probability*** *is given by*

$$\phi(u) = \Pr(U_t \geq 0 \text{ for all } t \geq 0 | U_0 = u).$$

Here we continuously check the surplus and demand that the portfolio be solvent forever. Both continuous checking and a requirement that the portfolio survive forever are unrealistic. In practice, it is more likely that surplus is checked at regular intervals. While we would like our portfolio to last forever, it is too much to ask that our model be capable of forecasting infinitely far into the future. A more useful quantity follows.

Definition 6.6 *The* ***discrete time, finite horizon survival probability*** *is given by*

$$\tilde{\phi}(u, \tau) = \Pr(U_t \geq 0 \text{ for all } t = 0, 1, \ldots, \tau | U_0 = u).$$

Here the portfolio is required to survive τ periods (usually years) and we only check at the end of each period. There are two intermediate cases.

Definition 6.7 *The* ***continuous time, finite horizon survival probability*** *is given by*

$$\phi(u, \tau) = \Pr(U_t \geq 0 \text{ for all } 0 \leq t \leq \tau | U_0 = u)$$

and the ***discrete time, infinite horizon survival probability*** *is given by*

$$\tilde{\phi}(u) = \Pr(U_t \geq 0 \text{ for all } t = 0, 1, \ldots | U_0 = u).$$

It should be clear that the following inequalities hold.

$$\tilde{\phi}(u, \tau) \geq \tilde{\phi}(u) \geq \phi(u)$$

and

$$\tilde{\phi}(u, \tau) \geq \phi(u, \tau) \geq \phi(u).$$

There are also some limits that should be equally obvious. They are

$$\lim_{\tau \to \infty} \phi(u, \tau) = \phi(u)$$

and

$$\lim_{\tau \to \infty} \tilde{\phi}(u, \tau) = \tilde{\phi}(u).$$

In many cases, convergence is rapid. This means that the choice of a finite or infinite horizon should depend as much on the ease of calculation as on the appropriateness of the model. We will find that if the Poisson process holds, infinite horizon probabilities are easier to obtain. For other cases, the finite horizon calculation may be easier.

Although we have not defined notation to express them, there is another pair of limits. As the frequency with which surplus is checked increases (that is, the number of times per year), the discrete time survival probabilities converge to their continuous time counterparts.

As this subsection refers to ruin, we close by defining the probability of ruin.

Definition 6.8 *The* ***continuous time, infinite horizon ruin probability*** *is given by*

$$\psi(u) = 1 - \phi(u)$$

and the other three ruin probabilities are defined and denoted in a similar manner.

6.3 DISCRETE, FINITE-TIME RUIN PROBABILITIES

6.3.1 The discrete time process

Let P_t be the premium collected in the tth period and let S_t be the losses paid in the tth period. We also add one generalization. Let C_t be any cash flow other than the collection of premiums and the payment of losses. The most significant cash flow is the earning of investment income on the surplus available at the beginning of the period. The surplus at the end of the tth period is then

$$U_t = u + \sum_{j=1}^{t}(P_j + C_j - S_j) = U_{t-1} + P_t + C_t - S_t.$$

The final assumption is that given U_{t-1}, the random variable $W_t = P_t + C_t - S_t$ depends only upon U_{t-1} and not upon any other previous experience. This makes $\{U_t; t = 1, 2, \ldots\}$ a **Markov process.**

In order to evaluate ruin probabilities, we consider a second process defined as follows. First, define

$$\begin{aligned} W_t^* &= \begin{cases} 0, & U_{t-1}^* < 0 \\ W_t, & U_{t-1}^* \geq 0 \end{cases} \\ &\text{and} \\ U_t^* &= U_{t-1}^* + W_t^* \end{aligned} \tag{6.1}$$

where the new process starts with $U_0^* = u$. In this case, the finite horizon survival probability is

$$\tilde{\phi}(u, \tau) = \Pr(U_\tau^* \geq 0).$$

The reason we need only check U_t^* at time τ is that once ruined, this process is not allowed to become non-negative. The following example illustrates this distinction and is a preview of the method presented in detail in Subsection 6.3.2.1.

Example 6.3 *Consider a process with an initial surplus of 2, a fixed annual premium of 3, and losses of either 0 or 6 with probabilities 0.6 and 0.4, respectively. There are no other cash flows. Determine* $\tilde{\phi}(2,2)$.

U_1 can assume only two values, 5 and -1, with probabilities 0.6 and 0.4. In each year, W_t takes the values 3 and -3 with probabilities 0.6 and 0.4. For year 2, there are four possible ways for the process to end. The are listed in the following table.

Case	U_1	W_2	W_2^*	U_2^*	Probability
1	5	3	3	8	0.36
2	5	-3	-3	2	0.24
3	-1	3	0	-1	0.24
4	-1	-3	0	-1	0.16

Then, $\tilde{\phi}(2,2) = 0.36 + 0.24 = 0.60$. Note that for U_2 the process would continue for cases 3 and 4 producing values of 2 and -4. But our process is not allowed to recover from ruin and so case three must be forced to remain negative. □

6.3.2 Evaluating the probability of ruin

There are three ways to evaluate the ruin probability. One way that is always available is simulation. Just as the aggregate loss distribution can be simulated, the progress of surplus can also be simulated. For extremely complicated models (for example, one encompassing medical benefits including hospitalization, prescription drugs, and outpatient visits, as well as including random inflation, interest rates, and utilization rates) this may be the only way to proceed. For more modest settings the other two methods will work well. The first is a brute force method that has few restrictions, and the second is an inversion method which has some restrictions.

6.3.2.1 Evaluation by convolutions For any practical use of this method, the distributions of all the random variables involved should be discrete and have finite support. If they are not, some discrete approximation should be constructed. The calculation is done recursively, using (6.1). For notational purposes, suppose we have obtained the discrete pf of U_{t-1}^*. Then the ruin probability is $\tilde{\psi}(u, t-1) = \Pr(U_{t-1}^* < 0)$ and the distribution of non-negative surplus is $f_j = \Pr(U_{t-1}^* = u_j)$, $j = 1, 2, \ldots, n$, where $u_j \geq 0$ for all j and u_n is the largest possible value of U_{t-1}^*. We have assumed that for each positive value of U_{t-1}^*, the distribution of W_t is known. Let $g_{j,k} = \Pr(W_t =$

$w_{j,k}|U^*_{t-1} = u_j)$. We have left open the possibility that even the values W_t may depend on u_j. We then obtain the probabilities of U^*_t by convolution. First,

$$\begin{aligned}
\tilde{\psi}(u,t) &= \tilde{\psi}(u,t-1) + \Pr(U^*_{t-1} \geq 0 \text{ and } U^*_{t-1} + W_t < 0) \\
&= \tilde{\psi}(u,t-1) + \sum_{j=1}^{n} \Pr(U^*_{t-1} + W_t < 0 | U^*_{t-1} = u_j) \Pr(U^*_{t-1} = u_j) \\
&= \tilde{\psi}(u,t-1) + \sum_{j=1}^{n} \Pr(u_j + W_t < 0 | U^*_{t-1} = u_j) f_j \\
&= \tilde{\psi}(u,t-1) + \sum_{j=1}^{n} \sum_{w_{j,k} < -u_j} g_{j,k} f_j.
\end{aligned}$$

Then,

$$\begin{aligned}
\Pr(U^*_t = x) &= \Pr(U^*_{t-1} \geq 0 \text{ and } U^*_{t-1} + W_t = x) \\
&= \sum_{j=1}^{n} \Pr(U^*_{t-1} \geq 0 \text{ and } U^*_{t-1} + W_t = x | U^*_{t-1} = u_j) \\
&\quad \times \Pr(U^*_{t-1} = u_j) \\
&= \sum_{j=1}^{n} \Pr(u_j + W_t = x | U^*_{t-1} = u_j) f_j \\
&= \sum_{j=1}^{n} \sum_{w_{j,k} + u_j = x} g_{j,k} f_j.
\end{aligned}$$

Although these formulas look a bit intimidating, they are fairly easy to implement. Consider the following example.

Example 6.4 *Suppose that annual losses can assume the values 0, 2, 4, and 6, with probabilities 0.4, 0.3, 0.2, and 0.1, respectively. Further suppose that the initial surplus is 2, and a premium of 2.5 is collected at the beginning of each year. Interest is earned at 10% on any surplus available at the beginning of the year because losses are paid at the end of the year. In addition, a rebate of 0.5 is given in any year in which there are no losses. Determine the survival probability at the end of each of the first two years.*

First note that the rebate cannot be such that it is applied to the next year's premium. Doing so would require that we not only begin the year knowing the surplus, we must also know if a rebate was to be provided.

At time zero, $\tilde{\psi}(2,0) = 0$ and $f_1 = \Pr(U^*_0 = 2) = 1$. The possible values of $w_{1,k}$ are given in Table 6.1, along with the probabilities, $g_{1,k}$.

For example, $w_{1,1}$ is based on a premium of 2.5, interest of 0.45 (on the surplus after collection of the premium), a loss payment of 0, and a rebate of

Table 6.1 ws and gs for Example 6.4

k	$w_{1,k}$	$g_{1,k}$
1	2.45	0.4
2	0.95	0.3
3	−1.05	0.2
4	−3.05	0.1

0.5. To evaluate $\tilde{\psi}(2,1)$, observe that the only value of $w_{1,k}$ which is below $-u_1 = -2$ is $w_{1,4}$ and so $\tilde{\psi}(2,1) = 0.1$. It is also easy to see that the only values of x that will have positive probability are those that are $2+w_{1,k}$. This gives the values for the distribution of U_1^* as in Table 6.2.

Table 6.2 U_1^* for Example 6.4

j	u_j	f_j
1	0.95	0.2
2	2.95	0.3
3	4.45	0.4

The remaining probability is at $\Pr(U_1^* = -1.05) = 0.1$.

One way to visualize year 2 is with a two-way table providing all of the combinations of u_j and $w_{j,k}$. The entries in Table 6.3 are $u_j + w_{j,k}$, $g_{j,k}$. Only the sums need be presented, because they are the only interesting quantities.

Table 6.3 $u + w$s and gs for Example 6.4

			k			
j	u_j	f_j	1	2	3	4
1	0.95	0.2	3.295, 0.4	1.795, 0.3	−0.205, 0.2	−2.205, 0.1
2	2.95	0.3	5.495, 0.4	3.995, 0.3	1.995, 0.2	−0.005, 0.1
3	4.45	0.4	7.145, 0.4	5.645, 0.3	3.645, 0.2	1.645, 0.1

The joint probability for any cell is the product of f_j from that row and the probability in that cell. The addition to $\tilde{\psi}(2,1)$ is the probability for all the cells with negative entries, that is,

$$\tilde{\psi}(2,2) = 0.1 + 0.2(0.2) + 0.2(0.1) + 0.3(0.1) = 0.19.$$

There are no duplicate values for $w_{i,k}$, so the best we can do is put the values in order and note that they are the new u_j values for the beginning of year 3. They are listed in Table 6.4.

Table 6.4 us for Example 6.4

j	u_j	f_j
1	1.645	0.04
2	1.795	0.06
3	1.995	0.06
4	3.295	0.08
5	3.645	0.08
6	3.995	0.09
7	5.495	0.12
8	5.645	0.12
9	7.145	0.16

The probabilities total 0.81, the complement of $\tilde{\psi}(2,2)$. By the earlier definition, the remaining 0.19 probability is associated with $U_2^* < 0$. □

It should be easy to see that the number of possible u values as well as the number of decimal places can increase rapidly. At some point, rounding would seem to be a good idea. A simple way to do this is to demand that at each period, the only allowable u values are some multiple of h, a span that may need to increase from period to period. When probability is assigned to some value that is not a multiple of h, it is distributed to the two nearest values in a way that will preserve the mean (spreading to more values could preserve higher moments).

Example 6.5 (Example 6.4 continued) *Distribute the probabilities for the surplus at the end of year 2 using a span of $h = 2$.*

The probability of 0.04 at 1.645 must be distributed to the points 0 and 2. To preserve the mean, $0.355(0.04)/2 = 0.0071$ is placed at zero and the remaining 0.0329 is placed at 2. The expected value is $0.0071(0)+0.0329(2) = 0.0658$, which matches the original value of 0.04(1.645). The value 0.355 is the distance from the point in question (1.645) to the next span point (2), and the denominator is the span. The probability is then placed at the previous span point. The resulting approximate distribution is given in Table 6.5.

□

6.3.2.2 Evaluation by inversion One of the strengths of the inversion method is that the act of computing a convolution is reduced to a few multiplications. This is true, provided that the random variables are independent. In this case that means that W_t is independent of U_{t-1}. We use a different approach with regard to keeping track of ruin (earlier that was accomplished by freezing U_t^* upon ruin). This idea could also be applied to the direct

Table 6.5 Probabilities for Example 6.5

j	u_j	f_j
1	0	0.0134
2	2	0.189225
3	4	0.258975
4	6	0.2568
5	8	0.0916

convolution approach. This time, let U_t^{**} be U_t conditioned on $U_t \geq 0$. At the end of each period, all probability associated with ruin is redistributed over the outcomes producing non-negative surplus. The year-by-year analysis proceeds as follows.

1. Determine $\varphi_{1,t}(z) = E(e^{izU_{t-1}^{**}})$, the characteristic function of U_{t-1}^{**}.
2. Determine $\varphi_{2,t}(z) = E(e^{izW_t})$, the characteristic function of W_t.
3. Then $\varphi_{3,t}(z) = \varphi_{1,t}(z)\cdot\varphi_{2,t}(z)$ is the characteristic function of $U_{t-1}^{**}+W_t$.
4. Use inversion to determine $f_t(u)$, the pf of $U_{t-1}^{**} + W_t$.
5. Let $r_t = \Pr(U_{t-1}^{**} + W_t < 0)$. This is the probability that, given survival to time $t-1$, the portfolio is ruined at time t.
6. Then $f_t^{**}(u) = f_t(u)/(1 - r_t)$ for $u \geq 0$ is the pf of U_t^{**}.
7. The probability of ruin by time t is then $\tilde{\psi}(u, t) = \tilde{\psi}(u, t-1) + r_t[1 - \tilde{\psi}(u, t-1)]$.

The process is initiated by noting that the pf of U_1 can be obtained directly by observing that $U_1 = u + W_1$, so all that needs to be done is to shift the arguments of the pf of W_1 by u.

Example 6.6 *Aggregate losses for one year are 0, 2, 4, and 6 with probabilities 0.4, 0.3, 0.2, and 0.1, respectively. Premiums of 2.5 are collected at the beginning of the year and initial surplus is 2. Determine the probability of ruin within the first two years using the fast Fourier transform (FFT).*

The pf of W_t is the same in all years and is given in Table 6.6

With an initial surplus of 2 it is easy to obtain the distribution of U_1. It is given in Table 6.7

This immediately gives $\tilde{\psi}(2, 1) = 0.1$ and the distribution of U_1^{**} is given in Table 6.8

In order for the FFT to work in a simple manner, it is best to have all amounts be positive. This can be accomplished by adding 3.5 to each variable.

Table 6.6 pf of W_t for Example 6.6

w	$\Pr(W = w)$
−3.5	0.1
−1.5	0.2
0.5	0.3
2.5	0.4

Table 6.7 pf of U_1 for Example 6.6

u	$\Pr(U_1 = u)$
−1.5	0.1
0.5	0.2
2.5	0.3
4.5	0.4

The shifted distributions are given in the second and third columns of Table 6.9. Anticipating that the shifted $U_1^{**} + W_2$ will take on values from 0 to 14 with a span of 2, we observe that 8 values are required. This is already a power of 2, so no extra zeros need be added. In Table 6.9 the fourth and fifth columns provide the FFT of the two input variables. They are followed by the product of the two characteristic functions, and then ultimately by the inverse of this characteristic function. The last column is the pf we seek. Of course, in this case it would have been trivial to perform the convolutions, but this way we can also verify that the FFT and its inverse do what they are supposed to.

We must note that the probabilities in the last column are shifted by 7. The actual distribution of the sum is given in Table 6.10. We see that $9/90 = 1/10$ of the probability is associated with negative values and so $\tilde{\psi}(2, 2) = 0.1 + 0.9(0.1) = 0.19$. The conditional distribution, U_2^{**} is also given in Table 6.10.□

Table 6.8 pf of U_1^{**} for Example 6.6

u	$\Pr(U_1^{**} = u)$
0.5	2/9
2.5	3/9
4.5	4/9

Table 6.9 Year 2 ruin calculation for Example 6.6

u	$f_1^{**}(u)$	$f_W(u)$	$\varphi_{1,2}/8$	$\varphi_{2,2}/8$	$\varphi_{3,2}/64$	$f_2(u)$
0	0	1/10	0.125	0.125	0.01563	0
2	0	2/10	$-0.08502 - 0.05724i$	$-0.00518 - 0.09053i$	$-0.00474 + 0.00799i$	0
4	2/9	3/10	$0.02778 + 0.04167i$	$-0.025 + 0.025i$	$-0.00174 - 0.00035i$	2/90
6	3/9	4/10	$-0.02609 - 0.00169i$	$0.03018 - 0.01553i$	$-0.00081 + 0.00035i$	7/90
8	4/9	0	0.04167	-0.025	-0.00104	16/90
10	0	0	$-0.02609 + 0.00169i$	$0.03018 + 0.01553i$	$-0.00081 - 0.00035i$	25/90
12	0	0	$0.02778 - 0.04167i$	$-0.025 - 0.25i$	$-0.00174 + 0.00035i$	24/90
14	0	0	$-0.08502 + 0.05724i$	$-0.00518 + 0.09053i$	$-0.00474 - 0.00799i$	16/90

Table 6.10 Distribution of surplus after year two for Example 6.6

u	$\Pr(U_1^{**} + W_2 = u)$	$\Pr(U_2^{**} = u)$
-3	2/90	0
-1	7/90	0
1	16/90	16/81
3	25/90	25/81
5	24/90	24/81
7	16/90	16/81

6.4 CONTINUOUS TIME MODELS

For the remainder of this chapter we turn to models that examine surplus continuously over time. Because these models tend to be difficult to analyze, we begin by restricting attention to models in which the number of claims has a Poisson distribution. In the discrete time case we found that answers could be obtained by brute force. For the continuous case we find that exact, analytic solutions can be obtained for some situations and approximations and an upper bound can be obtained for many situations. In this section we introduce the Poisson process and the continuous time approach to ruin.

6.4.1 The Poisson process

We consider the basic properties of the Poisson process $\{N_t;\ t \geq 0\}$ representing the number of claims on a portfolio of business. Thus, N_t is the number of claims in $(0, t]$. A formal definition of a Poisson process is now given.

Definition 6.9 *The number of claims process* $\{N_t;\ t \geq 0\}$ *is a* ***Poisson process*** *with rate* $\lambda > 0$ *if*

1. $N_0 = 0$.
2. *The process has stationary and independent increments.*
3. *The number of claims in an interval of length* t *is Poisson distributed with mean* λt. *That is, for all* $s, t > 0$ *we have*

$$\Pr(N_{t+s} - N_s = n) = \frac{(\lambda t)^n e^{-\lambda t}}{n!}, \qquad n = 0, 1, 2, \ldots. \tag{6.2}$$

Stationary increments means that the distribution of the number of claims in a fixed interval depends only on the length of the interval and not on when the interval occurs, for example, there is no trend effect. **Independent increments** means that the number of claims in an interval is statistically independent of the number of claims in any previous interval (not overlapping the present interval). Together, **stationary and independent increments** imply that the process can be thought of intuitively as starting over at any point in time. Actually, the assumption of stationarity in condition 2 in the definition is not necessary since it is implied by condition 3, but it is stated for clarity.

Condition 3 of the definition of a Poisson process may actually be replaced by a more intuitive condition. We introduce the notation $o(h)$ and say "the function $f(x)$ is $o(h)$" if $\lim_{h\to 0} f(h)/h = 0$. Intuitively, $f(x)$ goes to 0 as x goes to 0 faster than $f(x) = x$ itself. Thus, $f(x) = x^2$ is $o(h)$ but $f(x) = \sqrt{x}$ is not $o(h)$. We note that $o(h)$ is simply notational and is not a mathematical quantity, so that $o(h) + o(h) = o(h)$, for example, is a valid statement.

We state, but do not prove, that condition (3) in the definition of a Poisson process may be replaced by

3′. $\Pr(N_{t+s} - N_s = 1) = \lambda t + o(t)$ *and* $\Pr(N_{t+s} - N_s > 1) = o(t)$.

Condition 3′ implies that for a sufficiently short time interval, the probability of exactly one claim occurring in that interval is roughly proportional to the length of that interval, and that multiple claims at a single point in time cannot occur. It is also clear that 3′ implies

$$\Pr(N_{t+s} - N_s = 0) = 1 - \lambda t + o(t).$$

An important property of the Poisson process is that the times between claims are independent and identically exponentially distributed, each with mean $1/\lambda$. To see this, let W_j be the time between the $(j-1)$st and jth claims for $j = 1, 2, 3, \ldots$. Then,

$$\Pr(W_1 > t) = \Pr(N_t = 0) = e^{-\lambda t}$$

and so W_1 is exponential with mean $1/\lambda$. Also,

$$\begin{aligned}
\Pr(W_2 > t | W_1 = s) &= \Pr(W_1 + W_2 > s + t | W_1 = s) \\
&= \Pr(N_{t+s} = 1 | N_s = 1) \\
&= \Pr(N_{t+s} - N_s = 0 | N_s = 1) \\
&= \Pr(N_{t+s} - N_s = 0)
\end{aligned}$$

since the increments are independent. From 3, we then have

$$\Pr(W_2 > t | W_1 = s) = e^{-\lambda t}.$$

Since this is true for all s, $\Pr(W_2 > t) = e^{-\lambda t}$ and W_2 is independent of W_1. Similarly, $W_3, W_4, W_5, \ldots$ are independent and exponentially distributed, each with mean $1/\lambda$.

Finally, we remark that from a fixed point in time $t_0 \geq 0$, the time until the next claim occurs is also exponentially distributed with mean λ^{-1} due to the memoryless property of the exponential distribution ($\Pr(W_{n+1} > t+s | W_{n+1} > s) = e^{-\lambda t}$ for all $s \geq 0$ and $n = 0, 1, 2, \ldots$).

6.4.2 The continuous time problem

The model for claims payments will now be based on the Poisson process. The formal definition follows.

Definition 6.10 *Let the number of claims process* $\{N_t;\ t \geq 0\}$ *be a Poisson process with rate* λ. *Let the individual losses* $\{X_1, X_2, \ldots\}$ *be independent and identically distributed positive random variables, independent of* N_t, *each with cumulative distribution function* $F(x)$ *and mean* $\mu < \infty$. *Thus* X_j *is the amount of the* j*th loss. Let* S_t *be the total loss in* $(0, t]$. *It is given by* $S_t = 0$ *if* $N_t = 0$ *and* $S_t = \sum_{j=1}^{N_t} X_j$ *if* $N_t > 0$. *Then for fixed* t, S_t *has a compound Poisson distribution. The process* $\{S_t;\ t \geq 0\}$ *is said to be a* ***compound Poisson process****. Since* $\{N_t;\ t \geq 0\}$ *has stationary and independent increments, so does* $\{S_t; t \geq 0\}$. *Also,* $E(S_t) = E(N_t)E(X_j) = (\lambda t)(\mu) = \mu\lambda t$.

We assume that premiums are payable continuously at constant rate c per unit time. That is, the total net premium in $(0, t]$ is ct and we ignore interest for mathematical simplicity. We further assume that net premiums have a positive loading, that is, $ct > E(S_t)$, which implies that $c > \lambda\mu$. Thus let

$$c = (1 + \theta)\lambda\mu \tag{6.3}$$

where $\theta > 0$ is called the **relative security loading**.

For our model, we have now specified the loss and premium processes. The surplus process is thus

$$U_t = U_0 + ct - S_t, \quad t \geq 0.$$

6.5 THE ADJUSTMENT COEFFICIENT AND LUNDBERG'S INEQUALITY

In this section we determine a special quantity and then show that it can be used to obtain a bound on the value of $\psi(u)$. While it is only a bound, it is easy to obtain, and as an upper bound it provides a conservative estimate.

6.5.1 The adjustment coefficient

It is difficult to motivate the definition of the adjustment coefficient, so we just state it.

Definition 6.11 *Let* κ *be the unique positive solution to the equation*

$$1 + (1 + \theta)\mu\kappa = E(e^{\kappa X}). \tag{6.4}$$

If such a value exists, it is called the ***adjustment coefficient****.*

To see that there may be a solution, consider the two lines in the (t, y) plane given by $y_1(t) = 1 + (1+\theta)\mu t$ and $y_2(t) = E(e^{tX})$, the moment-generating function, where X is the individual loss amount random variable. Now, $y_1(t)$ is a straight line with positive slope $(1+\theta)\mu$. The mgf may not exist at all, or may exist only for some values of t. Assume for this discussion that the mgf exists for all non-negative t. Then $y_2'(t) = E(Xe^{tX}) > 0$ and $y_2''(t) = E(X^2e^{tX}) > 0$. Since $y_1(0) = y_2(0) = 1$, the two curves intersect when $t = 0$. But $y_2'(0) = E(X) = \mu < (1+\theta)\mu = y_1'(0)$. Thus as t increases from 0 the curve $y_2(t)$ initially falls below $y_1(t)$, but since $y_2'(t) > 0$ and $y_2''(t) > 0$, eventually $y_2(t)$ will cross $y_1(t)$ at a point $\kappa > 0$. The point κ is the adjustment coefficient.

We remark that there may not be a positive solution to (6.4); for example, if the single claim amount distribution has no moment generating function (e.g., Pareto, lognormal).

Example 6.7 (Exponential claim amounts) *If X has an exponential distribution with mean μ, determine the adjustment coefficient.*

We have $F(x) = 1 - e^{-x/\mu}$, $x > 0$. Then, $E(e^{tX}) = (1-\mu t)^{-1}$, $t < \mu^{-1}$. Thus, from (6.4), κ satisfies

$$1 + (1+\theta)\mu\kappa = (1-\mu\kappa)^{-1}. \tag{6.5}$$

As predicted, $\kappa = 0$ is one solution and the positive solution is $\kappa = \theta/[\mu(1+\theta)]$. The graph in Figure 6.3 displays plots of the left- and right-hand sides of (6.5) for the case $\theta = 0.2$ and $\mu = 1$. They intersect at 0 and at the adjustment coefficient, $\kappa = 0.2/1.2 = 0.1667$. □

Example 6.8 (A gamma distribution) *Suppose that $\theta = 2$ and the gamma distribution has $\alpha = 2$. To avoid confusion, let β be the scale parameter. Determine the adjustment coefficient.*

The single claim size density is

$$f(x) = \beta^{-2}xe^{-x/\beta}, \qquad x > 0.$$

For the gamma distribution $\mu = 2\beta$ and

$$E(e^{tX}) = \int_0^\infty e^{tx}f(x)dx = (1-\beta t)^{-2}, \qquad \beta t < 1.$$

Then from (6.4) we obtain

$$1 + 6\kappa\beta = (1-\beta\kappa)^{-2}$$

which may be rearranged as

$$6\beta^3\kappa^3 - 11\beta^2\kappa^2 + 4\beta\kappa = 0.$$

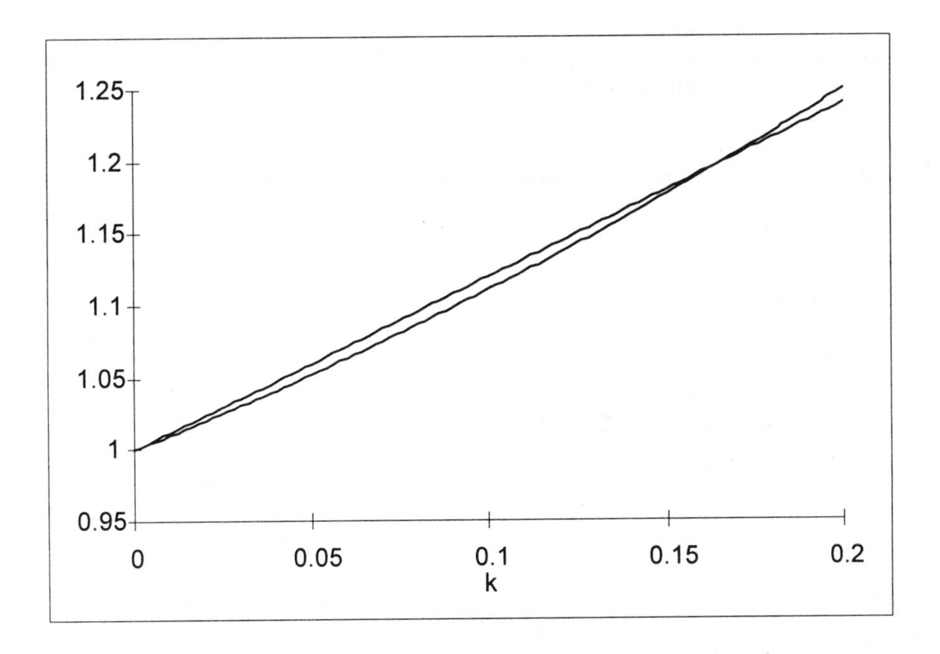

Fig. 6.3 Left and right sides of the adjustment coefficient equation

This is easily factored as

$$\kappa\beta(2\kappa\beta - 1)(3\kappa\beta - 4) = 0.$$

The adjustment coefficient is the only root that solves the original equation,[1] namely $\kappa = 1/(2\beta)$. □

For general claim amount distributions, it is not possible to explicitly solve for κ as was done in the above two examples. Normally, one must resort to numerical methods, many of which require an initial "guess" as to the value of κ. To find such a value, note that for (6.4) we may write

$$\begin{aligned} 1 + (1+\theta)\mu\kappa &= E(e^{\kappa X}) \\ &= E(1 + \kappa X + \kappa^2 X^2/2 + \cdots) \\ &> E(1 + \kappa X + \kappa^2 X^2/2) \\ &= 1 + \kappa\mu + \kappa^2 E(X^2)/2. \end{aligned}$$

Then subtraction of 1 from both sides of the inequality and division by κ results in

$$\kappa < 2\theta\mu/E(X^2). \tag{6.6}$$

[1] Of the two roots, the larger one, $4/(3\beta)$, is not a legitimate argument for the mgf. The mgf exists only for values less than $1/\beta$. When solving such equations, the adjustment coefficient will always be the smallest positive solution.

The right-hand side of (6.6) is usually a satisfactory initial value of κ. Other inequalities for κ are given in the exercises.

Example 6.9 *The aggregate loss random variable has variance equal to three times the mean. Determine a bound on the adjustment coefficient.*

For the compound Poisson distribution, $E(S_t) = \mu\lambda t$, $Var(S_t) = \lambda t E(X^2)$, and so $E(X^2) = 3\mu$. Hence, from (6.6), $\kappa < 2\theta/3$. □

Define

$$H(t) = 1 + (1+\theta)\mu t - E(e^{tX}). \tag{6.7}$$

Then the adjustment coefficient $\kappa > 0$ satisfies $H(\kappa) = 0$. To solve this equation, use the Newton–Raphson formula,

$$\kappa_{j+1} = \kappa_j - \frac{H(\kappa_j)}{H'(\kappa_j)}$$

where

$$H'(t) = (1+\theta)\mu - E(Xe^{tX})$$

beginning with an initial value κ_0. Since $H(0) = 0$, care must be taken so as not to converge to the value 0. Also note that it may be easier to obtain $E(Xe^{tX})$ by differentiating the mgf than by computing the expected value.

Example 6.10 *Suppose the Poisson parameter is $\lambda = 4$ and the premium rate is $c = 7$. Further suppose the individual loss amount distribution is given by*

$$\Pr(X = 1) = 0.6, \quad \Pr(X = 2) = 0.4.$$

Determine the adjustment coefficient.

We have

$$\mu = E(X) = (1)(0.6) + (2)(0.4) = 1.4$$

and

$$E(X^2) = (1)^2(0.6) + (2)^2(0.4) = 2.2.$$

Then $\theta = c(\lambda\mu)^{-1} - 1 = 7(5.6)^{-1} - 1 = 0.25$. From (6.6), we know that κ must be less than $\kappa_0 = 2(0.25)(1.4)/2.2 = 0.3182$. Now,

$$E(e^{tX}) = 0.6e^t + 0.4e^{2t}$$

and so from (6.7)

$$H(t) = 1 + 1.75t - 0.6e^t - 0.4e^{2t}.$$

We also have

$$E(Xe^{tX}) = (1e^t)(0.6) + (2e^{2t})(0.4)$$

and so

$$H'(t) = 1.75 - 0.6e^t - 0.8e^{2t}.$$

Our initial guess is $\kappa_0 = 0.3182$. Then $H(\kappa_0) = -0.02381$ and $H'(\kappa_0) = -0.5865$. Thus, an updated estimate of κ is

$$\kappa_1 = 0.3182 - (-0.02381)/(-0.5865) = 0.2776.$$

Then $H(0.2776) = -0.003091$, $H'(0.2776) = -0.4358$, and

$$\kappa_2 = 0.2776 - (-0.003091)/(-0.4358) = 0.2705.$$

Continuing on, we get $\kappa_3 = 0.2703$, $\kappa_4 = 0.2703$, and so the adjustment coefficient is $\kappa = 0.2703$. □

There is another form for the equation (6.4) which is often useful. We have, using integration by parts,

$$\int_0^\infty e^{\kappa x} dF(x) = -e^{\kappa x}[1 - F(x)]|_0^\infty + \kappa \int_0^\infty e^{\kappa x}[1 - F(x)]dx.$$

Now,

$$0 \le e^{\kappa x}[1 - F(x)] = e^{\kappa x} \int_x^\infty dF(y) \le \int_x^\infty e^{\kappa y} dF(y)$$

and $\lim_{x\to\infty} \int_x^\infty e^{\kappa y} dF(y) = 0$ since $E(e^{\kappa X}) = \int_0^\infty e^{\kappa y} dF(y) < \infty$. Thus, $\lim_{x\to\infty} e^{\kappa x}[1 - F(x)] = 0$ and so

$$E(e^{\kappa X}) = \int_0^\infty e^{\kappa x} dF(x) = 1 + \kappa \int_0^\infty e^{\kappa x}[1 - F(x)]dx.$$

Therefore, combining with (6.4), the following is an alternative definition of κ:

$$1 + \theta = \int_0^\infty e^{\kappa x} f_e(x)dx \tag{6.8}$$

where

$$f_e(x) = \frac{1 - F(x)}{\mu}, \qquad x > 0 \tag{6.9}$$

is a probability density function as was shown in Subsection 2.7.2.

6.5.2 Lundberg's inequality

The first main use of the adjustment coefficient lies in the following result.

Theorem 6.1 *Suppose $\kappa > 0$ satisfies (6.4). Then the probability of ruin $\psi(u)$ satisfies*

$$\psi(u) \le e^{-\kappa u}, \qquad u \ge 0. \tag{6.10}$$

Proof: Let $\psi_n(u)$ be the probability that ruin occurs on or before the nth claim for $n = 0, 1, 2, \ldots$. We will prove by induction on n that $\psi_n(u) \leq e^{-\kappa u}$. Obviously, $\psi_0(u) = 0 \leq e^{-\kappa u}$. Now assume that $\psi_n(u) \leq e^{-\kappa u}$ and we wish to show that $\psi_{n+1}(u) \leq e^{-\kappa u}$. Let us consider what happens on the first claim. The time until the first claim occurs is exponential with probability density function $\lambda e^{-\lambda t}$. If the claim occurs at time $t > 0$, the surplus available to pay the claim at time t is $u + ct$. Thus, ruin occurs on the first claim if the amount of the claim exceeds $u + ct$. The probability that this happens is $1 - F(u + ct)$. If the amount of the claim is x where $0 \leq x \leq u + ct$, ruin does not occur on the first claim. After payment of the claim, there is still a surplus of $u + ct - x$ remaining. Ruin can still occur on the next n claims. Since the surplus process has stationary and independent increments, this is the same probability as if we had started at the time of the first claim with initial reserve $u + ct - x$ and been ruined in the first n claims. Thus, by the law of total probability, we have the recursive equation

$$\psi_{n+1}(u) = \int_0^\infty \left[1 - F(u + ct) + \int_0^{u+ct} \psi_n(u + ct - x)dF(x)\right] \lambda e^{-\lambda t} dt.$$

Thus, using the inductive hypothesis,

$$\begin{aligned}
\psi_{n+1}(u) &= \int_0^\infty \left[\int_{u+ct}^\infty dF(x) + \int_0^{u+ct} \psi_n(u + ct - x)dF(x)\right] \lambda e^{-\lambda t} dt \\
&\leq \int_0^\infty \left[\int_{u+ct}^\infty e^{-\kappa(u+ct-x)} dF(x)\right. \\
&\quad \left. + \int_0^{u+ct} e^{-\kappa(u+ct-x)} dF(x)\right] \lambda e^{-\lambda t} dt
\end{aligned}$$

where we have also used the fact that $-\kappa(u + ct - x) > 0$ when $x > u + ct$. Combining the two inner integrals gives

$$\begin{aligned}
\psi_{n+1}(u) &\leq \int_0^\infty \left[\int_0^\infty e^{-\kappa(u+ct-x)} dF(x)\right] \lambda e^{-\lambda t} dt \\
&= \lambda e^{-\kappa u} \int_0^\infty e^{-\kappa ct} \left[\int_0^\infty e^{\kappa x} dF(x)\right] e^{-\lambda t} dt \\
&= \lambda e^{-\kappa u} \int_0^\infty e^{-(\lambda+\kappa c)t} \left[E(e^{\kappa X})\right] dt \\
&= \lambda E\left(e^{\kappa X}\right) e^{-\kappa u} \int_0^\infty e^{-(\lambda+\kappa c)t} dt \\
&= \frac{\lambda E\left(e^{\kappa X}\right)}{\lambda + \kappa c} e^{-\kappa u}.
\end{aligned}$$

But from (6.4) and (6.3)

$$\lambda E(e^{\kappa X}) = \lambda[1 + (1 + \theta)\kappa\mu] = \lambda + \kappa(1 + \theta)\lambda\mu = \lambda + \kappa c$$

and so $\psi_{n+1}(u) \le e^{-\kappa u}$. Therefore, $\psi_n(u) \le e^{-\kappa u}$ for all n and so $\psi(u) = \lim_{n\to\infty} \psi_n(u) \le e^{-\kappa u}$. □

This result is important since it may be used to examine the interplay between the level of surplus u and the premium loading θ, both parameters which are under the control of the insurer. Suppose one is willing to tolerate a probability of ruin of α (e.g., $\alpha = 0.01$) and a surplus of u is available. Then a loading of

$$\theta = \frac{u\left\{E\left[\exp\left(-\frac{\log\alpha}{u}X\right)\right]-1\right\}}{-\mu\log\alpha} - 1$$

ensures that (6.4) is satisfied by $\kappa = (-\log\alpha)/u$. Then, by Theorem 6.1, $\psi(u) \le e^{-\kappa u} = e^{\log\alpha} = \alpha$. On the other hand, if a specified loading of θ is desired, the surplus u required to ensure a ruin probability of no more than α is given by

$$u = (-\log\alpha)/\kappa$$

since $\psi(u) \le e^{-\kappa u} = e^{\log\alpha} = \alpha$ as before.

Also, (6.10) allows us to show that

$$\psi(\infty) = \lim_{u\to\infty} \psi(u) = 0. \tag{6.11}$$

Because the ruin probability is a probability, we have

$$0 \le \psi(u) \le e^{-\kappa u} \tag{6.12}$$

and thus

$$0 \le \lim_{u\to\infty} \psi(u) \le \lim_{u\to\infty} e^{-\kappa u} = 0$$

which establishes (6.11). We then have that for the survival probability,

$$\phi(\infty) = 1. \tag{6.13}$$

6.6 AN INTEGRODIFFERENTIAL EQUATION

We now consider the problem of finding an explicit formula for the ruin probability $\psi(u)$ or (equivalently) the survival probability $\phi(u)$. It is useful to consider a slightly more general function.

Definition 6.12 $G(u, y) = \Pr$*(ruin occurs with initial reserve u, and the deficit immediately after ruin occurs is at most y), $u \ge 0, y \ge 0$.*

Thus the surplus immediately after ruin is between 0 and $-y$. We then have

$$\psi(u) = \lim_{y\to\infty} G(u, y), \qquad u \ge 0. \tag{6.14}$$

We have the following result.

Theorem 6.2 *The function $G(u, y)$ satisfies the equation*

$$\frac{\partial}{\partial u}G(u,y) = \frac{\lambda}{c}G(u,y) - \frac{\lambda}{c}\int_0^u G(u-x,y)dF(x) - \frac{\lambda}{c}[F(u+y)-F(u)], \quad u \geq 0. \tag{6.15}$$

Proof: We consider what happens in the first h time units, where h is small. Because the number of claims process is a Poisson process, either 0 or 1 claims will occur (the probability of more than one claim is $o(h)$). If no claims occurs (with probability $1 - \lambda h$ (plus $o(h)$)), then, by stationary and independent increments, ruin with a deficit of at most y would then occur with probability $G(u + ch, y)$. If a claim occurs (with probability λh (plus $o(h)$) for amount x where $0 \leq x \leq u + ch$, ruin with a deficit of at most y has not yet occurred but would then occur with probability $G(u + ch - x, y)$, whereas if $u + ch < x \leq u + ch + y$, ruin occurs and the deficit is at most y (if $x > u + ch + y$ the deficit would exceed y). Thus, by the law of total probability

$$\begin{aligned} G(u,y) &= (1-\lambda h)G(u+ch,y) + \lambda h\int_0^{u+ch} G(u+ch-x,y)dF(x) \\ &\quad + \lambda h\left[F(u+ch+y) - F(u+ch)\right] + o(h). \end{aligned}$$

This may be rewritten as

$$\begin{aligned} c\frac{G(u+ch,y)-G(u,y)}{ch} &= \lambda G(u+ch,y) \\ &\quad -\lambda\int_0^{u+ch} G(u+ch-x,y)dF(x) \\ &\quad -\lambda\left[F(u+ch+y)-F(u+ch)\right] + \frac{o(h)}{h}. \end{aligned}$$

Let $h \to 0$ and divide by c to get

$$\frac{\partial}{\partial u}G(u,y) = \frac{\lambda}{c}G(u,y) - \frac{\lambda}{c}\int_0^u G(u-x,y)dF(x) - \frac{\lambda}{c}[F(u+y)-F(u)].$$

□

We now determine an explicit formula for $G(0, y)$.

Theorem 6.3 *The function $G(0, y)$ is given by*

$$G(0,y) = \frac{\lambda}{c}\int_0^y \{1-F(x)\}\,dx, \qquad y \geq 0. \tag{6.16}$$

Proof: First note that

$$0 \leq G(u,y) \leq \psi(u) \leq e^{-\kappa u}$$

and thus

$$0 \leq G(\infty,y) = \lim_{u\to\infty} G(u,y) \leq \lim_{u\to\infty} e^{-\kappa u} = 0,$$

and therefore $G(\infty,y)=0$. Also,

$$\int_0^\infty G(u,y)du \leq \int_0^\infty e^{-\kappa u}du = \kappa^{-1} < \infty.$$

Now let $\tau(y) = \int_0^\infty G(u,y)du$ and we know that $0 < \tau(y) < \infty$. Then, integrate (6.15) with respect to u from 0 to ∞ to get, using the above,

$$-G(0,y) = \frac{\lambda}{c}\tau(y) - \frac{\lambda}{c}\int_0^\infty \int_0^u G(u-x,y)dF(x)du - \frac{\lambda}{c}\int_0^\infty [F(u+y)-F(u)]du.$$

Interchanging the order of integration in the double integral yields

$$\begin{aligned} G(0,y) &= -\frac{\lambda}{c}\tau(y) + \frac{\lambda}{c}\int_0^\infty \int_x^\infty G(u-x,y)dudF(x) \\ &\quad +\frac{\lambda}{c}\int_0^\infty [F(u+y)-F(u)]du \end{aligned}$$

and changing the variable of integration from u to $v = u - x$ in the inner integral of the double integral results in

$$\begin{aligned} G(0,y) &= -\frac{\lambda}{c}\tau(y) + \frac{\lambda}{c}\int_0^\infty \int_0^\infty G(v,y)dvdF(x) \\ &\quad +\frac{\lambda}{c}\int_0^\infty [F(u+y)-F(u)]du \\ &= -\frac{\lambda}{c}\tau(y) + \frac{\lambda}{c}\int_0^\infty \tau(y)dF(x) + \frac{\lambda}{c}\int_0^\infty [F(u+y)-F(u)]du. \end{aligned}$$

Since $\int_0^\infty dF(x) = 1$, the first two terms on the right-hand side cancel, and so

$$\begin{aligned} G(0,y) &= \frac{\lambda}{c}\int_0^\infty [F(u+y)-F(u)]du \\ &= \frac{\lambda}{c}\int_0^\infty [1-F(u)]du - \frac{\lambda}{c}\int_0^\infty [1-F(u+y)]du. \end{aligned}$$

Then change the variable from u to $x = u$ is the first integral and from u to $x = u + y$ in the second integral. The result is

$$G(0,y) = \frac{\lambda}{c}\int_0^\infty [1-F(x)]dx - \frac{\lambda}{c}\int_y^\infty [1-F(x)]dx = \frac{\lambda}{c}\int_0^y [1-F(x)]dx.$$

□

We remark that (6.16) holds even if there is no adjustment coefficient. The function $G(0, y)$ is itself of interest later, but for now we shall return to the analysis of $\phi(u)$.

Theorem 6.4 *The survival probability with no initial reserve satisfies*

$$\phi(0) = \frac{\theta}{1+\theta}. \tag{6.17}$$

Proof: Recall that $\mu = \int_0^\infty [1-F(x)]dx$ and note that from (6.16),

$$\psi(0) = \lim_{y\to\infty} G(0,y) = \frac{\lambda}{c}\int_0^\infty [1-F(x)]dx = \frac{\lambda\mu}{c} = \frac{1}{1+\theta}.$$

Thus, $\phi(0) = 1 - \psi(0) = \theta/(1+\theta)$. □

The general solution to $\phi(u)$ may be obtained from the following integrodifferential equation subject to the initial condition (6.17).

Theorem 6.5 *The probability of ultimate survival $\phi(u)$ satisfies*

$$\phi'(u) = \frac{\lambda}{c}\phi(u) - \frac{\lambda}{c}\int_0^u \phi(u-x)dF(x), \qquad u \geq 0. \tag{6.18}$$

Proof: From (6.15) with $y \to \infty$ and (6.14),

$$\psi'(u) = \frac{\lambda}{c}\psi(u) - \frac{\lambda}{c}\int_0^u \psi(u-x)dF(x) - \frac{\lambda}{c}[1-F(u)], \qquad u \geq 0. \tag{6.19}$$

In terms of the survival probability $\phi(u) = 1 - \psi(u)$, (6.19) may be expressed as

$$\begin{aligned}
-\phi'(u) &= \frac{\lambda}{c}[1-\phi(u)] - \frac{\lambda}{c}\int_0^u [1-\phi(u-x)]dF(x) - \frac{\lambda}{c}[1-F(u)] \\
&= -\frac{\lambda}{c}\phi(u) - \frac{\lambda}{c}\int_0^u dF(x) + \frac{\lambda}{c}\int_0^u \phi(u-x)dF(x) + \frac{\lambda}{c}F(u) \\
&= -\frac{\lambda}{c}\phi(u) + \frac{\lambda}{c}\int_0^u \phi(u-x)dF(x)
\end{aligned}$$

since $F(u) = \int_0^u dF(x)$. The result then follows. □

It is largely a matter of taste whether one uses (6.18) or (6.19). We shall often use (6.18) since it is slightly simpler algebraically. Unfortunately, the solution for general $F(x)$ is rather complicated and we shall defer this general solution to Section 6.8. At this point we shall attempt to obtain the solution for some special choices of $F(x)$.

6.6.1 Analytic solutions for certain claim size distributions

In this section, we obtain analytic expressions for the ruin probabilities. That is, in certain cases we can eliminate the integral term on the right-hand side of (6.18) by differentiation. This results in a differential equation rather than an integrodifferential equation which is easier to solve. This is possible for certain claim size distributions.

Example 6.11 (The exponential distribution) *Suppose, as in Example 6.7, that* $F(x) = 1 - e^{-x/\mu}$, $x > 0$. *Determine* $\phi(u)$.

In this case (6.18) becomes

$$\phi'(u) = \frac{\lambda}{c}\phi(u) - \frac{\lambda}{\mu c}\int_0^u \phi(u-x)e^{-x/\mu}dx.$$

Change variables in the integral from x to $y = u - x$ to obtain

$$\phi'(u) = \frac{\lambda}{c}\phi(u) - \frac{\lambda}{\mu c}e^{-u/\mu}\int_0^u \phi(y)e^{y/\mu}dy. \tag{6.20}$$

We wish to eliminate the integral term in (6.20) so we differentiate with respect to u. This gives

$$\phi''(u) = \frac{\lambda}{c}\phi'(u) + \frac{\lambda}{\mu^2 c}e^{-u/\mu}\int_0^u \phi(y)e^{y/\mu}dy - \frac{\lambda}{\mu c}\phi(u).$$

The integral term can be eliminated using (6.20) to produce

$$\phi''(u) = \frac{\lambda}{c}\phi'(u) - \frac{\lambda}{\mu c}\phi(u) + \frac{1}{\mu}\left[\frac{\lambda}{c}\phi(u) - \phi'(u)\right]$$

which simplifies to

$$\phi''(u) = \left(\frac{\lambda}{c} - \frac{1}{\mu}\right)\phi'(u).$$

This may be rewritten as

$$\frac{\phi''(u)}{\phi'(u)} = -\frac{\theta}{\mu(1+\theta)}.$$

Integrating with respect to u gives

$$\log\phi'(u) = -\frac{\theta u}{\mu(1+\theta)} + K_1.$$

From (6.20) with $u = 0$ and using (6.17),

$$\begin{aligned}\phi'(0) &= \frac{\lambda}{c}\frac{\theta}{1+\theta}\\ &= \frac{\lambda}{\lambda\mu(1+\theta)}\frac{\theta}{1+\theta}\\ &= \frac{\theta}{\mu(1+\theta)^2}\end{aligned}$$

and so

$$K_1 = \log\left[\frac{\theta}{\mu(1+\theta)^2}\right].$$

Thus,

$$\phi'(u) = \frac{\theta}{\mu(1+\theta)^2}\exp\left[-\frac{\theta u}{\mu(1+\theta)}\right]$$

which may be integrated again to give

$$\phi(u) = -\frac{1}{1+\theta}\exp\left[-\frac{\theta u}{\mu(1+\theta)}\right] + K_2.$$

Now (6.17) gives $\phi(0) = \theta/(1+\theta)$, and so with $u = 0$, we have $K_2 = 1$. Thus

$$\phi(u) = 1 - \frac{1}{1+\theta}\exp\left[-\frac{\theta u}{\mu(1+\theta)}\right]$$

is the required probability. □

We next prove a result about differential equations upon which we will rely heavily in the remainder of this section.

Theorem 6.6 *Suppose that the function $k(x)$ satisfies*

$$k''(x) + bk'(x) + ck(x) = 0. \tag{6.21}$$

If the quadratic equation

$$r^2 + br + c = 0 \tag{6.22}$$

has real, distinct roots r_1 and r_2, then the solution to (6.21) is given by

$$k(x) = A_1 e^{r_1 x} + A_2 e^{r_2 x} \tag{6.23}$$

where A_1 and A_2 are arbitrary constants.

Proof: We introduce the function $m(x)$ defined by $k(x) = e^{r_1 x}m(x)$. Then

$$k'(x) = r_1 e^{r_1 x}m(x) + e^{r_1 x}m'(x)$$

and

$$k''(x) = r_1^2 e^{r_1 x}m(x) + 2r_1 e^{r_1 x}m'(x) + e^{r_1 x}m''(x).$$

Substitute into (6.21) to obtain

$$\begin{aligned} 0 &= e^{r_1 x}m''(x) + 2r_1 e^{r_1 x}m'(x) + r_1^2 e^{r_1 x}m(x) \\ &\quad + b[e^{r_1 x}m'(x) + r_1 e^{r_1 x}m(x)] + ce^{r_1 x}m(x). \end{aligned}$$

Divide by $e^{r_1 x}$ and rearrange to obtain

$$0 = m''(x) + (b + 2r_1)m'(x) + (r_1^2 + br_1 + c)m(x). \tag{6.24}$$

Since r_1 is a root of (6.22), we must have $r_1^2+br_1+c=0$ and the term involving $m(x)$ disappears. Also, $r^2+br+c=(r-r_1)(r-r_2)=r^2-(r_1+r_2)r+r_1r_2$, and so $b=-r_1-r_2$. Thus (6.24) becomes

$$m''(x)+(r_1-r_2)m'(x)=0.$$

We then have

$$m''(x)/m'(x)=r_2-r_1$$

which may be integrated to give

$$\log m'(x)=(r_2-r_1)x+\log C_1$$

or

$$m'(x)=C_1e^{(r_2-r_1)x}.$$

Integrating again gives

$$m(x)=A_1+A_2e^{(r_2-r_1)x}$$

where $A_2=C_1(r_2-r_1)^{-1}$. But $k(x)=e^{r_1x}m(x)$ and (6.23) follows. □

For notational purposes, it is convenient to define for any $n>0$ the function

$$r_n(u)=\int_0^u e^{ny}\phi(y)dy. \tag{6.25}$$

We will now illustrate the method of solution by way of example.

Example 6.12 (A mixture of exponential distributions) *Suppose* $\theta=4/11$ *and the single claim size density is*

$$F'(x)=e^{-3x}+\frac{10}{3}e^{-5x},\quad x>0.$$

Determine $\phi(u)$.

In this case, $F'(x)=\frac{1}{3}(3e^{-3x})+\frac{2}{3}(5e^{-5x})$ and the mean is $\mu=\frac{1}{3}(\frac{1}{3})+\frac{2}{3}(\frac{1}{5})=11/45$. Then

$$c/\lambda=\mu(1+\theta)=1/3.$$

Then (6.18) becomes

$$\begin{aligned}\phi'(u) &= 3\phi(u)-3\int_0^u\phi(u-x)\{e^{-3x}+\frac{10}{3}e^{-5x}\}dx\\ &= 3\phi(u)-3e^{-3u}r_3(u)-10e^{-5u}r_5(u).\end{aligned} \tag{6.26}$$

We have changed variables of integration from x to $y=u-x$ and used the definition (6.25). To eliminate the unknown functions $r_3(u)$ and $r_5(u)$, we differentiate (6.26). This gives

$$\phi''(u)=3\phi'(u)+9e^{-3u}r_3(u)-3\phi(u)+50e^{-5u}r_5(u)-10\phi(u).$$

We may eliminate $10e^{-5u}r_5(u)$ in this equation by solving for it in (6.26). Thus, we have

$$\begin{aligned}\phi''(u) &= 3\phi'(u) - 13\phi(u) + 9e^{-3u}r_3(u) \\ &\quad +5[-\phi'(u) + 3\phi(u) - 3e^{-3u}r_3(u)].\end{aligned}$$

In other words,

$$\phi''(u) = -2\phi'(u) + 2\phi(u) - 6e^{-3u}r_3(u). \tag{6.27}$$

To eliminate $r_3(u)$, differentiate (6.27) to obtain

$$\phi'''(u) = -2\phi''(u) + 2\phi'(u) + 18e^{-3u}r_3(u) - 6\phi(u).$$

Again we may use (6.27) to replace $6e^{-3u}r_3(u)$, giving

$$\begin{aligned}\phi'''(u) &= -2\phi''(u) + 2\phi'(u) - 6\phi(u) \\ &\quad +3[-\phi''(u) - 2\phi'(u) + 2\phi(u)].\end{aligned}$$

In other words,

$$\phi'''(u) + 5\phi''(u) + 4\phi'(u) = 0. \tag{6.28}$$

Equation (6.28) may be regarded as a differential equation of the form (6.21) for the function $\phi'(u)$. Since $0 = r^2 + 5r + 4 = (r+1)(r+4)$ is satisfied by the roots $r_1 = -1$ and $r_2 = -4$, it follows from Theorem 6.6 that the solution to (6.28) is

$$\phi'(u) = A_1e^{-u} + A_2e^{-4u} \tag{6.29}$$

where the constants A_1 and A_2 need to be determined from the boundary conditions. We may put $u = 0$ into (6.29) and (6.26) and use (6.17) to obtain

$$A_1 + A_2 = \phi'(0) = 3\phi(0) = 3\theta(1+\theta)^{-1} = 4/5. \tag{6.30}$$

We may then integrate (6.29) to get

$$\phi(u) = -A_1e^{-u} - (A_2/4)e^{-4u} + A_3. \tag{6.31}$$

But from (6.13), $\phi(\infty) = 1$, and thus $A_3 = 1$. Again, from (6.17), put $u = 0$ into (6.31) to get

$$A_1 + A_2/4 = 1 - \phi(0) = (1+\theta)^{-1} = 11/15. \tag{6.32}$$

The constants A_1 and A_2 must satisfy (6.30) and (6.32). Thus, $A_1 = 32/45$ and $A_2 = 4/45$. Substitution into (6.31) gives

$$\phi(u) = 1 - \frac{32}{45}e^{-u} - \frac{1}{45}e^{-4u}, \qquad u \geq 0$$

as the required solution. □

Example 6.13 (A gamma distribution). *As in Example 6.8, suppose that* $\theta = 2$ *and the single claim size density is*

$$F'(x) = \beta^{-2} x e^{-x/\beta}, \qquad x > 0.$$

Determine $\phi(u)$.

The mean is $\mu = 2\beta$ and so $c/\lambda = \mu(1+\theta) = 6\beta$. Then (6.18) becomes

$$6\beta\phi'(u) = \phi(u) - \beta^{-2} \int_0^u \phi(u-x) x e^{-x/\beta} dx. \tag{6.33}$$

Let us put the integral in (6.33) in a more convenient form. Change variables from x to $y = u - x$ to get

$$\int_0^u \phi(u-x) x e^{-x/\beta} dx = e^{-u/\beta} \int_0^u \phi(y) e^{y/\beta}(u-y) dy.$$

Recall the definition $r_{1/\beta}(u)$ from (6.25) and use integration by parts. This gives

$$\int_0^u \phi(y) e^{y/\beta}(u-y) dy = (u-y) r_{1/\beta}(y)\Big|_{y=0}^{u} + \int_0^u r_{1/\beta}(y) dy = \int_0^u r_{1/\beta}(y) dy.$$

Thus, (6.33) may be expressed as

$$6\beta\phi'(u) = \phi(u) - \beta^{-2} e^{-u/\beta} \int_0^u r_{1/\beta}(y) dy. \tag{6.34}$$

We wish to eliminate the integral term in (6.34) so we differentiate. This gives

$$6\beta\phi''(u) = \phi'(u) + \beta^{-3} e^{-u/\beta} \int_0^u r_{1/\beta}(y) dy - \beta^{-2} e^{-u/\beta} r_{1/\beta}(u).$$

We may replace $\beta^{-2} e^{-u/\beta} \int_0^u r_{1/\beta}(y) dy$ in this equation from (6.34), thus obtaining

$$6\beta\phi''(u) = \phi'(u) - \beta^{-2} e^{-u/\beta} r_{1/\beta}(u) + \beta^{-1}[\phi(u) - 6\beta\phi'(u)].$$

In other words,

$$6\beta\phi''(u) = -5\phi'(u) + \beta^{-1}\phi(u) - \beta^{-2} e^{-u/\beta} r_{1/\beta}(u). \tag{6.35}$$

We now wish to eliminate the term involving $r_{1/\beta}(u)$, so we differentiate a second time. This gives

$$6\beta\phi'''(u) = -5\phi''(u) + \beta^{-1}\phi'(u) + \beta^{-3} e^{-u/\beta} r_{1/\beta}(u) - \beta^{-2}\phi(u).$$

Replace $\beta^{-2} e^{-u/\beta} r_{1/\beta}(u)$ from (6.35) to get

$$\begin{aligned} 6\beta\phi'''(u) &= -5\phi''(u) + \beta^{-1}\phi'(u) - \beta^{-2}\phi(u) \\ &\quad + \beta^{-1}[-6\beta\phi''(u) - 5\phi'(u) + \beta^{-1}\phi(u)]. \end{aligned}$$

This may be rearranged as

$$\phi'''(u) + \frac{11}{6\beta}\phi''(u) + \frac{2}{3\beta^2}\phi'(u) = 0. \tag{6.36}$$

Equation (6.36) is a differential equation of the form (6.21) for the function $\phi'(u)$. Since

$$0 = r^2 + \frac{11}{6\beta}r + \frac{2}{3\beta^2} = \left(r + \frac{1}{2\beta}\right)\left(r + \frac{4}{3\beta}\right)$$

is satisfied by $r_1 = -1/(2\beta)$ and $r_2 = -4/(3\beta)$, it follows from Theorem 6.6 that the solution to (6.36) is

$$\phi'(u) = A_1 e^{-u/(2\beta)} + A_2 e^{-4u/(3\beta)} \tag{6.37}$$

where it remains to determine A_1 and A_2 from the boundary conditions. Put $u = 0$ simultaneously in (6.37) and (6.34) to get, using (6.17),

$$A_1 + A_2 = \phi'(0) = \phi(0)/(6\beta) = 1/(9\beta). \tag{6.38}$$

Now, integrate (6.37) to get

$$\phi(u) = -2\beta A_1 e^{-u/(2\beta)} - \frac{3}{4}\beta A_2 e^{-4u/(3\beta)} + A_3. \tag{6.39}$$

One must have $A_3 = 1$ since (6.13) implies that $\phi(\infty) = 1$. Also, we may use (6.17) with $u = 0$ in (6.39) to get

$$2\beta A_1 + \frac{3}{4}\beta A_2 = \frac{1}{3}. \tag{6.40}$$

Thus, A_1 and A_2 must satisfy (6.38) and (6.40). Solving them for A_1 and A_2 gives $A_1 = 1/(5\beta)$ and $A_2 = -4/(45\beta)$. Substitution into (6.39) gives

$$\phi(u) = 1 - \frac{2}{5}e^{-u/(2\beta)} + \frac{1}{15}e^{-4u/(3\beta)}, \qquad u \geq 0$$

as the required solution. □

In the above examples the method of solution is to eliminate integral terms by differentiation. This technique works for certain types of claim size distributions only, however, and the solutions involve combinations of exponential terms. While the general solution for arbitrary claim size densities is complicated and is considered in Section 6.8, the exponential form for the ruin probabilities holds fairly generally, at least in the tail. This idea is considered in the next section.

6.7 CRAMÉR'S ASYMPTOTIC RUIN FORMULA

In this section we derive one more piece of information about the ruin probability. We shall integrate the integrodifferential equation derived in the previous section to obtain a formula which is of considerable interest in its own right. First, we will show that

$$\int_0^t \int_0^u \psi(u-x)dF(x)du = \int_0^t \psi(t-x)F(x)dx. \tag{6.41}$$

To see (6.41), begin with the left hand side and reverse the order of integration. This gives

$$\int_0^t \int_0^u \psi(u-x)dF(x)du = \int_0^t \int_x^t \psi(u-x)dudF(x).$$

In the inner integral on the right-hand side, change variables from u to $y = u - x$. This gives

$$\int_0^t \int_x^t \psi(u-x)dudF(x) = \int_0^t \int_0^{t-x} \psi(y)dydF(x).$$

For notational convenience, define

$$\Lambda(x) = \int_0^x \psi(y)dy$$

and thus the double integral may be expressed as

$$\int_0^t \int_0^{t-x} \psi(y)dydF(x) = \int_0^t \Lambda(t-x)dF(x).$$

Then integration by parts gives

$$\begin{aligned}\int_0^t \Lambda(t-x)dF(x) &= \Lambda(t-x)F(x)|_0^t + \int_0^t F(x)\psi(t-x)dx \\ &= \int_0^t F(x)\psi(t-x)dx\end{aligned}$$

proving (6.41).

We are now going to integrate (6.19) over u from 0 to t. Using (6.41), we get

$$\psi(0) - \psi(t) = \frac{\lambda}{c}\int_0^t [1-F(u)]du - \frac{\lambda}{c}\int_0^t \psi(u)du + \frac{\lambda}{c}\int_0^t \psi(t-x)F(x)dx.$$

Change variables from u to x in the first integral and from u to $x = t - u$ in the second. Then combine the last two integrals to get

$$\psi(t) = \frac{\lambda}{c}\int_0^t \psi(t-x)[1-F(x)]dx + \psi(0) - \frac{\lambda}{c}\int_0^t [1-F(x)]dx.$$

Now,

$$\psi(0) = (1+\theta)^{-1} = \frac{\lambda\mu}{c} = \frac{\lambda}{c}\int_0^\infty [1 - F(x)]dx$$

and so the last two terms may be combined to give

$$\psi(t) = \frac{\lambda}{c}\int_0^t \psi(t-x)[1 - F(x)]dx + \frac{\lambda}{c}\int_t^\infty [1 - F(x)]dx. \tag{6.42}$$

Equation (6.42) has various uses. For general $F(\cdot)$, one can solve it numerically to obtain $\psi(x)$, although we will use a different approach to this problem in the next section. Another important use is in the derivation of the Cramér asymptotic ruin formula, given in the following theorem. Recall the notation $a(x) \sim b(x)$, $x \to \infty$ which means $\lim_{x\to\infty} a(x)/b(x) = 1$.

Theorem 6.7 *Suppose $\kappa > 0$ satisfies (6.4). Then the ruin probability satisfies*

$$\psi(u) \sim Ce^{-\kappa u}, \qquad u \to \infty \tag{6.43}$$

where

$$C = \frac{\theta\mu}{E(Xe^{\kappa X}) - \mu(1+\theta)}. \tag{6.44}$$

Proof: We begin by stating without proof a special case of the "renewal theorem." Suppose $g(x)$ is the pdf of a positive random variable, and $a(x)$ satisfies the renewal equation

$$a(x) = \int_0^x a(x-y)g(y)dy + b(x), \qquad x > 0. \tag{6.45}$$

Then

$$\lim_{x\to\infty} a(x) = \frac{\int_0^\infty b(y)dy}{\int_0^\infty yg(y)dy}. \tag{6.46}$$

Now, (6.42) may be expressed using (6.9) as

$$\psi(t) = \frac{1}{1+\theta}\int_0^t \psi(t-x)f_e(x)dx + \frac{1}{1+\theta}\int_t^\infty f_e(x)dx. \tag{6.47}$$

Equation (6.47) is "almost" a renewal equation of the form (6.45), except for the presence of the term $(1+\theta)^{-1}$ in front of the integral (and is actually called a defective renewal equation). Equation (6.8) provides a clue as to how to "turn it into" a renewal equation. Multiply both sides of (6.47) by $e^{\kappa t}$ to obtain

$$\psi_*(t) = \int_0^t \psi_*(t-x)g(x)dx + \frac{e^{\kappa t}\int_t^\infty f_e(x)dx}{1+\theta}$$

where $\psi_*(t) = e^{\kappa t}\psi(t)$, and $g(x) = e^{\kappa x}f_e(x)/(1+\theta)$. Clearly, (6.8) implies that $g(x)$ is a probability density function. Thus, $\psi_*(t)$ satisfies a renewal equation of the form (6.45) with $b(t) = e^{\kappa t} \int_t^\infty f_e(x)dx/(1+\theta)$. Therefore, from (6.46),

$$\lim_{t\to\infty} \psi_*(t) = \frac{\int_0^\infty b(y)dy}{\int_0^\infty yg(y)dy}. \tag{6.48}$$

To put (6.48) in a recognizable form, note that integration by parts gives

$$\begin{aligned}\int_0^\infty b(y)dy &= \frac{1}{1+\theta}\int_0^\infty e^{\kappa y}\left[\int_y^\infty f_e(x)dx\right]dy \\ &= \frac{1}{1+\theta}\left[\frac{1}{\kappa}e^{\kappa y}\int_y^\infty f_e(x)dx\Big|_{y=0}^\infty + \frac{1}{\kappa}\int_0^\infty e^{\kappa y} f_e(y)dy\right].\end{aligned}$$

Now, $0 \le e^{\kappa y}\int_y^\infty f_e(x)dx \le \int_y^\infty e^{\kappa x} f_e(x)dx$, and the right-hand side must go to 0 as $y \to \infty$ since (6.8) holds. The integral on the right-hand side is simply $1+\theta$ from (6.8), hence

$$\int_0^\infty b(y)dy = \frac{1}{1+\theta}\left(-\frac{1}{\kappa} + \frac{1+\theta}{\kappa}\right) = \frac{\theta}{\kappa(1+\theta)}.$$

Also,

$$\int_0^\infty yg(y)dy = \frac{1}{1+\theta}\int_0^\infty ye^{\kappa y} f_e(y)dy.$$

Integration by parts (with the "u" term given by $yf_e(y)$ and the "dv" term given by $e^{\kappa y}dy$) gives

$$\int_0^\infty ye^{\kappa y} f_e(y)dy = \frac{y}{\kappa} f_e(y)e^{\kappa y}\Big|_0^\infty - \frac{1}{\kappa}\int_0^\infty e^{\kappa y} f_e(y)dy + \frac{1}{\mu\kappa}\int_0^\infty ye^{\kappa y}dF(y).$$

As long as $E(Xe^{\kappa X}) = \int_0^\infty xe^{\kappa x}dF(x) < \infty$, then $\lim_{y\to\infty} ye^{\kappa y} f_e(y) = 0$ since

$$0 \le ye^{\kappa y} f_e(y) = \frac{ye^{\kappa y}}{\mu}\int_y^\infty dF(x) \le \frac{1}{\mu}\int_y^\infty xe^{\kappa x}dF(x).$$

Then, the first of the three terms above is 0 and the second is given by (6.8). We thus have

$$\int_0^\infty yg(y)dy = \frac{1}{1+\theta}\left[-\frac{1+\theta}{\kappa} + \frac{E(Xe^{\kappa X})}{\mu\kappa}\right] = \frac{E(Xe^{\kappa X}) - \mu(1+\theta)}{\mu\kappa(1+\theta)}.$$

Then (6.48) may be restated as

$$\begin{aligned}\lim_{u\to\infty} e^{\kappa u}\psi(u) &= \frac{\theta/[\kappa(1+\theta)]}{[E(Xe^{\kappa X}) - \mu(1+\theta)]/[\mu\kappa(1+\theta)]} \\ &= \frac{\mu\theta}{E(Xe^{\kappa X}) - \mu(1+\theta)}\end{aligned}$$

which proves the theorem. □

Thus, in addition to the exponential inequality given by Theorem 6.1, the ruin probability behaves like an exponential function for large u. Obviously, for Lundberg's inequality (6.10) to hold, it must be the case that C given by (6.44) must satisfy $C \leq 1$.

Example 6.14 *In Example 6.13, we found that the ruin probability satisfied*

$$\psi(u) = \frac{2}{5}e^{-u/(2\beta)} - \frac{1}{15}e^{-4u/(3\beta)}, \quad u \geq 0.$$

Determine the limit as $u \to \infty$ directly, and from the limit determine the values of κ and C.

We have

$$\lim_{u\to\infty} e^{u/(2\beta)}\psi(u) = \frac{2}{5} - \frac{1}{15}\lim_{u\to\infty} e^{-5u/(6\beta)} = \frac{2}{5}.$$

Thus, the adjustment coefficient κ is given by $1/(2\beta)$ (derived directly in Example 6.8), and $C = 2/5$. □

Example 6.15 (The exponential distribution) *If $F(x) = 1 - e^{-x/\mu}$, $x > 0$, determine the asymptotic ruin formula.*

We found in Example 6.7 that the adjustment coefficient was given by $\kappa = \theta/\{\mu(1+\theta)\}$ and $E(e^{tX}) = (1-\mu t)^{-1}$. Thus,

$$E(Xe^{tX}) = \frac{d}{dt}(1-\mu t)^{-1} = \mu(1-\mu t)^{-2}.$$

Also,

$$E(Xe^{\kappa X}) = \mu(1-\mu\kappa)^{-2} = \mu\{1-\theta(1+\theta)^{-1}\}^{-2} = \mu(1+\theta)^2.$$

Thus, from (6.44),

$$C = \frac{\mu\theta}{\mu(1+\theta)^2 - \mu(1+\theta)} = \frac{\theta}{(1+\theta)(1+\theta-1)} = \frac{1}{1+\theta}.$$

The asymptotic formula (6.43) becomes

$$\psi(u) \sim \frac{1}{1+\theta}\exp\left[-\frac{\theta u}{\mu(1+\theta)}\right], \qquad u \to \infty.$$

It turns out that this is the exact ruin probability as was demonstrated in Example 6.11. □

6.8 THE MAXIMUM AGGREGATE LOSS

We now derive the general solution to the integrodifferential equation (6.18) subject to the initial conditions (6.13) and (6.17).

Beginning with an initial reserve u, the probability that the surplus will ever fall below the initial level u is $\psi(0)$ since the surplus process has stationary and independent increments. Thus the probability of dropping below the initial level u is the same for all u, but we know that when $u = 0$ it is $\psi(0)$.

The key result is that, given that there is a drop below the initial level u, the random variable Y which represents the amount of this initial drop has probability density function $f_e(y)$ where $f_e(y)$ is given by (6.9).

Theorem 6.8 *Given that there is a drop below the initial level u, the random variable Y which represents the amount of this initial drop has probability density function $f_e(x) = [1 - F(x)]/E(X)$.*

Proof: Recall the function $G(u, y)$ from Definition 6.12. Because the surplus process has stationary and independent increments, $G(0, y)$ also represents the probability that the surplus drops below its initial level, and the amount of this drop is at most y. Thus, using Theorem 6.3, the amount of the drop, given that there is a drop, has cumulative distribution function

$$\begin{aligned}
\Pr(Y \leq y) &= G(0, y)/\psi(0) \\
&= \frac{\lambda}{c\psi(0)} \int_0^y [1 - F(u)]du \\
&= \frac{1}{E(X)} \int_0^y [1 - F(u)]du
\end{aligned}$$

and the result follows by differentiation. □

If there is a drop of y, the surplus immediately after the drop is $u - y$, and because the surplus process has stationary and independent increments, ruin occurs thereafter with probability $\psi(u - y)$, provided $u - y$ is positive, otherwise ruin would have already occurred. The probability of a second drop is $\psi(0)$, and the amount of the second drop has density $f_e(y)$ and is independent of the first drop. Due to the memoryless property of the Poisson process, the process "starts over" after each drop. Therefore, the total number of drops K is geometrically distributed, that is, $\Pr(K = 0) = 1 - \psi(0)$, $\Pr(K = 1) = [1 - \psi(0)]\psi(0)$, and more generally,

$$\Pr(K = k) = [1 - \psi(0)][\psi(0)]^k, \quad k = 0, 1, 2, \ldots .$$

After a drop, the surplus immediately begins to increase again. Thus, the lowest level of the surplus is $u - L$ where L, called the **maximum aggregate loss**, is the total of all the drop amounts. Let Y_j be the amount of the jth drop, and since the surplus process has stationary and independent increments, $\{Y_1, Y_2, ...\}$ is a sequence of independent and identically distributed random variables (each with density $f_e(y)$). Since the number of drops is K, it follows that

$$L = Y_1 + Y_2 + \cdots + Y_K$$

with $L = 0$ if $K = 0$. Thus, L is a compound geometric random variable with "claim size density" $f_e(y)$.

Clearly, ultimate survival beginning with initial reserve u occurs if the maximum loss L does not exceed u, that is,

$$\phi(u) = \Pr(L \le u), \quad u \ge 0.$$

Let $F_e^{*0}(y) = 0$ if $y < 0$ and 1 if $y \ge 0$. Also $F_e^{*k}(y) = \Pr\{Y_1 + Y_2 + \cdots + Y_k \le y\}$ is the cumulative distribution function of the k-fold convolution of the distribution of Y with itself. We then have the general solution, namely

$$\phi(u) = \sum_{k=0}^{\infty} \frac{\theta}{1+\theta}\left(\frac{1}{1+\theta}\right)^k F_e^{*k}(u), \qquad u \ge 0.$$

We may therefore compute ruin probabilities numerically by computing the cumulative distribution function of a compound geometric distribution using any of the techniques described in Chapter 4.

Example 6.16 *Suppose the individual loss distribution is Pareto with $\alpha = 3$ and $\theta = 1{,}000$. Let the security loading be $\theta = 0.2$. Determine $\phi(u)$ for $u = 100, 200, 300, \ldots$.*

We first require the cdf, $F_e(u)$. It can be found from its pdf

$$\begin{aligned} f_e(u) &= \frac{1 - F_X(u)}{E(X)} = \frac{1 - \left[1 - \left(\frac{1,000}{1,000 + u}\right)^3\right]}{500} \\ &= \frac{1}{500}\left(\frac{1{,}000}{1{,}000 + u}\right)^3 \end{aligned}$$

which happens to be the density function of a Pareto distribution with $\alpha = 2$ and $\theta = 1{,}000$. This new Pareto distribution is the severity distribution for a compound geometric distribution where the parameter is $\beta = 1/\theta = 5$. The compound geometric distribution can be evaluated using any of the techniques in Chapter 4. We used the recursive formula with a discretization which preserves the mean and a span of $h = 5$. The cumulative probabilities are then obtained by summing the discrete probabilities generated by the recursive formula. The values appear in Table 6.11.

Table 6.11 Survival probabilites, Pareto losses

u	$\phi(u)$	u	$\phi(u)$
100	0.193	5,000	0.687
200	0.216	7,500	0.787
300	0.238	10,000	0.852
500	0.276	15,000	0.923
1,000	0.355	20,000	0.958
2,000	0.473	25,000	0.975
3,000	0.561		

6.9 THE BROWNIAN MOTION RISK PROCESS

In this section, we study the relationship between Brownian motion (the Wiener process) and the surplus process $\{U_t;\ t \geq 0\}$ where

$$U_t = u + ct - S_t, \qquad t \geq 0 \tag{6.49}$$

and $\{S_t;\ t \geq 0\}$ is the total loss process defined by

$$S_t = X_1 + X_2 + \cdots + X_{N_t}, \qquad t \geq 0$$

where $\{N_t;\ t \geq 0\}$ is a Poisson process with rate λ and where $S_t = 0$ when $N_t = 0$. As in Sections 6.4 to 6.8 we assume that the individual losses $\{X_1, X_2, \ldots\}$ are independently and identically distributed positive random variables whose moment generating function exists. The surplus process $\{U_t;\ t \geq 0\}$ increases continuously with slope c, the premium rate per unit time, and has successive downward jumps of $\{X_1, X_2, \ldots\}$ at random jump times $\{T_1, T_2, \ldots\}$ as illustrated by Figure 6.4. In that figure, $u = 20$, $c = 35$, $\lambda = 3$, and X has an exponential distribution with mean 10.

Let

$$Z_t = U_t - u = ct - S_t, \qquad t \geq 0. \tag{6.50}$$

Then $Z_0 = 0$. Since S_t has a compound distribution, the process $\{Z_t;\ t \geq 0\}$ has mean

$$\begin{aligned} E(Z_t) &= ct - E(S_t) \\ &= ct - \lambda t E(X) \end{aligned}$$

and variance

$$Var(Z_t) = \lambda t E(X^2).$$

We now introduce the corresponding stochastic process based on Brownian motion.

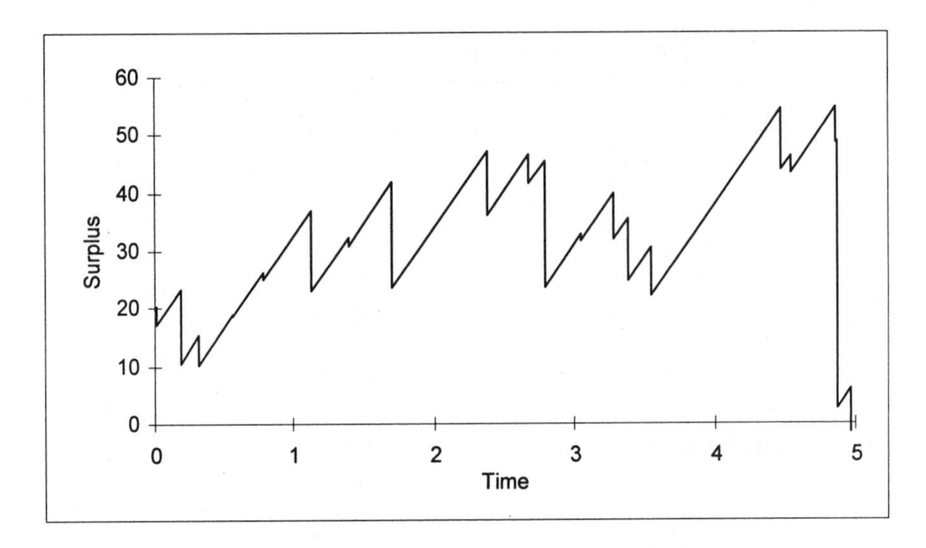

Fig. 6.4 Sample path for a Poisson surplus process

Definition 6.13 *A continuous time stochastic process* $\{W_t;\ t \geq 0\}$ *is a* ***Brownian motion*** *process if*

1. $W_0 = 0$;

2. $\{W_t,\ t \geq 0\}$ *has stationary and independent increments; and*

3. *for every* $t > 0$, W_t *is normally distributed with mean* 0 *and variance* $\sigma^2 t$.

The Brownian motion process, also called the **Wiener process** or **white noise**, has been used extensively in describing various physical phenomena. When $\sigma^2 = 1$, it is called **standard Brownian motion**. The English botanist Robert Brown discovered the process in 1827 and used it to describe the continuous irregular motion of a particle immersed in a liquid or gas. In 1905 Albert Einstein explained this motion by postulating perpetual collision of the particle with the surrounding medium. Norbert Wiener provided the analytical description of the process in a series of papers beginning in 1918. Since then it has been used in many areas of application from quantum mechanics to describing price levels on the stock market. It has become the key model underpinning modern financial theory.

Definition 6.14 *A continuous time stochastic process* $\{W_t;\ t \geq 0\}$ *is called a* ***Brownian motion with drift*** *process if it satisfies the properties of a Brownian motion process except that* W_t *has mean* μt *rather than* 0, *for some*

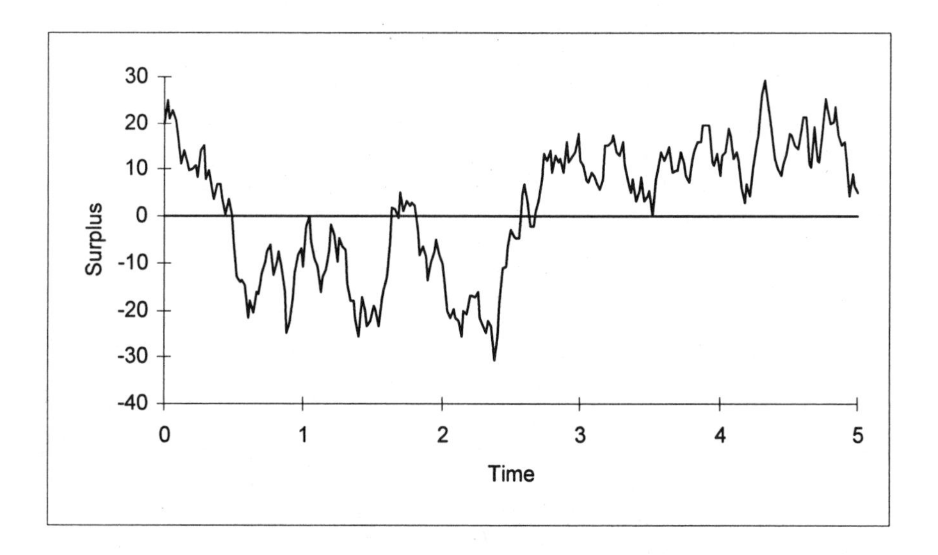

Fig. 6.5 Sample path for a Brownian motion with drift

$\mu > 0$. *A Brownian motion with drift is illustrated in Figure 6.5. This process has* $u = 20$, $\mu = 5$, *and* $\sigma^2 = 600$. *The illustrated process has an initial surplus of* 20, *so the mean of* W_t *is* $20+5t$. *Technically,* $W_t - 20$ *is a Brownian motion with drift process.*

We now show how the surplus process (6.50) based on the compound Poisson risk process is related to the Brownian motion with drift process. We will take a limit of the process (6.50) as the expected number of downward jumps becomes large, and simultaneously the size of the jumps becomes small. Since the Brownian motion with drift process is characterized by the infinitesimal mean μ and infinitesimal variance σ^2, we force the mean and variance functions to be the same for the two processes. In this way, the Brownian motion with drift can be thought of as an approximation to the compound Poisson-based surplus process. Similarly, the compound Poisson process can be used as an approximation for Brownian motion.

Let

$$\mu = c - \lambda E[X]$$

and

$$\sigma^2 = \lambda E[X^2]$$

denote the infinitesimal mean and variance of the Brownian motion with drift process. Then

$$\lambda = \frac{\sigma^2}{E[X^2]} \tag{6.51}$$

and

$$c = \mu + \sigma^2 \frac{E[X]}{E[X^2]}. \tag{6.52}$$

Now, in order to take limits, we can treat the jump size X as a scaled version of some other random variable Y, so that $X = \alpha Y$ where Y has fixed mean and variance. Then

$$\lambda = \frac{\sigma^2}{E[Y^2]} \cdot \frac{1}{\alpha^2}$$

and

$$c = \mu + \sigma^2 \frac{E[Y]}{E[Y^2]} \cdot \frac{1}{\alpha}.$$

Then, in order for $\lambda \to \infty$, we let $\alpha \to 0$.

Since the process $\{S_t;\ t \geq 0\}$ is a continuous time process with stationary and independent increments, so are the processes $\{U_t;\ t \geq 0\}$ and $\{Z_t;\ t \geq 0\}$. This will then also be the case for the limiting process. Since $Z_0 = 0$, we only need to establish that for every t, in the limit, Z_t is normally distributed with mean μt and variance $\sigma^2 t$ according to Definitions 6.13 and 6.14. We do this by looking at the moment generating function of Z_t.

$$\begin{aligned}
M_{Z_t}(r) &= M_{ct-S_t}(r) \\
&= E\{\exp[r(ct - S_t)]\} \\
&= \exp(t\{rc + \lambda[M_X(-r) - 1]\}).
\end{aligned}$$

Then

$$\begin{aligned}
\frac{\log M_{Z_t}(r)}{t} &= rc + \lambda[M_X(-r) - 1] \\
&= r[\mu + \lambda E(X)] \\
&\quad + \lambda \left[1 - rE(X) + \frac{r^2}{2!}E(X^2) - \frac{r^3}{3!}E(X^3) + \cdots - 1\right] \\
&= r\mu + \frac{r^2}{2}\lambda E(X^2) - \lambda\left[\frac{r^3}{3!}E(X^3) - \frac{r^4}{4!}E(X^4) + \cdots\right] \\
&= r\mu + \frac{r^2}{2}\sigma^2 - \lambda\alpha^2\left[\alpha\frac{r^3}{3!}E(Y^3) - \alpha^2\frac{r^4}{4!}E(Y^4) + \cdots\right] \\
&= r\mu + \frac{r^2}{2}\sigma^2 - \sigma^2\left[\alpha\frac{r^3}{3!}\frac{E(Y^3)}{E(Y^2)} - \alpha^2\frac{r^4}{4!}\frac{E(Y^4)}{E(Y^2)} + \cdots\right].
\end{aligned}$$

Since all terms except α are fixed, as $\alpha \to 0$ we have

$$\lim_{\alpha \to 0} M_{Z_t}(r) = \exp\{r\mu t + \frac{r^2}{2}\sigma^2 t\}$$

which is the mgf of the normal distribution with mean μt and $\sigma^2 t$. This establishes that the limiting process is Brownian motion with drift.

From Figure 6.4, it is clear that the process U_t is differentiable everywhere except at jump points. As the number of jump points increases indefinitely, the process becomes **nowhere differentiable**. Another property of a Brownian motion process is that **its paths are continuous functions of t with probability one**. Intuitively, this occurs because the jump sizes become small as $\alpha \to 0$.

Finally, the total distance traveled in $(0, t]$ by the process U_t is

$$\begin{aligned} D &= ct + S_t \\ &= ct + X_1 + \cdots + X_{N_t} \end{aligned}$$

which has expected value

$$\begin{aligned} E[D] &= ct + \lambda t E[X] \\ &= t\left[\mu + \sigma^2 \frac{E(Y)}{E(Y^2)} \frac{1}{\alpha} + \sigma^2 \frac{E(Y)}{E(Y^2)} \frac{1}{\alpha}\right] \\ &= t\left[\mu + 2\sigma^2 \frac{E(Y)}{E(Y^2)} \frac{1}{\alpha}\right]. \end{aligned}$$

This quantity becomes indefinitely large as $\alpha \to 0$. Hence, we have

$$\lim_{\alpha \to 0} E[D] = \infty.$$

This means that the expected distance traveled in a finite time interval is infinitely large! For a more rigorous discussion of the properties of the Brownian motion process, the text by Karlin and Taylor [74, Ch. 7] is recommended.

Since $Z_t = U_t - u$, we can just add u to the Brownian motion with drift process and then use equations (6.51) and (6.52) to develop an approximation for the process (6.50). Of course, the larger the value of λ and the smaller the jumps, the better will be the approximation. For a very large block of insurance policies (for example, for an entire company), this may be appropriate. In this case, the probability of ultimate ruin and the distribution of time until ruin are easily obtained from the approximating Brownian motion with drift process. This is done in the next section. Similarly, if a process is known to be Brownian motion with drift, a compound Poisson surplus process can be used as an approximation.

6.10 BROWNIAN MOTION AND THE PROBABILITY OF RUIN

Let $\{W_t;\ t \geq 0\}$ denote the Brownian motion with drift process with mean function μt and variance function $\sigma^2 t$. Let $U_t = u + W_t$ denote the Brownian motion with drift process with initial surplus $U_0 = u$.

We consider the probability of ruin in a finite time interval $(0, \tau)$ as well as the distribution of time until ruin if ruin occurs. Letting $\tau \to \infty$ will give ultimate ruin probabilities.

The probability of ruin before time τ can be expressed as

$$\begin{aligned}\psi(u,\tau) &= 1-\phi(u,\tau)\\ &= \Pr\{T<\tau\}\\ &= \Pr\left\{\min_{0<t<\tau} U_t < 0\right\}\\ &= \Pr\left\{\min_{0<t<\tau} W_t < -U_0\right\}\\ &= \Pr\left\{\min_{0<t<\tau} W_t < -u\right\}.\end{aligned}$$

Theorem 6.9 *For the process U_t described above, the ruin probability is given by*

$$\psi(u,\tau) = \Phi\left(-\frac{u+\mu\tau}{\sqrt{\sigma^2\tau}}\right) + \exp\left(-\frac{2\mu}{\sigma^2}u\right)\Phi\left(-\frac{u-\mu\tau}{\sqrt{\sigma^2\tau}}\right) \tag{6.53}$$

where $\Phi(\cdot)$ is the cdf of the standard normal distribution.

Proof: Any sample path of U_t with a final level $U_\tau < 0$ must have first crossed the barrier at some time $T < \tau$. For any such path U_t, we define a new path U_t^* which is the same as the original sample path for all $t < T$ but is the reflection about the barrier $U_t = 0$ of the original sample path for all $t > T$. Then

$$U_t^* = \begin{cases} U_t, & t \le T \\ -U_t, & t > T. \end{cases} \tag{6.54}$$

The reflected path U_t^* has final value $U_\tau^* = -U_\tau$. These are illustrated in Figure 6.6, which is based on the sample path in Figure 6.5.

Now consider any path that crosses the barrier $U_t = 0$ in $(0,\tau)$. Any such path is one of two possible types:

Type A: One that has a final value of $U_\tau < 0$

Type B: One that has a final value of $U_\tau > 0$.

Any path of Type B is a reflection of some other path of Type A. Hence, sample paths can be considered in reflecting pairs. The probability of ruin at some time in $(0,\tau)$ is the total probability of all such pairs:

$$\psi(u,\tau) = \Pr\{T<\tau\} = \Pr\left\{\min_{0<t<\tau} U_t < 0\right\}$$

where it is understood that all probabilities are conditional on $U_0 = u$. This probability is obtained by considering all original paths of Type A with final values $U_\tau = x < 0$. By adding all the corresponding reflecting paths U_t^* as well, all possible paths that cross the ruin barrier are considered. Note that the case $U_\tau = 0$ has been left out. The probability of this happening is zero and so this event can be ignored. In Figure 6.6 the original sample path ended at a positive surplus value and so is of Type B. The reflection is of Type A.

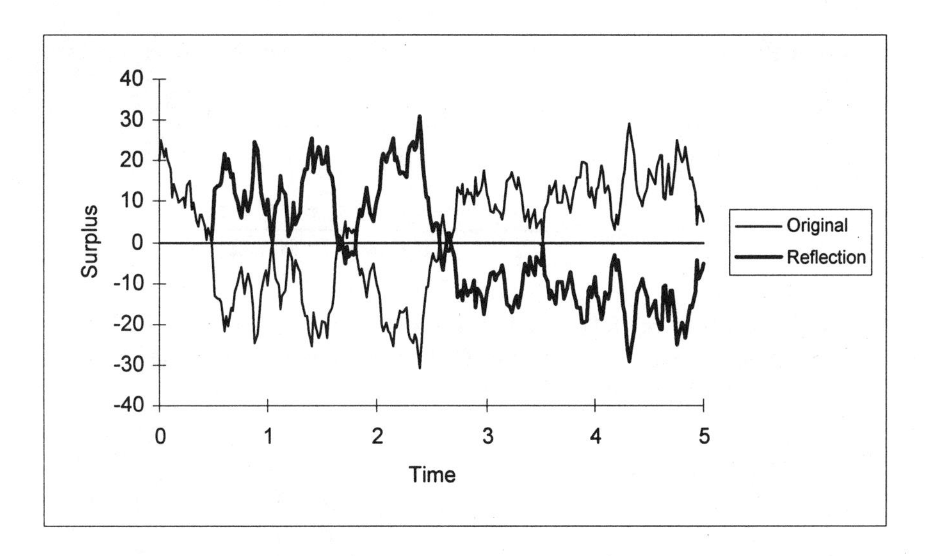

Fig. 6.6 Type B sample path and its reflection

Let A_x and B_x denote the sets of all possible paths of Types A and B which end at $U_\tau = x$ for Type A and $U_\tau = -x$ for Type B. Further let $\Pr\{A_x\}$ and $\Pr\{B_x\}$ denote the total probability associated with the paths in the sets.[2] Hence the probability of ruin is,

$$\Pr\{T < \tau\} = \int_{-\infty}^{0} \Pr\{U_\tau = x\} \frac{\Pr\{A_x\} + \Pr\{B_x\}}{\Pr\{U_\tau = x\}} dx. \tag{6.55}$$

Since A_x is the set of all possible paths ending at x

$$\Pr\{A_x\} = \Pr\{U_\tau = x\}$$

where the right-hand side is the pdf of U_τ. Then

$$\Pr\{T < \tau\} = \int_{-\infty}^{0} \Pr\{U_\tau = x\} \left[1 + \frac{\Pr\{B_x\}}{\Pr\{A_x\}}\right] dx. \tag{6.56}$$

Since $U_t - u$ is a Brownian motion process with drift, U_τ is normally distributed with mean $u + \mu\tau$ and variance $\sigma^2\tau$, and so

$$\Pr\{U_\tau = x\} = (2\pi\sigma^2\tau)^{-1/2} \exp\left[-\frac{(x - u - \mu\tau)^2}{2\sigma^2\tau}\right].$$

[2] We are abusing probability notation here. The actual probability of these events is zero. What is being called a probability is really a probability density which can be integrated to produce probabilities for sets which have positive probability.

To obtain $\Pr\{B_x\}/\Pr\{A_x\}$, we condition on all possible ruin times T. Then

$$\begin{aligned}\frac{\Pr\{B_x\}}{\Pr\{A_x\}} &= \frac{\int_0^\tau \Pr\{B_x|T=t\}\Pr\{T=t\}dt}{\int_0^\tau \Pr\{A_x|T=t\}\Pr\{T=t\}dt} \\ &= \frac{\int_0^\tau \Pr\{U_\tau=-x|T=t\}\Pr\{T=t\}dt}{\int_0^\tau \Pr\{U_\tau=x|T=t\}\Pr\{T=t\}dt}\end{aligned}$$

The conditional pdf of $U_\tau|T=t$ is the same as the pdf of $U_\tau - U_t$ because $T=t$ implies $U_t=0$. The process U_t has independent increments and so $U_\tau - U_t$ has a normal distribution. Then,

$$\begin{aligned}\Pr\{U_\tau = x|T=t\} &= \Pr\{U_\tau - U_t = x\} \\ &= \frac{\exp\left\{-\dfrac{[x-\mu(\tau-t)]^2}{2\sigma^2(\tau-t)}\right\}}{\sqrt{2\pi\sigma^2(\tau-t)}} \\ &= \frac{\exp\left\{-\dfrac{x^2-2x\mu(\tau-t)+\mu^2(\tau-t)^2}{2\sigma^2(\tau-t)}\right\}}{\sqrt{2\pi\sigma^2(\tau-t)}} \\ &= \exp\left(\frac{x\mu}{\sigma^2}\right)\frac{\exp\left\{-\dfrac{x^2+\mu^2(\tau-t)^2}{2\sigma^2(\tau-t)}\right\}}{\sqrt{2\pi\sigma^2(\tau-t)}}.\end{aligned}$$

Similarly, by replacing x with $-x$,

$$\Pr\{U_\tau = -x|T=t\} = \frac{\exp\left(-\frac{x\mu}{\sigma^2}\right)\exp\left\{-\dfrac{x^2+\mu^2(\tau-t)^2}{2\sigma^2(\tau-t)}\right\}}{\sqrt{2\pi\sigma^2(\tau-t)}}.$$

We then have

$$\begin{aligned}\frac{\Pr\{B_x\}}{\Pr\{A_x\}} &= \frac{\displaystyle\int_0^\tau \frac{\exp\left(-\frac{x\mu}{\sigma^2}\right)\exp\left\{-\dfrac{x^2+\mu^2(\tau-t)^2}{2\sigma^2(\tau-t)}\right\}}{\sqrt{2\pi\sigma^2(\tau-t)}}\Pr\{T=t\}dt}{\displaystyle\int_0^\tau \frac{\exp\left(\frac{x\mu}{\sigma^2}\right)\exp\left\{-\dfrac{x^2+\mu^2(\tau-t)^2}{2\sigma^2(\tau-t)}\right\}}{\sqrt{2\pi\sigma^2(\tau-t)}}\Pr\{T=t\}dt} \\ &= \exp\{-2\mu x/\sigma^2\}.\end{aligned}$$

Then from (6.55),

$$\begin{aligned}
\psi(u,\tau) &= \Pr\{T<\tau\}\\
&= \int_{-\infty}^{0} \Pr\{U_\tau = x\}\left[1+\frac{\Pr\{B_x\}}{\Pr\{A_x\}}\right]dx\\
&= \Phi\left(\frac{-u-\mu\tau}{\sqrt{\sigma^2\tau}}\right)\\
&\quad +\int_{-\infty}^{0}(2\pi\sigma^2\tau)^{-1/2}\exp\left[-\frac{(x-u-\mu\tau)^2}{2\sigma^2\tau}-\frac{2\mu x}{\sigma^2}\right]dx\\
&= \Phi\left(-\frac{u+\mu\tau}{\sqrt{\sigma^2\tau}}\right)\\
&\quad +\int_{-\infty}^{0}(2\pi\sigma^2\tau)^{-1/2}\exp\left[-\frac{(x-u+\mu\tau)^2+4\mu u\tau}{2\sigma^2\tau}\right]dx\\
&= \Phi\left(-\frac{u+\mu\tau}{\sqrt{\sigma^2\tau}}\right)+\exp\left(-\frac{2\mu}{\sigma^2}u\right)\Phi\left(-\frac{u-\mu\tau}{\sqrt{\sigma^2\tau}}\right).
\end{aligned}$$

□

Corollary 6.10 *The probability of ultimate ruin is given by*

$$\psi(u) = 1-\phi(u) = \Pr\{T<\infty\} = \exp\left(-\frac{2\mu}{\sigma^2}u\right). \tag{6.57}$$

By letting $\tau \to \infty$, this result follows from Theorem 6.9. It should be noted that the distribution (6.53) is a **defective distribution**, since the cdf does not approach 1 as $\tau \to \infty$. The corresponding proper distribution is obtained by conditioning on ultimate ruin.

Corollary 6.11 *The distribution of the time until ruin given that it occurs is given by*

$$\begin{aligned}
\frac{\psi(u,\tau)}{\psi(u)} &= Pr\{T<\tau|T<\infty\}\\
&= \exp\left(\frac{2\mu}{\sigma^2}u\right)\Phi\left(-\frac{u+\mu\tau}{\sqrt{\sigma^2\tau}}\right)+\Phi\left(-\frac{u-\mu\tau}{\sqrt{\sigma^2\tau}}\right), \qquad \tau > 0.
\end{aligned} \tag{6.58}$$

Corollary 6.12 *The probability density function of the time until ruin is given by*

$$f_T(\tau) = \frac{u}{\sqrt{2\pi\sigma^2}}\tau^{-3/2}\exp\left[-\frac{(u-\mu\tau)^2}{2\sigma^2\tau}\right], \qquad \tau > 0. \tag{6.59}$$

This is obtained by differentiation of (6.58) with respect to τ. By replacing u^2/σ^2 by θ, u/μ by μ (a different μ), and τ by x, (6.59) can be rewritten as

$$f_T(x) = \left(\frac{\theta}{2\pi x^3}\right)^{1/2}\exp\left[-\frac{\theta}{2x}\left(\frac{x-\mu}{\mu}\right)^2\right], \qquad x>0, \tag{6.60}$$

which is the usual form of the inverse Gaussian distribution (see Appendix A).

Then the time until ruin, if ruin occurs, is represented by an **inverse Gaussian distribution** with mean u/μ and variance $u\sigma^2/\mu^3$. With $\mu = 0$, ruin is certain and the time until ruin (6.59) has pdf

$$f_T(\tau) = \frac{u}{\sqrt{2\pi\sigma^2}}\tau^{-3/2}\exp\left(-\frac{u^2}{2\sigma^2\tau}\right), \qquad \tau > 0$$

and cdf (from (6.58))

$$F_T(\tau) = 2\Phi\left(-\frac{u}{\sigma\tau^{1/2}}\right), \qquad \tau > 0.$$

This distribution is the **one-sided stable law with index 1/2**.

These results can be used as approximations for the original surplus process (6.50) based on the compound Poisson model. We do this by substituting back using (6.51) and (6.52).

We then have for the process (6.50), from (6.53), (6.57) and (6.59),

$$\begin{aligned}
\psi(u,\tau) &\doteq \Phi\left[-\frac{u+\theta\lambda\tau E(X)}{\sqrt{\lambda\tau E(X^2)}}\right] \\
&\quad + \exp\left[-\frac{2E(X)}{E(X^2)}\theta u\right]\Phi\left[-\frac{u-\theta\lambda\tau E(X)}{\sqrt{\lambda\tau E(X^2)}}\right], \qquad u > 0,\ \tau > 0 \\
\psi(u) &\doteq \exp\left[-\frac{2E(X)}{E(X^2)}\theta u\right], \qquad u > 0,
\end{aligned}$$

and

$$f_T(\tau) \doteq \frac{u}{\sqrt{2\pi\lambda E(X^2)}}\tau^{-3/2}\exp\left\{-\frac{[u-\theta\lambda\tau E(X)]^2}{2\lambda\tau E(X^2)}\right\}, \qquad \tau > 0.$$

Here

$$c = (1+\theta)\lambda E(X)$$

and so θ is the relative premium loading. Then for any compound Poisson-based process, it is easy to get simple numerical approximations. For example, the expected time until ruin, if it occurs, is

$$E(T) = \frac{u}{\mu} = \frac{u}{\theta\lambda E(X)}. \tag{6.61}$$

Naturally, the accuracy of this approximation depends on the relative sizes of the quantities involved.

It should be noted from (6.61) that the expected time of ruin, given that it occurs, depends (as we might expect) on the four key quantities which describe the surplus process. A higher initial surplus (u) increases the time to ruin, while increasing any of the other components decreases the expected time. This may appear surprising at first, but, for example, increasing the loading increases the rate of expected growth in surplus, making ruin difficult. Therefore, if ruin should occur, it will have to happen soon, before the high loading leads to large gains. If λ is large, the company is essentially much larger and events happen more quickly. Therefore, ruin, if it happens, will occur sooner. Finally, a large value of $E(X)$ makes it easier for an early claim to wipe out the initial surplus.

All these are completely intuitive. However, formula (6.61) shows how each factor can have an influence on the expected ruin time.

On a final note, it is also possible to use the compound Poisson-based risk process Z_t as an approximation for a Brownian motion process. For known drift and variance parameters μ and σ^2, one can use (6.51) and (6.52) to obtain

$$\mu = \theta\lambda E(X) \tag{6.62}$$

$$\sigma^2 = \lambda E(X^2). \tag{6.63}$$

It is convenient to fix the jump sizes so that $E(X) = k$, say, and $E(X^2) = k^2$. Then we have

$$\lambda = \frac{\sigma^2}{k^2} \tag{6.64}$$

$$\theta = \frac{\mu}{\lambda k} = \frac{\mu}{\sigma^2}k. \tag{6.65}$$

When μ and σ^2 are fixed, choosing a value of k fixes λ and θ. Hence, the Poisson-based process can be used to approximate the Brownian motion with accuracy determined only by the parameter k. The smaller the value of k, the smaller the jump sizes and the larger the number of jumps per unit time.

Hence, simulation of the Brownian motion process can be done using the Poisson-based process. To simulate the Poisson-based process, the waiting times between successive events are generated first since these are exponentially distributed with mean $1/\lambda$. As k becomes small, λ becomes large and the mean waiting time becomes small.

6.11 EXERCISES

Section 6.3

6.1 The total claims paid in a year can be 0, 5, 10, 15, or 20 with probabilities 0.4, 0.3, 0.15, 0.1, and 0.05, respectively. Annual premiums of 6 are collected at the beginning of each year. Interest of 10% is earned on any funds available at the beginning of the year, and claims are paid at the end of the year.

(a) Determine $\tilde{\psi}(2,3)$ exactly.

(b) Determine $\tilde{\psi}(2,3)$ but after the premium is added and the interest credited, discretize the resulting distribution with a span of 5.

6.2 Repeat Exercise 6.1 using the FFT and re-discretizing to maintain a span of 5.

Section 6.4

6.3 Prove that equation (6.2) implies 3′.

6.4 For fixed $t_0 > 0$, prove that the time until the next claim occurs is exponential with mean λ^{-1}. *Hint*: Condition on the joint distribution of the time since the last claim occurred and the number of claims in $(0, t_0]$.

Section 6.5

6.5 Calculate the adjustment coefficient if $\theta = 0.32$ and the claim size distribution is the same as that of Example 6.8.

6.6 Calculate the adjustment coefficient if the individual loss size density is $f(x) = \sqrt{\beta/x}\, e^{-\beta x}/\Gamma(1/2),\ x > 0$.

6.7 Calculate the adjustment coefficient if $c = 3$, $\lambda = 4$, and the individual loss size density is $f(x) = e^{-2x} + \frac{3}{2}e^{-3x},\ x > 0$. Do not use an iterative numerical procedure.

6.8 If $c = 2.99$, $\lambda = 1$, and the individual loss size distribution is given by $\Pr(X = 1) = 0.2$, $\Pr(X = 2) = 0.3$, and $\Pr(X = 3) = 0.5$, use the Newton–Raphson procedure to numerically obtain the adjustment coefficient.

6.9 Repeat Exercise 6.7 using the Newton–Raphson procedure beginning with an initial estimate based on (6.6).

6.10 Suppose that $E(X^3)$ is known where X is a generic individual loss amount random variable. Prove that the adjustment coefficient κ satisfies

$$\kappa < \frac{-3E(X^2) + \sqrt{9\{E(X^2)\}^2 + 24\theta\mu E(X^3)}}{2E(X^3)}.$$

Also prove that the right-hand side of this inequality is strictly less than the bound given in (6.6), namely $2\theta\mu/E(X^2)$.

6.11 Recall that if $g''(x) \geq 0$, Jensen's inequality implies $E[g(Y)] \geq g[E(Y)]$. Also, from Subsection 2.7.2,

$$\int_0^\infty x f_e(x) dx = \frac{E(X^2)}{2\mu}$$

where $f_e(x)$ is defined by (6.9).

(a) Use equation (6.8) and the above to show that

$$\kappa \leq \frac{2\mu \log(1+\theta)}{E(X^2)}.$$

(b) Show that $\log(1+\theta) < \theta$ for $\theta > 0$ and thus the inequality in (a) is tighter than that from equation (6.6). *Hint*: Consider $h(\theta) = \theta - \log(1+\theta), \quad \theta \geq 0$.

(c) If there is a maximum claim size of m, show that equation (6.8) becomes

$$1 + \theta = \int_0^m e^{\kappa x} f_e(x) dx.$$

Show that the right hand side of this equality satisfies

$$\int_0^m e^{\kappa x} f_e(x) dx \leq e^{\kappa m}$$

and hence that

$$\kappa \geq \frac{1}{m} \log(1+\theta).$$

6.12 Recall from Exercise 2.90 that the claim size distribution is new worse than used in expectation (NWUE) if

$$\int_x^\infty f_e(y) dy \geq 1 - F(x), \qquad x \geq 0.$$

(a) Let Y have probability density function $f_e(y)$, $y \geq 0$, and let X have cumulative distribution function $F(x)$. Show that

$$\Pr(Y > y) \geq \Pr(X > y), \qquad y \geq 0$$

and hence that

$$\Pr(e^{\kappa Y} > t) \geq \Pr(e^{\kappa X} > t), \qquad t \geq 1.$$

(b) Use (a) to show that $E(e^{\kappa Y}) \geq E(e^{\kappa X})$.

(c) Use (b) to show that $\kappa \leq \dfrac{\theta}{\mu(1+\theta)}$.

(d) Prove that if the claim size distribution is NBUE

$$\kappa \geq \frac{\theta}{\mu(1+\theta)}.$$

6.13 Suppose that $\kappa > 0$ satisfies (6.4) and also that

$$\bar{F}(x) \leq \rho e^{-\kappa x} \int_x^{\infty} e^{\kappa y} dF(y) \tag{6.66}$$

for $0 < \rho \leq 1$, where $\bar{F}(x) = 1 - F(x)$. Prove that $\psi(u) \leq \rho e^{-\kappa u}$, $u \geq 0$. *Hint*: Use the method of Theorem 6.1.

6.14 Continue the previous exercise. Use integration by parts to show that

$$\int_x^{\infty} e^{\kappa y} dF(y) = e^{\kappa x}\bar{F}(x) + \kappa \int_x^{\infty} e^{\kappa y}\bar{F}(y)dy, \qquad x \geq 0.$$

6.15 Suppose $F(x)$ has a decreasing failure rate (Subsection 2.7.2). Prove that $\bar{F}(y) \geq \bar{F}(x)\bar{F}(y-x)$, $x \geq 0$, $y \geq x$. Then use Exercise 6.14 to show that (6.66) is satisfied with $\rho^{-1} = E(e^{\kappa X})$. Use (6.4) to conclude that

$$\psi(x) \leq [1 + (1+\theta)\kappa\mu]^{-1} e^{-\kappa x}, \qquad x \geq 0.$$

6.16 Suppose $F(x)$ has a failure rate $\mu(x) = -(d/dx)\log \bar{F}(x)$ which satisfies $\mu(x) \leq m < \infty$, $x \geq 0$. Use the result in Exercise 6.14 to show that (6.66) is satisfied with $\rho = 1 - \kappa/m$ and thus

$$\psi(x) \leq (1 - \kappa/m)e^{-\kappa x}, \qquad x \geq 0.$$

Hint: Show that for $y > x$, $\bar{F}(y) \geq \bar{F}(x)e^{-(y-x)m}$.

Section 6.6

6.17 Suppose that $\theta = 4/5$ and the individual loss amount density is given by

$$f(x) = (1+6x)e^{-3x}, \qquad x > 0.$$

Show that

$$\phi'''(u) + 5\phi''(u) + 4\phi'(u) = 0$$

and use this to obtain an expression for $\phi(u)$.

6.18 Suppose that $\theta = 3/11$ and the single claim amount density is given by

$$f(x) = 2e^{-4x} + \frac{7}{2}e^{-7x}, \qquad x > 0.$$

Show that

$$\phi'''(u) + 7\phi''(u) + 6\phi'(u) = 0$$

and use this to obtain an expression for $\phi(u)$.

6.19 Suppose that $\theta = 3/5$ and the single claim amount density is given by

$$f(x) = 3e^{-4x} + 0.5e^{-2x}, \qquad x > 0.$$

Show that

$$\phi'''(u) + 4\phi''(u) + 3\phi'(u) = 0$$

and use this to obtain an expression for $\phi(u)$.

6.20 Suppose that $\theta = 7/5$ and the individual loss density is the convolution of two exponential distributions given by

$$f(x) = \int_0^x 3e^{-3(x-y)}2e^{-2y}dy.$$

Show that

$$2\phi'''(u) + 9\phi''(u) + 7\phi'(u) = 0$$

and use this to obtain an expression for $\phi(u)$.

6.21 (a) Use integration by parts on (6.18) to show that

$$c\phi'(u) = \lambda\phi(0)\{1 - F(u)\} + \lambda \int_0^u \phi'(u - x)\{1 - F(x)\}dx.$$

(b) Let $m(u) = e^{\kappa u}\phi'(u)$ where κ is the adjustment coefficient and show that $m(u)$ satisfies

$$m(u) = \frac{\theta}{(1+\theta)^2}e^{\kappa u}f_e(u) + \frac{1}{1+\theta}\int_0^u m(u-x)\{e^{\kappa x}f_e(x)\}dx$$

where $f_e(x)$ is defined by (6.9). If $c = 1.7$, $\lambda = 3$, and $f(x) = \frac{2}{3}e^{-2x} + e^{-3x} + 2e^{-6x}$, $x \geq 0$, show that $\kappa = 1$ and use the above integral equation to find $m(u)$.

Section 6.7

6.22 You have determined the formula

$$\psi(u) = \frac{1}{4}e^{-u} + \frac{1}{12}e^{-2u}, \qquad u \geq 0.$$

(a) Obtain the adjustment coefficient and the loading.

(b) If $E(Xe^{\kappa X}) = 2$, find $E(e^{\kappa X})$.

6.23 The following formula holds:

$$\psi(u) = C_1 e^{-r_1 u} + C_2 e^{-r_2 u}, \qquad u \geq 0$$

where $0 < r_1 < r_2$. Show that $C_2 \geq -\phi(0)$.

6.24 Determine the adjustment coefficient κ (if it exists) and the asymptotic ruin formula for the following individual loss distributions. In each case assume that $\lambda = 10$ and $c = 15E(X)$.

(a) Gamma with $\alpha = 0.5$ and $\theta = 1{,}000$.

(b) $f(x) = F'(x) = 0.01,\ 0 < x < 100$.

(c) Geometric with $\beta = 9$. Then $\Pr(X = x) = 0.1(0.9)^x,\ x = 0, 1, \ldots$.

6.25 Use the renewal theorem (see the beginning of the proof of Theorem 6.7) to show that $\lim_{u\to\infty} m(u) = C\kappa$, where $m(u)$ is defined in Exercise 6.21.

6.26 Recall the function $G(u, y)$ given by Definition 6.12.

(a) Integrate (6.15) over u from 0 to t and use (6.16) to show that

$$G(t,y) = \frac{\lambda}{c}\int_0^t G(t-x,y)\{1-F(x)\}dx + \frac{\lambda}{c}\int_t^{y+t}\{1-F(x)\}dx.$$

(b) Use (a) and the renewal theorem to show that

$$G(u,y) \sim C(y)e^{-\kappa u}, \qquad u \to \infty$$

where

$$C(y) = \frac{\kappa \int_0^\infty e^{\kappa t} \int_t^{t+y} [1-F(x)]dx dt}{E\left(Xe^{\kappa X}\right) - \mu(1+\theta)}.$$

Section 6.8

6.27 Suppose the number of claims follows the Poisson process and the amount of an individual claim is exponential with mean 100. The relative security loading is $\theta = 0.1$. Determine $\psi(1{,}000)$ by using the method of this section. Use the method of rounding with a span of 50 to discretize the exponential distribution. Compare your answer to the exact ruin probability (see Example 6.11).

6.28 Consider the problem of Example 6.13 with $\beta = 50$. Use the method of this section (with discretization by rounding and a span of 1) to approximate $\psi(200)$. Compare your answer to the exact ruin probability.

6.29 Suppose that the claim severity pdf is given by

$$F'(x) = \sum_{k=1}^{r} q_k \frac{\beta^{-k} x^{k-1} e^{-x/\beta}}{(k-1)!}, \qquad x > 0$$

where $\sum_{k=1}^{r} q_k = 1$. Note that this is a mixture of gamma densities.

(a) Show that

$$f_e(x) = \sum_{k=1}^{r} q_k^* \frac{\beta^{-k} x^{k-1} e^{-x/\beta}}{(k-1)!}, \qquad x > 0$$

where

$$q_k^* = \sum_{j=k}^{r} q_j \Big/ \sum_{j=1}^{r} j q_j, \qquad k = 1, 2, ..., r$$

and also show that $\sum_{k=1}^{r} q_k^* = 1$.

(b) Define

$$Q^*(z) = \sum_{k=1}^{r} q_k^* z^k$$

and use the results of Exercise 4.36 to show that

$$\psi(u) = \sum_{n=1}^{\infty} c_n \sum_{j=0}^{n-1} \frac{(u/\beta)^j e^{-u/\beta}}{j!}, \qquad u \geq 0$$

where

$$C(z) = \{1 - \frac{1}{\theta}[Q^*(z) - 1]\}^{-1}$$

is a compound geometric pgf, with probabilities which may be computed recursively by $c_0 = \theta(1+\theta)^{-1}$ and

$$c_k = \frac{1}{1+\theta} \sum_{j=1}^{k} q_j^* c_{k-j}, \qquad k = 1, 2, ...$$

where $q_j^* = 0$ if $j \neq 1, 2, \ldots, r$.

(c) Use (b) to show that

$$\psi(u) = e^{-u/\beta} \sum_{j=0}^{\infty} \bar{C}_j \frac{(u/\beta)^j}{j!}, \qquad u \geq 0$$

where $\bar{C}_j = \sum_{k=j+1}^{\infty} c_k$, $j = 0, 1, \ldots$. Then use (b) to show that the $\bar{C}_n$s may be computed recursively from

$$\bar{C}_n = \frac{1}{1+\theta}\sum_{k=1}^{n} q_k^* \bar{C}_{n-k} + \frac{1}{1+\theta}\sum_{k=n+1}^{\infty} q_k^*, \qquad n = 1, 2, \ldots$$

beginning with $\bar{C}_0 = (1+\theta)^{-1}$.

(d) Show that (6.44) may be reexpressed as

$$C = \frac{\theta}{\kappa E(Ye^{\kappa Y})}$$

where Y has pf $f_e(y)$. Hence prove in this case that

$$\psi(u) \sim \frac{\theta}{\kappa\beta\sum_{j=1}^{r} jq_j^*(1-\beta\kappa)^{-j-1}} e^{-\kappa u}, \qquad u \to \infty$$

where $\kappa > 0$ satisfies

$$1+\theta = Q^*[(1-\beta\kappa)^{-1}] = \sum_{j=1}^{r} q_j^*(1-\beta\kappa)^{-j}.$$

6.12 CASE STUDY

In Chapter 4 a variety of reinsurance schemes were considered. For this part of the case study, assume that 100 actuaries are selected at random and so the total group produces claims with a Poisson distribution with $\lambda = 1.54578$. As determined earlier, all actuaries have a lognormal severity distribution with $\mu = 10.5430$ and $\sigma = 2.31315$. Further suppose all individual policies have no deductible and a per-claim policy limit of 10,000,000. The primary insurer adds 20% to the net premium to account for risk. You wish to offer individual excess-of-loss reinsurance with a deductible of 1,000,000. That is, KPWV will pay the excess over 1,000,000 on each claim. The reinsurance premium will be 125% of the net premium. One way to convince the primary insurer that this arrangement is beneficial is to show that $\tilde{\psi}(u, 5)$ will be smaller if reinsurance is purchased.

The premium collected by the primary insurer is

$$1.2(1.54578)E(X \wedge 10{,}000{,}000) = 697{,}980.$$

The distribution of aggregate losses was obtained via the recursive formula using a span of 10,000. Assuming no interest earnings and using the convolution approach of Section 6.3, the five-year ruin probability with an initial surplus of 1,000,000 is 0.268. The premium for the reinsurer is

$$1.25(1.54578)[E(X \wedge 10{,}000{,}000) - E(X \wedge 1{,}000{,}000)] = 378{,}851.$$

The premium retained by the primary insurer is the difference, 319,129. For the ruin probability, the severity distribution is limited to 1,000,000, recursions are done to obtain the aggregate distribution of retained claims, and then convolutions are used to obtain the ruin probability. It is reduced to 0.176.

Another view of the advantage of reinsurance is that the initial surplus could be reduced to 700,000. This would keep the ruin probability near the no reinsurance level of 0.271. This would likely be an argument against reinsurance. In order to reduce surplus requirements by 300,000 the primary insurer must give up 378,851 of premium income. This does not appear to be a good trade-off.

Appendix A: An Inventory of Continuous Distributions

A.1 INTRODUCTION

Descriptions of the models are given below. First a few mathematical preliminaries are presented that indicate how the various quantities can be computed.

The incomplete gamma function[1] is given by

$$\Gamma(\alpha; x) = \frac{1}{\Gamma(\alpha)} \int_0^x t^{\alpha-1} e^{-t} dt, \qquad \alpha > 0,\ x > 0$$

$$\text{with } \Gamma(\alpha) = \int_0^\infty t^{\alpha-1} e^{-t} dt, \qquad \alpha > 0.$$

When $\alpha \leq 0$ the integral does not exist. In that case, define

$$\Gamma(\alpha) G(\alpha; x) = \int_x^\infty t^{\alpha-1} e^{-t} dt, \qquad x > 0.$$

[1] Some references, such as [1], denote this integral $P(\alpha, x)$ and define $\Gamma(\alpha, x) = \int_x^\infty t^{\alpha-1} e^{-t} dt$. Note that this definition does not normalize by dividing by $\Gamma(\alpha)$. When using software to evaluate the incomplete gamma function, be sure to note how it is defined.

Integration by parts produces the relationship

$$\Gamma(\alpha)G(\alpha;x) = -\frac{x^\alpha e^{-x}}{\alpha} + \frac{\Gamma(\alpha+1)}{\alpha}G(\alpha+1;x)$$

which allows for recursive calculation because for $\alpha > 0$, $\Gamma(\alpha)G(\alpha;x) = \Gamma(\alpha)[1-\Gamma(\alpha;x)]$. When α is zero or a negative integer, this recursive formula requires the evaluation of

$$\Gamma(0)G(0;x) = \int_x^\infty t^{-1}e^{-t}dt = E_1(x)$$

which is called the **exponential integral**. A series expansion for this integral is

$$E_1(x) = -0.57721566490153 - \log x - \sum_{n=1}^{\infty}\frac{(-1)^n x^n}{n(n!)}.$$

When α is a positive integer, the incomplete gamma distribution can be evaluated exactly as given in the following theorem.

Theorem A.1 *For integer* α,

$$\Gamma(\alpha;x) = 1 - \sum_{j=0}^{\alpha-1}\frac{x^j e^{-x}}{j!}.$$

Proof: For $\alpha = 1$, $\Gamma(1;x) = \int_0^x e^{-t}dt = 1 - e^{-x}$ and so the theorem is true for this case. The proof is completed by induction. Assume it is true for $\alpha = 1, \ldots, n$. Then

$$\begin{aligned}
\Gamma(n+1;x) &= \frac{1}{n!}\int_0^x t^n e^{-t}dt \\
&= \frac{1}{n!}\left(-t^n e^{-t}\Big|_0^x + \int_0^x nt^{n-1}e^{-t}dt\right) \\
&= \frac{1}{n!}\left(-x^n e^{-x}\right) + \Gamma(n;x) \\
&= -\frac{x^n e^{-x}}{n!} + 1 - \sum_{j=0}^{n-1}\frac{x^j e^{-x}}{j!} \\
&= 1 - \sum_{j=0}^{n}\frac{x^j e^{-x}}{j!}.
\end{aligned}$$

□

The incomplete beta function is given by

$$\beta(a,b;x) = \frac{\Gamma(a+b)}{\Gamma(a)\Gamma(b)}\int_0^x t^{a-1}(1-t)^{b-1}dt, \qquad a>0,\ b>0,\ 0<x<1$$

and when $b < 0$ (but $a > 1 + [-b]$) repeated integration by parts produces

$$\begin{aligned}\Gamma(a)\Gamma(b)\beta(a,b;x) &= -\Gamma(a+b)\left[\frac{x^{a-1}(1-x)^b}{b}\right. \\ &+\frac{(a-1)x^{a-2}(1-x)^{b+1}}{b(b+1)}+\cdots \\ &\left.+\frac{(a-1)\cdots(a-r)x^{a-r-1}(1-x)^{b+r}}{b(b+1)\cdots(b+r)}\right] \\ &+\frac{(a-1)\cdots(a-r-1)}{b(b+1)\cdots(b+r)}\Gamma(a-r-1) \\ &\times\Gamma(b+r+1)\beta(a-r-1,b+r+1;x)\end{aligned}$$

where r is the smallest integer such that $b + r + 1 > 0$. The first argument must be positive (that is, $a - r - 1 > 0$).

Numerical approximations for both the incomplete gamma and the incomplete beta function are available in many statistical computing packages as well as in many spreadsheets because they are just the distribution functions of the gamma and beta distributions. The following approximations are taken from [1]. The suggestion regarding using different formulas for small and large x when evaluating the incomplete gamma function is from [103]. That reference also contains computer subroutines for evaluating these expressions. In particular, it provides an effective way of evaluating continued fractions.

For $x \le \alpha + 1$ use the following series expansion

$$\Gamma(\alpha;x) = \frac{x^\alpha e^{-x}}{\Gamma(\alpha)} \sum_{n=0}^{\infty} \frac{x^n}{\alpha(\alpha+1)\cdots(\alpha+n)}$$

while for $x > \alpha + 1$ use the following continued fraction expansion

$$1-\Gamma(\alpha;x) = \frac{x^\alpha e^{-x}}{\Gamma(\alpha)} \cfrac{1}{x+\cfrac{1-\alpha}{1+\cfrac{1}{x+\cfrac{2-\alpha}{1+\cfrac{2}{x+\cdots}}}}}.$$

The incomplete gamma function can also be used to produce cumulative probabilities from the standard normal distribution. Let $\Phi(z) = \Pr(Z \le z)$ where Z has the standard normal distribution. Then for $z \ge 0$, $\Phi(z) = 0.5 + \Gamma(0.5; z^2/2)/2$ while for $z < 0$, $\Phi(z) = 1 - \Phi(-z)$.

The incomplete beta function can be evaluated by the following series expansion

$$\begin{aligned}\beta(a, b; x) &= \frac{\Gamma(a+b)x^a(1-x)^b}{a\Gamma(a)\Gamma(b)} \\ &\times \left[1 + \sum_{n=0}^{\infty} \frac{(a+b)(a+b+1)\cdots(a+b+n)}{(a+1)(a+2)\cdots(a+n+1)} x^{n+1}\right]\end{aligned}$$

The gamma function itself can be found from

$$\begin{aligned}\log\Gamma(\alpha) &\doteq (\alpha - 1/2)\log\alpha - \alpha + \log(2\pi)/2 \\ &+ \frac{1}{12\alpha} - \frac{1}{360\alpha^3} + \frac{1}{1260\alpha^5} - \frac{1}{1680\alpha^7} + \frac{1}{1188\alpha^9} - \frac{691}{360,360\alpha^{11}} \\ &+ \frac{1}{156\alpha^{13}} - \frac{3617}{122,400\alpha^{15}} + \frac{43,867}{244,188\alpha^{17}} - \frac{174,611}{125,400\alpha^{19}}.\end{aligned}$$

For values of α above 10 the error is less than 10^{-19}. For values below 10 use the relationship

$$\log\Gamma(\alpha) = \log\Gamma(\alpha+1) - \log\alpha.$$

The distributions are presented in the following way. First the name is given along with the parameters. Next the density function ($f(x)$) and distribution function ($F(x)$) are given. These are followed by the kth moment, the kth limited expected value (when there is simplification, the formula is given for integer k), and the mode. These are followed by the derivatives of the distribution function with respect to the parameters. For distributions involving the incomplete gamma and beta functions, some of these derivatives are difficult to compute and so an approximation should be used. Finally, for some distributions, formulas for starting values are given. Within each family the distributions are presented in decreasing order with regard to the number of parameters. The Greek letters used are selected to be consistent. Any Greek letter that is not used in the distribution means that that distribution is a special case of one with more parameters, but with the missing parameters set equal to 1. Unless specifically indicated, all parameters must be positive.

Except for two distributions, inflation can be recognized by simply inflating the parameter θ. That is, if X has a particular distribution, then cX has the same distribution type, with all parameters unchanged except θ is changed to $c\theta$. For the lognormal distribution, μ changes to $\mu + \log(c)$ with σ unchanged, while for the inverse Gaussian both μ and θ are multiplied by c.

For several of the distributions, starting values are suggested. They are not necessarily good estimators, just places from which to start an iterative procedure to maximize the likelihood or other objective function. These are found by either the methods of moments or percentile matching. The quantities used are:

$$\text{Moments:} \qquad m = \frac{1}{n}\sum_{i=1}^{n} x_i, \qquad t = \frac{1}{n}\sum_{i=1}^{n} x_i^2$$

$$\text{Percentile matching:} \qquad p = \text{ 25th percentile}, \; q = \text{ 75th percentile}$$

For grouped data or data that have been truncated or censored, these quantities may have to be approximated. Because the purpose is to obtain starting values and not a useful estimate, it is often sufficient to just ignore modifications. For three- and four-parameter distributions, starting values can be obtained by using estimates from a special case, then making the new parameters equal to 1. An all-purpose starting value rule (for when all else fails) is to set the scale parameter (θ) equal to the mean and set all other parameters equal to 2.

A.2 TRANSFORMED BETA FAMILY

A.2.1 Four-parameter distribution

A.2.1.1 Transformed beta—$\alpha, \theta, \gamma, \tau$

$$\begin{aligned}
f(x) &= \frac{\Gamma(\alpha+\tau)}{\Gamma(\alpha)\Gamma(\tau)} \frac{\gamma(x/\theta)^{\gamma\tau}}{x[1+(x/\theta)^\gamma]^{\alpha+\tau}} \\
F(x) &= \beta(\tau,\alpha;u), \quad u = \frac{(x/\theta)^\gamma}{1+(x/\theta)^\gamma} \\
E[X^k] &= \frac{\theta^k\Gamma(\tau+k/\gamma)\Gamma(\alpha-k/\gamma)}{\Gamma(\alpha)\Gamma(\tau)}, \quad -\tau\gamma < k < \alpha\gamma \\
E[(X\wedge x)^k] &= \frac{\theta^k\Gamma(\tau+k/\gamma)\Gamma(\alpha-k/\gamma)}{\Gamma(\alpha)\Gamma(\tau)} \beta(\tau+k/\gamma, \alpha-k/\gamma; u) \\
&\quad + x^k[1-F(x)], \quad k > -\tau\gamma \\
\text{mode} &= \theta\left(\frac{\tau\gamma-1}{\alpha\gamma+1}\right)^{1/\gamma}, \quad \tau\gamma > 1, \text{ else } 0
\end{aligned}$$

$$\begin{aligned}
\frac{\partial F}{\partial \alpha} &= \beta_2(\tau,\alpha;u) \\
\frac{\partial F}{\partial \theta} &= -f(x)\frac{x}{\theta} \\
\frac{\partial F}{\partial \gamma} &= f(x)\frac{x}{\gamma}\log\left(\frac{x}{\theta}\right) \\
\frac{\partial F}{\partial \tau} &= \beta_1(\tau,\alpha;u).
\end{aligned}$$

A.2.2 Three-parameter distributions

A.2.2.1 Generalized Pareto—α, θ, τ

$$
\begin{aligned}
f(x) &= \frac{\Gamma(\alpha+\tau)}{\Gamma(\alpha)\Gamma(\tau)} \frac{\theta^\alpha x^{\tau-1}}{(x+\theta)^{\alpha+\tau}} \\
F(x) &= \beta(\tau, \alpha; u), \quad u = \frac{x}{x+\theta} \\
E[X^k] &= \frac{\theta^k \Gamma(\tau+k)\Gamma(\alpha-k)}{\Gamma(\alpha)\Gamma(\tau)}, \quad -\tau < k < \alpha \\
E[X^k] &= \frac{\theta^k \tau(\tau+1)\cdots(\tau+k-1)}{(\alpha-1)\cdots(\alpha-k)}, \quad \text{if } k \text{ is an integer} \\
E[(X \wedge x)^k] &= \frac{\theta^k \Gamma(\tau+k)\Gamma(\alpha-k)}{\Gamma(\alpha)\Gamma(\tau)} \beta(\tau+k, \alpha-k; u) \\
&\quad + x^k[1-F(x)], \quad k > -\tau \\
\text{mode} &= \theta \frac{\tau-1}{\alpha+1}, \quad \tau > 1, \text{ else } 0
\end{aligned}
$$

$$
\begin{aligned}
\frac{\partial F}{\partial \alpha} &= \beta_2(\tau, \alpha; u) \\
\frac{\partial F}{\partial \theta} &= -f(x)\frac{x}{\theta} \\
\frac{\partial F}{\partial \tau} &= \beta_1(\tau, \alpha; u).
\end{aligned}
$$

A.2.2.2 Burr—α, θ, γ

$$
\begin{aligned}
f(x) &= \frac{\alpha\gamma(x/\theta)^\gamma}{x[1+(x/\theta)^\gamma]^{\alpha+1}} \\
F(x) &= 1-u^\alpha, \quad u = \frac{1}{1+(x/\theta)^\gamma} \\
E[X^k] &= \frac{\theta^k \Gamma(1+k/\gamma)\Gamma(\alpha-k/\gamma)}{\Gamma(\alpha)}, \quad -\gamma < k < \alpha\gamma \\
E[(X \wedge x)^k] &= \frac{\theta^k \Gamma(1+k/\gamma)\Gamma(\alpha-k/\gamma)}{\Gamma(\alpha)} \beta(1+k/\gamma, \alpha-k/\gamma; 1-u) \\
&\quad + x^k u^\alpha, \quad k > -\gamma \\
\text{mode} &= \theta\left(\frac{\gamma-1}{\alpha\gamma+1}\right)^{1/\gamma}, \quad \gamma > 1, \text{ else } 0
\end{aligned}
$$

$$
\frac{\partial F}{\partial \alpha} = -u^\alpha \log u
$$

$$\begin{aligned}
\frac{\partial F}{\partial \theta} &= -f(x)\frac{x}{\theta} \\
\frac{\partial F}{\partial \gamma} &= f(x)\frac{x}{\gamma}\log\left(\frac{x}{\theta}\right).
\end{aligned}$$

A.2.2.3 Inverse Burr—τ, θ, γ

$$\begin{aligned}
f(x) &= \frac{\tau\gamma(x/\theta)^{\gamma\tau}}{x[1+(x/\theta)^\gamma]^{\tau+1}} \\
F(x) &= u^\tau, \quad u = \frac{(x/\theta)^\gamma}{1+(x/\theta)^\gamma} \\
E[X^k] &= \frac{\theta^k\Gamma(\tau+k/\gamma)\Gamma(1-k/\gamma)}{\Gamma(\tau)}, \quad -\tau\gamma < k < \gamma \\
E[(X\wedge x)^k] &= \frac{\theta^k\Gamma(\tau+k/\gamma)\Gamma(1-k/\gamma)}{\Gamma(\tau)}\beta(\tau+k/\gamma, 1-k/\gamma; u) \\
&\quad +x^k[1-u^\tau], \quad k > -\tau\gamma \\
\text{mode} &= \theta\left(\frac{\tau\gamma-1}{\gamma+1}\right)^{1/\gamma}, \quad \tau\gamma > 1, \text{ else } 0
\end{aligned}$$

$$\begin{aligned}
\frac{\partial F}{\partial \tau} &= u^\tau \log u \\
\frac{\partial F}{\partial \theta} &= -f(x)\frac{x}{\theta} \\
\frac{\partial F}{\partial \gamma} &= f(x)\frac{x}{\gamma}\log\left(\frac{x}{\theta}\right).
\end{aligned}$$

A.2.3 Two-parameter distributions

A.2.3.1 Pareto—α, θ

$$\begin{aligned}
f(x) &= \frac{\alpha\theta^\alpha}{(x+\theta)^{\alpha+1}} \\
F(x) &= 1-\left(\frac{\theta}{x+\theta}\right)^\alpha \\
E[X^k] &= \frac{\theta^k\Gamma(k+1)\Gamma(\alpha-k)}{\Gamma(\alpha)}, \quad -1 < k < \alpha \\
E[X^k] &= \frac{\theta^k k!}{(\alpha-1)\cdots(\alpha-k)}, \quad \text{if } k \text{ is an integer} \\
E[X\wedge x] &= \frac{\theta}{\alpha-1}\left[1-\left(\frac{\theta}{x+\theta}\right)^{\alpha-1}\right], \quad \alpha \neq 1 \\
E[X\wedge x] &= -\theta\log\left(\frac{\theta}{x+\theta}\right), \quad \alpha = 1
\end{aligned}$$

$$\begin{aligned}
E[(X \wedge x)^k] &= \frac{\theta^k \Gamma(k+1)\Gamma(\alpha-k)}{\Gamma(\alpha)} \beta[k+1, \alpha-k; x/(x+\theta)] \\
&\quad + x^k \left(\frac{\theta}{x+\theta}\right)^\alpha, \quad \text{all } k \\
\text{mode} &= 0
\end{aligned}$$

$$\begin{aligned}
\frac{\partial F}{\partial \alpha} &= -\left(\frac{\theta}{x+\theta}\right)^\alpha \log\left(\frac{\theta}{x+\theta}\right) \\
\frac{\partial F}{\partial \theta} &= -f(x)\frac{x}{\theta}
\end{aligned}$$

$$\hat{\alpha} = 2\frac{t-m^2}{t-2m^2}, \quad \hat{\theta} = \frac{mt}{t-2m^2}.$$

A.2.3.2 Inverse Pareto—τ, θ

$$\begin{aligned}
f(x) &= \frac{\tau\theta x^{\tau-1}}{(x+\theta)^{\tau+1}} \\
F(x) &= \left(\frac{x}{x+\theta}\right)^\tau \\
E[X^k] &= \frac{\theta^k \Gamma(\tau+k)\Gamma(1-k)}{\Gamma(\tau)}, \quad -\tau < k < 1 \\
E[X^k] &= \frac{\theta^k (-k)!}{(\tau-1)\cdots(\tau+k)}, \quad \text{if } k \text{ is a negative integer} \\
E[(X \wedge x)^k] &= \theta^k \tau \int_0^{x/(x+\theta)} y^{\tau+k-1}(1-y)^{-k} dy \\
&\quad + x^k \left[1 - \left(\frac{x}{x+\theta}\right)^\tau\right], \quad k > -\tau \\
\text{mode} &= \theta\frac{\tau-1}{2}, \quad \tau > 1, \text{ else } 0
\end{aligned}$$

$$\begin{aligned}
\frac{\partial F}{\partial \tau} &= \left(\frac{x}{x+\theta}\right)^\tau \log\left(\frac{x}{x+\theta}\right) \\
\frac{\partial F}{\partial \theta} &= -f(x)\frac{x}{\theta}.
\end{aligned}$$

A.2.3.3 Loglogistic—γ, θ

$$f(x) = \frac{\gamma(x/\theta)^\gamma}{x[1+(x/\theta)^\gamma]^2}$$

$$\begin{aligned}
F(x) &= u, \quad u = \frac{(x/\theta)^\gamma}{1+(x/\theta)^\gamma} \\
E[X^k] &= \theta^k \Gamma(1+k/\gamma)\Gamma(1-k/\gamma), \quad -\gamma < k < \gamma \\
E[(X \wedge x)^k] &= \theta^k \Gamma(1+k/\gamma)\Gamma(1-k/\gamma)\beta(1+k/\gamma, 1-k/\gamma; u) \\
&\quad + x^k(1-u), \quad k > -\gamma \\
\text{mode} &= \theta\left(\frac{\gamma-1}{\gamma+1}\right)^{1/\gamma}, \quad \gamma > 1, \text{ else } 0
\end{aligned}$$

$$\begin{aligned}
\frac{\partial F}{\partial \gamma} &= f(x)\frac{x}{\gamma}\log\left(\frac{x}{\theta}\right) \\
\frac{\partial F}{\partial \theta} &= -f(x)\frac{x}{\theta}
\end{aligned}$$

$$\hat{\gamma} = \frac{2\log(3)}{\log(q)-\log(p)}, \quad \hat{\theta} = \exp\left(\frac{\log(q)+\log(p)}{2}\right).$$

A.2.3.4 Paralogistic—α, θ

This is a Burr distribution with $\gamma = \alpha$.

$$\begin{aligned}
f(x) &= \frac{\alpha^2 (x/\theta)^\alpha}{x[1+(x/\theta)^\alpha]^{\alpha+1}} \\
F(x) &= 1-u^\alpha, \quad u = \frac{1}{1+(x/\theta)^\alpha} \\
E[X^k] &= \frac{\theta^k \Gamma(1+k/\alpha)\Gamma(\alpha-k/\alpha)}{\Gamma(\alpha)}, \quad -\alpha < k < \alpha^2 \\
E[(X \wedge x)^k] &= \frac{\theta^k \Gamma(1+k/\alpha)\Gamma(\alpha-k/\alpha)}{\Gamma(\alpha)}\beta(1+k/\alpha, \alpha-k/\alpha; 1-u) \\
&\quad + x^k u^\alpha, \quad k > -\alpha \\
\text{mode} &= \theta\left(\frac{\alpha-1}{\alpha^2+1}\right)^{1/\alpha}, \quad \alpha > 1, \text{ else } 0
\end{aligned}$$

$$\begin{aligned}
\frac{\partial F}{\partial \alpha} &= -u^\alpha[\log u + \alpha(1-u)\log(\theta/x)] \\
\frac{\partial F}{\partial \theta} &= -f(x)\frac{x}{\theta}.
\end{aligned}$$

Starting values can use estimates from the loglogistic (use γ for α) or Pareto (use α) distributions.

A.2.3.5 *Inverse paralogistic*—τ, θ

This is an inverse Burr distribution with $\gamma = \tau$.

$$
\begin{aligned}
f(x) &= \frac{\tau^2 (x/\theta)^{\tau^2}}{x[1+(x/\theta)^\tau]^{\tau+1}} \\
F(x) &= u^\tau, \quad u = \frac{(x/\theta)^\tau}{1+(x/\theta)^\tau} \\
E[X^k] &= \frac{\theta^k \Gamma(\tau + k/\tau)\Gamma(1-k/\tau)}{\Gamma(\tau)}, \quad -\tau^2 < k < \tau \\
E[(X \wedge x)^k] &= \frac{\theta^k \Gamma(\tau + k/\tau)\Gamma(1-k/\tau)}{\Gamma(\tau)} \beta(\tau + k/\tau, 1-k/\tau; u) \\
&\quad + x^k[1-u^\tau], \quad k > -\tau^2 \\
\text{mode} &= \theta \, (\tau - 1)^{1/\tau}, \quad \tau > 1, \text{ else } 0
\end{aligned}
$$

$$
\begin{aligned}
\frac{\partial F}{\partial \tau} &= u^\tau [\log u - \tau(1-u)\log(\theta/x)] \\
\frac{\partial F}{\partial \theta} &= -f(x)\frac{x}{\theta}.
\end{aligned}
$$

Starting values can use estimates from the loglogistic (use γ for τ) or inverse Pareto (use τ) distributions.

A.3 TRANSFORMED GAMMA FAMILY

A.3.1 Three-parameter distributions

A.3.1.1 *Transformed gamma*—α, θ, τ

$$
\begin{aligned}
f(x) &= \frac{\tau u^\alpha e^{-u}}{x\Gamma(\alpha)}, \quad u = (x/\theta)^\tau \\
F(x) &= \Gamma(\alpha; u) \\
E[X^k] &= \frac{\theta^k \Gamma(\alpha + k/\tau)}{\Gamma(\alpha)}, \quad k > -\alpha\tau \\
E[(X \wedge x)^k] &= \frac{\theta^k \Gamma(\alpha + k/\tau)}{\Gamma(\alpha)} \Gamma(\alpha + k/\tau; u) \\
&\quad + x^k[1-\Gamma(\alpha; u)], \quad k > -\alpha\tau \\
\text{mode} &= \theta \left(\frac{\alpha\tau - 1}{\tau}\right)^{1/\tau}, \quad \alpha\tau > 1, \text{ else } 0
\end{aligned}
$$

$$\frac{\partial F}{\partial \alpha} = \Gamma'(\alpha; u)$$
$$\frac{\partial F}{\partial \theta} = -f(x)\frac{x}{\theta}$$
$$\frac{\partial F}{\partial \tau} = f(x)\frac{x}{\tau}\log\left(\frac{x}{\theta}\right).$$

A.3.1.2 Inverse transformed gamma—α, θ, τ

$$f(x) = \frac{\tau u^\alpha e^{-u}}{x\Gamma(\alpha)}, \quad u = (\theta/x)^\tau$$
$$F(x) = 1 - \Gamma(\alpha; u)$$
$$E[X^k] = \frac{\theta^k \Gamma(\alpha - k/\tau)}{\Gamma(\alpha)}, \quad k < \alpha\tau$$
$$E[(X \wedge x)^k] = \frac{\theta^k \Gamma(\alpha - k/\tau)}{\Gamma(\alpha)}[1 - \Gamma(\alpha - k/\tau; u)] + x^k \Gamma(\alpha; u)$$
$$= \frac{\theta^k \Gamma(\alpha - k/\tau)}{\Gamma(\alpha)} G(\alpha - k/\tau; u) + x^k \Gamma(\alpha; u), \quad \text{all } k$$
$$\text{mode} = \theta\left(\frac{\tau}{\alpha\tau + 1}\right)^{1/\tau}$$

$$\frac{\partial F}{\partial \alpha} = -\Gamma'(\alpha; u)$$
$$\frac{\partial F}{\partial \theta} = -f(x)\frac{x}{\theta}$$
$$\frac{\partial F}{\partial \tau} = f(x)\frac{x}{\tau}\log\left(\frac{x}{\theta}\right).$$

A.3.2 Two-parameter distributions

A.3.2.1 Gamma—α, θ

$$f(x) = \frac{(x/\theta)^\alpha e^{-x/\theta}}{x\Gamma(\alpha)}$$
$$F(x) = \Gamma(\alpha; x/\theta)$$
$$E[X^k] = \frac{\theta^k \Gamma(\alpha + k)}{\Gamma(\alpha)}, \quad k > -\alpha$$
$$E[X^k] = \theta^k(\alpha + k - 1)\cdots\alpha, \quad \text{if } k \text{ is an integer}$$

$$
\begin{aligned}
E[(X \wedge x)^k] &= \frac{\theta^k \Gamma(\alpha+k)}{\Gamma(\alpha)} \Gamma(\alpha+k; x/\theta) + x^k[1-\Gamma(\alpha; x/\theta)], \quad k > -\alpha \\
&= \alpha(\alpha+1)\cdots(\alpha+k-1)\theta^k \Gamma(\alpha+k; x/\theta) \\
&\quad + x^k[1-\Gamma(\alpha; x/\theta)], \quad k \text{ an integer} \\
\text{mode} &= \theta(\alpha-1), \quad \alpha > 1, \text{ else } 0
\end{aligned}
$$

$$
\begin{aligned}
\frac{\partial F}{\partial \alpha} &= \Gamma'(\alpha; x/\theta) \\
\frac{\partial F}{\partial \theta} &= -f(x)\frac{x}{\theta}
\end{aligned}
$$

$$
\hat{\alpha} = \frac{m^2}{t-m^2}, \quad \hat{\theta} = \frac{t-m^2}{m}.
$$

A.3.2.2 Inverse gamma—α, θ

$$
\begin{aligned}
f(x) &= \frac{(\theta/x)^\alpha e^{-\theta/x}}{x\Gamma(\alpha)} \\
F(x) &= 1-\Gamma(\alpha; \theta/x) \\
E[X^k] &= \frac{\theta^k \Gamma(\alpha-k)}{\Gamma(\alpha)}, \quad k < \alpha \\
E[X^k] &= \frac{\theta^k}{(\alpha-1)\cdots(\alpha-k)}, \quad \text{if } k \text{ is an integer} \\
E[(X \wedge x)^k] &= \frac{\theta^k \Gamma(\alpha-k)}{\Gamma(\alpha)}[1-\Gamma(\alpha-k; \theta/x)] + x^k \Gamma(\alpha; \theta/x) \\
&= \frac{\theta^k \Gamma(\alpha-k)}{\Gamma(\alpha)} G(\alpha-k; \theta/x) + x^k \Gamma(\alpha; \theta/x), \text{ all } k \\
&= \frac{\theta^k}{(\alpha-1)\cdots(\alpha-k)} G(\alpha-k; \theta/x) \\
&\quad + x^k \Gamma(\alpha; \theta/x), \quad k \text{an integer} \\
\text{mode} &= \theta/(\alpha+1)
\end{aligned}
$$

$$
\begin{aligned}
\frac{\partial F}{\partial \alpha} &= -\Gamma'(\alpha; \theta/x) \\
\frac{\partial F}{\partial \theta} &= -f(x)\frac{x}{\theta}
\end{aligned}
$$

$$
\hat{\alpha} = \frac{2t-m^2}{t-m^2}, \quad \hat{\theta} = \frac{mt}{t-m^2}.
$$

A.3.2.3 Weibull—θ, τ

$$
\begin{aligned}
f(x) &= \frac{\tau(x/\theta)^\tau e^{-(x/\theta)^\tau}}{x} \\
F(x) &= 1 - e^{-(x/\theta)^\tau} \\
E[X^k] &= \theta^k \Gamma(1 + k/\tau), \quad k > -\tau \\
E[(X \wedge x)^k] &= \theta^k \Gamma(1 + k/\tau)\Gamma[1 + k/\tau; (x/\theta)^\tau] + x^k e^{-(x/\theta)^\tau}, \quad k > -\tau \\
\text{mode} &= \theta \left(\frac{\tau - 1}{\tau}\right)^{1/\tau}, \quad \tau > 1, \text{ else } 0
\end{aligned}
$$

$$
\begin{aligned}
\frac{\partial F}{\partial \tau} &= f(x)\frac{x}{\tau}\log\left(\frac{x}{\theta}\right) \\
\frac{\partial F}{\partial \theta} &= -f(x)\frac{x}{\theta}
\end{aligned}
$$

$$
\hat{\theta} = \exp\left(\frac{g\log(p) - \log(q)}{g - 1}\right), \quad g = \frac{\log(\log(4))}{\log(\log(4/3))}, \quad \hat{\tau} = \frac{\log(\log(4))}{\log(q) - \log(\hat{\theta})}
$$

A.3.2.4 Inverse Weibull—θ, τ

$$
\begin{aligned}
f(x) &= \frac{\tau(\theta/x)^\tau e^{-(\theta/x)^\tau}}{x} \\
F(x) &= e^{-(\theta/x)^\tau} \\
E[X^k] &= \theta^k \Gamma(1 - k/\tau), \quad k < \tau \\
E[(X \wedge x)^k] &= \theta^k \Gamma(1 - k/\tau)\{1 - \Gamma[1 - k/\tau; (\theta/x)^\tau]\} \\
&\quad + x^k \left[1 - e^{-(\theta/x)^\tau}\right], \quad \text{all } k \\
&= \theta^k \Gamma(1 - k/\tau) G[1 - k/\tau; (\theta/x)^\tau] + x^k \left[1 - e^{-(\theta/x)^\tau}\right] \\
\text{mode} &= \theta \left(\frac{\tau}{\tau + 1}\right)^{1/\tau}
\end{aligned}
$$

$$
\begin{aligned}
\frac{\partial F}{\partial \tau} &= f(x)\frac{x}{\tau}\log\left(\frac{x}{\theta}\right) \\
\frac{\partial F}{\partial \theta} &= -f(x)\frac{x}{\theta}.
\end{aligned}
$$

$$
\hat{\theta} = \exp\left(\frac{g\log(q) - \log(p)}{g - 1}\right), \quad g = \frac{\log(\log(4))}{\log(\log(4/3))}, \quad \hat{\tau} = \frac{\log(\log(4))}{\log(\hat{\theta}) - \log(p)}
$$

A.3.3 One-parameter distributions

A.3.3.1 Exponential—θ

$$\begin{aligned} f(x) &= \frac{e^{-x/\theta}}{\theta} \\ F(x) &= 1-e^{-x/\theta} \\ E[X^k] &= \theta^k\Gamma(k+1), \quad k>-1 \\ E[X^k] &= \theta^k k!, \quad \text{if } k \text{ is an integer} \\ E[X\wedge x] &= \theta(1-e^{-x/\theta}) \\ E[(X\wedge x)^k] &= \theta^k\Gamma(k+1)\Gamma(k+1;x/\theta)+x^k e^{-x/\theta}, \quad k>-1 \\ &= \theta^k k!\Gamma(k+1;x/\theta)+x^k e^{-x/\theta}, \quad k \text{ an integer} \\ \text{mode} &= 0 \end{aligned}$$

$$\frac{\partial F}{\partial \theta} = -f(x)\frac{x}{\theta}$$

$$\hat{\theta} = m.$$

A.3.3.2 Inverse exponential—θ

$$\begin{aligned} f(x) &= \frac{\theta e^{-\theta/x}}{x^2} \\ F(x) &= e^{-\theta/x} \\ E[X^k] &= \theta^k\Gamma(1-k), \quad k<1 \\ E[(X\wedge x)^k] &= \theta^k\Gamma(1-k)G(1-k;\theta/x)+x^k(1-e^{-\theta/x}), \quad \text{all } k \\ \text{mode} &= \theta/2 \end{aligned}$$

$$\frac{\partial F}{\partial \theta} = -f(x)\frac{x}{\theta}$$

$$\hat{\theta} = -q\log(3/4).$$

A.4 OTHER DISTRIBUTIONS

A.4.1.1 Lognormal—μ,σ (μ can be negative)

$$\begin{aligned} f(x) &= \frac{1}{x\sigma\sqrt{2\pi}}\exp(-z^2/2) = \phi(z)/(\sigma x), \quad z=\frac{\log x-\mu}{\sigma} \\ F(x) &= \Phi(z) \end{aligned}$$

$$
\begin{aligned}
E[X^k] &= \exp(k\mu + k^2\sigma^2/2) \\
E[(X \wedge x)^k] &= \exp(k\mu + k^2\sigma^2/2)\Phi\left(\frac{\log x - \mu - k\sigma^2}{\sigma}\right) + x^k[1 - F(x)] \\
\text{mode} &= \exp(\mu - \sigma^2)
\end{aligned}
$$

$$
\begin{aligned}
\frac{\partial F}{\partial \mu} &= -\phi(z)/\sigma \\
\frac{\partial F}{\partial \sigma} &= -z\phi(z)/\sigma
\end{aligned}
$$

$$
\hat{\sigma} = \sqrt{\log(t) - 2\log(m)}, \quad \hat{\mu} = \log(m) - \hat{\sigma}^2/2.
$$

A.4.1.2 Inverse Gaussian—μ, θ

$$
\begin{aligned}
f(x) &= \left(\frac{\theta}{2\pi x^3}\right)^{1/2} \exp\left\{-\frac{\theta z^2}{2x}\right\}, \quad z = \frac{x - \mu}{\mu} \\
F(x) &= \Phi\left[z\left(\frac{\theta}{x}\right)^{1/2}\right] + \exp(2\theta/\mu)\Phi\left[-y\left(\frac{\theta}{x}\right)^{1/2}\right], \quad y = \frac{x + \mu}{\mu} \\
E[X] &= \mu, \quad Var[X] = \mu^3/\theta \\
E[X \wedge x] &= x - \mu z\Phi\left[z\left(\frac{\theta}{x}\right)^{1/2}\right] - \mu y \exp(2\theta/\mu)\Phi\left[-y\left(\frac{\theta}{x}\right)^{1/2}\right]
\end{aligned}
$$

$$
\begin{aligned}
\frac{\partial F}{\partial \mu} &= -\phi\left[z\left(\frac{\theta}{x}\right)^{1/2}\right]\left(\frac{\theta}{x}\right)^{1/2}\frac{x}{\mu^2} - \frac{2\theta}{\mu^2}\exp(2\theta/\mu)\Phi\left[-y\left(\frac{\theta}{x}\right)^{1/2}\right] \\
&\quad + \exp(2\theta/\mu)\phi\left[-y\left(\frac{\theta}{x}\right)^{1/2}\right]\left(\frac{\theta}{x}\right)^{1/2}\frac{x}{\mu^2} \\
\frac{\partial F}{\partial \theta} &= z\phi\left[z\left(\frac{\theta}{x}\right)^{1/2}\right]\frac{z}{2(\theta x)^{1/2}} + \frac{2}{\mu}\exp(2\theta/\mu)\Phi\left[-y\left(\frac{\theta}{x}\right)^{1/2}\right] \\
&\quad - \exp(2\theta/\mu)\phi\left[-y\left(\frac{\theta}{x}\right)^{1/2}\right]\frac{y}{2(\theta x)^{1/2}}
\end{aligned}
$$

$$
\hat{\mu} = m, \quad \hat{\theta} = \frac{m^3}{t - m^2}.
$$

A.4.1.3 Single-parameter Pareto—α, θ

$$
\begin{aligned}
f(x) &= \frac{\alpha\theta^\alpha}{x^{\alpha+1}}, \quad x > \theta \\
F(x) &= 1 - (\theta/x)^\alpha, \quad x > \theta \\
E[X^k] &= \frac{\alpha\theta^k}{\alpha - k}, \quad k < \alpha \\
E[(X \wedge x)^k] &= \frac{\alpha\theta^k}{\alpha - k} - \frac{k\theta^\alpha}{(\alpha - k)x^{\alpha-k}} \\
\text{mode} &= \theta
\end{aligned}
$$

$$
\frac{\partial F}{\partial \alpha} = -\left(\frac{\theta}{x}\right)^\alpha \log\left(\frac{\theta}{x}\right)
$$

$$
\hat{\alpha} = \frac{m}{m - \theta}.
$$

Note: Although there appears to be two parameters, only α is a true parameter. The value of θ must be set in advance.

A.5 DISTRIBUTIONS WITH FINITE SUPPORT

For these two distributions, the scale parameter θ is assumed known.

A.5.1.1 Generalized beta—a, b, θ, τ

$$
\begin{aligned}
f(x) &= \frac{\Gamma(a+b)}{\Gamma(a)\Gamma(b)} u^a (1-u)^{b-1} \frac{\tau}{x}, \quad 0 < x < \theta, \quad u = (x/\theta)^\tau \\
F(x) &= \beta(a, b; u) \\
E[X^k] &= \frac{\theta^k \Gamma(a+b)\Gamma(a+k/\tau)}{\Gamma(a)\Gamma(a+b+k/\tau)}, \quad k > -a\tau \\
E[(X \wedge x)^k] &= \frac{\theta^k \Gamma(a+b)\Gamma(a+k/\tau)}{\Gamma(a)\Gamma(a+b+k/\tau)} \beta(a+k/\tau, b; u) + x^k[1 - \beta(a, b; u)]
\end{aligned}
$$

$$
\begin{aligned}
\frac{\partial F}{\partial a} &= \beta_1(a, b; u) \\
\frac{\partial F}{\partial b} &= \beta_2(a, b; u) \\
\frac{\partial F}{\partial \tau} &= -f(x) u \log \theta.
\end{aligned}
$$

A.5.1.2 beta—a, b, θ

$$
\begin{aligned}
f(x) &= \frac{\Gamma(a+b)}{\Gamma(a)\Gamma(b)} u^a (1-u)^{b-1} \frac{1}{x}, \quad 0 < x < \theta, \quad u = x/\theta \\
F(x) &= \beta(a, b; u) \\
E[X^k] &= \frac{\theta^k \Gamma(a+b)\Gamma(a+k)}{\Gamma(a)\Gamma(a+b+k)}, \quad k > -a \\
E[X^k] &= \frac{\theta^k a(a+1)\cdots(a+k-1)}{(a+b)(a+b+1)\cdots(a+b+k-1)}, \quad \text{if } k \text{ is an integer} \\
E[(X \wedge x)^k] &= \frac{\theta^k a(a+1)\cdots(a+k-1)}{(a+b)(a+b+1)\cdots(a+b+k-1)} \beta(a+k, b; u) \\
&\quad + x^k [1 - \beta(a, b; u)]
\end{aligned}
$$

$$
\begin{aligned}
\frac{\partial F}{\partial a} &= \beta_1(a, b; u) \\
\frac{\partial F}{\partial b} &= \beta_2(a, b; u)
\end{aligned}
$$

$$
\hat{a} = \frac{\theta m^2 - mt}{\theta t - \theta m^2}, \quad \hat{b} = \frac{(\theta m - t)(\theta - m)}{\theta t - \theta m^2}.
$$

Appendix B: An Inventory of Discrete Distributions

B.1 INTRODUCTION

The sixteen models fall into three classes. The divisions are based on the algorithm by which the probabilities are computed. For some of the more familiar distributions these formulas will look different from the ones you may have learned, but they produce the same probabilities. After each name, the parameters are given. All parameters are positive unless otherwise indicated. In all cases, p_k is the probability of observing k losses.

For finding moments, the most convenient form is to give the factorial moments. The jth factorial moment is $\mu_{(j)} = E[N(N-1)\cdots(N-j+1)]$. We have $E[N] = \mu_{(1)}$ and $Var(N) = \mu_{(2)} + \mu_{(1)} - \mu_{(1)}^2$.

The estimators which are presented are not intended to be useful estimators but rather for providing starting values for maximizing the likelihood (or other) function. For determining starting values, the following quantities are used [where n_k is the observed frequency at k (if, for the last entry, n_k represents the number of observations at k or more, assume it was at exactly k)

and n is the sample size]:

$$\hat{\mu} = \frac{1}{n}\sum_{k=1}^{\infty} kn_k, \quad \hat{\sigma}^2 = \frac{1}{n}\sum_{k=1}^{\infty} k^2 n_k - \hat{\mu}^2.$$

When the method of moments is used to determine the starting value, a circumflex (e.g., $\hat{\lambda}$) is used. For any other method, a tilde (e.g., $\tilde{\lambda}$) is used. When the starting value formulas do not provide admissible parameter values, a truly crude guess is to set the product of all λ and β parameters equal to the sample mean and set all other parameters equal to 1. If there are two λ and/or β parameters, an easy choice is to set each to the square root of the sample mean.

The last item presented is the probability generating function,

$$P(z) = E[z^N].$$

B.2 THE $(a, b, 0)$ CLASS

The distributions in this class have support on 0, 1, ... For this class, a particular distribution is specified by setting p_0 and then using $p_k = (a + b/k)p_{k-1}$. Specific members are created by setting p_0, a, and b. For any member, $\mu_{(1)} = (a+b)/(1-a)$ and for higher j, $\mu_{(j)} = (aj+b)\mu_{(j-1)}/(1-a)$. The variance is $(a+b)/(1-a)^2$.

B.2.1.1 Poisson—λ

$$\begin{aligned}
p_0 &= e^{-\lambda}, \quad a = 0, \quad b = \lambda \\
p_k &= \frac{e^{-\lambda}\lambda^k}{k!} \\
E[N] &= \lambda, \quad Var[N] = \lambda \\
\hat{\lambda} &= \hat{\mu} \\
P(z) &= e^{\lambda(z-1)}
\end{aligned}$$

B.2.1.2 Geometric—β

$$\begin{aligned}
p_0 &= 1/(1+\beta), \quad a = \beta/(1+\beta), \quad b = 0 \\
p_k &= \frac{\beta^k}{(1+\beta)^{k+1}} \\
E[N] &= \beta, \quad Var[N] = \beta(1+\beta) \\
\hat{\beta} &= \hat{\mu} \\
P(z) &= [1 - \beta(z-1)]^{-1}.
\end{aligned}$$

This is a special case of the negative binomial with $r = 1$.

B.2.1.3 Binomial—$q, m, (0 < q < 1, m$ an integer)

$$
\begin{aligned}
p_0 &= (1-q)^m, \quad a = -q/(1-q), \quad b = (m+1)q/(1-q) \\
p_k &= \binom{m}{k} q^k (1-q)^{m-k}, \quad k = 0, 1, \ldots, m \\
E[N] &= mq, \quad Var[N] = mq(1-q) \\
\hat{q} &= \hat{\mu}/m \\
P(z) &= [1 + q(z-1)]^m.
\end{aligned}
$$

B.2.1.4 Negative binomial—β, r

$$
\begin{aligned}
p_0 &= (1+\beta)^{-r}, \quad a = \beta/(1+\beta), \quad b = (r-1)\beta/(1+\beta) \\
p_k &= \frac{r(r+1)\cdots(r+k-1)\beta^k}{k!(1+\beta)^{r+k}} \\
E[N] &= r\beta, \quad Var[N] = r\beta(1+\beta) \\
\hat{\beta} &= \frac{\hat{\sigma}^2}{\hat{\mu}} - 1, \quad \hat{r} = \frac{\hat{\mu}^2}{\hat{\sigma}^2 - \hat{\mu}} \\
P(z) &= [1 - \beta(z-1)]^{-r}.
\end{aligned}
$$

B.3 THE $(a, b, 1)$ CLASS

To distinguish this class from the $(a, b, 0)$ class, the probabilities are denoted $\Pr(N = k) = p_k^M$ or $\Pr(N = k) = p_k^T$ depending on which subclass is being represented. For this class, p_0^M is arbitrary (that is, it is a parameter) and then p_1^M or p_1^T is a specified function of the parameters a and b. Subsequent probabilities are obtained recursively as in the $(a, b, 0)$ class: $p_k^M = (a + b/k)p_{k-1}^M$, $k = 2, 3, \ldots$ with the same recursion for p_k^M. There are two subclasses of this class. When discussing their members, we often refer to the "corresponding" member of the $(a, b, 0)$ class. This refers to the member of that class with the same values for a and b. The notation p_k will continue to be used for probabilities for the corresponding $(a, b, 0)$ distribution.

B.3.1 The zero-truncated subclass

The members of this class have $p_0^T = 0$ and therefore it need not be estimated. These distributions should only be used when a value of zero is impossible. The first factorial moment is $\mu_{(1)} = (a+b)/[(1-a)(1-p_0)]$, where p_0 is the value for the corresponding member of the $(a, b, 0)$ class. For the logarithmic distribution (which has no corresponding member), $\mu_{(1)} = \beta/\log(1+\beta)$. Higher factorial moments are obtained recursively with the same formula as with the $(a, b, 0)$ class. The variance is

$$(a+b)[1-(a+b+1)p_0]/[(1-a)(1-p_0)]^2.$$

For those members of the subclass which have corresponding $(a, b, 0)$ distributions, $p_k^T = p_k/(1-p_0)$.

B.3.1.1 Zero-truncated Poisson—λ

$$
\begin{aligned}
p_1^T &= \lambda/(e^\lambda - 1), \quad a = 0, \quad b = \lambda \\
p_k^T &= \frac{\lambda^k}{k!(e^\lambda - 1)} \\
E[N] &= \lambda/(1-e^{-\lambda}), \quad Var[N] = \lambda[1-(\lambda+1)e^{-\lambda}]/(1-e^{-\lambda})^2 \\
\tilde{\lambda} &= \log(n\hat{\mu}/n_1) \\
P(z) &= \frac{e^{\lambda z} - 1}{e^\lambda - 1}.
\end{aligned}
$$

B.3.1.2 Zero-truncated geometric—β

$$
\begin{aligned}
p_1^T &= 1/(1+\beta), \quad a = \beta/(1+\beta), \quad b = 0 \\
p_k^T &= \frac{\beta^{k-1}}{(1+\beta)^k} \\
E[N] &= 1+\beta, \quad Var[N] = \beta(1+\beta) \\
\hat{\beta} &= \hat{\mu} - 1 \\
P(z) &= \frac{[1-\beta(z-1)]^{-1} - (1+\beta)^{-1}}{1-(1+\beta)^{-1}}.
\end{aligned}
$$

This is a special case of the zero-truncated negative binomial with $r = 1$.

B.3.1.3 Logarithmic—β

$$
\begin{aligned}
p_1^T &= \beta/[(1+\beta)\log(1+\beta)], \quad a = \beta/(1+\beta), \quad b = -\beta/(1+\beta) \\
p_k^T &= \frac{\beta^k}{k(1+\beta)^k \log(1+\beta)} \\
E[N] &= \beta/\log(1+\beta), \quad Var[N] = \beta[1+\beta-\beta/\log(1+\beta)]/\log(1+\beta) \\
\tilde{\beta} &= \frac{n\hat{\mu}}{n_1} - 1 \quad \text{or} \quad \frac{2(\hat{\mu}-1)}{\hat{\mu}} \\
P(z) &= 1 - \frac{\log[1-\beta(z-1)]}{\log(1+\beta)}.
\end{aligned}
$$

This is a limiting case of the zero-truncated negative binomial as $r \to 0$.

B.3.1.4 Zero-truncated binomial—$q, m, (0 < q < 1, m$ an integer$)$

$$
\begin{aligned}
p_1^T &= m(1-q)^{m-1}q/[1-(1-q)^m] \\
a &= -q/(1-q), \quad b = (m+1)q/(1-q)
\end{aligned}
$$

$$
\begin{aligned}
p_k^T &= \frac{\binom{m}{k} q^k (1-q)^{m-k}}{1-(1-q)^m}, \quad k = 1, 2, \ldots, m \\
E[N] &= mq/[1-(1-q)^m] \\
Var[N] &= \frac{mq[(1-q)-(1-q+mq)(1-q)^m]}{[1-(1-q)^m]^2} \\
\tilde{q} &= \hat{\mu}/m \\
P(z) &= \frac{[1+q(z-1)]^m - (1-q)^m}{1-(1-q)^m}.
\end{aligned}
$$

B.3.1.5 Zero-truncated negative binomial—$\beta, r, (r > -1)$

$$
\begin{aligned}
p_1^T &= r\beta/[(1+\beta)^{r+1} - (1+\beta)], \\
a &= \beta/(1+\beta), \quad b = (r-1)\beta/(1+\beta) \\
p_k^T &= \frac{r(r+1)\cdots(r+k-1)}{k![(1+\beta)^r - 1]} \left(\frac{\beta}{1+\beta}\right)^k \\
E[N] &= r\beta/[1-(1+\beta)^{-r}] \\
Var[N] &= \frac{r\beta[(1+\beta)-(1+\beta+r\beta)(1+\beta)^{-r}]}{[1-(1+\beta)^{-r}]^2} \\
\tilde{\beta} &= \frac{\hat{\sigma}^2}{\hat{\mu}} - 1, \quad \tilde{r} = \frac{\hat{\mu}^2}{\hat{\sigma}^2 - \hat{\mu}} \\
P(z) &= \frac{[1-\beta(z-1)]^{-r} - (1+\beta)^{-r}}{1-(1+\beta)^{-r}}.
\end{aligned}
$$

This distribution is sometimes called the extended truncated negative binomial distribution because the parameter r can extend below 0.

B.3.2 The zero-modified subclass

A zero-modified distribution is created by starting with a truncated distribution and then placing an arbitrary amount of probability at zero. This probability, p_0^M, is a parameter. The remaining probabilities are adjusted accordingly. Values of p_k^M can be determined from the corresponding zero-truncated distribution as $p_k^M = (1-p_0^M)p_k^T$ or from the corresponding $(a, b, 0)$ distribution as $p_k^M = (1-p_0^M)p_k/(1-p_0)$. The same recursion used for the zero-truncated subclass applies.

The mean is $1 - p_0^M$ times the mean for the corresponding zero-truncated distribution. The variance is $1 - p_0^M$ times the zero-truncated variance plus $p_0^M(1-p_0^M)$ times the square of the zero-truncated mean. The probability-generating function is $P^M(z) = p_0^M + (1-p_0^M)P(z)$, where $P(z)$ is the probability-generating function for the corresponding zero-truncated distribution.

The maximum likelihood estimator of q_0 is always the sample relative frequency at 0.

B.4 THE COMPOUND CLASS

Members of this class are obtained by compounding one distribution with another. That is, let N be a discrete distribution, called the **primary distribution** and let $M_1, M_2, \ldots$ be identically and independently distributed with another discrete distribution, called the **secondary distribution**. The compound distribution is $S = M_1 + \cdots + M_N$. The probabilities for the compound distributions are found from

$$p_k = \sum_{y=1}^{k} (a + by/k) f_y p_{k-y} / (1 - af_0)$$

for $n = 1, 2, \ldots$, where a and b are the usual values for the primary distribution (which must be a member of the $(a, b, 0)$ class) and f_y is p_y for the secondary distribution. The only two primary distributions used here are Poisson (for which $p_0 = \exp[-\lambda(1 - f_0)]$) and geometric (for which $p_0 = 1/[1 + \beta - \beta f_0]$). Because this information completely describes these distributions, only the names and starting values are given below.

The moments can be found from the moments of the individual distributions:

$$E[S] = E[N]E[M] \quad \text{and} \quad Var[S] = E[N]Var[M] + Var[N]E[M]^2.$$

The probability-generating function is $P(z) = P_{\text{primary}}(P_{\text{secondary}}(z))$.

In the following list the primary distribution is always named first. For the first, second, and fourth distributions, the secondary distribution is the $(a, b, 0)$ class member with that name. For the third and the last three distributions (the Poisson–ETNB and its two special cases) the secondary distribution is the zero-truncated version.

B.4.1 Some compound distributions

B.4.1.1 *Poisson–binomial*—$\lambda, q, m (0 < q < 1,$ *m* ***an integer***)

$$\hat{q} = \frac{\hat{\sigma}^2/\hat{\mu} - 1}{m - 1}, \quad \hat{\lambda} = \hat{\mu}/(m\hat{q}) \quad \text{or} \quad \tilde{q} = 0.5,\ \tilde{\lambda} = 2\hat{\mu}/m.$$

B.4.1.2 *Poisson–Poisson*—λ_1, λ_2

The parameter λ_1 is for the primary Poisson distribution, and λ_2 is for the secondary Poisson distribution. This distribution is also called the **Neyman Type A**.

$$\tilde{\lambda}_1 = \tilde{\lambda}_2 = \sqrt{\hat{\mu}}.$$

B.4.1.3 *Geometric–extended truncated negative binomial*—$\beta_1, \beta_2, r\ (r > -1)$

The parameter β_1 is for the primary geometric distribution. The last two parameters are for the secondary distribution. The truncated version is used so that the extension of r is available.

$$\tilde{\beta}_1 = \tilde{\beta}_2 = \sqrt{\hat{\mu}}.$$

B.4.1.4 Geometric–Poisson—β, λ

$$\tilde{\beta} = \tilde{\lambda} = \sqrt{\hat{\mu}}.$$

B.4.1.5 Poisson–extended truncated negative binomial—λ, β, r

$$\tilde{r} = \frac{\hat{\mu}(K - 3\hat{\sigma}^2 + 2\hat{\mu}) - 2(\hat{\sigma}^2 - \hat{\mu})^2}{\hat{\mu}(K - 3\hat{\sigma}^2 + 2\hat{\mu}) - (\hat{\sigma}^2 - \hat{\mu})^2}, \quad \tilde{\beta} = \frac{\hat{\sigma}^2 - \hat{\mu}}{\hat{\mu}(1 + \hat{r})}, \quad \tilde{\lambda} = \frac{\hat{\mu}}{\hat{r}\hat{\beta}}, \text{ or}$$

$$\tilde{r} = \frac{\hat{\sigma}^2 n_1/n - \hat{\mu}^2 n_0/n}{(\hat{\sigma}^2 - \hat{\mu}^2)(n_0/n)\log(n_0/n) - \hat{\mu}(\hat{\mu}n_0/n - n_1/n)},$$

$$\tilde{\beta} = \frac{\hat{\sigma}^2 - \hat{\mu}}{\hat{\mu}(1 + \hat{r})}, \quad \tilde{\lambda} = \frac{\hat{\mu}}{\hat{r}\hat{\beta}}$$

where $K = \frac{1}{n}\sum_{k=0}^{\infty} k^3 n_k - 3\hat{\mu}\frac{1}{n}\sum_{k=0}^{\infty} k^2 n_k + 2\hat{\mu}^3$

This distribution is also called the **generalized Poisson–Pascal.**

B.4.1.6 Polya–Aeppli—λ, β

$$\hat{\beta} = \frac{\hat{\sigma}^2 - \hat{\mu}}{2\hat{\mu}}, \; \hat{\lambda} = \frac{\hat{\mu}}{1 + \hat{\beta}}.$$

This is a special case of the Poisson–extended truncated negative binomial with $r = 1$. It is actually a Poisson–truncated geometric.

B.4.1.7 Poisson–inverse Gaussian—λ, β

$$\tilde{\lambda} = -\log(n_0/n), \; \tilde{\beta} = 4(\hat{\mu} - \tilde{\lambda})/\hat{\mu}$$

This is a special case of the Poisson–extended truncated negative binomial with $r = -0.5$.

B.5 A HIERARCHY OF DISCRETE DISTRIBUTIONS

The following table indicates which distributions are special or limiting cases of others. For the special cases, one parameter is set equal to a constant to create the special case. For the limiting cases, two parameters go to infinity or zero in some special way.

Distribution	is a special case of	is a limiting case of
Poisson	ZM Poisson	negative binomial, Poisson–binomial, Poisson–inverse Gaussian, Polya–Aeppli, Neyman– A
ZT Poisson	ZM Poisson	ZT negative binomial
ZM Poisson		ZM negative binomial
geometric	negative binomial, ZM geometric	geometric–Poisson
ZT geometric	ZT negative binomial	
ZM geometric	ZM negative binomial	
logarithmic		ZT negative binomial
ZM logarithmic		ZM negative binomial
binomial	ZM binomial	
negative binomial	ZM negative binomial	Poisson–ETNB
Poisson–inverse Gaussian	Poisson–ETNB	
Polya–Aeppli	Poisson–ETNB	
Neyman–A		Poisson–ETNB

Appendix C: The Simplex Method

The method (which is not related to the simplex method from operations research) was introduced for use with maximum likelihood estimation by Nelder and Mead in 1965 [95]. An excellent reference (and the source of the particular version presented here) is *Sequential Simplex Optimization* by Walters, Parker, Morgan, and Deming [123].

Let $\mathbf{x}$ be a $k \times 1$ vector and $f(\mathbf{x})$ be the function in question. The iterative step begins with $k+1$ vectors, $\mathbf{x}_1, \ldots, \mathbf{x}_{k+1}$, and the corresponding functional values, $f_1, \ldots, f_{k+1}$. At any iteration the points will be ordered so that $f_2 < \cdots < f_{k+1}$. When starting, also arrange for $f_1 < f_2$. Three of the points have names: $\mathbf{x}_1$ is called *worstpoint*, $\mathbf{x}_2$ is called *secondworstpoint*, and $\mathbf{x}_{k+1}$ is called *bestpoint*. It should be noted that after the first iteration these names may not perfectly describe the points. Now identify five new points. The first one, $\mathbf{y}_1$, is the center of $\mathbf{x}_2, \ldots, \mathbf{x}_{k+1}$, That is, $\mathbf{y}_1 = \sum_{j=2}^{k+1} \mathbf{x}_j / k$, and is called *midpoint*. The other four points are found as follows:

$$
\begin{aligned}
\mathbf{y}_2 &= 2\mathbf{y}_1 - \mathbf{x}_1, && \textit{refpoint} \\
\mathbf{y}_3 &= 2\mathbf{y}_2 - \mathbf{x}_1, && \textit{doublepoint} \\
\mathbf{y}_4 &= (\mathbf{y}_1 + \mathbf{y}_2)/2, && \textit{halfpoint} \\
\mathbf{y}_5 &= (\mathbf{y}_1 + \mathbf{x}_1)/2, && \textit{centerpoint.}
\end{aligned}
$$

Then let $g_2, \ldots, g_5$ be the corresponding functional values (the value at $\mathbf{y}_1$ is never used). The key is to replace *worstpoint* ($\mathbf{x}_1$) with one of these points. The decision process proceeds as follows:

1. If $f_2 < g_2 < f_{k+1}$, then replace it with *refpoint*.

2. If $g_2 \geq f_{k+1}$ and $g_3 > f_{k+1}$, then replace it with *doublepoint*.

3. If $g_2 \geq f_{k+1}$ and $g_3 \leq f_{k+1}$, then replace it with *refpoint*.

4. If $f_1 < g_2 \leq f_2$, then replace it with *halfpoint*.

5. If $g_2 \leq f_1$, then replace it with *centerpoint*.

After the replacement has been made, the old *secondworstpoint* becomes the new *worstpoint*. The remaining k points are then ordered. The one with the smallest functional value becomes the new *secondworstpoint*, and the one with the largest functional value becomes the new *bestpoint*. In practice, there is no need to compute $\mathbf{y}_3$ and g_3 until you have reached step 2. Also note that at most one of the pairs $(\mathbf{y}_4, g_4)$ and $(\mathbf{y}_5, g_5)$ needs to be obtained, depending on which (if any) of the conditions in steps 4 and 5 hold.

The following graph indicates the progress of the simplex. The problem is to maximize the function $f(x, y) = -a^2 - 2b^2$. The starting simplex is $(1, 1)$, $(1, 2)$, $(2, 1)$ and the maximum is at $(0, 0)$.

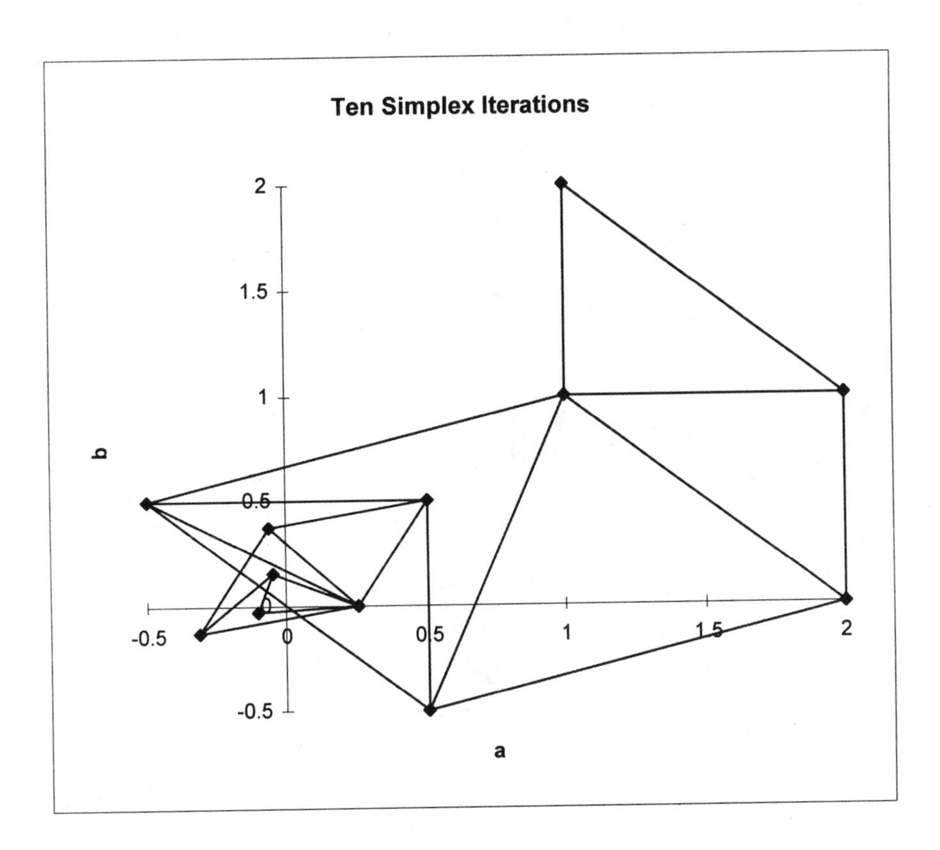
Ten Simplex Iterations
2
1.5
1
0.5
0
-0.5
b
-0.5
0
0.5
1
1.5
2
a

Appendix D: Frequency and Severity Relationships

Let N be the number of losses random variable and let X be the severity random variable. If there is a deductible of d imposed, there are two ways to modify X. One is to create Y, the amount paid per loss:

$$\begin{aligned} Y &= 0, & X \leq d \\ &= X - d, & X > d. \end{aligned}$$

In this case the appropriate frequency distribution continues to be N.

An alternative approach is to create Y^*, the amount paid per payment:

$$\begin{aligned} Y^* &= \text{undefined}, & X \leq d \\ &= X - d, & X > d. \end{aligned}$$

In this case the frequency random variable must be altered to reflect the number of payments. Let this variable be N^*. Assume that for each loss the probability is $v = 1 - F_X(d)$ that a payment will result. Further assume that the incidence of making a payment is independent of the number of losses. Then $N^* = L_1 + L_2 + \cdots + L_N$ where L_i is 0 with probability $1 - v$ and is 1 with probability v. Probability generating functions yield the following relationships:

N	Parameters for N^*
Poisson	$\lambda^* = v\lambda$
ZM Poisson	$p_0^{M*} = \dfrac{p_0^M - e^{-\lambda} + e^{-v\lambda} - p_0^M e^{-v\lambda})}{1 - e^{-\lambda}}, \ \lambda^* = v\lambda$
binomial	$q^* = vq$
ZM binomial	$p_0^{M*} = \dfrac{p_0^M - (1-q)^m + (1-vq)^m - p_0^M(1-vq)^m}{1-(1-q)^m}$ $q^* = vq$
negative binomial	$\beta^* = v\beta, \ r^* = r$
ZM negative binomial	$p_0^{M*} = \dfrac{p_0^M - (1+\beta)^{-r} + (1+v\beta)^{-r} - p_0^M(1+v\beta)^{-r}}{1-(1+\beta)^{-r}}$ $\beta^* = v\beta, \ r^* = r$
ZM logarithmic	$p_0^{M*} = 1 - (1 - p_0^M)\log(1+v\beta)/\log(1+\beta)$ $\beta^* = v\beta$

The geometric distribution is not presented as it is a special case of the negative binomial with $r = 1$. For zero-truncated distributions, the above is still used as the distribution for N^* will now be zero-modified. For compound distributions, modify only the secondary distribution. For ETNB secondary distributions the parameter for the primary distribution is multiplied by $1 - p_0^{M*}$ as obtained above while the secondary distribution remains zero-truncated (however, $\beta^* = v\beta$).

There are occasions in which frequency data are collected which provide a model for N^*. There would have to have been a deductible d in place and therefore v is available. It is possible to recover the distribution for N, although there is no guarantee that reversing the process will produce a legitimate probability distribution. The solutions are the same as above, only now $v = 1/[1 - F_X(d)]$.

Now suppose the current frequency model is N^*, which is appropriate for a deductible of d. Now suppose the deductible is to be changed to d'. The new frequency for payments is N' and is of the same type. Then use the table with $v = [1 - F_X(d')]/[1 - F_X(d)]$.

Finally, suppose a truncating limit is imposed. In this case

$$
\begin{aligned}
Y^* &= \text{undefined}, & X \le d \\
&= X - d, & d < X < u \\
&= \text{undefined}, & X \ge u.
\end{aligned}
$$

Then proceed as above, only now $v = F_X(u) - F_X(d)$.

Appendix E: Limited Expected Value Calculations

The limited expected value is defined as

$$E[X \wedge x] = \int_0^x t f(t)dt + x[1 - F(x)].$$

It can be used to obtain the expected payment when the loss X or aggregate loss S is modified by a deductible (d), a limit (u), or a coinsurance(α). Suppose all three are present, so that the amount paid, Y, is related to X by

$$\begin{aligned} Y &= 0, & X < d \\ &= \alpha(X - d), & d \leq X \leq u \\ &= \alpha(u - d), & X > u. \end{aligned}$$

Then $E[Y] = \alpha\{E[X\wedge u] - E[X\wedge d]\}$. If there is inflation so that $X^* = (1+r)X$ and Y is based on X^*, then $E[Y] = \alpha(1+r)\{E[X\wedge u/(1+r)] - E[X\wedge d/(1+r)]\}$.

The limited expected squared value is defined as

$$E[(X \wedge x)^2] = \int_0^x t^2 f(t)dt + x^2[1 - F(x)].$$

We then have $E[Y^2] = \alpha^2\{E[(X \wedge u)^2] - E[(X \wedge d)^2] - 2dE[X \wedge u] + 2dE[X \wedge d]\}$. The variance can be found by subtracting the square of the mean. These calculations can apply to either the individual or the aggregate losses.

Appendix F: The Recursive Formula

The recursive formula is

$$g_i = \frac{\left\{[p_1 - (a+b)p_0]f_i + \sum_{j=1}^{i}(a+bj/i)f_j g_{i-j}\right\}}{(1 - af_0)}, \quad i = 1, 2, \ldots$$

where $g_i = \Pr(S = i)$, $p_i = \Pr(N = i)$, and $f_j = \Pr(X = j)$. The formula must be initialized with the value of g_0. These values are given in Table F.1. It should be noted that if N is a member of the $(a, b, 0)$ class, $p_1 - (a+b)p_0 = 0$ and so the first term will vanish. If N is a member of the compound class, the recursion must be run twice. The first pass uses the secondary distribution for p_i, a, and b. The second pass uses the output from the first pass as f_j and uses the primary distribution for p_i, a, and b.

Table F.1 Starting values (g_0) for recursions

Distribution	g_0
Poisson	$\exp[\lambda(f_0 - 1)]$
geometric	$[1 + \beta(1 - f_0)]^{-1}$
binomial	$[1 + q(f_0 - 1)]^m$
negative binomial	$[1 + \beta(1 - f_0)]^{-r}$
ZM Poisson	$p_0^M + (1 - p_0^M)\dfrac{\exp(\lambda f_0) - 1}{\exp(\lambda) - 1}$
ZM geometric	$p_0^M + (1 - p_0^M)\dfrac{f_0}{1 + \beta(1 - f_0)}$
ZM binomial	$p_0^M + (1 - p_0^M)\dfrac{[1 + q(f_0 - 1)]^m - (1 - q)^m}{1 - (1 - q)^n}$
ZM negative binomial	$p_0^M + (1 - p_0^M)\dfrac{[1 + \beta(1 - f_0)]^{-r} - (1 + \beta)^{-r}}{1 - (1 + \beta)^{-r}}$
ZM logarithmic	$p_0^M + (1 - p_0^M)\left\{1 - \dfrac{\log[1 + \beta(1 - f_0)]}{\log(1 + \beta)}\right\}$

Appendix G: Discretization of the Severity Distribution

There are two ways to discretize the severity distribution that are used in the program. One is the method of rounding and the other is a mean-preserving method.

G.1 THE METHOD OF ROUNDING

This method has two features: All probabilities are positive, and the probabilities add to one. Let h be the span and let Y be the discretized version of X. If there are no modifications, then

$$\begin{aligned} f_j &= \Pr(Y = jh) = \Pr[(j - 1/2)h \leq X < (j + 1/2)h] \\ &= F_X[(j + 1/2)h] - F_X[(j - 1/2)h]. \end{aligned}$$

Suppose a deductible of d, limit of u, and coinsurance of α are to be applied. If the modifications are to be applied before the discretization, then

$$g_0 = \frac{F_X(d + h/2) - F_X(d)}{1 - F_X(d)}$$

$$\begin{aligned} g_j &= \frac{F_X[d+(j+1/2)h] - F_X[d-(j-1/2)h]}{1-F_X(d)}, \\ & \qquad j = 1, \ldots, \frac{u-d}{h} - 1 \\ g_{(u-d)/h} &= \frac{1 - F_X(u-h/2)}{1-F_X(d)}. \end{aligned}$$

where $g_j = \Pr(Z = j\alpha h)$ where Z is the modified distribution. This method does not require that the limits be multiples of h but does require that $u-d$ be a multiple of h. This method gives the probabilities of payments per payment.

Finally, if there is truncation from above at u, change all denominators to $F_X(u) - F_X(d)$ and also change the numerator of $g_{(u-d)/h}$ to $F_X(u) - F_X(u - h/2)$.

G.2 MEAN-PRESERVING

This method ensures that the discretized distribution has the same mean as the original severity distribution. With no modifications the discretization is

$$\begin{aligned} f_0 &= 1 - E[X \wedge h]/h \\ f_j &= (2E[X \wedge jh] - E[X \wedge (j-1)h] - E[X \wedge (j+1)h])/h, \quad j = 1, 2, \ldots. \end{aligned}$$

For the modified distribution,

$$\begin{aligned} g_0 &= 1 - \frac{E[X \wedge d+h] - E[X \wedge d]}{h[1-F_X(d)]} \\ gj &= \frac{2E[X \wedge d+jh] - E[X \wedge d+(j-1)h] - E[X \wedge d+(j+1)h]}{h[1-F_X(d)]}, \\ & \qquad j = 1, \ldots, \frac{u-d}{h} - 1 \\ g_{(u-d)/h} &= \frac{E[X \wedge uh] - E[X \wedge (u-1)h]}{h[1-F_X(d)]}. \end{aligned}$$

To incorporate truncation from above, change the denominators to $h[F_X(u) - F_X(d)]$. Also, add $h[1 - F_X(u)]$ to the numerator of g_0 and subtract it from the numerator for $g_{(u-d)/h}$.

G.3 UNDISCRETIZATION OF A DISCRETIZED DISTRIBUTION

Assume we have $\Pr(S = 0)$, the true probability that the random variable is zero. Let $g_j = \Pr(S^* = jh)$, where S^* is a discretized distribution and h is the span. The following are approximations for the cdf and LEV of S, the true distribution which was discretized as S^*. They are all based on the assumption that S has a uniform distribution over the interval from $(j-1/2)h$ to $(j+1/2)h$ for integral j. The first interval is from 0 to $h/2$, and the probability $p_0 - \Pr(S=0)$ is assumed to be uniformly distributed over it. Let S^{**} be the random variable with this approximate continuous distribution. Then the approximate distribution function can be found by interpolation as follows. First, let

$$F_j = F_{S^{**}}[(j+1/2)h] = \sum_{i=0}^{j} g_i, \qquad j = 0, 1, \ldots.$$

Then, for x in the interval $(j-1/2)h$ to $(j+1/2)h$,

$$\begin{aligned} F_{S^{**}}(x) &= F_{j-1} + \int_{(j-1/2)h}^{x} h^{-1} g_j dt = F_{j-1} + [x-(j-1/2)h]h^{-1}g_j \\ &= F_{j-1} + [x-(j-1/2)h]h^{-1}(F_j - F_{j-1}) \\ &= (1-w)F_{j-1} + wF_j, \quad w = x/h - j + 1/2. \end{aligned}$$

Because the first interval is only half as wide, the formula for $0 \le x \le h/2$ is

$$F_{S^{**}}(x) = (1-w)\Pr(S=0) + wp_0, \qquad w = 2x/h.$$

It is also possible to express these formulas in terms of the discrete probabilities:

$$\begin{aligned} F_{S^{**}}(x) &= \Pr(S=0) + \frac{2x}{h}[g_0 - \Pr(S=0)], \qquad 0 < x \le h/2 \\ &= \sum_{i=0}^{j-1} g_i + \frac{x-(j-1/2)h}{h} g_j, \qquad (j-1/2)h < x \le (j+1/2)h, \end{aligned}$$

With regard to the limited expected value, expressions for the first and kth LEVs are

$$\begin{aligned} E(S^{**} \wedge x) &= x[1-\Pr(S=0)] - \frac{x^2}{h}[g_0 - \Pr(S=0)], \quad 0 < x \le h/2 \\ &= \frac{h}{4}[g_0 - \Pr(S=0)] + \sum_{i=1}^{j-1} ihg_i + \frac{x^2 - [(j-1/2)h]^2}{2h} g_j \\ &\quad + x[1 - F_{S^{**}}(x)], \quad (j-1/2) < x \le (j+1/2)h\ , \end{aligned}$$

and

$$
\begin{aligned}
E[(S^{**} \wedge x)^k] &= \frac{2x^{k+1}}{h(k+1)}[g_0 - \Pr(S=0)] + x^k[1 - F_{S^{**}}(x)], \quad 0 < x \leq h/2 \\
&= \frac{(h/2)^k[g_0 - \Pr(S=0)]}{k+1} + \sum_{i=1}^{j-1} \frac{h^k[(i+1/2)^{k+1} - (i-1/2)^{k+1}]}{k+1} g_i \\
&\quad + \frac{x^{k+1} - [(j-1/2)h]^{k+1}}{h(k+1)} g_j \\
&\quad + x^k[1 - F_{S^{**}}(x)], \qquad (j-1/2) < x \leq (j+1/2)h \,.
\end{aligned}
$$

Appendix H: Simulation

Far from being a comprehensive survey of simulation methods, this Appendix presents just two methods. One is an all-purpose method for simulating observations from a univariate distribution. The second is a method for simulating observations from a multivariate normal distribution. In both cases it is assumed that an algorithm for simulating observations from the uniform distribution on the interval (0,1) is available. One such formula is given below.

A good introduction to the issues involved in simulation as well as the methods presented here is the book by Ripley [105]. This source, as do many others, provides efficient methods that are specific for particular distributions, such as the Poisson and gamma.

H.1 UNIFORM (0,1) SIMULATION

A popular method of generating pseudouniform numbers is the multiplicative congruential generator. In 1984 Fishman and Moore [42] tested a large number of generators, and among those recommended was the following one. Begin with x_0 an integer between 1 and $2^{31} - 2$ inclusive. This is known as the **seed**. Then calculate $x_1 = 742{,}938{,}285 x_0 \bmod(2^{31} - 1)$. The first generated pseudouniform number is $u_1 = x_1/(2^{31} - 1)$. This process can be repeated indefinitely by letting $x_j = 742{,}938{,}285x_{j-1} \bmod (2^{31}-1)$ and $u_j = x_j/(2^{31}-1)$.

The only remaining problem is obtaining the seed. Sometimes it is a good idea to specify the number. This makes it possible to repeat the simulation with the same sequence of pseudouniform numbers. This can be especially useful when debugging a simulation program. Most computer packages allow for a mechanical means of setting the seed, perhaps by looking at the time on the system clock.

H.2 UNIVARIATE SIMULATION

The inversion method requires access to the distribution function. If u is a simulated observation from the uniform (0,1) distribution, then, provided the solution is unique, the simulated value for the random variable X is the solution to $F_X(x) = u$. For continuous random variables the solution will always be unique.

Example H.1 *Simulate an observation from the Pareto distribution with* $\alpha = 4$ *and* $\theta = 1{,}000$ *based on a uniform random number* $u = 0.774$.

The equation to solve is

$$1 - \left(\frac{1{,}000}{1{,}000 + x}\right)^4 = 0.774$$

and the solution is $x = 450.35$. □

For discrete random variables there will either be no solution or multiple solutions. If there is no solution, select the smallest value of x for which $F_X(x) > u$. If there are multiple solutions, again select the smallest value of x for which $F_X(x) > u$. This last statement may be surprising because you are asked to select none of the solutions. The following example may clarify this.

Example H.2 *Simulate two observations from a discrete random variable that has probabilities 0.1, 0.3, 0.4, and 0.2 for the values 0, 1, 2, and 3. The uniform random numbers are 0.37 and 0.80.*

The distribution function is a step function which is 0 up to (but not including) 0, 0.1 from 0 to 1, 0.4 from 1 to 2, 0.8 from 2 to 3, and 1.0 from 3 on. It does not take on the value 0.37. The next largest value is 0.4 and the smallest x that has that value is $x = 1$. The value 0.80 is taken on from 2 to 3, but not at 3. The next largest value is 1.0, which is first achieved at 3 and so $x = 3$ is the simulated value.

Extending this analysis reveals that uniform numbers from 0.00 to 0.09 will simulate a 0, those from 0.10 to 0.39 will simulate a 1, those from 0.40 to 0.79 will simulate a 2, and those from 0.80 to 0.99 will simulate a 3. The reason for the step up at exact values (such as 0.80) is that most uniform random number generators will have 0 as a possible simulated outcome, but 1 will not be possible. □

H.3 MULTIVARIATE NORMAL

While univariate normal variables can be simulated by the inversion method, there are better approaches. One popular method is called the polar method. Begin with uniform (0,1) random numbers U_1 and U_2. Next compute $V_1 = 2U_1 - 1$ and $V_2 = 2U_2 - 1$. Then compute $S = V_1^2 + V_2^2$. If $S \geq 1$, then discard the two numbers and start again. If $S < 1$, complete the simulation by computing

$$Z_1 = V_1\sqrt{\frac{-2\log S}{S}} \quad \text{and} \quad Z_2 = V_2\sqrt{\frac{-2\log S}{S}}.$$

Then Z_1 and Z_2 will be simulated independent standard normal random variables.

For an arbitrary univariate normal variable with mean μ and standard deviation σ, all that need be done is to multiply Z by σ and then add μ.

A multivariate normal random variable of dimension n is completely specified by the means of the n marginal variables $\boldsymbol{\mu} = (\mu_1, \ldots, \mu_n)'$, the variances of the n marginal variables σ_{ii}, $i = 1, \ldots, n$, and the covariances of the various pairs of variables σ_{ij}, $i \neq j$. The collection of variances and covariances can be displayed in the matrix $\Sigma = (\sigma_{ij})$. The first step in the simulation process is to factor the covariance matrix as $\Sigma = CC'$. While any factorization is acceptable, the Choleski factorization is the most efficient (and must be available because Σ is positive definite). This factorization requires that C be lower triangular with positive elements on the diagonal. Factorizations are directly provided by many computer programs, and the algorithms appear in all numerical analysis texts (e.g., Burden and Faires [21]).

The next step is to simulate n independent standard normal variables. Let $\mathbf{z}$ be the vector of simulated numbers. The final step is to compute the multivariate normal value from $\mathbf{x} = C\mathbf{z} + \boldsymbol{\mu}$.

Example H.3 *Simulate one observation from the bivariate normal distribution with means 2 and 5, standard deviations 3 and 6, and correlation 0.6. Use the uniform random numbers 0.3 and 0.4.*

The polar method produces

$$V_1 = 2(0.3) - 1 = -0.4, \quad V_2 = 2(0.4) - 1 = -0.2,$$

$$
\begin{aligned}
S &= (-0.4)^2 + (-0.2)^2 = 0.2, \\
Z_1 &= -0.4\sqrt{\frac{2\log 0.2}{0.2}} = -1.6047, \quad Z_2 = -0.2\sqrt{\frac{2\log 0.2}{0.2}} = -0.80236.
\end{aligned}
$$

The variances are 9 and 36 while the covariance is $0.6(3)(6) = 10.8$. The covariance matrix factors as

$$
\begin{bmatrix} 9 & 10.8 \\ 10.8 & 36 \end{bmatrix} = \begin{bmatrix} 3 & 0 \\ 3.6 & 4.8 \end{bmatrix} \begin{bmatrix} 3 & 3.6 \\ 0 & 4.8 \end{bmatrix}
$$

and therefore the simulated value is

$$
\begin{bmatrix} 3 & 0 \\ 3.6 & 4.8 \end{bmatrix} \begin{bmatrix} -1.6047 \\ -0.80236 \end{bmatrix} + \begin{bmatrix} 2 \\ 5 \end{bmatrix} = \begin{bmatrix} -2.8141 \\ -4.628248 \end{bmatrix}.
$$

□

H.4 POISSON

There is a way to simulate a Poisson random variable without computing the Poisson probabilities. Begin by setting $a = e^{-\lambda}$. Obtain a sequence of uniform $(0, 1)$ random numbers, $U_1, U_2, \ldots$. Then let $T_k = U_1 U_2 \cdots U_k$ for $k = 1, 2, \ldots$. Let n be the smallest value such that $T_n > a$. Then the simulated random value is $N = n - 1$.

Example H.4 *Simulate two observations from the Poisson distribution with mean 4. Use the uniform random numbers 0.285, 0.249, 0.336, 0.504, 0.880, 0.783, 0.672, 0.669, 0.459, 0.063, and 0.664.*

We have $a = e^{-4} = 0.0183$. For the first four uniform random numbers, $T_1 = 0.285$, $T_2 = 0.0710$, $T_3 = 0.0238$, and $T_4 = 0.0120$. Therefore, the simulated value is 3. For the second simulated Poisson value, $T_1 = 0.880$, $T_2 = 0.6890$, $T_3 = 0.4630$, $T_4 = 0.3098$, $T_5 = 0.1422$, and $T_6 = 0.0090$. Therefore, the simulated value is 5. □

Appendix I: Answers to Selected Exercises

I.1 CHAPTER 2

2.1 (a) $\hat{\mu} = 204{,}900$. $\hat{\mu}'_2 = 1.4134 \times 10^{11}$. $\hat{\sigma} = 325{,}807$. $\hat{\mu}'_3 = 1.70087 \times 10^{17}$, $\hat{\mu}_3 = 9.62339 \times 10^{16}$, $\hat{c} = 1.590078$, $\hat{\gamma}_1 = 2.78257$.
(b) $\hat{\pi}_{0.5} = 59{,}917$. $\hat{\pi}_{0.75} = 227{,}338$. $\hat{\pi}_{0.9} = 627{,}622$. $\hat{\pi}_{0.95} = 1{,}018{,}705$.
(c) The CI is (103,217,000, 513,586,000).
(d) $E_n(500{,}000) = 153{,}139$. $E_n^{(2)}(500{,}000) = 53{,}732{,}687{,}032$.
2.4 The median is 1.4142. The mode is 1.
2.6 The probability is approximately $1 - \Phi(1.25) = 0.106$.
2.8 38.6.
2.10 (a) 13.75.
(b) The ogive connects the points $(0.5, 0)$, $(2.5, 0.35)$, $(8.5, 0.65)$, $(15.5, 0.85)$,, and $(29.5, 1)$.
(c) The histogram has height $.35/2 = 0.175$ on the interval $(0.5, 2.5)$, height $0.3/6 = 0.05$ on the interval $(2.5, 8.5)$, height $0.2/7 = 0.028571$ on the interval $(8.5, 15.1)$, and height $0.15/14 = 0.010714$ on the interval $(15.5, 29.5)$.
2.12 (a) $(79, 121)$.
(b) Using (2.8), $(72, 128)$. Using Example 2.8, $(79, 137)$.
2.13 $\Pr(Z > 1.625) = 0.052$.

2.17 0.625.
2.19 $bias = -1/3$.
2.21 $\alpha = 0.6122$, $\beta = 0.5102$.
2.23 $\text{bias}(\hat{\theta}_1) = 0.2$, $Var(\hat{\theta}_1) = 0.16$, $\text{MSE}(\hat{\theta}_1) = 0.2$. $\text{bias}(\hat{\theta}_2) = -0.04$, $Var(\hat{\theta}_2) = 0.3184$, $\text{MSE}(\hat{\theta}_2) = 0.32$. The relative efficiency is 0.625.
2.25 $\hat{\theta} = n/\Sigma x_j^2$.
2.27 $\hat{\gamma} = 2$ and $\hat{\theta} = 200$.
2.29 (a) $\hat{p} = -n/\Sigma \log x_j$.
(b) $\hat{p} = \bar{x}/(1-\bar{x})$.
2.31 (a) $\hat{\mu} = 7.39817$, $\hat{\sigma} = 0.636761$. $\Pr(X > 4{,}500) = 1 - \Phi(1.5919) = 0.056$.
(b) $\hat{\mu} = 7.33429$, $\hat{\sigma} = 0.753263$. $\Pr(X > 4{,}500) = 1 - \Phi(1.4305) = 0.076$.
2.33 (a) $\hat{\beta} = 3.35112$.
(b) $\hat{\beta} = 3.20031$.
2.35 (a) $\hat{\alpha} = 4.141$.
(b) $\hat{\alpha} = 3.8629$.
2.37 $\hat{\theta} = 4.374$.
2.39 The minimum modified chi-square estimate is $\hat{\theta} = 89.976$ while the loglikelihood is maximized at $\hat{\theta} = 93.188$. The denominator requested is 0.02701.
2.41 (a) $Var(\hat{\theta}) \doteq \theta^2/n$.
(b) The CI is $\bar{x} \pm 1.96\bar{x}/\sqrt{n}$.
(c) The CI is $\bar{x}^2 \pm 1.96(2\bar{x}^2)/\sqrt{n}$.
2.43 (a) $\hat{\lambda} = (1/4)\log(4/3)$.
(b) $\widehat{Var(\hat{\lambda})} = 1/192$.
2.46 The CI is 0.076 ± 0.179.
2.48 The CI is 239.88 ± 133.50.
2.50 $Var(\hat{\theta}) \doteq 0.02701$, $\hat{\theta} = 93.188$. The CI is 0.00928 ± 0.00003.
2.53 (a) A percentage increase of 8.46%.
(b) With a deductible of 1,000 the percentage decrease is 2.87%.
2.55 Y is inverse Gaussian with parameters $c\mu$ and $c\theta$, it is a scale family. Because both μ and θ change there is no scale parameter.
2.58 $e_{\hat{X}}(1{,}000) = 4{,}500$, $e_{\hat{X}}(3{,}000) = 4{,}475$, $e_{\hat{X}}(5{,}000) = 4{,}507$, $e_{\hat{X}}(7{,}000) = 4{,}520$, $e_{\hat{X}}(9{,}000) = 4{,}500$. $e_{\hat{X}}(d)$ is essentially constant. The Pareto distribution has a straight line with positive slope, so the data are not consistent.
2.60 The percentage increase is 8%.
2.62 $e_{\hat{X}}(250) = 50$.
2.64 LER = 0.076.
2.66 From Exercise 2.38, $\hat{\alpha} = 2.848$. Also, $\theta = 100$.
(a) $E(X) = \alpha\theta/(\alpha - 1) = 154.11$.
(b) LER = 0.9025.
(c) $E(\text{payment/payment}) = 216.45$. From the data, there were no losses above 500, so the empirical average payment/loss is zero and the average payment/payment is undefined.
(d) $E(Y) = 161.82$.

(e) LER $= 0.8933$.
2.68 (a) The increased limits factor is 2,829/1,982 $= 1.427$.
(b) The increased limits factor is 3,459/2,570 $= 1.346$.
(c) The increased limits factor is 2,877/2,007 $= 1.433$.
(d) For model (b) versus model (c) the test statistic is 363.16 There are 4 degrees of freedom and the null hypothesis of model (b) is rejected. For model (a) versus model (c) the test statistic is 63.72 with 4 degrees of freedom and the null hypothesis of model (c) is rejected. Conclusion: choose the most complicated model where each group has its own model.
2.70 $F_Y(y) = 1 - \left(\frac{\theta}{\theta + y}\right)^\alpha$ a Pareto distribution. $f_Y(y) = dF_Y(y)/dy = \frac{\alpha\theta^\alpha}{(\theta + y)^{\alpha+1}}$.
2.72 $F_Y(y) = \frac{(y\theta)^\gamma}{1 + (y\theta)^\gamma}$, loglogistic with γ unchanged and $\theta = 1/\theta$.
2.74 $F_Y(y) = \frac{1}{2}e^{\ln y/\theta}$, $0 < y \le 1$, and $F_Y(y) = 1 - \frac{1}{2}e^{-\ln y/\theta}$, $y \ge 1$.
2.76 The limiting distribution is then transformed gamma with $\tau = 1$, which is a gamma distribution. The gamma parameters are $\alpha = \tau$ and $\theta = \xi$.
2.79 For a single exponential distribution, $\hat{\theta} = 32{,}874.8$ and NLL $= 548.722$. For the mixture, mles are $\hat{\theta}_1 = 78{,}615.8$, $\hat{\theta}_2 = 9{,}209.40$, and $\hat{p} = 0.35651$ with NLL $= 500.537$. The likelihood ratio test statistic is 96.37. With two degrees of freedom, the p-value is very small. The mixture distribution is preferred.
2.81 $\Pr(Y > 2.2) = 0.125$.
2.84 The limiting distribution is gamma with $\alpha = 3$ and $\theta = 50$.
2.92 (a) The expected outcome is $\frac{200\pi(100) + 50\pi(50)}{\pi(100) + \pi(50)}$.
(b) If switch, the expected outcome is 100.
(c) Switch if $x < 2\log 2$.
2.94 $\hat{\alpha}_{\text{Bayes}} = (10 + \gamma)(\theta^{-1} - 10\log 100 + \Sigma \log x_j)^{-1}$. $\hat{\alpha}_{\text{mle}} = 10(\Sigma \log x_j - 10\log 100)^{-1}$. The two estimators are equal when $\gamma = 0$ and $\theta = \infty$. This corresponds to $\pi(\alpha) = \alpha^{-1}$, an improper prior.
2.96 (a) $\pi(\mu, \sigma|\mathbf{x}) \propto \sigma^{-n} \exp\left[-\Sigma \frac{1}{2}\left(\frac{\log x_j - \mu}{\sigma}\right)^2\right] \sigma^{-1}$
(b) $\hat{\mu} = \frac{1}{n}\Sigma \log x_j$, $\hat{\sigma} = \left[\frac{1}{n+1}\Sigma(\log x_j - \hat{\mu})^2\right]^{1/2}$.
(c) The 95% HPD interval is $\hat{\mu} \pm 1.96\hat{\sigma}/\sqrt{n}$.
2.98 For the lognormal model, 3,961. For the Pareto model it is 4,787.
2.100 The test statistic is 9.504. With two degrees of freedom, the critical value at $\alpha = 0.01$ is 9.210 and at $\alpha = 0.005$ it is 10.597. Accept at $\alpha = 0.005$.
2.101 The test statistic is 0.320. The critical value is $1.36/\sqrt{5} = 0.608$ and so the null hypothesis and therefore the proposed distribution are accepted.
2.103 The likelihood ratio test and the p-values favor the inverse Weibull distribution ($\hat{\tau} = 0.600639$ and $\hat{\theta} = 49.8671$) for the refinancing model and the inverse gamma distribution ($\hat{\alpha} = 0.0783721$ and $\hat{\theta} = 4.52473$) for the

original mortgages. If a common model is to be used, the total NLL favors the inverse Weibull by a small margin over the inverse gamma. Either model would be reasonable as a common model for the two populations.

2.105 The test statistic is 0.172 and even at $\alpha = 0.2$ where the critical value is $1.07/\sqrt{10} = 0.338$, the null hypothesis, and therefore the gamma distribution, is accepted.

2.107 The best one parameter model is the inverse exponential with NLL $= 659.76$ and a p-value of 0.3241. No higher parameter model can be justified.

2.109 For the 10,000 limit the transformed gamma is slightly better than the Weibull. For the 25,000 limit the Burr is better than the lognormal. When a common model is used, the Burr is significantly better than the lognormal (twice the difference is 12.36 and there are two degrees of freedom). The common (Burr) model adds only 0.77 to the NLL so it is reasonable to use a common model. The Burr is not a bad second choice to the transformed gamma for the 10,000 limit.

2.111 The SBC for the model with common parameters is
$54{,}260.96 + 2\log(26{,}655/2\pi) = 54{,}277.67$.

With a common τ it is
$54{,}079.38 + 6\log(26{,}655/2\pi) = 54{,}129.50$.

With distinct parameters it is
$54{,}047.52 + 10\log(26{,}655/2\pi) = 54{,}131.05$.

The smallest value is for the model with common τ but distinct θ.

2.113 The solution by percentage matching is $\hat{\tau} = 0.87308$ and $\hat{\theta} = 1{,}338.8$. The likelihood is maximized at $\hat{\tau} = 0.778187$ and $\hat{\theta} = 1{,}343.67$.

2.115 (a) $\Pr(\text{Loss} > 25) = 0.4724$.
(b) $\Pr(Y < 25) = 0.7135$.

2.117 $F(x) = 1 - e^{-\lambda x}$. $L(\lambda) = \dfrac{(1 - e^{-2{,}000\lambda})^2(e^{-2{,}000\lambda} - e^{-5{,}000\lambda})^4}{(1 - e^{-5{,}000\lambda})^6}$.

2.119 The inverse Gaussian is the best choice. The parameter estimates are $\hat{\mu} = 203{,}500$ and $\hat{\theta} = 38{,}060.3$.

2.121 The best model is the lognormal with $\hat{\mu} = 11.0447$ and $\hat{\sigma} = 1.65859$. For the truncated model in Exercise 2.117 the expected payment (in thousands) is 244,839. For the shifted model from this problem the expected payment per payment (in thousands) is 305,339.

2.123 $L = \alpha^8 (125)^{8\alpha} (\prod x_i)^{-\alpha-1} (125/200)^{3\alpha}$. The maximum occurs at $\hat{\alpha} = 2.9217$.

2.125 The best model is the transformed gamma with $\hat{\alpha} = 0.386573$, $\hat{\theta} = 90.0735$, and $\hat{\tau} = 2.83224$. The expected payment per payment is 51.395.

2.128 $f_K(x;\alpha) = \dfrac{1}{2}\displaystyle\sum_{j=1}^{n} \frac{\alpha[(\alpha-1)x_j]^\alpha}{[x + (\alpha-1)x_j]^{\alpha+1}}$.

2.130 Separate fitting of the marginal distributions produces scale parameter estimates of 6,365 and 216.2 for the loss and ALAE respectively. The combined NLL is 496.58. Fitting Frank's copula produces scale parameter estimates of 5,886 and 218.2 as well as $\hat{\alpha} = 0.3035$. The NLL is 495.81. With one degree of freedom, the reduction in the NLL of 0.77 is not significant. Assuming the inverse exponential distribution is correct, the association is not significant.

I.2 CHAPTER 3

3.1

$$P_N^{(k)}(z) = \sum_{j=0}^{\infty} p_j \frac{d^k}{dz^k} z^j = \sum_{j=k}^{\infty} p_j j(j-1)\cdots(j-k+1)z^{j-k},$$

$$P_N^{(k)}(0) = p_k k!.$$

3.3 (a) The mle is 0.1001.
(b) (0.0939, 0.1063).
(c) The test statistic is 0.0008. There is one degree of freedom, the 5% critical value is 3.84 and the null hypothesis (and therefore the Poisson model) is accepted.
3.5 (a) The sample means are: underinsured, 0.109; insured, 0.057.
(b) For underinsured the test statistic is 0.75. With one degree of freedom, the 5% critical value is 3.84 and the null hypothesis (and therefore the Poisson model) is accepted. For uninsured the last two groups must be combined. The test statistic is 0.11 and there are zero degrees of freedom, so the test cannot be completed.
(c) The Poisson parameter is 0.166.
3.7 (a) $\hat{\lambda} = 15.688$. The NLL is 3,578.58.
(b) The CI is 15.688 ± 0.359.
(c) The test statistic is 4,558.40. The Poisson model is rejected.
3.9 (a) $\hat{\beta} = 0.352072$, $\hat{r} = 0.47149$.
(b) The test statistic is 4.785. With one degree of freedom, the critical value is 3.84 and the negative binomial distribution is rejected.
(c) $\hat{r} = 0.656060$ and $\hat{\beta} = 0.253026$.
(d) The test statistic is 23.568 while the critical value is 3.84. Reject the Poisson in favor of the negative binomial.
(e) Yes, it could be a mixture of different independent Poissons.
(f) The information matrix is $\begin{bmatrix} 328.738 & 798.067 \\ 798.067 & 2{,}069.260 \end{bmatrix}$ and its inverse is $\begin{bmatrix} 0.047754 & -0.018418 \\ -0.018418 & 0.0075865 \end{bmatrix}$.
(g) For r: 0.65606 ± 0.42831. For β: 0.25303 ± 0.17072.
3.11 Because $r = 1$, $\hat{\beta} = \bar{X}$. $E(l'') = n/[\beta(1+\beta)]$. The reciprocal matches the true variance of the mle.
3.13 (a) geometric: $\hat{\beta} = 19.146$, negative binomial: $\hat{r} = 0.564181$, $\hat{\beta} = 37.9032$.
(b) For the geometric, the test statistic is 146.84, there are 7 df and the 5% critical value is 14.07, and so the model is rejected. For the negative binomial,

the test statistic is 30.16, there are 6 df and the 5% critical value is 12.59, and so that model is also rejected.
(c) For Poisson vs. geometric the geometric is preferred. For the negative binomial vs. geometric, the test statistic is 67.22, which is clearly significant. Choose the negative binomial.
(d) The CI is 19.145 ± 1.878.
(e) The expected benefit is $10\left[\beta - \dfrac{\beta^{21}}{(1+\beta)^{20}}\right]$. At $\beta = 19.145$ it is 122.30. The CI is 122.30 ± 5.26.
3.15 (a) $\hat{q} = 0.025025$.
(b) There is one degree of freedom. The test statistic is 3.699. At 5% the critical value is 3.841 and the binomial model is accepted.
(c) From part (d): (0.023494, 0.026556). From part (e): (0.023540, 0.026601).
(d) The *mle* is at $m = \infty$ because the sample variance exceeds the sample mean.
(e) The Poisson is to be preferred.
(f) Because the binomial does not beat the Poisson, use the Poisson as indicated in the solution to Exercise 3.12.
3.17 For Exercise 3.15 the values at $k = 1, 2, 3$ are 0.1000, 0.0994, and 0.1333, which are nearly constant. The Poisson distribution is recommended. For Exercise 3.16 the values at $k = 1, 2, 3, 4$ are 0.1405, 0.2149, 0.6923, and 1.3333, which is increasing. The geometric/negative binomial is recommended.
3.19 (a) For $k = 1, 2, 3, 4, 5, 6, 7$ the values are 0.2760, 0.2315, 0.2432, 0.2891, 0.4394, 0.2828, and 0.4268, which are nearly constant. The Poisson model may work well.
(b) The values appear below. Because the sample variance exceeds the sample mean, there is no mle for the binomial distribution.

Model	Parameters	NLL	Chi-square	df
Poisson	$\hat{\lambda} = 1.74128$	2,532.86	1,080.80	5
Geometric	$\hat{\beta} = 1.74128$	2,217.71	170.72	7
Negative binomial	$\hat{r} = .867043,\ \hat{\beta} = 2.00830$	2,216.07	165.57	6

(c) The geometric is better than the Poisson by both likelihood and chi-square measures. The negative binomial distribution is not an improvement over the geometric because the NLL decreases by only 1.64. When doubled, 3.28 does not exceed the critical value of 3.841. The best choice is geometric, but it does not pass the goodness-of-fit test.
3.23 The pgf goes to $1 - (1-z)^{-r}$ as $\beta \to \infty$. The derivative with respect to z is $-r(1-z)^{-r-1}$. The expected value is this derivative evaluated at $z = 1$, which is infinite due to the negative exponent.
3.25 The covariance matrix is diagonal. The diagonal elements are $Var(\hat{q}_0) = 2.51884 \times 10^{-7}$ and $Var(\hat{\beta}) = 9.36090 \times 10^{-6}$.

For the goodness-of-fit test, the test statistic is 38.562. There are 3 degrees of freedom, the critical value at 5% significance is 7.815, and so the zero-modified logarithmic distribution is rejected.

3.26 For Exercise 3.3, the Poisson is the clear choice.

For Exercise 3.6 an argument could be made for the ZM negative binomial, but the simpler geometric still looks to be a good choice.

For Exercise 3.7, the best distribution is the ZM geometric which passes the goodness-of-fit test.

For Exercise 3.19, the ZM negative binomial is the best choice, but it does not look very promising.

3.28 (a) $p_k/p_{k-1} = \left(\frac{k-1}{k}\right)^{\rho+1} \neq a + b/k$ for any choices of a, b, and ρ and for all k.

(b) The mle is $\hat{\rho} = 3.0416$.

(c) The test statistic is 785.18 and with 3 degrees of freedom the model is clearly not acceptable.

3.32 The covariance matrix is obtained by inverting the information matrix:

$$\begin{bmatrix} 13{,}820.89 & 104{,}633.0 \\ 104{,}633.0 & 821{,}647.1 \end{bmatrix}^{-1} = \begin{bmatrix} 0.0020147 & -0.00025656 \\ -0.00025656 & 0.000033889 \end{bmatrix}$$

3.33 For Exercise 3.3, the Poisson cannot be topped.

For Exercise 3.6, the geometric model, which easily passed the goodness–of-fit test, still looks good.

For Exercise 3.7, none of the models improved the loglikelihood over the ZM geometric.

For Exercise 3.19, none of the models have a superior loglikelihood versus the ZM negative binomial.

3.39 The coefficient discussed in the section is $\frac{(\mu_3 - 3\sigma^2 + 2\mu)\mu}{(\sigma^2 - \mu)^2}$. The five coefficient estimates are (a) -0.85689, (b) $-1{,}817.27$, (c) -5.47728, (d) 1.48726, (e) -0.42125. For all but data set (d) it appears that a compound Poisson model will not be appropriate. For data set (d) it appears that Polya–Aeppli model will do well.

(a) The simpler negative binomial is an excellent choice.

(b) The Poisson is clearly acceptable.

(c) The Poisson is acceptable and should be our choice.

(d) Use the negative binomial, which passes the goodness-of-fit test. The moment analysis supported the Polya–Aeppli, which was acceptable but not as good as the negative binomial.

(e) The best is Poisson-binomial with $m = 2$, though the simpler and more popular negative binomial is a close alternative.

3.43 Then N is mixed Poisson. The mixing distribution has pgf $P(z) = \prod_{i=1}^{n} P_i(z)$.

3.46 EXCEL solver reports the following mles to four decimals: $\hat{p} = 0.9312$, $\hat{\lambda}_1 = 0.1064$, and $\hat{\lambda}_2 = 0.6560$. The negative loglikelihood is 10,221.9. Round-

ing these numbers to two decimals produces a negative loglikelihod of 10,223.3 while Tröbliger's solution is superior at 10,222.1. A better two decimal solution is (0.94,0.11,0.69) which gives 10,222.0. The negative binomial distribution was found to have a negative loglihood of 10,223.4. The extra parameter for the two-point mixture cannot be justified (using the likelihood ratio test).
3.48 (a) For frequency the probability of a claim is 0.03, and for severity the probability of a 10,000 claim is 1/3 and that of a 20,000 claim is 2/3.
(b) $\Pr(X = 0) = 1/3$ and $\Pr(X = 10{,}000) = 2/3$.
(c) For frequency, the probability of a claim is 0.02 and the severity distribution places probability one at 10,000.
3.50 The value of p_0^M will be negative and so there is no appropriate frequency distribution to describe effect of lowering the deductible.

I.3 CHAPTER 4

4.1 $(c) + (d) + (e)$ is:

For $X \leq 1{,}000$, $\quad 0 + 0 + X = X$,

For $1{,}000 < X \leq 63{,}500$, $\quad 0.8(X - 1{,}000) + 800 + 0.2X = X$,

For $63{,}500 < X \leq 126{,}000$, $\quad 0.8(X - 63{,}500) + 50{,}000 + 800 + 0.2X = X$,

For $X > 126{,}000$, $\quad 50{,}000 + 50{,}000 + X - 100{,}000 = X$.

4.3 The Poisson and all compound distributions with a Poisson primary distribution have a pgf of the form $P(z) = \exp\{\lambda[P_2(z) - 1]\} = [Q(z)]^\lambda$, where $Q(z) = \exp[P_2(z) - 1]$.

The negative binomial and geometric distributions and all compound distributions with a negative binomial or geometric primary distribution have $P(z) = \{1 - \beta[P_2(z) - 1]\}^{-r} = [Q(z)]^r$, where $Q(z) = \{1 - \beta[P_2(z) - 1]\}^{-1}$.

The same is true for the binomial distribution and binomial-X compound distributions with $\alpha = m$ and $Q(z) = 1 + q[P_2(z) - 1]$.

The zero-truncated and zero-modified distributions cannot be written in this form.
4.5 $E(S) = 5{,}600$, $Var(S) = 9{,}710{,}400$.
4.7 $Var(N) = 4.5$.
4.9 $p = 0.8$.
4.11 Let C be the number of cigarettes smoked. Then $E(C) = 33.6$ and $Var(C) = 370.88$.
4.13 The loading is 0.5885.
4.15 $F(80) = 0.75$.
4.17 $E(\text{benefits}) = 41{,}960$.
4.19 The difference is 7.44.
4.21 $E(Z) = 19.44$.

4.23 The total cost is 5,499.50.
4.26 $\lambda = 0.75568$.
4.28 Replace β by $\beta^* = \beta[1 - f_X(0)]$.
4.30 The answer is the sum of
$0.80R_{100}$ pays 80% of all in excess of 100,
$0.10R_{1100}$ pays an additional 10% in excess of 1,100, and
$0.10R_{2100}$ pays an additional 10% in excess of 2,100.
4.32 $E(S) = \frac{1}{2}(\log 4 - \log 2)$.
4.34 $E[(X - 100)_+] = 177$.
4.38 The expected compensation is 5,419,344.
4.40 $f_X(2) = \frac{2}{5}$.
4.44 The probability of claims being 800 or less is 0.2384.
4.46 $F_X^{*4}(6) = 0.94922$.
4.48 $\lambda = 0.8$.
4.50 In units of 20, $Var(S) = 82.5$. In payment units it is 33,000.
4.52 Using program CR, the result is 0.016.
4.53 Using program CR, the result is 0.055. Using a normal approximation with a mean of 250 and a variance 8,000 the probability is 0.047.
4.54 The answers are the same as for Exercises 4.52 and 4.53.
4.55 The answers are about the same as for Exercises 4.52 and 4.53.
4.57 (a) The recursive formula was used with a discretization interval of 25 and a mean-preserving discretization. The answers are 6,192.69, 4,632.13, and 12,800.04.
(b) The individual deductible of 100 requires a change in the frequency distribution. The probability of exceeding 100 under the gamma distribution is 0.7772974 and so the new Poisson parameter is 5 times this probability or 3.886487. The results are 5,773.24, 4,578.78, and 12,073.35.
(c) The frequency distribution is altered as in part (b). The results are 148.27, 909.44, and 0.
4.61 $B = 375.136$. $A = 378.515$ and $A/B = 1.009$.
4.63 $F_S^{*n}(5) = 0.78$.
4.67 $x = 10$.

I.4 CHAPTER 5

5.1 $f(x|y) = \dfrac{\binom{n_1}{x}\binom{n_2}{y-x}}{\binom{n_1+n_2}{y}}$. This is the hypergeometric distribution.

5.2(a) $f_X(0) = 0.3,\ f_X(1) = 0.4,\ f_X(2) = 0.3$.
$f_Y(0) = 0.25,\ f_Y(1) = 0.3,\ f_Y(2) = 0.45$.
(b) The following array presents the values for $x = 0, 1, 2$

$$f(x|Y=0) \quad = \quad 0.2/0.25 = 0.8,\ 0/0.25 = 0,\ 0.05/0.25 = 0.2$$

$$\begin{aligned} f(x|Y=1) &= 0/0.3 = 0,\ 0.15/0.3 = 0.5,\ 0.15/0.3 = 0.5 \\ f(x|Y=2) &= 0.1/0.45 = 0.22,\ 0.25/0.45 = 0.56,\ 0.1/0.45 = 0.22. \end{aligned}$$

(c)

$$\begin{aligned} E(X|Y=0) &= 0(0.8) + 1(0) + 2(0.2) = 0.4 \\ E(X|Y=1) &= 0(0) + 1(0.5) + 2(0.5) = 1.5 \\ E(X|Y=2) &= 0(0.22) + 1(0.56) + 2(0.22) = 1 \\ E(X^2|Y=0) &= 0(0.8) + 1(0) + 4(0.2) = 0.8 \\ E(X^2|Y=1) &= 0(0) + 1(0.5) + 4(0.5) = 2.5 \\ E(X^2|Y=2) &= 0(0.22) + 1(0.56) + 4(0.22) = 1.44 \\ Var(X|Y=0) &= 0.8 - 0.4^2 = 0.64 \\ Var(X|Y=1) &= 2.5 - 1.5^2 = 0.25 \\ Var(X|Y=2) &= 1.44 - 1^2 = 0.44. \end{aligned}$$

(d)

$$\begin{aligned} E(X) &= 0.4(0.25) + 1.5(0.3) + 1(0.45) = 1 \\ E[Var(X|Y)] &= 0.64(0.25) + 0.25(0.3) + 0.44(0.45) = 0.433 \\ Var[E(X|Y)] &= 0.16(0.25) + 2.25(0.3) + 1(0.45) - 1^2 = 0.165 \\ Var(X) &= 0.433 + 0.165 = 0.598. \end{aligned}$$

5.7 (b) The posterior distribution is

$$\pi(\theta|\mathbf{x}) = \frac{\Gamma(a+b+\Sigma K_j)}{\Gamma(a+\Sigma x_j)\Gamma(b+\Sigma K_j+\Sigma x_j)}\theta^{a+\Sigma x_j-1}(1-\theta)^{b+\Sigma K_j-\Sigma x_j-1}$$

with mean

$$E(\theta|\mathbf{x}) = \frac{a+\Sigma x_j}{a+b+\Sigma K_j}.$$

5.9 (b) $\pi(\theta|\mathbf{x})$ is beta with parameters

$$\begin{aligned} a^* &= \alpha + nr \quad \text{and} \quad b^* = b + \sum x_j \\ E(\theta|\mathbf{x}) &= \frac{a^*}{a^*+b^*} = \frac{a+nr}{a+nr+b+\sum x_j} \end{aligned}$$

5.11 (b) Because the mean of Θ_1 given Θ_2 and $\mathbf{x}$ does not depend on θ_2 it is also the mean of Θ_1 given just $\mathbf{x}$, which is μ_*. The mean of Θ_2 given $\mathbf{x}$ is the ratio of the parameters or

$$\frac{\alpha+n/2}{\beta+\frac{1}{2}\sum(x_j-\overline{x})^2+\dfrac{n(\bar{x}-\mu)^2}{2(1+n\sigma^2)}}.$$

5.13 $\theta = \log(1+\beta)$, $p(x) = \dfrac{\Gamma(\alpha+x)}{\Gamma(\alpha)x!}$, and $q(\theta) = \left(\dfrac{1}{1-e^{-\theta}}\right)^{\alpha}$.

5.21 $\Pr(D=2|2,3,4,1,4) = \frac{3}{4}$.

5.23 $\Pr(H = 1/4|d = 1) = \frac{2}{3}$.
5.25 $\pi(\theta|y = 1) = 4\theta^2 e^{-2\theta}$.
5.27 $\pi(\theta|r = 1) = 60\theta^2(1 - \theta)^3$.
5.29 $\pi(\lambda|y = 3) = 27\lambda^3 e^{-3\lambda/2}/32$.
5.31 $\pi(t|x = 5) = 108t^2 e^{-6t}$.
5.33 $\alpha = 3.7647$, $\beta = 26.5625$.
5.35 The standard for full credibility is

$$nr\beta = 1{,}082.41\left(1 + \beta + \frac{\sigma_Y^2}{\theta_Y^2}\right).$$

Partial credibility is obtained by taking the square root of ratio of the number of claims to the standard for full credibility.
5.37 $Z = \sqrt{1{,}082/1{,}443.21} = 0.86586$.
5.39 $Z = 0.55408$. The credibility estimate is 16,001,328.
5.41 The coefficient of variation is 1.2661.
5.43 $K = 0.056527$
5.45 $k = 0.075$ or 7.5%.
5.47 (a) $\pi(\theta_{ij}) = 1/6$ for die i and spinner j.
(b)(c)

i	1	1	1	2	2	2
j	1	2	3	1	2	3
$\mu(\theta_{ij})$	35/30	30/30	20/30	140/30	120/30	80/30
$v(\theta_{ij})$	6,725/900	5,400/900	2,600/900	12,200/900	10,800/900	5,600/900

(d) $\Pr(X_1 = 3) = 35/180$.
(e)

i	1	1	1	2	2	2
j	1	2	3	1	2	3
$\Pr(\Theta = \theta_{ij}\|X_1 = 3)$	1/35	2/35	4/35	4/35	8/35	16/35

(f) $\Pr(X_2 = 0|X_1 = 3) = 455/1{,}050$, $\Pr(X_2 = 3|X_1 = 3) = 357/1{,}050$, $\Pr(X_2 = 8|X_1 = 3) = 238/1{,}050$,
(g) $E(X_2|X_1 = 3) = \dfrac{2{,}975}{1{,}050}$
(h) $\Pr(X_2 = 0,\ X_1 = 3) = \dfrac{455}{5{,}400}\Pr(X_2 = 3|X_1 = 3) = \dfrac{357}{5,400}$. $\Pr(X_2 = 8|X_1 = 3) = \dfrac{238}{5,400}$.
(i) Divide answers to (h) by 35/180 to obtain the answers to (f).
(j) $E(X_2|X_1 = 3) = \dfrac{2,975}{1,050}$.

(k) $\mu = \frac{425}{180}$, $v = \frac{43{,}325}{5{,}400}$, $a = \frac{76{,}925}{32{,}400}$.
(l) $Z = 0.228349$, $P_c = 2.507$.
5.49 $Z = \frac{4}{4 + 36.083} = 0.10$, $P_c = 0.10(125) + 0.90(50) = 57.50$.
5.55 (a) The Bayesian premium is $E(X_{n+1}|\mathbf{x}) = \frac{mn}{mn+k}\bar{x} + \frac{k}{mn+k}\mu$.
(b) The Bühlmann premium must be the same because the Bayesian premium is linear in the observations.
(c) Replace θ with m and μ with $(2\theta)^{-1/2}$ to obtain

$$f(x) = \left(\frac{m}{2\pi x^3}\right)^{1/2} \exp\left[-m\theta x + m(2\theta)^{1/2} - \frac{m}{2x}\right].$$

Now let $p(m, x) = \left(\frac{m}{2\pi x^3}\right)^{1/2} \exp\left(\frac{m}{2x}\right)$ and $q(\theta) = \exp[-(2\theta)^{1/2}]$.
5.57 (c) This is a negative binomial distribution.
5.58 (a) 2.9.
(b) 3.10709.
5.60 $Z = 6/7$.
5.62 (a) 11/9..
(b) 2/3.
5.64 (a) 9/7.
(b) 9/7.
5.66 (a) $\Pr(N = 0) = 0.159046$.
(b) 13/7.
(c) 1.8505.
5.68 $E(X_2|X_1 = T) = 10.6$.
5.70 (a) 1.5.
(b) 1.5.
5.72 The Bayesian estimate is $\mu(A) = 0.1$.
5.74 $Z = 0.6$.
5.76 $Z = 0.37333$.
5.78 (a) 3,493.53.
(b) 3,592.34.
5.84 The three estimates are 734.67, 640.66, and 891.34.
5.86 496.23.
5.88 The premiums are 0.09770, 0.27996, 0.46222, 0.64448, and 0.82674 for 0, 1, 2, 3, and 4, claims respectively.
5.90 The equation to be solved is

$$r\mu = \sum_{i=1}^{r} \frac{t_i + 1}{\mu^{-1} + m_i}.$$

5.92 0.79192.

I.5 CHAPTER 6

6.1 (a) $\tilde{\psi}(2,3) = 0.425$.
(b) $\tilde{\psi}(2,3) = 0.4334222 > 0.425$.

6.5 $\kappa = \dfrac{1}{6\beta}$.

6.7 $\kappa = 1$.

6.17 $\phi(u) = 1 - \dfrac{16}{27}e^{-u} + \dfrac{1}{27}e^{-4u}$.

6.19 $\phi(u) = 1 - \dfrac{9}{16}e^{-u} - \dfrac{1}{16}^{-3u}$.

6.21 (c) $m(u) = \dfrac{70}{129} + \dfrac{A_1}{r_1}e^{r_1 u} + \dfrac{A_2}{r_2}e^{r_2 u}$ where $r_1 = \dfrac{-53 - 2\sqrt{154}}{17}$ and $r_2 = \dfrac{-53 + 2\sqrt{154}}{17}$.

6.22 (a) $\kappa = 1$ (in which case $C = 1/4$), and $\dfrac{1}{1+\theta} = \psi(0) = \dfrac{1}{3}$ and $\theta = 2$.
(b) $E\left(e^{\kappa X}\right) = \dfrac{17}{11}$.

6.24 (a) $\kappa = 0.000425$, $\psi(u) \sim 0.630e^{-0.000425u}$, $u \to \infty$.
(b) $\kappa = 0.0110923$, $\psi(u) \sim 0.7535e^{-0.0110923u}$, $u \to \infty$.
(c) $\kappa = 0.0351$, $\psi(u) \sim 0.667e^{-0.0351u}$, $u \to \infty$.

6.27 $\psi(1{,}000) = 0.36626$ (exact). With a span of 50, $\psi(1{,}000) = 0.362858$.

6.28 $\psi(200) = 0.053812$ (exact). With a span of 25, $\psi(200) = 0.0541275$.

References

1. Abramowitz, M. and Stegun, I. (1964), *Handbook of Mathematical Functions with Formulas, Graphs, and Mathematical Tables*, New York: Wiley.

2. Accomando, F. and Weissner, E. (1988), "Report Lag Distributions: Estimation and Application to IBNR Counts," *Transcripts of the 1988 Casualty Loss Reserve Seminar*, Arlington, VA: Casualty Actuarial Society, 1038–1133.

3. *American Heritage Electronic Dictionary* (1992), Houghton-Mifflin (also, 1993, Wordstar International).

4. Arnold, B. (1983), *Pareto Distributions (Statistical Distributions in Scientific Work, Vol. 5)*, Fairland, MD: International Co-operative Publishing House.

5. Bailey, A. (1942), "Sampling Theory in Casualty Insurance, Parts I and II," *Proceedings of the Casualty Actuarial Society*, **XXIX**, 50–95.

6. Bailey, A. (1943), "Sampling Theory in Casualty Insurance, Parts III through VII," *Proceedings of the Casualty Actuarial Society*, **XXX**, 31–65.

7. Bailey, A. (1950), "Credibility procedures," *Proceedings of the Casualty Actuarial Society*, **XXXVII**, 7–23 and 94–115.

8. Bailey, W. (1992), "A Method for Determining Confidence Intervals for Trend," *Transactions of the Society of Actuaries*, **XLIV**, 11–54.

9. Baker, C. (1977), *The Numerical Treatment of Integral Equations*, Oxford: Clarendon Press.

10. Beard, R., Pentikainen, T., and Pesonen, E. (1984), *Risk Theory 3rd ed.*, London: Chapman and Hall.

11. Beirlant, J., Teugels, J., and Vynckier, P. (1996), *Practical Analysis of Extreme Values*, Leuven, Belgium: Leuven University Press.

12. Berger, J. (1985), *Bayesian Inference in Statistical Analysis, 2nd ed.*, New York: Springer-Verlag.

13. Bertram, J. (1981), "Numerische Berechnumg von Gesamtschadenverteilungen," *Blätter der deutschen Gesellschaft Versicherungsmathematik*, Band **15.2**, 175–194.

14. Bevan, J. (1963), "Comprehensive Medical Insurance—Statistical Analysis for Ratemaking," *Proceedings of the Casualty Actuarial Society*, **L**, 111–128.

15. Bondy, M. (1958), "Auto B.I. Liability Rates—Use of 10/20 Experience in the Establishment of Territorial Relativities," *Proceedings of the Casualty Actuarial Society*, **XLV**, 1–8.

16. Bowers, N., Gerber, H., Hickman, J., Jones, D., and Nesbitt, C. (1986), *Actuarial Mathematics*, Schaumburg, IL: Society of Actuaries.

17. Brockett, P. (1991), "Information Theoretic Approach to Actuarial Science: A Unification and Extension of Relevant Theory and Applications," with discussion, *Transactions of the Society of Actuaries*, **XLIII**, 73–135.

18. Bühlmann, H. (1967), "Experience rating and credibility," *ASTIN Bulletin*, **4**, 199–207.

19. Bühlmann, H. (1970), *Mathematical Methods in Risk Theory*, New York: Springer-Verlag.

20. Bühlmann, H. and Straub, E. (1970), "Glaubwürdigkeit für Schadensätze (credibility for loss ratios)," *Mitteilungen der Vereinigung Schweizerischer Versicherungs-Mathematiker*, **70**, 111–133.

21. Burden, R. and Faires, J. (1993), *Numerical Analysis, 5th ed.*, Boston: PWS-Kent.

22. Cahill, J. (1940), "Excess Coverage (per Accident Basis) for Self-Insurers: Workmen's Compensation—New York," *Proceedings of the Casualty Actuarial Society*, **XXVII**, 77–111.

23. Carlin, B. and Klugman, S. (1993) "Hierarchical Bayesian Whitaker Graduation," *Scandinavian Actuarial Journal*, 183–196.

24. Carriere, J. (1993), "Nonparametric Estimators of a Distribution Function Based on Mixtures of Gamma Distributions," *Actuarial Research Clearing House*, **1993.3**, 1–11.

25. Crane, H. (1935), "Commerical Accident and Health Insurance from the Standpoint of the Reinsurance Company," *Proceedings of the Casualty Actuarial Society*, **XXI**, 303–312.

26. Casualty Actuarial Society (1990), *Foundations of Casualty Actuarial Science*, Arlington, VA: Casualty Actuarial Society.

27. Daykin, C., Pentaikainen, T., and Pesonen, M. (1994), *Practical Risk Theory for Actuaries*, London: Chapman and Hall.

28. DePril, N. (1986), "On the Exact Computation of the Aggregate Claims Distribution in the Individual Life Model," *ASTIN Bulletin*, **16**, 109–112.

29. DePril, N. (1988), "Improved Approximations for the Aggregate Claims Distribution of a Life Insurance Portfolio," *Scandinavian Actuarial Journal*, 61–68.

30. DePril, N. (1989), "The Aggregate Claim Distribution in the Individual Model with Arbitrary Positive Claims," *ASTIN Bulletin*, **19**, 9–24.

31. Douglas, J. (1980), *Analysis with Standard Contagious Distributions*, Fairland, Maryland: International Co-operative Publishing House.

32. Downey, E. and G. Kelly (1918), "The Revision of Pennsylvania Compensation Insurance Rates, 1918, *Proceedings of the Casualty Actuarial Society*, **V**, 243–278.

33. Dropkin, L. (1959), "Some Considerations on Automobile Rating Systems Utilizing Individual Driving Records," *Proceedings of the Casualty Actuarial Society*, **XLVI**, 165–176.

34. Dropkin, L. (1964), "Size of Loss Distributions in Workmen's Compensation Insurance," *Proceedings of the Casualty Actuarial Society*, **LI**, 198–259.

35. Efron, B. (1986), "Why Isn't Everyone a Bayesian?" *The American Statistician*, **40**, 1–11 (including comments and reply).

36. Efron, B. and Tibshirani, R. (1993), *An Introduction to the Bootstrap*, New York: Chapman and Hall.

37. Ericson, W. (1969), "A Note on the Posterior Mean of a Population Mean," *Journal of the Royal Statistical Society, Series B*, **31**, 332–334.

38. Feller, W. (1950), *An Introduction to Probability Theory and Its Applications, Vol. 1, 3rd ed.*, New York: Wiley.

39. Feller, W. (1971), *An Introduction to Probability Theory and Its Applications, Vol. 2, 2nd ed.*, New York: Wiley.

40. Finger, R. (1976), "Estimating Pure Premiums by Layer—An Approach," *Proceedings of the Casualty Actuarial Society*, **LXIII**, 34–52.

41. Fisher, A. (1915), "Note on the Application of Recent Mathematical-Statistical Methods to Coal Mine Accidents, with Special Reference to Catastrophes in Coal Mines in the United States, *Proceedings of the Casualty Actuarial Society*, **II**, 70–78.

42. Fishman, G. and Moore, L. (1984), "An Exhaustive Analysis of Multiplicative Congruential Generators with Modulus $2^{31} - 1$," *SIAM Journal Sci. Stat. Comput.*, **7**, 24–45.

43. Fisz, M. (1963), *Probability Theory and Mathematical Statistics*, New York: Wiley.

44. Frees, E., Carriere, J., and Valdez, E. (1995), "Annuity Valuation with Dependent Mortality," working paper.

45. Gavin, J., Haberman, S., and Verrall, R. (1993), "Moving Weighted Average Graduation Using Kernel Methods," *Insurance: Mathematics and Economics*, **12**, 113–126.

46. Gavin, J., Haberman, S., and Verrall, R. (1994), "On the Choice of Bandwidth for Kernel Graduation," *Journal of the Institute of Actuaries*, **121**, 119–134.

47. Genest, C. (1987), "Frank's Family of Bivariate Distributions," *Biometrika*, **74**, 549–555.

48. Genest, C., and McKay, J. (1986), "The Joy of Copulas: Bivariate Distributions with Uniform Marginals," *The American Statistician*, **40**, 280–283.

49. Gerber, H. (1982), "On the Numerical Evaluation of the Distribution of Aggregate Claims and its Stop-Loss Premiums," *Insurance: Mathematics and Economics*, **1**, 13–18.

50. Gerber, H. and D. Jones (1976), "Some Practical Considerations in Connection with the Calculation of Stop-Loss Premiums," *Transactions of the Society of Actuaries*, **XXVIII**, 215–231.

51. Gillam, W. (1992), "Parametrizing the Workers Compensation Experience Rating Plan," *Proceedings of the Casualty Actuarial Society*, **LXXIX**, 21–56.

52. Goovaerts, M.J., and Hoogstad, W.J. (1987), *Credibility Theory, Surveys of Actuarial Studies No. 4*, Rotterdam: Nationale-Nederlanden.

53. Hachemeister, C.A. (1975), "Credibility for Regression Models With Application to Trend," in *Credibility: Theory and Applications* (P. Kahn, ed.), New York: Academic Press, 129–163.

54. Harwayne, F. (1958), "Estimating Ultimate Incurred Losses in Auto Liability Insurance," *Proceedings of the Casualty Actuarial Society*, **XLV**, 63–87.

55. Harwayne, F. (1959), "Merit Rating in Private Passenger Automobile Liability Insurance and the California Driver Record Study," *Proceedings of the Casualty Actuarial Society*, **XLVI**, 189–195.

56. Hayne, R. (1989), "Application of Collective Risk Theory to Estimate Variability in Loss Reserves," *Proceedings of the Casualty Actuarial Society*, **LXXVI**, 77–109.

57. Hayne, R. (1994), "Extended Service Contracts," *Proceedings of the Casualty Actuarial Society*, **LXXXI**, 243–302.

58. Heckman, P. and G. Meyers (1983), "The Calculation of Aggregate Loss Distributions from Claim Severity and Claim Count Distributions," *Proceedings of the Casualty Actuarial Society*, **LXX**, 22–61.

59. Herzog, T.N. (1994), *Introduction to Credibility Theory*, Winsted, CT: ACTEX.

60. Herzog, T. and Laverty, J. (1995), "Experience of Refinanced FHA Section 203(b) Single Family Mortages," *Actuarial Research Clearing House*, **1995.1**, 97–129.

61. Hewitt, C., Jr. (1966), "Distribution of Workmens' Compensation Plans by Annual Premium Size," *Proceedings of the Casualty Actuarial Society*, **LIII**, 106–121.

62. Hewitt, C., Jr. (1967), "Loss Ratio Distributions—A Model," *Proceedings of the Casualty Actuarial Society*, **LIV**, 70–88.

63. Hewitt, C., Jr. (1979), "Methods for Fitting Distributions to Insurance Loss Data," *Proceedings of the Casualty Actuarial Society*, **LXVI**, 139–160.

64. Hipp, G. (1938), "Special Funds Under the New York Workmen's Compensation Law," *Proceedings of the Casualty Actuarial Society*, **XXIV**, 247–275.

65. Hogg, R. and Craig, A. (1978), *Introduction to Mathematical Statistics, 4th ed.*, New York: Macmillan.

66. Hogg, R. and Klugman, S. (1984), *Loss Distributions*, New York: Wiley.

67. Holgate, P. (1970), "The Modality of Some Compound Poisson Distributions," *Biometrika*, **57**, 666–667.

68. Hossack, I., Pollard, J. and Zehnwirth, B. (1983), *Introductory Statistics with Applications in General Insurance*, Cambridge: Cambridge University Press.

69. Hutchinson, T. and Lai, C. (1990), *Continuous Bivariate Distributions, Emphasizing Applications*, Adelaide: Rumsby.

70. Hyndman, R. and Fan, Y. (1996), "Sample Quantiles in Statistical Packages," *The American Statistician*, **50**, 361–365.

71. Jewell, W. (1974), "Credibility is Exact Bayesian for Exponential Families," *ASTIN Bulletin*, **8**, 77–90.

72. Johnson, N. and Kotz, A. (1972), *Distributions in Statistics, Vol. 4*, New York: Wiley.

73. Johnson, N., Kotz, A., and Kemp, A. (1992), *Univariate Discrete Distributions*, New York: Wiley.

74. Karlin, S. and Taylor, H. (1975), *A First Course in Stochastic Processes, 2nd ed.*, New York: Academic Press.

75. Klugman, S. (1987), "Credibility for Classification Ratemaking Via the Hierarchical Linear Model," *Proceedings of the Casualty Actuarial Society*, **LXXIV**, 272–321.

76. Klugman, S. (1992), *Bayesian Statistics in Actuarial Science with Emphasis on Credibility*, Boston: Kluwer.

77. Klugman, S. and Parsa, A. (1993), "Minimum Distance Estimation of Loss Distributions," *Proceedings of the Casualty Actuarial Society*, **LXXX**, 250–270.

78. Kormes, M. (1961), "Patterns of Serious Illness," *Proceedings of the Casualty Actuarial Society*, **XLVIII**, 121–130.

79. Kornya, P. (1983), "Distribution of Aggregate Claims in the Individual Risk Model," *Transactions of the Society of Actuaries*, **XXXV**, 837–858.

80. Lemaire, J. (1995), *Automobile Insurance: Actuarial Models, 2nd ed.*, Boston: Kluwer.

81. Lindley, D. (1987), "The Probability Approach to the Treatment of Uncertainty in Artificial Intelligence and Expert Systems," *Statistical Science*, **2**, 17–24 (also related articles in that issue).

82. London, D. (1988), *Survival Models and Their Estimation*, Winsted, CT: ACTEX Publications.

83. Longley-Cook, L. (1952), "A Statistical Study of Large Fire Losses with Applications to a Problem in Catastrophe Insurance," *Proceedings of the Casualty Actuarial Society*, **XXXIX**, 77–83.

84. Longley-Cook, L. (1958), "The Employment of Property and Casualty Actuaries," *Proceedings of the Casualty Actuarial Society*, **XLV**, 9–10.

85. Longley-Cook, L.H. (1962), "An Introduction to Credibility Theory," *Proceeding of the Casualty Actuarial Society*, **XLIX**, 194–221.

86. Luong, A. and Doray, L. (1996), "Goodness of Fit Test Statistics for the Zeta Family," *Insurance: Mathematics and Economics*, **10**, 45–53.

87. Mardia, K. (1970), *Families of Bivariate Distributions*, London: Griffin.

88. Meyers, G. (1984), "Empirical Bayesian Credibility for Workers' Compensation Classification Ratemaking," *Proceedings of the Casualty Actuarial Society*, **LXXI**, 96–121.

89. Meyers, G. (1994), "Quantifying the Uncertainty in Claim Severity Estimates for an Excess Layer When Using the Single Parameter Pareto," *Proceedings of the Casualty Actuarial Society*, **LXXXI**, 91–122 (including discussion).

90. Miccolis, R. (1977), "On the Theory of Increased Limits and Excess of Loss Pricing," *Proceedings of the Casualty Actuarial Society*, **LXIV**, 27–59.

91. Moore, D. (1978), "Chi-square Tests" in Hogg, R., ed., *Studies in Statistics, Vol. 19*, Washington, DC: Mathematical Association of America, 453–463.

92. Moore, D. (1986), "Tests of Chi-Squared Type," in D'Agostino, R. and Stephens, M., eds., *Goodness-of-Fit Techniques*, New York: Marcel Dekker, 63–95.

93. Mowbray, A.H. (1914), "How Extensive a Payroll Exposure is Necessary to Give a Dependable Pure Premium?" *Proceedings of the Casualty Actuarial Society* **I**, 24–30.

94. Mowbray, A. (1923), "Legal Limits of Weekly Compensation in their Bearing on Ratemaking for Workmen's Compensation Insurance," *Proceedings of the Casualty Actuarial Society*, **IX**, 208–241.

95. Nelder, J. and Mead, U. (1965), "A Simplex Method for Function Minimization," *The Computer Journal*.

96. Norberg, R. (1979), "The Credibility Approach to Experience Rating," *Scandinavian Actuarial Journal*, 181–221.

97. Outwater, O. (1920), "An American Accident Table," *Proceedings of the Casualty Actuarial Society*, **VII**, 57–77.

98. Patrik, G. (1980), "Estimating Casualty Insurance Loss Amount Distributions," *Proceedings of the Casualty Actuarial Society*, **LXVII**, 57–109.

99. Panjer, H. and Lutek, B. (1983), "Practical Aspects of Stop-Loss Calculations," *Insurance: Mathematics and Economics*, **2**, 159–177.

100. Panjer, H. and Wang, S. (1993), "On the Stability of Recursive Formulas," *ASTIN Bulletin*, **23**, 227–258.

101. Panjer, H. and Willmot, G. (1986), "Computational Aspects of Recursive Evaluation of Compound Distributions," *Insurance: Mathematics and Economics*, **5**, 113–116.

102. Panjer, H. and Willmot, G. (1992), *Insurance Risk Models*, Chicago: Society of Actuaries.

103. Press, W., Flannery, B., Teukolsky, S., and Vetterling, W. (1988), *Numerical Recipes in C*, Cambridge: Cambridge University Press.

104. Rao, C. (1965), *Linear Statistical Inference and Its Applications*, New York: Wiley.

105. Ripley, B. (1987), *Stochastic Simulation*, New York: Wiley.

106. Roberts, L. (1959), "Credibility of 10/20 Experience as Compared with 5/10 Experience," *Proceedings of the Casualty Actuarial Society*, **XLVI**, 235–250.

107. Robertson, J. (1992), "The Computation of Aggregate Loss Distributions," *Proceedings of the Casualty Actuarial Society*, **LXXIX**, 57–133.

108. Robertson, J. (1994), "It's a Puzzlement," *The Actuarial Review*, **21**, No. 3, 20.

109. Rohatgi, V. (1976), *An Introduction to Probability Theory and Mathematical Statistics*, New York: Wiley.

110. Self, S. and Liang, K. (1987), "Asymptotic Properties of Maximum Likelihood Estimators and Likelihood Ratio Tests Under Nonstandard Conditions," *Journal of the American Statistical Association*, **82**, 605–610.

111. Schwartz, G. (1978), "Estimating the Dimension of a Model," *Annals of Statistics*, **6**, 461–464.

112. Simon, L. (1961), "Fitting Negative Binomial Distributions by the Method of Maximum Likelihood," *Proceedings of the Casualty Actuarial Society*, **XLVIII**, 45–53.

113. Society of Actuaries Committee on Actuarial Principles (1992), "Principles of Actuarial Science," *Transactions of the Society of Actuaries*, **XLIV**, 565–628.

114. Society of Actuaries Committee on Actuarial Principles (1994), "Principles Regarding Provisions for Life Risks, Exposure Draft," Chicago: Society of Actuaries.

115. Stanard, J., and R. John (1990), "Evaluating the Effect of Reinsurance Contract Terms," *Proceedings of the Casualty Actuarial Society*, **LXXVII**, 1–41.

116. Stephens, M. (1986), "Tests Based on EDF Statistics," in *Goodness-of-Fit Techniques*, D'Agostino, R. and Stephens, M., eds., New York: Marcel Dekker, 97–193.

117. Sundt, B. (1986), *Special issue on credibility theory, Insurance: Abstracts and Reviews*, **2**.

118. Sundt, B. (1991), *An Introduction to Non-Life Insurance Mathematics, 2nd ed.*, Karlsruhe: Springer-Verlag.

119. Thyrion, P. (1961), "Contribution a l'Etude du Bonus pour non Sinstre en Assurance Automobile," *ASTIN Bulletin*, **1**, 142–162.

120. Tröbliger, A. (1961),"Mathematische Untersuchungen zur Beitragsruckgewahr in der Kraftfahrversicherung," *Blatter der Deutsche Gesellschaft fur Versicherungsmathematik*, **5**, 327–348.

121. Vaughan, E. (1992), *Fundamentals of Risk and Insurance, 6th ed.*, New York: Wiley.

122. Venter, G. (1983), "Transformed Beta and Gamma Distributions and Aggregate Losses," *Proceedings of the Casualty Actuarial Society*, **LXX**, 156–193.

123. Walters, F., Parker, L., Morgan, S. and Deming, S. (1991), *Sequential Simplex Optimization*, Boca Raton, FL: CRC Press.

124. Walters, M. (1974), "Homeowners Insurance Ratemaking," *Proceedings of the Casualty Actuarial Society*, **LXI**, 15–57.

125. Waters, H.R. (1993), *Credibility Theory*, Edinburgh: Department of Actuarial Mathematics & Statistics, Heriot-Watt University.

126. Weber, D. (1970), "A Stochastic Approach to Automobile Compensation," *Proceedings of the Casualty Actuarial Society*, **LVII**, 27–61.

127. Whitney, A.W. (1917), "The Theory of Experience Rating," *Proceedings of the Casualty Actuarial Society*, **IV**, 274–292.

128. Woll, R. (1979), "A Study of Risk Assessment," *Proceedings of the Casualty Actuarial Society*, **LXVI**, 84–138.

Index

WILEY SERIES IN PROBABILITY AND STATISTICS
ESTABLISHED BY WALTER A. SHEWHART AND SAMUEL S. WILKS

Probability and Statistics Section

*ANDERSON · The Statistical Analysis of Time Series
ARNOLD, BALAKRISHNAN, and NAGARAJA · A First Course in Order Statistics
BACCELLI, COHEN, OLSDER, and QUADRAT · Synchronization and Linearity: An Algebra for Discrete Event Systems
BASILEVSKY · Statistical Factor Analysis and Related Methods: Theory and Applications
BERNARDO and SMITH · Bayesian Statistical Concepts and Theory
BILLINGSLEY · Convergence of Probability Measures
BOROVKOV · Asymptotic Methods in Queuing Theory
BRANDT, FRANKEN, and LISEK · Stationary Stochastic Models
CAINES · Linear Stochastic Systems
CAIROLI and DALANG · Sequential Stochastic Optimization
CONSTANTINE · Combinatorial Theory and Statistical Design
COVER and THOMAS · Elements of Information Theory
CSÖRGŐ and HORVÁTH · Weighted Approximations in Probability Statistics
CSÖRGŐ and HORVÁTH · Limit Theorems in Change Point Analysis
DETTE and STUDDEN · The Theory of Canonical Moments with Applications in Statistics, Probability, and Analysis
*DOOB · Stochastic Processes
DRYDEN and MARDIA · Statistical Analysis of Shape
DUPUIS and ELLIS · A Weak Convergence Approach to the Theory of Large Deviations
ETHIER and KURTZ · Markov Processes: Characterization and Convergence
FELLER · An Introduction to Probability Theory and Its Applications, Volume 1, *Third Edition*, Revised; Volume II, *Second Edition*
FULLER · Introduction to Statistical Time Series, *Second Edition*
FULLER · Measurement Error Models
GELFAND and SMITH · Bayesian Computation
GHOSH, MUKHOPADHYAY, and SEN · Sequential Estimation
GIFI · Nonlinear Multivariate Analysis
GUTTORP · Statistical Inference for Branching Processes
HALL · Introduction to the Theory of Coverage Processes
HAMPEL · Robust Statistics: The Approach Based on Influence Functions
HANNAN and DEISTLER · The Statistical Theory of Linear Systems
HUBER · Robust Statistics
IMAN and CONOVER · A Modern Approach to Statistics
JUREK and MASON · Operator-Limit Distributions in Probability Theory
KASS and VOS · Geometrical Foundations of Asymptotic Inference
KAUFMAN and ROUSSEEUW · Finding Groups in Data: An Introduction to Cluster Analysis

*Now available in a lower priced paperback edition in the Wiley Classics Library.

Probability and Statistics (Continued)

KELLY · Probability, Statistics, and Optimization
LINDVALL · Lectures on the Coupling Method
McFADDEN · Management of Data in Clinical Trials
MANTON, WOODBURY, and TOLLEY · Statistical Applications Using Fuzzy Sets
MARDIA and JUPP · Statistics of Directional Data, *Second Edition*
MORGENTHALER and TUKEY · Configural Polysampling: A Route to Practical Robustness
MUIRHEAD · Aspects of Multivariate Statistical Theory
OLIVER and SMITH · Influence Diagrams, Belief Nets and Decision Analysis
*PARZEN · Modern Probability Theory and Its Applications
PRESS · Bayesian Statistics: Principles, Models, and Applications
PUKELSHEIM · Optimal Experimental Design
RAO · Asymptotic Theory of Statistical Inference
RAO · Linear Statistical Inference and Its Applications, *Second Edition*
*RAO and SHANBHAG · Choquet-Deny Type Functional Equations with Applications to Stochastic Models
ROBERTSON, WRIGHT, and DYKSTRA · Order Restricted Statistical Inference
ROGERS and WILLIAMS · Diffusions, Markov Processes, and Martingales, Volume I: Foundations, *Second Edition;* Volume II: Îto Calculus
RUBINSTEIN and SHAPIRO · Discrete Event Systems: Sensitivity Analysis and Stochastic Optimization by the Score Function Method
RUZSA and SZEKELY · Algebraic Probability Theory
SCHEFFE · The Analysis of Variance
SEBER · Linear Regression Analysis
SEBER · Multivariate Observations
SEBER and WILD · Nonlinear Regression
SERFLING · Approximation Theorems of Mathematical Statistics
SHORACK and WELLNER · Empirical Processes with Applications to Statistics
SMALL and McLEISH · Hilbert Space Methods in Probability and Statistical Inference
STAPLETON · Linear Statistical Models
STAUDTE and SHEATHER · Robust Estimation and Testing
STOYANOV · Counterexamples in Probability
TANAKA · Time Series Analysis: Nonstationary and Noninvertible Distribution Theory
THOMPSON and SEBER · Adaptive Sampling
WELSH · Aspects of Statistical Inference
WHITTAKER · Graphical Models in Applied Multivariate Statistics
YANG · The Construction Theory of Denumerable Markov Processes

Applied Probability and Statistics Section

ABRAHAM and LEDOLTER · Statistical Methods for Forecasting
AGRESTI · Analysis of Ordinal Categorical Data
AGRESTI · Categorical Data Analysis
ANDERSON, AUQUIER, HAUCK, OAKES, VANDAELE, and WEISBERG · Statistical Methods for Comparative Studies
ARMITAGE and DAVID (editors) · Advances in Biometry
*ARTHANARI and DODGE · Mathematical Programming in Statistics
ASMUSSEN · Applied Probability and Queues
*BAILEY · The Elements of Stochastic Processes with Applications to the Natural Sciences
BARNETT and LEWIS · Outliers in Statistical Data, *Third Edition*

*Now available in a lower priced paperback edition in the Wiley Classics Library.

Applied Probability and Statistics (Continued)

BARTHOLOMEW, FORBES, and McLEAN · Statistical Techniques for Manpower Planning, *Second Edition*
BATES and WATTS · Nonlinear Regression Analysis and Its Applications
BECHHOFER, SANTNER, and GOLDSMAN · Design and Analysis of Experiments for Statistical Selection, Screening, and Multiple Comparisons
BELSLEY · Conditioning Diagnostics: Collinearity and Weak Data in Regression
BELSLEY, KUH, and WELSCH · Regression Diagnostics: Identifying Influential Data and Sources of Collinearity
BHAT · Elements of Applied Stochastic Processes, *Second Edition*
BHATTACHARYA and WAYMIRE · Stochastic Processes with Applications
BIRKES and DODGE · Alternative Methods of Regression
BLOOMFIELD · Fourier Analysis of Time Series: An Introduction
BOLLEN · Structural Equations with Latent Variables
BOULEAU · Numerical Methods for Stochastic Processes
BOX · Bayesian Inference in Statistical Analysis
BOX and DRAPER · Empirical Model-Building and Response Surfaces
BOX and DRAPER · Evolutionary Operation: A Statistical Method for Process Improvement
BUCKLEW · Large Deviation Techniques in Decision, Simulation, and Estimation
BUNKE and BUNKE · Nonlinear Regression, Functional Relations and Robust Methods: Statistical Methods of Model Building
CHATTERJEE and HADI · Sensitivity Analysis in Linear Regression
CHOW and LIU · Design and Analysis of Clinical Trials
CLARKE and DISNEY · Probability and Random Processes: A First Course with Applications, *Second Edition*
*COCHRAN and COX · Experimental Designs, *Second Edition*
CONOVER · Practical Nonparametric Statistics, *Second Edition*
CORNELL · Experiments with Mixtures, Designs, Models, and the Analysis of Mixture Data, *Second Edition*
*COX · Planning of Experiments
CRESSIE · Statistics for Spatial Data, *Revised Edition*
DANIEL · Applications of Statistics to Industrial Experimentation
DANIEL · Biostatistics: A Foundation for Analysis in the Health Sciences, *Sixth Edition*
DAVID · Order Statistics, *Second Edition*
*DEGROOT, FIENBERG, and KADANE · Statistics and the Law
DODGE · Alternative Methods of Regression
DOWDY and WEARDEN · Statistics for Research, *Second Edition*
DUNN and CLARK · Applied Statistics: Analysis of Variance and Regression, *Second Edition*
ELANDT-JOHNSON and JOHNSON · Survival Models and Data Analysis
EVANS, PEACOCK, and HASTINGS · Statistical Distributions, *Second Edition*
FLEISS · The Design and Analysis of Clinical Experiments
FLEISS · Statistical Methods for Rates and Proportions, *Second Edition*
FLEMING and HARRINGTON · Counting Processes and Survival Analysis
GALLANT · Nonlinear Statistical Models
GLASSERMAN and YAO · Monotone Structure in Discrete-Event Systems
GNANADESIKAN · Methods for Statistical Data Analysis of Multivariate Observations, *Second Edition*
GOLDSTEIN and LEWIS · Assessment: Problems, Development, and Statistical Issues
GREENWOOD and NIKULIN · A Guide to Chi-Squared Testing
*HAHN · Statistical Models in Engineering
HAHN and MEEKER · Statistical Intervals: A Guide for Practitioners

*Now available in a lower priced paperback edition in the Wiley Classics Library.

Applied Probability and Statistics (Continued)

HAND · Construction and Assessment of Classification Rules
HAND · Discrimination and Classification
HEIBERGER · Computation for the Analysis of Designed Experiments
HINKELMAN and KEMPTHORNE: · Design and Analysis of Experiments, Volume 1: Introduction to Experimental Design
HOAGLIN, MOSTELLER, and TUKEY · Exploratory Approach to Analysis of Variance
HOAGLIN, MOSTELLER, and TUKEY · Exploring Data Tables, Trends and Shapes
HOAGLIN, MOSTELLER, and TUKEY · Understanding Robust and Exploratory Data Analysis
HOCHBERG and TAMHANE · Multiple Comparison Procedures
HOCKING · Methods and Applications of Linear Models: Regression and the Analysis of Variables
HOGG and KLUGMAN · Loss Distributions
HOLLANDER and WOLFE · Nonparametric Statistical Methods
HOSMER and LEMESHOW · Applied Logistic Regression
HØYLAND and RAUSAND · System Reliability Theory: Models and Statistical Methods
HUBERTY · Applied Discriminant Analysis
JACKSON · A User's Guide to Principle Components
JOHN · Statistical Methods in Engineering and Quality Assurance
JOHNSON · Multivariate Statistical Simulation
JOHNSON and KOTZ · Distributions in Statistics
Continuous Multivariate Distributions
JOHNSON, KOTZ, and BALAKRISHNAN · Continuous Univariate Distributions, Volume 1, *Second Edition*
JOHNSON, KOTZ, and BALAKRISHNAN · Continuous Univariate Distributions, Volume 2, *Second Edition*
JOHNSON, KOTZ, and BALAKRISHNAN · Discrete Multivariate Distributions
JOHNSON, KOTZ, and KEMP · Univariate Discrete Distributions, *Second Edition*
JUREČKOVÁ and SEN · Robust Statistical Procedures: Aymptotics and Interrelations
KADANE · Bayesian Methods and Ethics in a Clinical Trial Design
KADANE AND SCHUM · A Probabilistic Analysis of the Sacco and Vanzetti Evidence
KALBFLEISCH and PRENTICE · The Statistical Analysis of Failure Time Data
KELLY · Reversability and Stochastic Networks
KHURI, MATHEW, and SINHA · Statistical Tests for Mixed Linear Models
KLUGMAN, PANJER, and WILLMOT · Loss Models: From Data to Decisions
KLUGMAN, PANJER, and WILLMOT · Solutions Manual to Accompany Loss Models: From Data to Decisions
KOVALENKO, KUZNETZOV, and PEGG · Mathematical Theory of Reliability of Time-Dependent Systems with Practical Applications
LAD · Operational Subjective Statistical Methods: A Mathematical, Philosophical, and Historical Introduction
LANGE, RYAN, BILLARD, BRILLINGER, CONQUEST, and GREENHOUSE · Case Studies in Biometry
LAWLESS · Statistical Models and Methods for Lifetime Data
LEE · Statistical Methods for Survival Data Analysis, *Second Edition*
LePAGE and BILLARD · Exploring the Limits of Bootstrap
LINHART and ZUCCHINI · Model Selection
LITTLE and RUBIN · Statistical Analysis with Missing Data
MAGNUS and NEUDECKER · Matrix Differential Calculus with Applications in Statistics and Econometrics
MALLER and ZHOU · Survival Analysis with Long Term Survivors
MANN, SCHAFER, and SINGPURWALLA · Methods for Statistical Analysis of Reliability and Life Data

*Now available in a lower priced paperback edition in the Wiley Classics Library.

Applied Probability and Statistics (Continued)

McLACHLAN and KRISHNAN · The EM Algorithm and Extensions
McLACHLAN · Discriminant Analysis and Statistical Pattern Recognition
McNEIL · Epidemiological Research Methods
MILLER · Survival Analysis
MONTGOMERY and PECK · Introduction to Linear Regression Analysis, *Second Edition*
MYERS and MONTGOMERY · Response Surface Methodology: Process and Product in Optimization Using Designed Experiments
NELSON · Accelerated Testing, Statistical Models, Test Plans, and Data Analyses
NELSON · Applied Life Data Analysis
OCHI · Applied Probability and Stochastic Processes in Engineering and Physical Sciences
OKABE, BOOTS, and SUGIHARA · Spatial Tesselations: Concepts and Applications of Voronoi Diagrams
PANKRATZ · Forecasting with Dynamic Regression Models
PANKRATZ · Forecasting with Univariate Box-Jenkins Models: Concepts and Cases
PIANTADOSI · Clinical Trials: A Methodologic Perspective
PORT · Theoretical Probability for Applications
PUTERMAN · Markov Decision Processes: Discrete Stochastic Dynamic Programming
RACHEV · Probability Metrics and the Stability of Stochastic Models
RÉNYI · A Diary on Information Theory
RIPLEY · Spatial Statistics
RIPLEY · Stochastic Simulation
ROUSSEEUW and LEROY · Robust Regression and Outlier Detection
RUBIN · Multiple Imputation for Nonresponse in Surveys
RUBINSTEIN · Simulation and the Monte Carlo Method
RUBINSTEIN, MELAMED, and SHAPIRO · Modern Simulation and Modeling
RYAN · Statistical Methods for Quality Improvement
SCHUSS · Theory and Applications of Stochastic Differential Equations
SCOTT · Multivariate Density Estimation: Theory, Practice, and Visualization
*SEARLE · Linear Models
SEARLE · Linear Models for Unbalanced Data
SEARLE, CASELLA, and McCULLOCH · Variance Components
STOYAN, KENDALL, and MECKE · Stochastic Geometry and Its Applications, *Second Edition*
STOYAN and STOYAN · Fractals, Random Shapes and Point Fields: Methods of Geometrical Statistics
THOMPSON · Empirical Model Building
THOMPSON · Sampling
TIJMS · Stochastic Modeling and Analysis: A Computational Approach
TIJMS · Stochastic Models: An Algorithmic Approach
TITTERINGTON, SMITH, and MAKOV · Statistical Analysis of Finite Mixture Distributions
UPTON and FINGLETON · Spatial Data Analysis by Example, Volume 1: Point Pattern and Quantitative Data
UPTON and FINGLETON · Spatial Data Analysis by Example, Volume II: Categorical and Directional Data
VAN RIJCKEVORSEL and DE LEEUW · Component and Correspondence Analysis
WEISBERG · Applied Linear Regression, *Second Edition*
WESTFALL and YOUNG · Resampling-Based Multiple Testing: Examples and Methods for p-Value Adjustment
WHITTLE · Systems in Stochastic Equilibrium
WOODING · Planning Pharmaceutical Clinical Trials: Basic Statistical Principles
WOOLSON · Statistical Methods for the Analysis of Biomedical Data
*ZELLNER · An Introduction to Bayesian Inference in Econometrics

*Now available in a lower priced paperback edition in the Wiley Classics Library.

Texts and References Section

AGRESTI · An Introduction to Categorical Data Analysis
ANDERSON · An Introduction to Multivariate Statistical Analysis, *Second Edition*
ANDERSON and LOYNES · The Teaching of Practical Statistics
ARMITAGE and COLTON · Encyclopedia of Biostatistics: Volumes 1 to 6 with Index
BARTOSZYNSKI and NIEWIADOMSKA-BUGAJ · Probability and Statistical Inference
BERRY, CHALONER, and GEWEKE · Bayesian Analysis in Statistics and Econometrics: Essays in Honor of Arnold Zellner
BHATTACHARYA and JOHNSON · Statistical Concepts and Methods
BILLINGSLEY · Probability and Measure, *Second Edition*
BOX · R. A. Fisher, the Life of a Scientist
BOX, HUNTER, and HUNTER · Statistics for Experimenters: An Introduction to Design, Data Analysis, and Model Building
BOX and LUCEÑO · Statistical Control by Monitoring and Feedback Adjustment
BROWN and HOLLANDER · Statistics: A Biomedical Introduction
CHATTERJEE and PRICE · Regression Analysis by Example, *Second Edition*
COOK and WEISBERG · An Introduction to Regression Graphics
COX · A Handbook of Introductory Statistical Methods
DILLON and GOLDSTEIN · Multivariate Analysis: Methods and Applications
DODGE and ROMIG · Sampling Inspection Tables, *Second Edition*
DRAPER and SMITH · Applied Regression Analysis, *Third Edition*
DUDEWICZ and MISHRA · Modern Mathematical Statistics
DUNN · Basic Statistics: A Primer for the Biomedical Sciences, *Second Edition*
FISHER and VAN BELLE · Biostatistics: A Methodology for the Health Sciences
FREEMAN and SMITH · Aspects of Uncertainty: A Tribute to D. V. Lindley
GROSS and HARRIS · Fundamentals of Queueing Theory, *Third Edition*
HALD · A History of Probability and Statistics and their Applications Before 1750
HALD · A History of Mathematical Statistics from 1750 to 1930
HELLER · MACSYMA for Statisticians
HOEL · Introduction to Mathematical Statistics, *Fifth Edition*
JOHNSON and BALAKRISHNAN · Advances in the Theory and Practice of Statistics: A Volume in Honor of Samuel Kotz
JOHNSON and KOTZ (editors) · Leading Personalities in Statistical Sciences: From the Seventeenth Century to the Present
JUDGE, GRIFFITHS, HILL, LÜTKEPOHL, and LEE · The Theory and Practice of Econometrics, *Second Edition*
KHURI · Advanced Calculus with Applications in Statistics
KOTZ and JOHNSON (editors) · Encyclopedia of Statistical Sciences: Volumes 1 to 9 wtih Index
KOTZ and JOHNSON (editors) · Encyclopedia of Statistical Sciences: Supplement Volume
KOTZ, REED, and BANKS (editors) · Encyclopedia of Statistical Sciences: Update Volume 1
KOTZ, REED, and BANKS (editors) · Encyclopedia of Statistical Sciences: Update Volume 2
LAMPERTI · Probability: A Survey of the Mathematical Theory, *Second Edition*
LARSON · Introduction to Probability Theory and Statistical Inference, *Third Edition*
LE · Applied Survival Analysis
MALLOWS · Design, Data, and Analysis by Some Friends of Cuthbert Daniel
MARDIA · The Art of Statistical Science: A Tribute to G. S. Watson
MASON, GUNST, and HESS · Statistical Design and Analysis of Experiments with Applications to Engineering and Science
MURRAY · X-STAT 2.0 Statistical Experimentation, Design Data Analysis, and Nonlinear Optimization

*Now available in a lower priced paperback edition in the Wiley Classics Library.

Texts amd References (Continued)

PURI, VILAPLANA, and WERTZ · New Perspectives in Theoretical and Applied Statistics
RENCHER · Methods of Multivariate Analysis
RENCHER · Multivariate Statistical Inference with Applications
ROSS · Introduction to Probability and Statistics for Engineers and Scientists
ROHATGI · An Introduction to Probability Theory and Mathematical Statistics
RYAN · Modern Regression Methods
SCHOTT · Matrix Analysis for Statistics
SEARLE · Matrix Algebra Useful for Statistics
STYAN · The Collected Papers of T. W. Anderson: 1943–1985
TIERNEY · LISP-STAT: An Object-Oriented Environment for Statistical Computing and Dynamic Graphics
WONNACOTT and WONNACOTT · Econometrics, *Second Edition*

WILEY SERIES IN PROBABILITY AND STATISTICS

ESTABLISHED BY WALTER A. SHEWHART AND SAMUEL S. WILKS

Editors
Robert M. Groves, Graham Kalton, J. N. K. Rao, Norbert Schwarz, Christopher Skinner

Survey Methodology Section

BIEMER, GROVES, LYBERG, MATHIOWETZ, and SUDMAN · Measurement Errors in Surveys
COCHRAN · Sampling Techniques, *Third Edition*
COX, BINDER, CHINNAPPA, CHRISTIANSON, COLLEDGE, and KOTT (editors) · Business Survey Methods
*DEMING · Sample Design in Business Research
DILLMAN · Mail and Telephone Surveys: The Total Design Method
GROVES · Survey Errors and Survey Costs
GROVES, BIEMER, LYBERG, MASSEY, NICHOLLS, and WAKSBERG · Telephone Survey Methodology
*HANSEN, HURWITZ, and MADOW · Sample Survey Methods and Theory, Volume 1: Methods and Applications
*HANSEN, HURWITZ, and MADOW · Sample Survey Methods and Theory, Volume II: Theory
KASPRZYK, DUNCAN, KALTON, and SINGH · Panel Surveys
KISH · Statistical Design for Research
*KISH · Survey Sampling
LESSLER and KALSBEEK · Nonsampling Error in Surveys
LEVY and LEMESHOW · Sampling of Populations: Methods and Applications
LYBERG, BIEMER, COLLINS, de LEEUW, DIPPO, SCHWARZ, TREWIN (editors) · Survey Measurement and Process Quality
SKINNER, HOLT, and SMITH · Analysis of Complex Surveys

*Now available in a lower priced paperback edition in the Wiley Classics Library.